Worldwide Multivariable Calculus

David B. Massey

in memory of my father, Robert Brian Massey (1934-2012), who taught me to love all things mathematical and scientific

©2012-2016, Worldwide Center of Mathematics, LLC

v. 0517160752

ISBN 978-0-9842071-3-8

Contents

CONTENTS

0.1 Preface

Multivariable Calculus refers to Calculus involving functions of more than one variable, i.e., multivariable functions. Not surprisingly, it is important that the reader have a good command of one-variable Calculus, both differential and integral Calculus, before diving into multivariable Calculus.

Several aspects of multivariable Calculus are quite simple. Partial derivatives are just one-variable derivatives, in which you treat all other independent variables as constants. Iterated integrals are the analogous concept for integration; the integrals involved are "partial integrals" (though no one calls them that).

However, the complexity comes in when you consider the different directions in which you can ask for the rates of change of a multivariable function. For a one-variable function, $f(x)$, you are interested in the instantaneous rate of change in f as x moves to the right (i.e., increases) or as x moves to the left (i.e., decreases). For a function of even two variables, $f(x, y)$, there are an infinite number of directions in which (x, y) can move and in which you would want the corresponding rate of change of f.

The fact that you want to look at rates of change in an infinite number of directions means that the derivative, at a given point, of a multivariable function is itself a function of the direction in which the point moves. Once the derivative has to be a function, it is nicest to let the derivative incorporate not only the direction of movement of the point, but also the speed. This leads us to consider the derivative, at a point, as a function that can be applied to arbitrary *vectors*, for vectors are things which have both direction and magnitude.

This point of view of the derivative as a vector function is extremely beautiful, and a large part of its beauty stems from the fact that the derivative is then a *linear transformation*, the fundamental type of function considered in *linear algebra*. For this reason, many statements and results in multivariable Calculus look nicest when given in the language of linear algebra.

Directions and vectors also arise in the most complicated aspects of multivariable integration problems, in which you want, for various reasons, to integrate a *vector field*. The theorems and applications involving integration of vector fields are certainly the most difficult parts of multivariable Calculus.

And so, the question arises of how to best present both the easy and the difficult aspects of multivariable Calculus. To give the rigorous, technical definitions or hypotheses would make even reasonably simple results look difficult, and make the difficult results look nightmarish. The proofs would also complicate the presentation. Finally, there is the dilemma of whether or not to include serious linear algebra in the discussion.

We deal with these issues in a variety of ways.

- First, most sections are divided into subsections, two of which are labeled **Basics** and **More Depth**, so that the "easier" material is separated from the "more difficult" material. Each subsection of **Basics** has an associated video lecture, which can be viewed online by clicking the play button in the margin below the **Basics** box.

- Material in a given section that can be presented nicely in terms of linear transformations and matrices is found in a third subsection, which is labeled + **Linear Algebra**. These subsections can be easily omitted from a course syllabus.

If the + **Linear Algebra** subsections are included in the syllabus, it should be noted that, while vectors are discussed in the body of the textbook, linear transformations and matrices are not. We assume that either the reader already knows linear algebra, or is willing to learn it along the way.

To aid the student in the latter case, rather than include an appendix on this material, we have taken the more modern, but somewhat worrisome, approach of putting in links, in green, to the relevant Wikipedia articles. We say that this is "somewhat worrisome" since Wikipedia articles are open for anyone to edit and, typically, are not a good rigorous reference for deep mathematics. However, we have vetted the linear algebra articles, and they seem to be very good and free from errors. Nonetheless, the Wikipedia articles should be used as an introduction or a refresher, but not as a substitute for a serious linear algebra textbook.

- In those sections which are divided into subsections, the exercises are also divided into **Basics**, **More Depth**, and + **Linear Algebra** for ease in assigning appropriate problems.

- Proofs, other than short ones, which illuminate the material, are **not** contained in this textbook. Instead, the reader is pointed to the appropriate external references. In particular, we refer as often as possible to the excellent, free, pdf textbook of Trench, [8], and provide a link to that pdf. This use of external references, with links, should increase the readability of this textbook, and shorten the book for possible printing, while at the same time providing fully rigorous mathematics. We believe that external links to technical proofs is the future of high school and undergraduate mathematics textbooks.

The background required to read this book is a good understanding of single-variable Calculus, and we assume that you have had courses in differential and integral Calculus. Ideally, you would also be familiar with infinite series, but that material rarely comes up in multivariable Calculus.

Basic references for technical results, and results beyond the scope of this textbook, are Rudin, [7], and Trench, [8].

David B. Massey
February 2011

Chapter 1

Multivariable Spaces and Functions

This chapter is an introduction to Euclidean space, of arbitrarily high dimension, with emphasis on the 2-dimensional case of the xy-plane, $\mathbb{R}^2$, and the 3-dimensional case of xyz-space, $\mathbb{R}^3$.

We define higher-dimensional analogs of concepts that you are familiar with in the real line, $\mathbb{R}$; we generalize such notions as: intervals, open intervals, closed bounded intervals, directions, absolute value/magnitude, functions, continuity, graphs, etc.

We discuss vectors, angles, lines, and planes. This leads us to define two special product operations: the dot product and the cross product.

Finally, this chapter contains a small amount of Calculus. We present the relatively easy, but important, case of derivatives of a function of a single variable, which takes values in a higher-dimensional Euclidean space.

1.1   Euclidean Space

Multivariable Calculus, as the name implies, deals with the Calculus of functions of more than one variable. In this section, we define and discuss basic notions and terminology concerning n-dimensional Euclidean space, where n could be any natural number. We focus on the cases of $\mathbb{R}^2$, the xy-plane, and $\mathbb{R}^3$, xyz-space.

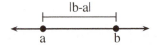

Basics:

The set of real numbers $\mathbb{R}$, or $(-\infty, \infty)$, is frequently referred to, and pictured as, the *real line*.

We assume that you are familiar with the notions of *open intervals*, (a, b), closed intervals, $[a, b]$, and intervals, in general, in the real line.

Also, if a and b are real numbers, you should know that the *distance* between a and b is the absolute value of the difference, i.e.,

$$\text{dist}(a, b) \;=\; |b - a| \;=\; \sqrt{(b - a)^2},$$

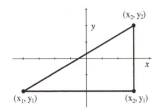

Figure 1.1.1: The distance between points in the real line.

where we have written this last complicated form for the absolute value because it generalizes nicely. The set $\mathbb{R}$, together with its distance function, is known as 1-*dimensional Euclidean space*.

You are also familiar with the xy-plane, in which points are described by pairs of real numbers (x, y). The set of pairs of real numbers is denoted by $\mathbb{R}^2$. Distance in the xy-plane is computed via the Pythagorean Theorem; the distance between points (x_1, y_1) and (x_2, y_2) is given by

$$\text{dist}\big((x_1, y_1), (x_2, y_2)\big) \;=\; \sqrt{(x_2 - x_1)^2 + (y_2 - y_1)^2}.$$

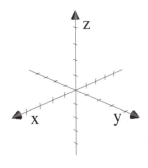

Figure 1.1.2: The distance between points in $\mathbb{R}^2$.

You may already be familiar with $\mathbb{R}^3$, 3-*dimensional Euclidean space*, also known as xyz-space. As a set, this consists of ordered triples of real numbers (x, y, z). This is frequently thought of as representing the space we live in. You could imagine that the floor is the xy-plane, and we have a third axis, the z-axis, which is perpendicular to the floor, i.e., perpendicular to the xy-plane, with the positive z-axis being above the floor, and the negative z-axis being below the floor. Then, z would measure height above the floor, so that a negative z value would indicate that you're below the floor.

In Figure 1.1.3, we've drawn a typical sketch of the x-, y-, and z-axes (in perspective), with the arrows pointing in the positive directions. Sometimes, to get a better view of some of our later graphs in $\mathbb{R}^3$, it will be convenient to rotate the axes, such as in Figure 1.1.4. Note, however, that, even though it's convenient to rotate the axes, we

Figure 1.1.3: The x-, y-, and z-axes.

will always use *right-handed axes*; this means that, if you take your right hand, and point your index finger in the direction of the positive x-axis, while pointing your middle finger in the direction of the positive y-axis, then your thumb will point in the direction of the positive z-axis. See Figure 1.1.5. Of course, the other choice of axes is left-handed, as in Figure 1.1.6; we won't use left-handed axes in this book.

In the xy-plane, you have four quadrants, corresponding to the four choices of positive/negative for the x- and y-coordinates. In $\mathbb{R}^3$, there are eight *octants*, corresponding to the eight choices of positive/negative for the x-, y-, and z-coordinates. However, only one of these octants is given a name; the *1st octant* consists of those points where x, y, and z are ≥ 0.

In three dimensions, by dropping a perpendicular line to the xy-plane, and using the Pythagorean Theorem twice, we can determine the distance between two points (x_1, y_1, z_1) and (x_2, y_2, z_2); see Figure 1.1.7. We find that

$$\text{dist}\left((x_1, y_1, z_1), (x_2, y_2, z_2)\right) \;=\; \sqrt{(x_2 - x_1)^2 + (y_2 - y_1)^2 + (z_2 - z_1)^2}.$$

Okay, that's $\mathbb{R}^2$ and $\mathbb{R}^3$, but how do we define $\mathbb{R}^n$, n-dimensional Euclidean space, where n could be any natural number?

It's simple, really. The set $\mathbb{R}^n$, *n-dimensional Euclidean space*, consists of *points* which are n real numbers in order; we call such a point an *ordered n-tuple*. Thus, examples of points in $\mathbb{R}^4$ are the ordered 4-tuples $(2, -1, 0, 3)$ and $(\pi, \sqrt{2}, 7, -e)$. Of course, in place of 2-tuple and 3-tuple, we say pair and triple, respectively. The numbers in the different positions in the n-tuple are called the *components* or *coordinates* of the point. We frequently write a point in $\mathbb{R}^n$ by $\mathbf{x} = (x_1, x_2, \ldots, x_n)$, where the boldface is used to indicate that $\mathbf{x}$ is a point with more than one component.

How do you picture $\mathbb{R}^4$, or $\mathbb{R}^n$ for $n > 4$? You don't. You picture $\mathbb{R}^2$ and $\mathbb{R}^3$, and hope that gives you some intuition for higher-dimensional $\mathbb{R}^n$. In fact, the context of dealing with $\mathbb{R}^n$ might mean that it's completely unreasonable to think of the different coordinates as specifying position. For instance, a company might produce four different types of liquid cleaning products, and keep track of how much of each product they have in stock simply by listing the number of thousands of liters of each, in order, e.g.,

$$(102.7, 34.1, 86.0, 385.4).$$

In this context, it wouldn't be reasonable to think of this point in $\mathbb{R}^4$ as specifying a position; it's just four real numbers in order. Note that we say that this point in $\mathbb{R}^4$ has units of thousands of liters only if every coordinate has the same units of thousands of liters.

We referred to $\mathbb{R}^n$ as *Euclidean space*, but, technically, Euclidean space refers to $\mathbb{R}^n$ with a specific notion of *distance*.

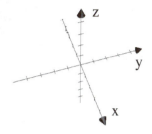

Figure 1.1.4: Rotated x-, y-, and z-axes.

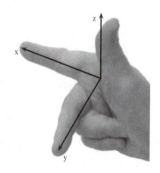

Figure 1.1.5: Right-handed axes.

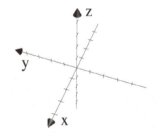

Figure 1.1.6: Left-handed axes.

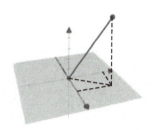

Figure 1.1.7: The Pythagorean Theorem in 3d.

Definition 1.1.1. Euclidean n-space, *or* **n-dimensional Euclidean space,** *is the set $\mathbb{R}^n$ of ordered n-tuples of real numbers, together with the notion of the* **distance** *between two points* $\mathbf{a} = (a_1, a_2, \ldots, a_n)$ *and* $\mathbf{b} = (b_1, b_2, \ldots, b_n)$ *given by*

$$\text{dist}(\mathbf{a}, \mathbf{b}) \;=\; \sqrt{(a_1 - b_1)^2 + (a_2 - b_2)^2 + \cdots (a_n - b_n)^2}.$$

The **origin in $\mathbb{R}^n$** *is the ordered n-tuple of all zeroes; we usually write $\mathbf{0}$ for the origin (regardless of what dimension we are using).*

Note that when $n = 1$, 2, or 3, the notion of distance in Euclidean space is exactly what we already discussed.

We need to discuss how to generalize standard notions from the real line to higher dimensions; we need some generalizations/replacements for intervals, open intervals, closed and bounded intervals, etc.

First, we need some even more basic terminology and notation. We sometimes use braces, curly brackets, $\{\cdots\}$, to enclose the elements of a set. We use a vertical line, $|$, as shorthand for the phrase "such that", when describing sets. So, for instance, the set of real numbers greater than 4 is written

$$\{x \in \mathbb{R} \mid x > 4\},$$

which you read as "the set of those x in $\mathbb{R}$ such that $x > 4$". Of course, this is a subset of $\mathbb{R}$, and is the same as the interval $(4, \infty)$.

A *subset E* of $\mathbb{R}^n$ is a collection of some (including, possibly, all or none) of the points in $\mathbb{R}^n$. We write $E \subseteq \mathbb{R}^n$ to indicate that E is a subset of $\mathbb{R}^n$, and we write $\mathbf{p} \in E$ to indicate that $\mathbf{p}$ is a point in, or element of, E. If A and B are both subsets of $\mathbb{R}^n$ (or, are sets, in general), then A is a subset of B, written $A \subseteq B$, if and only if every element of A is also in B.

Now, we can define the most basic generalizations of open and closed intervals in $\mathbb{R}$: open and closed balls in $\mathbb{R}^n$.

Technically, there is a difference between an ordered set with a single real number in one component and the real number itself. We shall not worry about this distinction.

Definition 1.1.2. *Suppose that r is a positive real number, and that $\mathbf{p}$ is a point in $\mathbb{R}^n$. Then, the n-dimensional* **open** *(respectively,* **closed***) ball, $B_r^n(\mathbf{p})$ (respectively, $\overline{B_r^n}(\mathbf{p})$), centered at $\mathbf{p}$, of radius r is the set of points in $\mathbb{R}^n$ whose distance from the center is less than (respectively, less than or equal to) r.*

Thus, the open ball is given, in set notation, by

$$B_r^n(\mathbf{p}) = \{\mathbf{x} \in \mathbb{R}^n \mid \mathrm{dist}(\mathbf{x}, \mathbf{p}) < r\},$$

while the closed ball is given by

$$\overline{B_r^n}(\mathbf{p}) = \{\mathbf{x} \in \mathbb{R}^n \mid \mathrm{dist}(\mathbf{x}, \mathbf{p}) \leq r\},$$

The $(n-1)$-**dimensional sphere**, $S_r^{n-1}(\mathbf{p})$, of radius $r > 0$, centered at $\mathbf{p}$, is the boundary of the n-dimensional ball. Thus,

$$S_r^{n-1}(\mathbf{p}) = \{\mathbf{x} \in \mathbb{R}^n \mid \mathrm{dist}(\mathbf{x}, \mathbf{p}) = r\}.$$

This means that, if $\mathbf{p} = (p_1, \ldots, p_n)$, then $S_r^{n-1}(\mathbf{p})$ is the set of points $\mathbf{x} = (x_1, \ldots, x_n)$ in $\mathbb{R}^n$ which satisfy the equation

$$(x_1 - p_1)^2 + (x_2 - p_2)^2 + \cdots + (x_n - p_n)^2 = r^2.$$

Note that a 2-dimensional sphere, which lies in $\mathbb{R}^3$, is what is usually referred to as just "a sphere".

You may have noticed that this definition means that, what we normally call a circle in $\mathbb{R}^2$, is also referred to as a 1-dimensional sphere. It's also true that n can be 1, which means that a 0-dimensional sphere in the real line consists of two points.

Example 1.1.3. For $r > 0$, the set of points (x, y, z) such that

$$x^2 + y^2 + z^2 = r^2,$$

describes a sphere of radius r, centered at the origin.

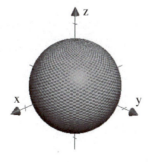

Figure 1.1.8: A sphere of radius $r > 0$, centered at the origin.

Obviously, the set $\mathbb{R}^n$ is an open subset of itself. Perhaps not so obvious is that the empty set is also open, because the condition is *vacuously* satisfied. This means that the condition holds, since there are no points in the empty set that can make it fail.

Definition 1.1.4. *A subset E of $\mathbb{R}^n$ is **open** if and only if, for all $\mathbf{p} \in E$, every point in $\mathbb{R}^n$ that is sufficiently close to $\mathbf{p}$ is also in E. More precisely, E is an open subset of $\mathbb{R}^n$ if and only if, for all $\mathbf{p} \in E$, there exists $r > 0$ such that $B_r^n(\mathbf{p}) \subseteq E$.*

Example 1.1.5. The subset E of $\mathbb{R}^2$ of those (x, y) such that $x > 0$ is an open set. How do we show this?

Suppose that (a, b) is a point in E. This means that $a > 0$. We need to show that there is an open ball $B_r^2(a, b)$ in $\mathbb{R}^2$, centered at (a, b), for some positive r, such that $B_r^2(a, b) \subseteq E$.

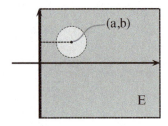

Figure 1.1.9: An open 2-dimensional ball of radius $a/2$, centered at (a, b), where $a > 0$. Each point inside the ball has a positive x-coordinate.

Intuitively, this is easy; we can, for instance, take an open ball of radius equal to half the distance to the y-axis, i.e., we can take $r = a/2$.

To show it rigorously, we suppose that (x, y) is in the open ball, i.e., that

$$\sqrt{(x-a)^2 + (y-b)^2} < \frac{a}{2},$$

and we need to show that this implies that $x > 0$.

As $(y - b)^2 \geq 0$, we have

$$|x - a| = \sqrt{(x-a)^2} \leq \sqrt{(x-a)^2 + (y-b)^2} < \frac{a}{2}.$$

Therefore,

$$-\frac{a}{2} < x - a < \frac{a}{2},$$

and so,

$$0 < \frac{a}{2} = a - \frac{a}{2} < x,$$

which is what we wanted to show.

Definition 1.1.6. *An* **open neighborhood** *of a point* **p** *in* $\mathbb{R}^n$ *is an open subset of* $\mathbb{R}^n$ *which contains* **p**.

Definition 1.1.7. *Suppose that E is a subset of* $\mathbb{R}^n$. *Then, the largest open subset of* $\mathbb{R}^n$ *which is contained in E is called the* **interior** *of E, and denoted* $\text{int}(E)$. *Equivalently,*

$$\text{int}(E) = \{\mathbf{p} \in E \mid \text{there exists } r > 0 \text{ such that } B_r^n(\mathbf{p}) \subseteq E\}.$$

Note that it is possible for $\text{int}(E)$ to be empty.

Example 1.1.8. For instance, the interior of the closed interval $[2,7]$ in $\mathbb{R}$ is the open interval $(2,7)$.

The interior of a circle (not the disk inside the circle) in $\mathbb{R}^2$ is empty; see Figure 1.1.10

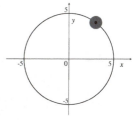

Figure 1.1.10: Any open ball around a point on the circle extends outside the circle.

Definition 1.1.9. *Suppose that E is a subset of* $\mathbb{R}^n$. *A point* **p** *in* $\mathbb{R}^n$ *(which may be in E or not) is a* **boundary point** *of E if and only if every open ball around* **p** *in* $\mathbb{R}^n$ *contains both a point in E and a point not in E.*
The set of boundary points of E is simply call the **boundary of E**.

Example 1.1.10. In most of the examples that we deal with, the boundary of the set will be what you would intuitively call the boundary. In Figure 1.1.11, we show two filled-in ellipses; for both, the boundaries are the precisely the "bounding" ellipses.

We dotted the boundary of the green filled-in ellipse to indicate that we mean that the boundary is **not** included in the subset of $\mathbb{R}^2$. For the blue filled-in ellipse, we drew a solid ellipse for the boundary to indicate that we mean that the boundary **is** included in the subset of $\mathbb{R}^2$ that we drew.

Figure 1.1.11: Subsets of $\mathbb{R}^2$, which don't include and do include their boundaries.

> It is important that a subset of $\mathbb{R}^n$ which contains a single point is a closed subset of $\mathbb{R}^n$.

Definition 1.1.11. *A subset E of $\mathbb{R}^n$ is* **closed** *if and only if E contains all of its boundary points. This is equivalent to saying that E is closed if and only if the subset of points in $\mathbb{R}^n$ that are* **not** *in E is an open set.*

Note that, if E is a subset of $\mathbb{R}^n$, then every point in E is either an interior point of E or a boundary point of E; but, keep in mind, if a subset is not closed, then it has boundary points which do not lie in the set itself.

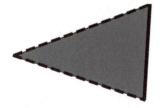

Figure 1.1.12: A filled-in triangle which is neither open nor closed.

Example 1.1.12. In Example 1.1.10, the blue, filled-in ellipse, which contains its boundary, is closed, while the green filled-in ellipse is open.

However, be warned: a subset of $\mathbb{R}^n$ needn't be open or closed. For instance, the filled-in triangle in Figure 1.1.12 contains only some of its boundary points, and is neither open nor closed.

There are also precisely two subsets of R^n that are **both** open and closed, but they are exceptional: the empty set and all of $\mathbb{R}^n$ are both open and closed.

A subset of $\mathbb{R}^n$ which doesn't extend out "infinitely far" is called *bounded*; the easiest way to say this in a technically precise way is to say that the subset is contained in a ball, which we can assume is centered at the origin.

Definition 1.1.13. *A subset E of $\mathbb{R}^n$ is* **bounded** *if and only if there exists $r > 0$ such that $E \subseteq B_r^n(\mathbf{0})$.*

Subsets of $\mathbb{R}^n$ which are closed and bounded play an important role in this textbook, and in mathematics in general.

Definition 1.1.14. *A subset E of $\mathbb{R}^n$ is* **compact** *if and only if it is closed and bounded.*

We have used this as our definition of *compact* because the actual definition from topology is not something we want to discuss. The fact that, for subsets of $\mathbb{R}^n$, the topologically compact subsets are precisely the closed and bounded subsets is the *Heine-Borel Theorem*; see Rudin, [7], 2.14.

Example 1.1.15. A closed ball of positive radius in $\mathbb{R}^n$ is compact, while an open ball is not.

The set of (x, y) in $\mathbb{R}^2$ such that $y \geq 0$ is closed, because it contains its boundary, the x-axis (where $y = 0$). However, this set is not bounded, as it contains points arbitrarily far from the origin; thus, the set is not compact.

More Depth:

An interval in $\mathbb{R}$ is precisely a subset of $\mathbb{R}$ which is in one, connected piece. Earlier in this section, we mentioned that open and closed balls are the most basic generalization in $\mathbb{R}^n$ of open and closed intervals in the real line. However, for many purposes, open and closed balls are too special; all we really want is a good notion of what it means for a set to be *connected*. First, we need some more set notation.

The empty set, the set with no elements, is denoted by $\emptyset$. For two sets (we care mainly about two subsets of the same $\mathbb{R}^n$) A and B, the *union* of A and B, denoted $A \cup B$, is the set which contains all of the elements of A and all of the elements of B. That is,

$$x \in A \cup B = \{x \mid x \in A \text{ or } x \in B\}.$$

The *intersection* of A and B, denoted $A \cap B$, is the set which contains those elements of A which are also elements of B, i.e.,

$$A \cap B = \{x \mid x \in A \text{ and } x \in B\}.$$

In mathematics, "or" always means the non-exclusive "or", i.e., we allow for both conditions to hold. Thus, here, if x is in A and x is in B, then $x \in A \cup B$.

Now we can define *connected subsets* in $\mathbb{R}^n$. A subset of R^n is connected if it does not consist of two (or more) separated pieces. Technically:

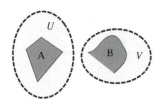

Figure 1.1.13: A separation of the set $E = A \cup B$.

Definition 1.1.16. *Suppose that $E \subseteq \mathbb{R}^n$. A **separation** of E consists of two open subsets $\mathcal{U}$ and $\mathcal{V}$ of $\mathbb{R}^n$ which separate E, i.e., such that, if we let $A = \mathcal{U} \cap E$ and $B = \mathcal{V} \cap E$, then*

1. *$A \neq \emptyset$ and $B \neq \emptyset$,*

2. *$E = A \cup B$, and*

3. *$A \cap B = \emptyset$.*

*A subset E of $\mathbb{R}^n$ is **connected** if and only if no separation of E exists.*

Remark 1.1.17. In $\mathbb{R}$, it is a theorem that the connected subsets are precisely the intervals.

Throughout this book, the notion of connectedness should not cause you any problems; the connected subsets of $\mathbb{R}^2$ and $\mathbb{R}^3$ that we look at will be the ones that you would intuitively refer to as connected.

Online answers to select exercises are here.

1.1.1   Exercises

Basics:

In the following exercises, find the distance between the given pairs of points.

1. $(1, 2, 3)$ and $(-3, -2, -1)$

2. $(1, 0, 1, 0)$ and $(0, 1, 0, 1)$

3. $(\sqrt{2}, 1, -3, 5, -1)$ and $(0, 2, -2, 6, 1)$

4. $(2, 2, 0)$ and $(1, 5, -1)$

Are the following points inside, outside, or on the sphere of radius 2, centered at $(1, 1, 1)$ in $\mathbb{R}^3$?

5. $(2, 2, 2)$

6. $(3, 3, 3)$

7. $(1, 3, 1)$

8. $(-1, 0, -1)$

Suppose that $\mathbf{x} = (x_1, x_2, x_3, x_4) = (2.3, 8.9, 10, 7.5) \in \mathbb{R}^4$. Explain how to interpret this point, in each of the following scenarios.

9. x_i represents the grade of student number i on a 10-point quiz.

10. x_i represents the temperature, in degrees Celsius, on a given day, i hours after 6 am.

11. x_i represents the number of thousands of dollars in bank account number i.

12. x_i represents the squirrel population-density, in hundreds of squirrels per square mile, at a distance of i miles from some point $\mathbf{p}$.

For each of the following subsets of $\mathbb{R}^n$, state whether the given subset is open, closed, bounded, compact, or none of these. Note that a given subset may possess more than one of these properties.

13. The open ball in $\mathbb{R}^2$ of radius 3, centered at $(-1, 2)$.

14. The filled-in square in $\mathbb{R}^2$, where $0 \le x \le 2$ and $0 \le y \le 2$.

15. The boundary of the filled-in square in Exercise 14.

16. The interior of the filled-in square in Exercise 14.

17. The subset of all (x, y) in $\mathbb{R}^2$ such that $|y| \le 2$ and $|x| = 7$.

18. The subset of all (x, y) in $\mathbb{R}^2$ such that $|y| > 2$ and $|x| = 7$.

19. The subset of all (x, y, z) in $\mathbb{R}^3$ such that $z \ge 0$.

20. The subset of all (x, y, z) in $\mathbb{R}^3$ such that $x^2 + y^2 + z^2 \le 1$ and $z \ge 0$ (the top half of a ball).

21. The subset of all (x, y, z) in $\mathbb{R}^3$ such that $x^2 + y^2 + z^2 \le 1$ and $z > 0$ (the top half of a ball).

22. The subset of all (x, y, z) in $\mathbb{R}^3$ such that $x^2 + y^2 + z^2 < 1$ and $z > 0$ (the top half of a ball).

In each of the following problems, you are told which way two of the positive axes point (on the sheet of paper, as you look at them) in $\mathbb{R}^3$. Sketch these positive axes, and then include the missing positive axis, given that the coordinate system is to be right-handed. You may have to experiment to produce a reasonable perspective.

23. The positive x-axis points upward, and the positive y-axis points to the right.

24. The positive x-axis points upward, and the positive z-axis points to the right.

25. The positive y-axis points upward, and the positive x-axis points to the right.

26. The positive y-axis points upward, and the positive z-axis points to the right.

More Depth: Sketch each of the following subsets of $\mathbb{R}^2$, and observe that they appear to be disconnected. Then, produce an explicit separation of the set, and verify the properties in Definition 1.1.16 to show that you, indeed, found a separation, and conclude the subset is not connected.

27. The set of points (x, y) such that $xy = 1$.

28. The set of points (x, y) such that $|x| > 1$.

29. The set of points (x, y) such that $x^2 - y^2 = 1$.

30. The set of points (x, y) such that $y^2 = 1$.

31. Prove that a subset of $\mathbb{R}^n$ which contains a single point is closed.

32. Prove that an arbitrary union of open subsets in $\mathbb{R}^n$ is open. That is, suppose that I is some set, possibly with an infinite number of elements, and that for each i in I, $\mathcal{U}_i$ is an open subset of $\mathbb{R}^n$. Show that the union $\bigcup_{i \in I} \mathcal{U}_i$ is an open subset of $\mathbb{R}^n$. Note that, by definition, a point $\mathbf{p}$ is in this union if and only if there exists an $i \in I$ such $\mathbf{p} \in I_i$.

33. Prove that an finite intersection of open subsets in $\mathbb{R}^n$ is open. That is, suppose that $\mathcal{U}_1, \ldots, \mathcal{U}_k$ is a finite collection of open subsets of $\mathbb{R}^n$. Show that

$$\mathcal{U}_1 \cap \cdots \cap \mathcal{U}_k$$

is open in $\mathbb{R}^n$.

34. Give an explicit example of a infinite collection of open subsets of some $\mathbb{R}^n$ whose intersection is not open.

35. Prove that an arbitrary intersection of closed subsets in $\mathbb{R}^n$ is closed. That is, suppose that I is some set, possibly with an infinite number of elements, and that for each i in I, C_i is an closed subset of $\mathbb{R}^n$. Show that the intersection $\bigcap_{i \in I} C_i$ is a closed subset of $\mathbb{R}^n$. Note that, by definition, a point $\mathbf{p}$ is in this intersection if and only if, for all $i \in I$, $\mathbf{p} \in I_i$.

36. Prove that an finite union of closed subsets in $\mathbb{R}^n$ is closed. That is, suppose that $C_1, \ldots, C_k$ is a finite collection of closed subsets of $\mathbb{R}^n$. Show that

$$C_1 \cup \cdots \cup C_k$$

is closed in $\mathbb{R}^n$.

37. Give an explicit example of a infinite collection of closed subsets of some $\mathbb{R}^n$ whose union is not closed.

38. Suppose that E and F are connected subsets of $\mathbb{R}^n$, and that $E \cap F$ is non-empty. Prove that $E \cup F$ is connected.

1.2 $\mathbb{R}^n$ as a Vector Space

Physically, a vector is a quantity that has both magnitude and direction. Typical vector quantities are displacement, velocity, acceleration, and force.

Mathematically, vectors in $\mathbb{R}^n$ can be identified with points in $\mathbb{R}^n$, together with two vector operations: addition and scalar multiplication.

Basics:

A *vector* is frequently described in physics and engineering classes as a quantity which has magnitude and direction. Typical vector quantities are things such as "move north 3 miles", "travel west at 5 meters per second", and "a force of 10 pounds, pushing south"; respectively, these are a displacement vector, a velocity vector, and a force vector.

Physical quantities which have only magnitude, but no direction, are called *scalar quantities*; examples of scalar quantities include mass, speed, and energy.

A vector in the real line, $\mathbb{R}$, the xy-plane, $\mathbb{R}^2$, or xyz-space, $\mathbb{R}^3$, is frequently represented by an arrow, which points in the direction of the vector and has length equal to the magnitude of the vector (even if vector itself does not have length units).

Notice that "move north 3 miles" does not specify a starting point or an ending point; of course, if you knew the starting point, you'd know the ending point. Similarly, "travel west at 5 meters per second" doesn't specify where you begin, and a "force of 10 pounds, pushing south" doesn't specify the point at which the force is acting. The arrow representing a given vector has no natural, fixed starting point; parallel arrows which have the same length and which have their tails and heads at corresponding ends (i.e., point in the same direction) represent the same vector. We say that these arrows, which have the same direction and magnitude, all *represent* the same vector.

If we do wish to specify a particular "starting point" for a vector, we will refer to the vector as being **based** at the given point, and refer to the point as the **base point**. For a moving object, at any given time, it is standard to base its velocity vector at the location of the object. When we talk about a force acting on an object, which is located at the point $\mathbf{p}$, it is natural to base the force vector at $\mathbf{p}$. Later, when we discuss tangents vectors to curves and surfaces at a point, it will be natural to consider those tangent vectors as being based at the point at which they're tangent.

If we have a vector, and consider it as starting at the origin $\mathbf{0}$, then the direction and magnitude of the vector are completely determined by the endpoint of the vector. In this way, we can identify points in $\mathbb{R}^n$ with vectors. For instance, we can consider the point $(1,2)$ in $\mathbb{R}^2$, and identify it with the vector represented by an arrow from $(0,0)$ out to $(1,2)$. In fact, it is common to use the same notation for a point and the vector, based at the origin, determined by the point; thus, we may talk about the point

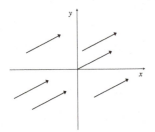

Figure 1.2.1:   Arrows which all represent the same vector.

$(1,2)$ and the vector $(1,2)$. We shall not place little arrows over variables representing vectors; instead, as with points, we shall use boldface variables to denote vectors, e.g., the vector $\mathbf{v} = (1,2)$.

Specifying a vector by starting it at the origin and specifying the ending point is called *giving the vector in coordinates*. The coordinates of a vector are also referred to as the *components*. The displacement vector in $\mathbb{R}^3$ given by $(2, -1, 3)$ meters means a displacement such that the x-coordinate increases by 2 meters, the y-coordinate decreases by 1 meter, and the z-coordinate increases by 3 meters. The velocity vector in $\mathbb{R}^2$ given by $(4, 3)$ meters per second means that the x-coordinate is changing at a rate of 4 m/s and the y-coordinate is changing at a rate of 3 m/s.

Two points $\mathbf{a} = (a_1, a_2, \ldots, a_n)$ and $\mathbf{b} = (b_1, b_2, \ldots, b_n)$ determine a *displacement vector from* $\mathbf{a}$ *to* $\mathbf{b}$, denoted $\overrightarrow{\mathbf{ab}}$; this vector gives the change in each coordinate, and so

$$\overrightarrow{\mathbf{ab}} = (b_1 - a_1, b_2 - a_2, \ldots, b_n - a_n),$$

that is, you take the coordinates of the final point and subtract the coordinates of the initial point. For instance, $\overrightarrow{(1,2)(0,5)} = (-1, 3)$ and $\overrightarrow{(2,1)(1,4)} = (-1, 3)$. You can see the equality of these displacement vectors in Figure 1.2.2.

For a displacement vector, we want its magnitude to equal the length from the starting point to the ending point. Hence, if we have a displacement $\mathbf{v} = (v_1, v_2, \ldots, v_n)$, which is represented by an arrow from the origin to the point $\mathbf{v}$, then we use the Euclidean distance from the previous section to define the magnitude of $\mathbf{v}$ to be

$$|\mathbf{v}| = \sqrt{v_1^2 + v_2^2 + \cdots + v_n^2}.$$

In fact, we define the magnitude of a vector in this manner, whether or not the vector is a displacement vector.

> **Definition 1.2.1.** *Given a vector* $\mathbf{v} = (v_1, v_2, \ldots, v_n)$ *in* $\mathbb{R}^n$, *we define the* **magnitude** *of* $\mathbf{v}$ *to be*
> $$|\mathbf{v}| = \sqrt{v_1^2 + v_2^2 + \cdots + v_n^2}.$$

As you probably know already:

> *The magnitude of the velocity (vector) of an object has a special name; it is called* **speed**.

Some books write little arrows over vector variables and enclose vectors, given in coordinates, by $\langle \rangle$. We shall not use these notations, for there are too many times when we need to consider a point as a vector and vice-versa.

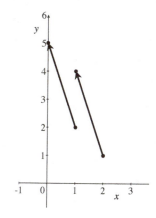

Figure 1.2.2: The same displacement vector.

Remark 1.2.2. If each component of the vector **v** has the same units as the other components, then the magnitude also has those common units. If the components have different units, then the magnitude is just considered as a unitless number.

You may wonder about the use of "absolute value signs" to denote magnitude. You might think that confusion would arise when we look at a vector in $\mathbb{R}$, i.e., when we simply look at a single real number. However, notice that there's no conflicting notation, for, if x is in $\mathbb{R}$, then the absolute value of x is precisely $\sqrt{x^2}$.

Example 1.2.3. Consider the force vector $\mathbf{F} = (3, -2, 4)$ Newtons. Find the magnitude of **F**.

Solution:

The magnitude of **F** is

$$|\mathbf{F}| \;=\; \sqrt{3^2 + (-2)^2 + 4^2} \;=\; \sqrt{29} \;\text{ Newtons.}$$

Example 1.2.4. An object is moving in space such a way that its velocity, at time t seconds, is given by

$$\mathbf{v}(t) \;=\; (2\cos t,\; 2\sin t,\; \sqrt{5}) \;\text{ m/s.}$$

Show that, even though the velocity is changing, the speed of the object is constant, and determine that constant speed.

Solution:

Speed is the magnitude of velocity. We find

$$|\mathbf{v}(t)| \;=\; \sqrt{(2\cos t)^2 + (2\sin t)^2 + (\sqrt{5})^2} \;=\; \sqrt{4(\cos^2 t + \sin^2 t) + 5} \;=\; \sqrt{9} \;=\; 3 \;\text{ m/s.}$$

Saying that the zero vector has every direction is also very convenient; it will make some of our later results easier to state if the zero vector is parallel to every other vector and is also perpendicular to every other vector.

The *zero vector* is the vector all of whose coordinates are zero; thus, we denote the zero vector the same way that we denote the origin: $\mathbf{0} = (0, 0, \ldots, 0)$. The context will make it clear when we are considering $\mathbf{0}$ as the origin and when we are considering it to be the zero vector. Clearly, the zero vector has magnitude 0, and is the only vector (in a given $\mathbb{R}^n$) with magnitude 0, but what's the direction of the zero vector? We usually say that the zero vector $\mathbf{0}$ has **every** direction. Why is this reasonable? Thinking in terms of displacement, moving 0 meters north is the same as moving 0 meters south, or 0 meters in any other direction.

We want to define *vector addition*. This is easiest to motivate by considering displacement. Suppose that an object is moving in the *xy*-plane, where we think of the positive *y* direction as north and the positive *x* direction as east. If the object first moves 2 units north and 3 units east, and then moves 1 unit north and -2 units east (i.e., 2 units west), then what is the total displacement of the object? This isn't difficult; the total displacement is 3 units north and 1 unit east, where we obtained this simply by adding the x-coordinates of the two displacements and adding the y-coordinates of the two displacements. We want for this final, total displacement to be called the sum of the two individual displacements, i.e., we want to define vector addition coordinate-wise (or component-wise), so that

$$(3, 2) + (-2, 1) \;=\; (1, 3).$$

Hence, we make the definition:

Definition 1.2.5. *Given vectors* $\mathbf{a} = (a_1, a_2, \ldots, a_n)$ *and* $\mathbf{b} = (b_1, b_2, \ldots, b_n)$ *in* $\mathbb{R}^n$, *we define the sum of* $\mathbf{a}$ *and* $\mathbf{b}$ *to be the coordinate-wise sum*

$$\mathbf{a} + \mathbf{b} \;=\; (a_1 + b_1, \, a_2 + b_2, \, \ldots, \, a_n + b_n).$$

Remark 1.2.6. It is easy to picture vector addition geometrically in $\mathbb{R}^2$ and $\mathbb{R}^3$, and to see the commutativity of vector addition. Suppose, for instance, that you have vectors $\mathbf{a}$ and $\mathbf{b}$ in $\mathbb{R}^2$. How can you picture the sum $\mathbf{a} + \mathbf{b}$? You should think: "first, you do $\mathbf{a}$, and then you do $\mathbf{b}$".

So, you draw an arrow representing the vector $\mathbf{a}$, and then you start the arrow representing the vector $\mathbf{b}$ where $\mathbf{a}$ ends; that is, you make the starting point of $\mathbf{b}$ the ending point of $\mathbf{a}$. Then, $\mathbf{a} + \mathbf{b}$ is represented by the arrow from the starting point of $\mathbf{a}$ to the ending point of $\mathbf{b}$; see Figure 1.2.3. Figure 1.2.3 also shows that you end up with the same resultant vector if you first start with $\mathbf{b}$ and then add $\mathbf{a}$, i.e., you see geometrically that $\mathbf{a} + \mathbf{b} = \mathbf{b} + \mathbf{a}$.

Because $\mathbf{a} + \mathbf{b}$ is represented by the diagonal of the parallelogram in Figure 1.2.3, this geometric way of looking at vector addition is sometimes referred to as the "parallelogram rule" for adding vectors.

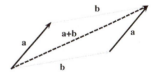

Figure 1.2.3: The "parallelogram rule" for adding vectors.

Example 1.2.7. An important example of vector addition is the *net force*, or the total force, acting on an object.

If $\mathbf{F}_1$, $\mathbf{F}_2$, ..., $\mathbf{F}_k$ are forces acting on an object in space, then the net force acting on the object is

$$\mathbf{F}_{\text{net}} \ = \ \mathbf{F}_1 + \mathbf{F}_2 + \cdots + \mathbf{F}_k.$$

Suppose, for instance, that forces $\mathbf{F}_1 = (2, 3, 5)$ and $\mathbf{F}_2 = (-3, 1, -2)$ Newtons act on an object. Let's determine the magnitude of the net force acting on the object.

We find

$$|\mathbf{F}_{\text{net}}| \ = \ |(2, 3, 5) + (-3, 1, -2)| \ = \ |(-1, 4, 3)| \ = \ \sqrt{(-1)^2 + 4^2 + 3^2} \ = \ \sqrt{26} \ \text{Newtons}.$$

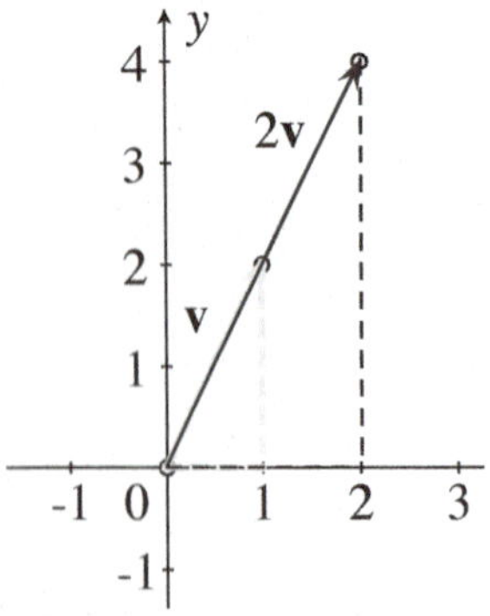

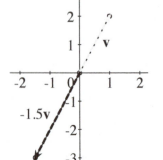

Figure 1.2.4: The vectors $\mathbf{v}$ and $2\mathbf{v}$ point in the same direction.

We need to consider another vector operation. Consider the vector $\mathbf{v} = (1, 2)$ in $\mathbb{R}^2$. If we double the x and y components of $\mathbf{v}$, we get $(2, 4)$; let's denote this vector by $2\mathbf{v}$.

If we draw both $\mathbf{v}$ and $2\mathbf{v}$ based at the origin, they point in the same direction. You can think of this in terms of similar triangles; since we doubled both the x- and y-coordinates, the arrow representing $2\mathbf{v} = (2, 4)$ lies on top of the arrow representing $\mathbf{v} = (1, 2)$. See Figure 1.2.4.

Note that the vectors point in the same direction, instead of opposite directions, because we multiplied the components by the same **positive** number.

What if we had taken $\mathbf{v} = (1, 2)$ and multiplied each of its components by the same **negative** number? Let $-1.5\mathbf{v} = (-1.5, -3)$, i.e., let $-1.5\mathbf{v}$ be the vector obtained from $\mathbf{v}$ by multiplying each component by -1.5. See Figure 1.2.5, in which we see that $-1.5\mathbf{v}$ is parallel to $\mathbf{v}$, but points in the opposite direction.

Before we discuss this further, let's go ahead and give the definition of *scalar multiplication*.

Figure 1.2.5: The vectors $\mathbf{v}$ and $-1.5\mathbf{v}$ point in opposite directions.

Definition 1.2.8. *Given a vector* $\mathbf{v} = (v_1, v_2, \ldots, v_n)$ *and a real number* r, *which in this context is referred to as a* **scalar**, *the* **scalar multiplication** *of* $\mathbf{v}$ *by* r *is denoted by* $r\mathbf{v}$, *and is given by multiplying each component of* $\mathbf{v}$ *by* r, *i.e.,*

$$r\mathbf{v} \ = \ (rv_1, rv_2, \ldots, rv_n).$$

If $r \neq 0$, *we frequently write* $\dfrac{\mathbf{v}}{r}$ *in place of* $\left(\dfrac{1}{r}\right)\mathbf{v}$. *We also write* $\mathbf{a} - \mathbf{b}$ *in place of* $\mathbf{a} + (-\mathbf{b})$, *and* $-\mathbf{v}$ *in place of* $-1\mathbf{v}$.

We are omitting the easy proof that $-1\mathbf{v}$ is, in fact, the additive inverse of $\mathbf{v}$, which is what $-\mathbf{v}$ really means.

It's easy to show that:

Theorem 1.2.9. *If you scalar multiply* **v** *by* r, *then the magnitude is multiplied (so, the vector is "scaled") by the absolute value of* r, *i.e.,*

$$|r\mathbf{v}| = |r||\mathbf{v}|.$$

Proof. We simply calculate:

$$|r\mathbf{v}| = |(rv_1, rv_2, \ldots, rv_n)| = \sqrt{r^2 v_1^2 + r^2 v_2^2 + \cdots + r^2 v_n^2} =$$

$$\sqrt{r^2} \cdot \sqrt{v_1^2 + v_2^2 + \cdots + v_n^2} = |r||\mathbf{v}|.$$

$\square$

Example 1.2.10. Perhaps the most famous of all scalar multiplications is contained in *Newton's 2nd Law of Motion*, which is usually stated simply as

$$\mathbf{F} = m\mathbf{a},$$

where **F** is the net force (vector) acting on an object, m is the object's mass (a scalar, which is assumed to be constant), and **a** is the acceleration (vector) of the object.

For instance, suppose that an object of mass 10 kg is accelerating at $\mathbf{a} = (\sqrt{6}, -3, 1)$ (m/s)/s. Then, the net force acting of the object is

$$\mathbf{F} = m\mathbf{a} = 10(\sqrt{6}, -3, 1) = (10\sqrt{6}, -30, 10) \text{ Newtons.}$$

The magnitude of this force is

$$|\mathbf{F}| = |m\mathbf{a}| = m|\mathbf{a}| = 10\sqrt{(\sqrt{6})^2 + (-3)^2 + 1^2} = 10\sqrt{6 + 9 + 1} = 40 \text{ Newtons.}$$

Example 1.2.11. Suppose that $\mathbf{v} \neq \mathbf{0}$, so that $|\mathbf{v}| \neq 0$. Let

$$\mathbf{u} = \left(\frac{1}{|\mathbf{v}|}\right)\mathbf{v} = \frac{\mathbf{v}}{|\mathbf{v}|}.$$

Show that the magnitude of **u** is 1.

Solution:

Using Theorem 1.2.9, we find

$$|\mathbf{u}| = \left|\left(\frac{1}{|\mathbf{v}|}\right)\mathbf{v}\right| = \frac{1}{|\mathbf{v}|} \cdot |\mathbf{v}| = 1.$$

The properties of scalar multiplication that we looked at in the examples before Definition 1.2.8 should convince you that, in $\mathbb{R}$, $\mathbb{R}^2$, and $\mathbb{R}^3$, two non-zero vectors are parallel if and only if they are scalar multiples of each other; the vectors point in the same direction if the scalar is positive, and in opposite directions if the scalar is negative.

However, what does "direction" mean in $\mathbb{R}^n$, where $n \geq 4$? For that matter, what's the technical definition of *direction* even in $\mathbb{R}$, $\mathbb{R}^2$, and $\mathbb{R}^3$?

Our definition is motivated by the intuition that we develop by looking at vectors in $\mathbb{R}^2$ and $\mathbb{R}^3$. Intuitively, two vectors have the same direction if and only if, after we scale them (by a positive scalar) so that the vectors have the same magnitude, we obtain the same vector. Thus, we pick some "natural" magnitude, greater than zero, and want to look at vectors after they're scaled to have this natural magnitude. The obvious choice for a "natural" positive magnitude is 1. As we saw in Example 1.2.11, if $\mathbf{v} \neq 0$, then $\mathbf{v}/|\mathbf{v}|$ has magnitude 1.

Therefore, we make the following definitions:

Definition 1.2.12. *A **unit vector** is a vector which has magnitude 1.*
*The **direction** of a non-zero vector $\mathbf{v}$ is defined to be the unit vector $\mathbf{v}/|\mathbf{v}|$.*
*We say that the zero vector $\mathbf{0}$ **has every direction**.*

Suppose that $\mathbf{v} \neq \mathbf{0}$ and $r > 0$. Then, if we calculate the direction of $r\mathbf{v}$, we find

$$\frac{r\mathbf{v}}{|r\mathbf{v}|} = \frac{r}{|r|} \cdot \frac{\mathbf{v}}{|\mathbf{v}|} = \frac{\mathbf{v}}{|\mathbf{v}|},$$

since $r > 0$. Hence, any positive scalar multiple of $\mathbf{v}$ has the same direction as $\mathbf{v}$.

Conversely, if $\mathbf{v}$ and $\mathbf{w}$ are non-zero vectors with the same direction, then

$$\frac{\mathbf{w}}{|\mathbf{w}|} = \frac{\mathbf{v}}{|\mathbf{v}|},$$

so that

$$\mathbf{w} = \left(\frac{|\mathbf{w}|}{|\mathbf{v}|} \right) \mathbf{v};$$

thus, $\mathbf{w}$ is a positive scalar multiple of $\mathbf{v}$.

Suppose that $\mathbf{v} \neq \mathbf{0}$ and $r < 0$. Then, if we calculate the direction of $r\mathbf{v}$, we find

$$\frac{r\mathbf{v}}{|r\mathbf{v}|} = \frac{r}{|r|} \cdot \frac{\mathbf{v}}{|\mathbf{v}|} = -\frac{\mathbf{v}}{|\mathbf{v}|},$$

since $r < 0$. Hence, any negative scalar multiple of $\mathbf{v}$ has the negative, or opposite, direction of that of $\mathbf{v}$. Conversely, it is easy to show that if two non-zero vectors have opposite direction, then each is a negative scalar multiple of the other.

Two non-zero vectors are parallel, by definition, if and only if they have the same or opposite directions. Therefore, we conclude:

Theorem 1.2.13. *Two non-zero vectors* **v** *and* **w** *have the same direction (we also say "point in the same direction") if and only if they are positive scalar multiples of each other.*

Two non-zero vectors **v** *and* **w** *have the opposite direction if and only if they are negative scalar multiples of each other.*

Two non-zero vectors **v** *and* **w** *are parallel if and only if they are (non-zero) scalar multiples of each other.*

Example 1.2.14. Find the magnitude and direction of the displacement vector from the point $\mathbf{p} = (1, 2, 3)$ to the point $\mathbf{q} = (-3, 0, 4)$, where all distances are in meters.

Solution:

We will also treat $\mathbf{p}$ and $\mathbf{q}$ as vectors, so that we can use vector operations to make sense of $\mathbf{q} - \mathbf{p}$.

The displacement vector is

$$\mathbf{d} = \overrightarrow{\mathbf{pq}} = \mathbf{q} - \mathbf{p} = (-4, -2, 1) \text{ meters.}$$

Its magnitude is

$$|\mathbf{d}| = \sqrt{(-4)^2 + (-2)^2 + 1^2} = \sqrt{21} \text{ meters.}$$

Thus, its direction, the unique vector in the direction of $\mathbf{d}$ is

$$\frac{\mathbf{d}}{|\mathbf{d}|} = \frac{1}{\sqrt{21}}(-4, -2, 1).$$

Note that the direction is **unitless**, the meters cancel out.

For many calculations, it is simpler to deal with this unit vector as we have written it, rather than multiplying the scalar times each component.

Remark 1.2.15. Something that arose in the previous example is worth noting on its own: the distance between two points is the magnitude of their difference, if you treat the points as vectors, i.e.,

$$\text{dist}(\mathbf{p}, \mathbf{q}) = |\mathbf{q} - \mathbf{p}| = |\mathbf{p} - \mathbf{q}|.$$

Example 1.2.16. Decide if any of the following vectors are parallel to any of the others:

$$(2, -3, 0, 4), \quad (-3, 4, 0, -6), \quad (-1.5, 2, 0, -3), \quad (8, -12, 1, 7).$$

For parallel vectors, decide if they point in the same or opposite directions.

Solution:

We need to see which vectors are (necessarily, non-zero) scalar multiples of other vectors, and also note whether the (possible) scalars are positive or negative.

First, note that the first three vectors all have a 0 for their third component, while the fourth vector has a 1 in its third component. Thus, the fourth vector cannot possibly be a scalar multiple of any of the first three.

So, now we ask: is there a scalar r such that $(2, -3, 0, 4) = r(-3, 4, 0, -6)$? This would mean $(2, -3, 0, 4) = (-3r, 4r, 0r, -6r)$. From the 1st component, we see that we would have to have $2 = -3r$, i.e., r would have to be $-2/3$. Then you just check if $r = -2/3$ works in each of the other components. It doesn't, because the 2nd component would require $-3 = 4r$, and $4(-2/3)$ is not -3. Therefore, there is no such r, and $(2, -3, 0, 4)$ and $(-3, 4, 0, -6)$ are **not** parallel.

Is there an r such that $(2, -3, 0, 4) = r(-1.5, 2, 0, -3)$? Again, looking at the 1st component tells you what r would have to be; we would have to have $r = 2/(-1.5) = -4/3$. But, again, this r doesn't work in the 2nd component. Hence, there is no such r, and $(2, -3, 0, 4)$ and $(-1.5, 2, 0, -3)$ are not parallel.

Finally, is there an r such that $(-3, 4, 0, -6) = r(-1.5, 2, 0, -3)$? From the 1st component, r would have to be 2, and you check and see that $r = 2$ also works for the other three components. Thus, $(-3, 4, 0, -6) = 2(-1.5, 2, 0, -3)$, so the vectors $(-3, 4, 0, -6)$ and $(-1.5, 2, 0, -3)$ are parallel and, in fact, point in the same direction, since the scalar is positive.

Example 1.2.17. Find the unique vector $\mathbf{v}$ which has the same direction as $\mathbf{w} = (1, 3, 2)$, but which has magnitude 7.

Solution:

The unit vector with the same direction as $\mathbf{w}$ is

$$\mathbf{u} = \frac{\mathbf{w}}{|\mathbf{w}|} = \frac{(1, 3, 2)}{\sqrt{14}}.$$

We want a vector with 7 times this magnitude, but the same direction; we simply scalar multiply by 7. Thus,

$$\mathbf{v} = \frac{7}{\sqrt{14}}(1, 3, 2).$$

Example 1.2.18. Suppose we have two objects which we assume to be *point-masses*, i.e., which occupy so little volume that we assume that they are located at individual points. Suppose that the point-masses have masses M and m and are located at points $\mathbf{p}_M$ and $\mathbf{p}_m$, respectively. Let $\mathbf{r} = \mathbf{p}_M - \mathbf{p}_m$, so that $\mathbf{r}$ is a vector which points from m to M, and whose magnitude is the distance between the masses. We assume that the masses are located at different points, so that $\mathbf{r} \neq 0$.

Then, *Newton's Law of Universal Gravitation* states that each mass exerts a gravitational attraction force on the other, along the line between masses, and that the magnitude of the force is proportional to the product of the masses and inversely proportional to the square of the distance between the masses. The proportionality constant, the *universal gravitational constant*, is denoted by G, and its value, in metric units, is approximately 6.67384×10^{-11} m^3/(kg·s^2).

We can write this very succinctly using vector notation. The force $\mathbf{F}_{M \leftarrow m}$ that M exerts on m is

$$\mathbf{F}_{M \leftarrow m} \;=\; \frac{GMm}{|\mathbf{r}|^2}\left(\frac{\mathbf{r}}{|\mathbf{r}|}\right).$$

The force $\mathbf{F}_{m \leftarrow M}$ that m exerts on M is

$$\mathbf{F}_{m \leftarrow M} \;=\; \frac{GMm}{|-\mathbf{r}|^2}\left(\frac{-\mathbf{r}}{|-\mathbf{r}|}\right) \;=\; -\mathbf{F}_{M \leftarrow m}.$$

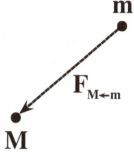

Figure 1.2.6: The gravitational force that M exerts on m.

We frequently combine vectors by scalar multiplying and adding, e.g,

$$3(1, -1, 2) + 7(4, 3, -5).$$

We give a name to such expressions:

Definition 1.2.19. *A* **linear combination** *of the vectors* $\mathbf{v}_1$, $\mathbf{v}_2$, ..., $\mathbf{v}_k$ *in* $\mathbb{R}^n$ *is any expression of the form*

$$t_1\mathbf{v}_1 + t_2\mathbf{v}_2 + \cdots + t_k\mathbf{v}_k,$$

where the t_i's are real numbers (i.e., scalars).

Example 1.2.20. Let $\mathbf{a} = (-1, 2, 0, 5)$, $\mathbf{b} = (3, 1, -3, 4)$, and $\mathbf{c} = (0, 1, 4, -7)$. Calculate the linear combinations

$$2\mathbf{a} - \mathbf{b} + 3\mathbf{c} \quad \text{and} \quad -\frac{3}{2}\mathbf{a} + 2\mathbf{b} - \frac{1}{2}\mathbf{c}.$$

Solution:

There's nothing tricky to do here; you simply have to calculate.

We find:

$$2\mathbf{a} - \mathbf{b} + 3\mathbf{c} = 2(-1, 2, 0, 5) - (3, 1, -3, 4) + 3(0, 1, 4, -7) =$$

$$(-2, 4, 0, 10) - (3, 1, -3, 4) + (0, 3, 12, -21) = (-5, 6, 15, -15)$$

and

$$-\frac{3}{2}\mathbf{a} + 2\mathbf{b} - \frac{1}{2}\mathbf{c} = -\frac{3}{2}(-1, 2, 0, 5) + 2(3, 1, -3, 4) - \frac{1}{2}(0, 1, 4, -7) =$$

$$\left(\frac{3}{2}, -3, 0, -\frac{15}{2}\right) + (6, 2, -6, 8) + \left(0, -\frac{1}{2}, -2, \frac{7}{2}\right) = \left(\frac{15}{2}, -\frac{3}{2}, -8, 4\right) =$$

$$\frac{1}{2}(15, -3, -16, 8).$$

There are special unit vectors in $\mathbb{R}^n$ that we give names to.

Definition 1.2.21. *In $\mathbb{R}^2$, we let $\mathbf{i} = (1, 0)$, which we frequently think of as pointing "east". and we let $\mathbf{j} = (0, 1)$, which we frequently think of as "north". The vectors $\mathbf{i}$ and $\mathbf{j}$ form the* **standard basis** *for $\mathbb{R}^2$.*

In $\mathbb{R}^3$, we let $\mathbf{i} = (1, 0, 0)$, which we frequently think of as pointing "east". and we let $\mathbf{j} = (0, 1, 0)$, which we frequently think of as "north". We let $\mathbf{k} = (0, 0, 1)$, which we frequently think of as "up". The vectors $\mathbf{i}$, $\mathbf{j}$, and $\mathbf{k}$ form the **standard basis** *for $\mathbb{R}^3$.*

More generally, in $\mathbb{R}^n$, for $1 \leq i \leq n$, we let $\mathbf{e}_i$ denote the vector which has zeroes in every component, except for having a 1 in the i-th component. The ordered collection of the unit vectors $\mathbf{e}_i$ is the **standard basis** *for $\mathbb{R}^n$.*

Remark 1.2.22. Many books prefer to write essentially all vectors in $\mathbb{R}^2$ and $\mathbb{R}^3$ in terms of $\mathbf{i}$, $\mathbf{j}$, and $\mathbf{k}$. This is certainly easy, but usually a bit cumbersome.

For instance, how do you write $(3, -5, 7)$ in terms of $\mathbf{i}$, $\mathbf{j}$, and $\mathbf{k}$? It's simple:

$$(3, -5, 7) \;=\; 3\mathbf{i} - 5\mathbf{j} + 7\mathbf{k},$$

that is, you just scalar multiply the components of the vector times the corresponding standard basis vector and add.

Why? Because of how scalar multiplication and vector addition are defined. If you work it out, it looks like this:

$$(3, -5, 7) \;=\; (3, 0, 0) + (0, -5, 0) + (0, 0, 7) \;=$$

$$3(1, 0, 0) - 5(0, 1, 0) + 7(0, 0, 1) \;=\; 3\mathbf{i} - 5\mathbf{j} + 7\mathbf{k}.$$

Of course, there are times when it's definitely preferable to write vectors in terms of the standard basis. For instance, the force of gravity on an object of mass m is frequently written as mg, where g denotes the (positive) acceleration to to gravity; this, however, is just the **magnitude** of the force vector. The direction is straight down, i.e., in the direction of $-\mathbf{k}$. Thus, the force of gravity, acting on the mass is actually

$$\mathbf{F}_{\text{grav}} \;=\; -mg\mathbf{k}.$$

Vector addition is common when discussing the velocities of boats in moving water, or airplanes in moving air.

> **More Depth:**

Consider an airplane which is flying through (possibly) moving air. There are three related velocities that are important here. There's $\mathbf{v}_{\text{air,rel.ground}}$, the velocity at which the air is moving, relative to the fixed ground. There's also $\mathbf{v}_{\text{plane,rel.ground}}$, the velocity at which the airplane is moving, relative to the fixed ground; you want the direction of this to be towards your destination, and its magnitude determines how long it will take to get there.

Finally, there's $\mathbf{v}_{\text{plane,rel.air}}$, the velocity at which the plane is moving, relative to the air that it's in. If you ignored the ground, and just used the air itself as your frame of reference, the velocity you'd measure would be $\mathbf{v}_{\text{plane,rel.air}}$. For instance, if the plane is simply flowing along with the air, we would have $\mathbf{v}_{\text{plane,rel.air}} = 0$. The velocity $\mathbf{v}_{\text{plane,rel.air}}$ is what is actually controlled by the pilots.

What's the relationship between these three velocities? Well...it's easy to see in special cases. For instance, if the wind is blowing due north at 20 mph, and the plane is moving 300 mph north relative to the air, then this *tail wind* makes the plane move at 320 mph relative to the ground. If you think about other cases, you should conclude more generally that

$$\mathbf{v}_{\text{plane,rel.ground}} \;=\; \mathbf{v}_{\text{plane,rel.air}} + \mathbf{v}_{\text{air,rel.ground}}.$$

Example 1.2.23. Suppose that an airplane maintains a constant altitude, so that the direction of its motion is determined by an xy-grid on the ground.

Suppose that the wind is blowing at 30 mph in the direction of the vector $(1,2)$, and the plane wishes to travel at 400 mph, relative to the ground, in the direction of $\mathbf{j} = (0,1)$, i.e., due north. What vector should the plane use for its velocity, relative to the air?

Solution:

We know that

$$\mathbf{v}_{\text{plane,rel.air}} \;=\; \mathbf{v}_{\text{plane,rel.ground}} - \mathbf{v}_{\text{air,rel.ground}}.$$

We're told that $\mathbf{v}_{\text{plane,rel.ground}}$ should equal $400\mathbf{j} = (0,400)$ mph. We're also told $\mathbf{v}_{\text{air,rel.ground}}$, but some calculation is involved; what we're told is that

$$\mathbf{v}_{\text{air,rel.ground}} \;=\; 30 \cdot \frac{(1,2)}{|(1,2)|} \;=\; \frac{30}{\sqrt{5}}(1,2) \;=\; 6\sqrt{5}(1,2) \;=\; (6\sqrt{5}, 12\sqrt{5}) \text{ mph.}$$

Therefore, the vector that the plane should use, relative to the air, is

$$(0,400) - (6\sqrt{5}, 12\sqrt{5}) \;=\; (-6\sqrt{5}, 400 - 12\sqrt{5}) \;\approx\; (-13.4164, 373.1672) \text{ mph.}$$

From the definitions of vector addition and scalar multiplication in $\mathbb{R}^n$, there are many algebraic properties which obviously hold; properties that seem so natural that you'd probably just assume they are true, even if we didn't state them explicitly.

It's easy to verify that vector addition and scalar multiplication in $\mathbb{R}^n$ satisfy the following properties, where all vectors are in $\mathbb{R}^n$:

1. for all vectors $\mathbf{a}$ and $\mathbf{b}$, $\mathbf{a} + \mathbf{b} \;=\; \mathbf{b} + \mathbf{a}$;

2. for all vectors $\mathbf{a}$, $\mathbf{b}$, and $\mathbf{c}$, $(\mathbf{a} + \mathbf{b}) + \mathbf{c} = \mathbf{a} + (\mathbf{b} + \mathbf{c})$;

3. there is an additive identity element, the zero vector $\mathbf{0}$, such that, for all vectors $\mathbf{a}$, $\mathbf{0} + \mathbf{a} = \mathbf{a} = \mathbf{a} + \mathbf{0}$;

4. every vector $\mathbf{b}$ has a unique additive inverse, $-\mathbf{b}$, such that $\mathbf{b} + (-\mathbf{b}) = \mathbf{0} = (-\mathbf{b}) + \mathbf{b}$;

5. for all vectors $\mathbf{a}$ and $\mathbf{b}$ and all scalars r, $r(\mathbf{a} + \mathbf{b}) = r\mathbf{a} + r\mathbf{b}$;

6. for all vectors $\mathbf{a}$ and all scalars r and s, $(r + s)\mathbf{a} = r\mathbf{a} + s\mathbf{a}$;

7. for all vectors $\mathbf{a}$ and all scalars r and s, $r(s\mathbf{a}) = (rs)\mathbf{a}$; and

8. for all vectors $\mathbf{a}$, $1\mathbf{a} = \mathbf{a}$.

In fact, we make the following general definition:

Definition 1.2.24. *A **vector space over** $\mathbb{R}$ is a set of objects, called **vectors**, and two operations called **vector addition** and **scalar multiplication** that take two vectors (respectively, a real number and a vector), and return a vector, such that the eight **axioms** above are satisfied.*

You can use scalars other than the real numbers, and the axioms for a vector space remain the same; you need for the scalars to form what is known as a field. Examples of fields other than $\mathbb{R}$ are the rational numbers $\mathbb{Q}$ and the complex numbers $\mathbb{C}$.

You may wonder why certain, desirable algebraic properties are omitted from the axioms, such as:
$$r\mathbf{a} = \mathbf{0} \quad \text{if and only if} \quad r = 0 \text{ or } \mathbf{a} = 0.$$
The answer is: you can prove that this, and other properties, follow from the given eight axioms.

An example of another important vector space, where the vectors are not in $\mathbb{R}^n$, is the vector space of functions $f : \mathbb{R}^n \to \mathbb{R}^m$, where we define vector addition by defining $f + g$ to mean the function from $\mathbb{R}^n$ to $\mathbb{R}^m$ given by

$$(f + g)(\mathbf{x}) = f(\mathbf{x}) + g(\mathbf{x}),$$

and defining rf to mean the function given by

$$(rf)(\mathbf{x}) = rf(\mathbf{x}).$$

+ Linear Algebra:

In linear algebra, matrices (the plural form of *matrix*) play a central role. A matrix is usually described as rectangular collection of numbers (usually, for us, real numbers), the horizontal lines are *rows*, while the vertical lines are *columns*. For instance, the matrix

$$\begin{bmatrix} 3 & -5 & 7 \\ -1 & 0 & 2 \end{bmatrix}$$

has 2 rows and 3 columns. We say that this is a 2×3 (read: 2 by 3) matrix.

It is then standard to consider a vector in $\mathbb{R}^n$ as a *column vector*, a matrix with 1 column and n rows.

For instance, if $\mathbf{v} = (2, 5, -1, 7)$, and we want to write $\mathbf{v}$ as a column vector, then we put square brackets around $\mathbf{v}$ and write the corresponding 1×4 matrix, i.e.,

$$[\mathbf{v}] = \begin{bmatrix} 2 \\ 5 \\ -1 \\ 7 \end{bmatrix}.$$

Example 1.2.25. Calculate

$$3 \begin{bmatrix} 2 \\ 5 \\ -1 \\ 7 \end{bmatrix} - 4 \begin{bmatrix} 1 \\ 0 \\ 3 \\ -2 \end{bmatrix}.$$

Solution:

This is just written in a different form, but it's not any more difficult to calculate:

$$3 \begin{bmatrix} 2 \\ 5 \\ -1 \\ 7 \end{bmatrix} - 4 \begin{bmatrix} 1 \\ 0 \\ 3 \\ -2 \end{bmatrix} = \begin{bmatrix} 6 \\ 15 \\ -3 \\ 21 \end{bmatrix} - \begin{bmatrix} 4 \\ 0 \\ 12 \\ -8 \end{bmatrix} = \begin{bmatrix} 2 \\ 15 \\ -15 \\ 29 \end{bmatrix}.$$

Finally, in Definition 1.2.21, we defined the standard basis for $\mathbb{R}^n$. What does the term *basis* mean? A basis for $\mathbb{R}^n$ is any collection of vectors $\mathbf{b}_1$, $\mathbf{b}_2$, ..., $\mathbf{b}_m$ such that, for every vector $\mathbf{v}$ in $\mathbb{R}^n$, there exist unique real numbers t_1, t_2, ..., t_m (which depend on $\mathbf{v}$) such that

$$\mathbf{v} = t_1 \mathbf{b}_1 + t_2 \mathbf{b}_2 + \cdots + t_m \mathbf{b}_m.$$

The standard basis is obviously a basis for $\mathbb{R}^n$ since, given $\mathbf{v} = (v_1, v_2, \ldots, v_n)$, we have the unique "expansion"

$$(v_1, v_2, \ldots, v_n) = v_1 \mathbf{e}_1 + v_2 \mathbf{e}_2 + \cdots + v_n \mathbf{e}_n.$$

It can be shown that all bases for the same vector space have the same number of basis elements. As there are n standard basis vectors for $\mathbb{R}^n$, every basis for $\mathbb{R}^n$ has exactly n vectors. Thus, above, when we used m for the number of basis vectors for $\mathbb{R}^n$, it was, in fact, true that m had to be equal to n.

1.2.1 Exercises

Online answers to select exercises are here.

Basics:

In each of the following exercises, determine the magnitude and direction of the given vector. The direction should be a unit vector.

1. $\mathbf{v} = (3, 4)$ feet per second.

2. $\mathbf{p} = (5, 12)$ meters.

3. $\mathbf{v} = (-6, 1, 6)$ meters per second.

4. $\mathbf{F} = (2, -1, 3)$ Newtons.

5. $\mathbf{a} = (1, -1, 1, -1)$ (m/s)/s.

6. $\mathbf{b} = (0, 1, 2, 0, -2, -1)$ feet.

7. $2\mathbf{i} - 3\mathbf{j} + \mathbf{k}$ pounds

8. $5\mathbf{i} - \mathbf{j} + \sqrt{10}\,\mathbf{k}$ meters per second

In each of the following problems, you are given two vectors $\mathbf{v}$ and $\mathbf{w}$, and a scalar a. Sketch the vectors $\mathbf{v}$, $\mathbf{w}$, and $\mathbf{v} + \mathbf{w}$, as in the parallelogram rule in Example 1.2.6. Also sketch $a\mathbf{v}$, $a\mathbf{w}$, and $a\mathbf{v} + a\mathbf{w}$

9. $\mathbf{v} = (1, 2)$, $\mathbf{w} = (1, 1)$, $a = 2$.

10. $\mathbf{v} = (1, 2)$, $\mathbf{w} = (1, 1)$, $a = -2$.

11. $\mathbf{v} = (1, 2)$, $\mathbf{w} = (1, 1)$, $a = 0.5$.

12. $\mathbf{v} = (1, 2)$, $\mathbf{w} = (1, 1)$, $a = -3$.

In each of the following problems, you are given a pair of vectors. Determine whether the given vectors are parallel to each other and, if they are, determine whether the vectors have the same or opposite directions. Do this without explicitly finding the unit vectors for the directions.

13. $(1, 2)$ and $\left(-\frac{1}{2}, -1\right)$

14. $(3, 4)$ and $(-6, -7)$

15. $(1, -2, 3)$ and $(2, -4, 5)$

16. $(1, -2, 3)$ and $(3, -6, 9)$

17. $(e^{-x}x^2, -x^3)$ and $(-2x^2, 2e^x x^3)$

18. $(ab, b^2 + 1)$ and $(a^3b, a^2b^2 + 1)$ (Here, your answer may include conditions on a and/or b.)

In each of the following problems, you are given a collection of force vectors, in Newtons, which are all of the forces acting on an object which has a mass of 2 kilograms. Determine the acceleration a of the object, and the magnitude of the acceleration.

19. $\mathbf{F}_1 = (1, 2)$ and $\mathbf{F}_2 = (-1, 2)$.

20. $\mathbf{F}_1 = 2(1, 2, -1)$ and $\mathbf{F}_2 = 3(-1, 2, 3)$.

21. $\mathbf{F}_1 = \mathbf{i} + 2\mathbf{j} - \mathbf{k}$ and $\mathbf{F}_2 = -\mathbf{j} + 2\mathbf{k}$.

22. $\mathbf{F}_1 = 3\mathbf{i} - 2\mathbf{j}$, $\mathbf{F}_2 = \mathbf{i} + 4\mathbf{j}$, and $\mathbf{F}_3 = -2\mathbf{i} + 5\mathbf{j}$.

In each of the following problems, you are given the velocity, in feet per second, of an object. You are given this velocity as a linear combination of other velocity vectors. Determine the speed of the object.

23. $5(1, 2) - 7(0, 1) + 4(2, -3)$

24. $3(1, 0, -2) + 4(1, 1, 1)$

25. $2(3\mathbf{i} - \mathbf{j}) - 5(\mathbf{i} + \mathbf{j})$

26. $7(\mathbf{i} + \mathbf{j} - \mathbf{k}) + 2(-\mathbf{i} - \mathbf{j} + 3\mathbf{k})$

In each of the following problems, you are given points a and b representing locations in $\mathbb{R}^n$, measured in meters. Determine the displacement vector from a to b, and find its magnitude and direction.

27. $\mathbf{a} = (1, 2)$ and $\mathbf{b} = (-2, -3)$.

28. $\mathbf{a} = (\pi, 1)$ and $\mathbf{b} = (-7 + \pi, 1)$.

29. $\mathbf{a} = (1, 2, 0)$ and $\mathbf{b} = (0, -2, -3)$.

30. $\mathbf{a} = (7, 11, -13)$ and $\mathbf{b} = (5, 10, -15)$.

31. $\mathbf{a} = (1, 2, 0, 1)$ and $\mathbf{b} = (0, -2, -3, 2)$.

32. $\mathbf{a} = (7, 11, -13, 15)$ and $\mathbf{b} = (5, 10, -15, 17)$.

In each of the following problems, you are given a vector. Determine the vectors of magnitudes 3 and 7 that have the same direction as the given vector.

33. $(3, 4)$

34. $(5, 12)$

35. $(-6, 1, 6)$

36. $(2, -1, 3)$

37. $(1, -1, 1, -1)$

38. $(0, 1, 2, 0, -2, -1)$

39. $2\mathbf{i} - 3\mathbf{j} + \mathbf{k}$

40. $5\mathbf{i} - \mathbf{j} + \sqrt{10}\,\mathbf{k}$

41. Suppose that $|\mathbf{v}| = 3$. What is the magnitude of $-5\mathbf{v}$?

42. Suppose that $\mathbf{v} \neq \mathbf{0}$. What is the magnitude of $\dfrac{13\mathbf{v}}{|\mathbf{v}|}$?

43. Suppose that $|\mathbf{v}| = 3$ and that $\mathbf{v}$ has the same direction as $(1, 1)$. What is $\mathbf{v}$?

44. Suppose that $|\mathbf{v}| = 2$ and that $\mathbf{v}$ has the same direction as $(-1, 1, \sqrt{7})$. What is $\mathbf{v}$?

45. A mass of 3 kilograms is acted on by gravity (near the Earth's surface). The only other force acting on the mass is a force of 30 Newtons, upward. Assume that the Earth's gravitational constant is $g = 9.81$ (m/s)/s. Determine the acceleration vector of the mass. Is the mass accelerating upward or downward, or is it impossible to decide? Is the mass moving upward or downward, or is it impossible to decide?

46. A point-mass A of mass $m_A = 3$ kg is located at $\mathbf{p}_A = (1, 2, 3)$ meters. A point-mass B of mass $m_b = 5$ kg is located at $\mathbf{p}_B = (2, 5, 7)$ meters. What gravitational force (vector) does A exert on B?

47. A point-mass A of mass $m_A = 3$ kg is located at $\mathbf{p}_A = (1, 2, 3)$ meters. A point-mass B is located at $\mathbf{p}_B = (2, 3, 4)$ meters, and exerts a gravitational force on A given by

$$\mathbf{F}_{B \leftarrow A} = 0.0001\,\mathbf{u} \quad \text{Newtons},$$

where $\mathbf{u}$ is a unit vector. What is $\mathbf{u}$, and what is the mass of B?

More Depth:

48. Suppose that an airplane maintains a constant altitude, so that the direction of its motion is determined by an xy-grid on the ground.

    Suppose that the wind is blowing at 15 m/s in the direction of the vector $(1,1)$ (northeast), and the plane wishes to travel at 200 m/s, relative to the ground, in the direction of $(1,-1)$ (southeast). What vector should the plane use for its velocity, relative to the air?

49. Suppose that the wind is blowing at 15 m/s in the direction of the vector $(1,1,2)$, and a plane wishes to travel at 200 m/s, relative to the ground, in the direction of $(1,-1,0)$ (southeast, at a constant altitude). What vector should the plane use for its velocity, relative to the air?

50. Prove that $\mathbb{R}^2$, with its standard vector addition and scalar multiplication, satisfies the 8 axioms for a vector space over $\mathbb{R}$.

51. Suppose that $\mathbf{v}$ is a vector in a vector space over $\mathbb{R}$, and $\mathbf{v} + \mathbf{v} = \mathbf{v}$. Prove that $\mathbf{v} = \mathbf{0}$, the zero vector.

52. Suppose that $\mathbf{v}$ is a vector in a vector space over $\mathbb{R}$. Prove that $0\mathbf{v} = \mathbf{0}$, the zero vector.

53. Suppose that r is a real number, and that $\mathbf{0}$ is the zero vector in some vector space over $\mathbb{R}$. Prove that $r\mathbf{0} = \mathbf{0}$.

54. Suppose that $\mathbf{v}$ is a vector in a vector space over $\mathbb{R}$, and that r is a real number such that $r\mathbf{v} = \mathbf{0}$. Prove that $r = 0$ or $\mathbf{v} = \mathbf{0}$.

+ Linear Algebra:

55. Calculate

$$3\begin{bmatrix} 2 \\ 5 \\ 7 \end{bmatrix} - 4\begin{bmatrix} 1 \\ 3 \\ -2 \end{bmatrix} + 2\begin{bmatrix} -1 \\ 6 \\ 0 \end{bmatrix}.$$

56. Calculate the magnitude of

$$2\begin{bmatrix} 1 \\ 3 \\ -1 \\ 0 \end{bmatrix} - \begin{bmatrix} 2 \\ 0 \\ -3 \\ 5 \end{bmatrix}.$$

57. Suppose that $\mathbf{b}_1$, $\mathbf{b}_2$, ..., $\mathbf{b}_n$ is a basis for $\mathbb{R}^n$. Prove that the $\mathbf{b}_i$'s are *linearly independent*, that is, prove that, if

$$t_1\mathbf{b}_1 + t_2\mathbf{b}_2 + \cdots + t_n\mathbf{b}_n = \mathbf{0},$$

then $t_1 = t_2 = \cdots = t_n = 0$.

1.3 Dot Product, Angles, and Orthogonal Projection

The dot product on $\mathbb{R}^n$ is an easy-to-calculate operation that you perform on pairs of vectors and which gives you back a real number, not a vector. The dot product is important because, in 2 and 3 dimensions, the dot product gives us an easy way of computing the angle between vectors. In higher dimensions, the dot product is used to define the angle between two vectors. A fundamental application of the dot product is in calculating the work done by a constant force as an object undergoes a displacement.

The dot product also arises when dealing with orthogonal projection. Orthogonal projection gives us a simple way to decompose a vector into a sum of two vectors, one of which is parallel to a prescribed vector **b** and the other of which is perpendicular, or orthogonal, to **b**.

Basics:

If we have two vectors in $\mathbb{R}^2$ or $\mathbb{R}^3$, and we draw them as starting at the same point, then we may talk about the *angle between the vectors*, which could be anywhere from 0 degrees or radians, when the vectors point in the same direction, to 180 degrees, or π radians, when the vectors point in opposite directions.

Given two vectors in coordinates, we would like to find an easy way to calculate the angle between them.

Consider two vectors in $\mathbb{R}^2$, $\mathbf{a} = (x_1, y_1)$ and $\mathbf{b} = (x_2, y_2)$. We draw these vectors based at the origin, and we let $\mathbf{c} = \mathbf{b} - \mathbf{a}$; see Figure 1.3.3.

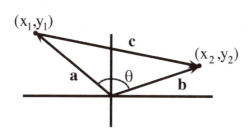

Figure 1.3.3: Vectors **a** and **b**, and the angle between them.

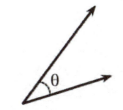

Figure 1.3.1: An acute angle between vectors.

Recall the *Law of Cosines* from trigonometry, which, in terms of our vectors, says

$$|\mathbf{c}|^2 \;=\; |\mathbf{a}|^2 + |\mathbf{b}|^2 - 2|\mathbf{a}||\mathbf{b}|\cos\theta.$$

Noting that

$$|\mathbf{a}|^2 = x_1^2 + y_1^2, \qquad |\mathbf{b}|^2 = x_2^2 + y_2^2, \qquad \text{and} \qquad |\mathbf{c}|^2 = (x_2 - x_1)^2 + (y_2 - y_1)^2,$$

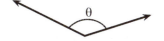

Figure 1.3.2: An obtuse angle between vectors.

we find that the Law of Cosines becomes

$$x_2^2 - 2x_1x_2 + x_1^2 + y_2^2 - 2y_1y_2 + y_1^2 \ = \ x_1^2 + y_1^2 + x_2^2 + y_2^2 - 2|\mathbf{a}||\mathbf{b}|\cos\theta.$$

Subtracting $x_1^2 + y_1^2 + x_2^2 + y_2^2$ from both sides, and dividing by -2, we arrive at

$$x_1x_2 + y_1y_2 \ = \ |\mathbf{a}||\mathbf{b}|\cos\theta.$$

Note how simple the left-hand side of the above formula is; it's just the product of the x-coordinates of the vectors $\mathbf{a}$ and $\mathbf{b}$ added to the product of the y-coordinates.

You can perform the analogous calculations for two vectors $\mathbf{p} = (x_1, y_1, z_1)$ and $\mathbf{q} = (x_2, y_2, z_2)$ in $\mathbb{R}^3$, and what you find is

$$x_1x_2 + y_1y_2 + z_1z_2 \ = \ |\mathbf{p}||\mathbf{q}|\cos\theta,$$

where θ is the angle between $\mathbf{p}$ and $\mathbf{q}$. Again, the left-hand side is simply the sum of the product of the corresponding coordinates of the two vectors.

This motivates us to define:

> It is a common, but horrible, mistake to think that the dot product of two vectors yields another vector. You add together the products of the corresponding coordinates, so **you end up with a number, a scalar, not a vector.**

> **Definition 1.3.1.** *Let* $\mathbf{v} = (v_1, v_2, \ldots, v_n)$ *and* $\mathbf{w} = (w_1, w_2, \ldots, w_n)$ *be vectors in* $\mathbb{R}^n$*. Then, the* **dot product** $\mathbf{v} \cdot \mathbf{w}$ *of* $\mathbf{v}$ *and* $\mathbf{w}$ *is the real number given by adding together the product to the corresponding coordinates of the two vectors, i.e.,*
>
> $$\mathbf{v} \cdot \mathbf{w} \ = \ v_1w_1 + v_2w_2 + \cdots + v_nw_n.$$

The important properties of the dot product, which we shall use throughout the remainder of this book are:

Theorem 1.3.2. *Let* **a**, **b**, *and* **c** *be vectors in* $\mathbb{R}^n$, *and let* r *and* s *be real numbers. Then,*

1. **(commutativity)** $\mathbf{a} \cdot \mathbf{b} = \mathbf{b} \cdot \mathbf{a}$;

2. **(distributivity)** $\mathbf{a} \cdot (\mathbf{b} + \mathbf{c}) = (\mathbf{a} \cdot \mathbf{b}) + (\mathbf{a} \cdot \mathbf{c})$;

3. **(scalar extraction)** $(r\mathbf{a}) \cdot (s\mathbf{b}) = (rs)(\mathbf{a} \cdot \mathbf{b})$;

4. $\mathbf{a} \cdot \mathbf{a} = |\mathbf{a}|^2$;

5. **(Cauchy-Schwarz Inequality)** *the absolute value of* $\mathbf{a} \cdot \mathbf{b}$ *satisfies*

$$|\mathbf{a} \cdot \mathbf{b}| \leq |\mathbf{a}||\mathbf{b}|,$$

   and the equality holds if and only if **a** *and* **b** *are parallel; and*

6. *the dot product is related to the angles between vectors by*

$$\mathbf{a} \cdot \mathbf{b} = |\mathbf{a}||\mathbf{b}|\cos\theta,$$

   where θ *is the angle between the vectors* **a** *and* **b**.

> Note that, if one of the vectors in the dot product is the zero vector, then there is no "the" angle between the vectors, because we allow the zero vector to have **every** direction. Still, we go ahead and write that $\mathbf{a} \cdot \mathbf{b} = |\mathbf{a}||\mathbf{b}|\cos\theta$, since if **a** or **b** equals **0**, the equality holds for all θ.

Remark 1.3.3. We leave the verification of properties 1-4 of Theorem 1.3.2 as exercises; they follow easily from the definition and corresponding properties of real numbers.

We will prove the Cauchy-Schwarz Inequality in the More Depth portion of this section.

We derived the formula $\mathbf{a} \cdot \mathbf{b} = |\mathbf{a}||\mathbf{b}|\cos\theta$ for vectors in $\mathbb{R}^2$; the proof in $\mathbb{R}^3$ is essentially identical. What happens in $\mathbb{R}^n$, where $n \geq 4$?

The answer may seem like cheating. If **a** and **b** are non-zero vectors in $\mathbb{R}^n$, then the Cauchy-Schwarz Inequality tells us that

$$-1 \leq \frac{\mathbf{a} \cdot \mathbf{b}}{|\mathbf{a}||\mathbf{b}|} \leq 1,$$

and we **define** the angle between **a** and **b** to be

$$\theta = \cos^{-1}\left(\frac{\mathbf{a} \cdot \mathbf{b}}{|\mathbf{a}||\mathbf{b}|}\right).$$

Example 1.3.8. Sometimes, given a non-zero vector (a, b) in $\mathbb{R}^2$, it is desirable to produce a non-zero vector which is perpendicular to (a, b). How do you do this? It's easy.

Just swap a and b and negate one them. That is, take the vector $(b, -a)$ or $(-b, a)$. It is trivial to verify that the dot product of either of these with (a, b) is 0:

$$(a, b) \cdot (b, -a) \ = \ ab + (b)(-a) \ = \ 0 \quad \text{and} \quad (a, b) \cdot (-b, a) \ = \ (a)(-b) + ba \ = \ 0.$$

In many physical problems, you are given a vector $\mathbf{F}$, and a non-zero vector $\mathbf{v}$, and you want to consider the "part of $\mathbf{F}$ that is parallel to $\mathbf{v}$". What does this mean?

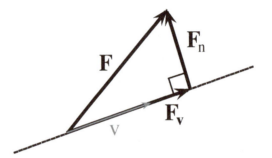

Figure 1.3.4: Writing $\mathbf{F}$ as a sum of vectors parallel and normal to $\mathbf{v}$.

What we mean is that we want to write $\mathbf{F}$ as the sum of two vectors $\mathbf{F_v}$ and $\mathbf{F}_n$, where $\mathbf{F_v}$ is a scalar multiple of $\mathbf{v}$ and $\mathbf{F}_n$ is orthogonal, or normal, to $\mathbf{v}$. Thus, we want

$$\mathbf{F} \ = \ \mathbf{F_v} + \mathbf{F}_n \ = \ t\mathbf{v} + \mathbf{F}_n,$$

for some t, where $\mathbf{F}_n$ is orthogonal to $\mathbf{v}$.

How do we determine t? We take the dot product of both sides of the previous equation with $\mathbf{v}$ and use that we are requiring that $\mathbf{F}_n$ is orthogonal to $\mathbf{v}$, so that $\mathbf{F}_n \cdot \mathbf{v} = 0$; we find

$$\mathbf{F} \cdot \mathbf{v} \ = \ t(\mathbf{v} \cdot \mathbf{v}) \ + \ 0.$$

Therefore, t would have to equal $(\mathbf{F} \cdot \mathbf{v})/(\mathbf{v} \cdot \mathbf{v})$, and so, if there exist $\mathbf{F_v}$ and $\mathbf{F}_n$ with the desired properties, we must have that

$$\mathbf{F_v} \ = \ t\mathbf{v} \ = \ \left(\frac{\mathbf{F} \cdot \mathbf{v}}{\mathbf{v} \cdot \mathbf{v}} \right) \mathbf{v} \ = \ \left(\frac{\mathbf{F} \cdot \mathbf{v}}{|\mathbf{v}|^2} \right) \mathbf{v},$$

and

$$\mathbf{F}_n = \mathbf{F} - \left(\frac{\mathbf{F} \cdot \mathbf{v}}{|\mathbf{v}|^2} \right) \mathbf{v}.$$

Moreover, it is easy to check that these $\mathbf{F}_\mathbf{v}$ and $\mathbf{F}_n$ do, indeed, satisfy the properties that we wanted; for clearly, $\mathbf{F} = \mathbf{F}_\mathbf{v} + \mathbf{F}_n$, $\mathbf{F}_\mathbf{v}$ is parallel to $\mathbf{v}$, and $\mathbf{F}_n$ is orthogonal to $\mathbf{v}$ because

$$\left(\mathbf{F} - \left(\frac{\mathbf{F} \cdot \mathbf{v}}{|\mathbf{v}|^2} \right) \mathbf{v} \right) \cdot \mathbf{v} = \mathbf{F} \cdot \mathbf{v} - \left(\frac{\mathbf{F} \cdot \mathbf{v}}{|\mathbf{v}|^2} \right) (\mathbf{v} \cdot \mathbf{v}) = \mathbf{F} \cdot \mathbf{v} - \mathbf{F} \cdot \mathbf{v} = 0.$$

We give names to $\mathbf{F}_\mathbf{v}$ and $\mathbf{F}_n$.

Definition 1.3.9. *Given a vector* $\mathbf{F}$ *and a non-zero vector* $\mathbf{v}$*, both in* $\mathbb{R}^n$*, we define the* **orthogonal projection of $\mathbf{F}$ onto $\mathbf{v}$** *to be the vector*

$$\mathbf{F}_\mathbf{v} = \operatorname{proj}_\mathbf{v} \mathbf{F} = \left(\frac{\mathbf{F} \cdot \mathbf{v}}{|\mathbf{v}|^2} \right) \mathbf{v} = \left(\mathbf{F} \cdot \frac{\mathbf{v}}{|\mathbf{v}|} \right) \frac{\mathbf{v}}{|\mathbf{v}|}.$$

This is also referred to as the **component of $\mathbf{F}$, parallel to $\mathbf{v}$**.
In this context, the vector $\mathbf{F}_n = \mathbf{F} - \mathbf{F}_\mathbf{v}$ *is referred to as the* **component of $\mathbf{F}$, normal to $\mathbf{v}$**

Remark 1.3.10. We shall usually use the notation $\mathbf{F}_\mathbf{v}$ for the orthogonal projection. However, the notation $\operatorname{proj}_\mathbf{v} \mathbf{F}$ is better if you're going to project multiple vectors onto $\mathbf{v}$, and so want to discuss the *orthogonal projection function* $\operatorname{proj}_\mathbf{v} : \mathbb{R}^n \to \mathbb{R}^n$ defined by

$$\operatorname{proj}_\mathbf{v}(\mathbf{F}) = \left(\frac{\mathbf{F} \cdot \mathbf{v}}{|\mathbf{v}|^2} \right) \mathbf{v}.$$

The vector $\mathbf{F}_n$, the component of $\mathbf{F}$ normal to $\mathbf{v}$ that we discussed earlier, is not usually given special notation or a special name; you simply write $\mathbf{F} - \mathbf{F}_\mathbf{v}$ for this normal component.

Note that the vector $\mathbf{u} = \mathbf{v}/|\mathbf{v}|$ is the unit vector in the direction of $\mathbf{v}$, so that

$$\mathbf{F}_\mathbf{v} = \left(\mathbf{F} \cdot \frac{\mathbf{v}}{|\mathbf{v}|} \right) \frac{\mathbf{v}}{|\mathbf{v}|} = (\mathbf{F} \cdot \mathbf{u})\mathbf{u} = \mathbf{F}_\mathbf{u},$$

which makes it clear that only the direction of $\mathbf{v}$ matters when calculating the orthogonal projection. In fact,

$$\mathbf{F}_{-\mathbf{v}} = \mathbf{F}_{-\mathbf{u}} = (\mathbf{F} \cdot -\mathbf{u})(-\mathbf{u}) = (\mathbf{F} \cdot \mathbf{u})\mathbf{u} = \mathbf{F}_\mathbf{u} = \mathbf{F}_\mathbf{v}$$

pick $+$ if $\mathbf{F_d}$ is in the direction of $\mathbf{d}$, and pick $-$ if the direction of $\mathbf{F_d}$ is opposite that of $\mathbf{d}$.

Therefore, the absolute value of the work is given by:

$$|\text{work}| \;=\; \left|\left(\frac{\mathbf{F}\cdot\mathbf{d}}{|\mathbf{d}|^2}\right)\mathbf{d}\right|\cdot|\mathbf{d}| \;=\; |\mathbf{F}\cdot\mathbf{d}|.$$

But notice that $\mathbf{F_d}$ points in the direction of $\mathbf{d}$ if and only if $\mathbf{F}\cdot\mathbf{d}\geq 0$, and points in the direction opposite $\mathbf{d}$ if and only if $\mathbf{F}\cdot\mathbf{d}\leq 0$. It follows that the work done, with the appropriate $\pm$ sign is simply given by $\mathbf{F}\cdot\mathbf{d}$.

The above discussion was our intuitive lead-in to making the following definition:

Definition 1.3.13. *The work done by a (constant) force $\mathbf{F}$ in $\mathbb{R}^n$, acting on an object, as the object is displaced along a line by a displacement vector $\mathbf{d}$ in $\mathbb{R}^n$ is given by*

$$\text{work} \;=\; \mathbf{F}\cdot\mathbf{d} \;=\; |\mathbf{F}||\mathbf{d}|\cos\theta,$$

where θ is the angle between $\mathbf{F}$ and $\mathbf{d}$.

Later, in Section 1.6, we shall see that the object does not need to move along a straight line and yet, still, to calculate the work, you dot the constant force with the total, straight, displacement vector. However, we make this more basic definition here, and show that the general case follows from integrating the case where you look at infinitesimal displacement in a straight line.

We will wait until Section 4.2 to address the case where the force is not constant.

Example 1.3.14. Suppose that a force of $\mathbf{F} = (-1, 3, 2)$ Newtons acts on an object as it is displaced along a line from the point $(0, 2, 5)$ to $(4, 0, -7)$, where all coordinates are measured in meters. How much work does the force do on the object?

Solution:

The displacement vector is

$$\mathbf{d} \;=\; (4, 0, -7) - (0, 2, 5) \;=\; (4, -2, -12) \text{ meters.}$$

Thus, the work done by $\mathbf{F}$ is

$$\mathbf{F}\cdot\mathbf{d} \;=\; (-1, 3, 2)\cdot(4, -2, -12) \;=\; -4 - 6 - 24 \;=\; -34 \text{ joules.}$$

More Depth:

Example 1.3.15. Find the angles in the triangle with vertices $(1, 1)$, $(2, 5)$, and $(4, 0)$.

Solution:

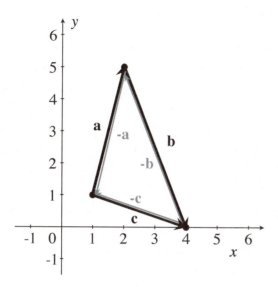

Figure 1.3.6: Vectors, and their negatives, between the vertices of a triangle.

We denote the angle inside the triangle, at a given vertex, by subscripting θ by the vertex, and thus want to calculate $\theta_{(1,1)}$, $\theta_{(2,5)}$, and $\theta_{(4,0)}$.

We calculate the displacement vectors between the vertices, as indicated in Figure 1.3.6. We find

$$\mathbf{a} = (2,5)-(1,1) = (1,4), \quad \mathbf{b} = (4,0)-(2,5) = (2,-5), \quad \text{and} \quad \mathbf{c} = (4,0)-(1,1) = (3,-1).$$

Now, remembering that we want the vectors to start at the same base point to determine the angle, we calculate

$$\theta_{(1,1)} = \cos^{-1}\left(\frac{\mathbf{a} \cdot \mathbf{c}}{|\mathbf{a}||\mathbf{c}|}\right) = \cos^{-1}\left(\frac{-1}{\sqrt{17}\sqrt{10}}\right) \approx 1.647568 \text{ radians} \approx 94.40°;$$

$$\theta_{(2,5)} = \cos^{-1}\left(\frac{-\mathbf{a} \cdot \mathbf{b}}{|-\mathbf{a}||\mathbf{b}|}\right) = \cos^{-1}\left(\frac{18}{\sqrt{17}\sqrt{29}}\right) \approx 0.625485 \text{ radians} \approx 35.84°;$$

and

$$\theta_{(4,0)} = \cos^{-1}\left(\frac{-\mathbf{b} \cdot -\mathbf{c}}{|-\mathbf{b}||-\mathbf{c}|}\right) = \cos^{-1}\left(\frac{11}{\sqrt{29}\sqrt{10}}\right) \approx 0.868539 \text{ radians} \approx 49.76°.$$

Note that, as they should, the angles add up to $180°$.

Example 1.3.16. Suppose that a constant force $\mathbf{F}$, in $\mathbb{R}^2$ or $\mathbb{R}^3$ (or, really, in $\mathbb{R}^n$), acts on an object as the object is displaced along a straight line from a point $\mathbf{p}_0$ to a point $\mathbf{p}_1$, then along a straight line from $\mathbf{p}_1$ to $\mathbf{p}_2$, then along a straight line from $\mathbf{p}_2$ to $\mathbf{p}_3$, and so on, and then finally along a straight line from $\mathbf{p}_{k-1}$ to $\mathbf{p}_k$.

Show that the total work done by the force is equal to simply the force dotted with the net displacement $\mathbf{d} = \mathbf{p}_k - \mathbf{p}_0$.

Solution

This is actually quite easy. For each i, where $1 \leq i \leq k$, let $\mathbf{d}_i = \mathbf{p}_i - \mathbf{p}_{i-1}$ be the displacement vector from $\mathbf{p}_{i-1}$ to $\mathbf{p}_i$. Then, the work done by $\mathbf{F}$, as the object is displaced by $\mathbf{d}_i$ is $W_i = \mathbf{F} \cdot \mathbf{d}_i$.

Thus, the total work is

$$W = \sum_{i=1}^{k} W_i = \sum_{i=1}^{k} \mathbf{F} \cdot \mathbf{d}_i = \mathbf{F} \cdot \left(\sum_{i=1}^{k} \mathbf{d}_i \right) = \mathbf{F} \cdot \left(\sum_{i=1}^{k} (\mathbf{p}_i - \mathbf{p}_{i-1}) \right) =$$

$$\mathbf{F} \cdot \big[(\mathbf{p}_1 - \mathbf{p}_0) + (\mathbf{p}_2 - \mathbf{p}_1) + (\mathbf{p}_3 - \mathbf{p}_2) + \cdots + (\mathbf{p}_k - \mathbf{p}_{k-1}) \big],$$

which "telescopes" to $\mathbf{F} \cdot (\mathbf{p}_k - \mathbf{p}_0) = \mathbf{F} \cdot \mathbf{d}$.

We would now like to prove the Cauchy-Schwarz Inequality from Theorem 1.3.2.

Let $\mathbf{a}$ and $\mathbf{b}$ be vectors in $\mathbb{R}^n$. Note that, if either $\mathbf{a}$ or $\mathbf{b}$ is the zero vector, then the equality holds, and the vectors are parallel, since the zero vector has every direction.

So, assume that $\mathbf{b} \neq \mathbf{0}$. Then, for all real numbers t

$$(\mathbf{a} + t\mathbf{b}) \cdot (\mathbf{a} + t\mathbf{b}) = |\mathbf{a} + t\mathbf{b}|^2 \geq 0,$$

and the equality holds if and only if $\mathbf{a} + t\mathbf{b} = \mathbf{0}$. Note that $\mathbf{a} + t\mathbf{b} = \mathbf{0}$ implies that $\mathbf{a}$ and $\mathbf{b}$ are parallel.

Now, expanding algebraically, using properties 1-4 in Theorem 1.3.2, we find

$$(\mathbf{a} + t\mathbf{b}) \cdot (\mathbf{a} + t\mathbf{b}) = \mathbf{a} \cdot \mathbf{a} + t(\mathbf{a} \cdot \mathbf{b}) + t(\mathbf{b} \cdot \mathbf{a}) + t^2 (\mathbf{b} \cdot \mathbf{b}) =$$

$$|\mathbf{a}|^2 + 2t(\mathbf{a} \cdot \mathbf{b}) + t^2 |\mathbf{b}|^2 = |\mathbf{b}|^2 \left[t^2 + 2 \left(\frac{\mathbf{a} \cdot \mathbf{b}}{|\mathbf{b}|^2} \right) t + \frac{|\mathbf{a}|^2}{|\mathbf{b}|^2} \right],$$

where $|\mathbf{b}|^2 \neq 0$, since $\mathbf{b} \neq \mathbf{0}$.

This shows that, for all t,

$$t^2 + 2 \left(\frac{\mathbf{a} \cdot \mathbf{b}}{|\mathbf{b}|^2} \right) t + \frac{|\mathbf{a}|^2}{|\mathbf{b}|^2} \geq 0,$$

where equality holds if and only if $\mathbf{a} + t\mathbf{b} = \mathbf{0}$.

We are now going to complete the square in the t variable, and then see what special value of t gives us the Cauchy-Schwarz Inequality.

Completing the square, we obtain, for all t

$$t^2 + 2\left(\frac{\mathbf{a} \cdot \mathbf{b}}{|\mathbf{b}|^2}\right)t + \frac{|\mathbf{a}|^2}{|\mathbf{b}|^2} = \left[t + \frac{\mathbf{a} \cdot \mathbf{b}}{|\mathbf{b}|^2}\right]^2 + \frac{|\mathbf{a}|^2|\mathbf{b}|^2 - (\mathbf{a} \cdot \mathbf{b})^2}{|\mathbf{b}|^4} \geq 0.$$

Therefore, if we let $t = -(\mathbf{a} \cdot \mathbf{b})/|\mathbf{b}|^2$, we conclude that

$$|\mathbf{a}|^2|\mathbf{b}|^2 - (\mathbf{a} \cdot \mathbf{b})^2 \geq 0,$$

and, if equality holds, then $\mathbf{a} = -t\mathbf{b}$, i.e., $\mathbf{a}$ is parallel to $\mathbf{b}$. Now note that, after taking square roots, the previous inequality is equivalent to the Cauchy-Schwarz Inequality: $|\mathbf{a} \cdot \mathbf{b}| \leq |\mathbf{a}||\mathbf{b}|$.

It remains to be shown that, if $\mathbf{a}$ and $\mathbf{b}$ are parallel, then $|\mathbf{a} \cdot \mathbf{b}| = |\mathbf{a}||\mathbf{b}|$. However, this is easy; if $\mathbf{a} = s\mathbf{b}$, then both sides of the equality equal $|s||\mathbf{b}|^2$.

We can now prove a "geometrically obvious" theorem; one which effectively says that the sum of the lengths of two sides of a triangle is greater than the length of the remaining side.

> **Theorem 1.3.17. (Triangle Inequality)** *Suppose that $\mathbf{a}$ and $\mathbf{b}$ are vectors in $\mathbb{R}^n$. Then,*
> $$|\mathbf{a} + \mathbf{b}| \leq |\mathbf{a}| + |\mathbf{b}|,$$
> *and equality holds if and only if $\mathbf{a}$ and $\mathbf{b}$ have the same direction.*

Proof. The inequality is equivalent to:

$$|\mathbf{a} + \mathbf{b}|^2 \leq |\mathbf{a}|^2 + 2|\mathbf{a}||\mathbf{b}| + |\mathbf{b}|^2.$$

Now, we use that

$$|\mathbf{a} + \mathbf{b}|^2 = (\mathbf{a} + \mathbf{b}) \cdot (\mathbf{a} + \mathbf{b}) = (\mathbf{a} \cdot \mathbf{a}) + 2(\mathbf{a} \cdot \mathbf{b}) + (\mathbf{b} \cdot \mathbf{b}) = |\mathbf{a}|^2 + 2(\mathbf{a} \cdot \mathbf{b}) + |\mathbf{b}|^2$$

to conclude that the Triangle Inequality is equivalent to

$$\mathbf{a} \cdot \mathbf{b} \leq |\mathbf{a}||\mathbf{b}|,$$

which follows at once from the Cauchy-Schwarz Inequality.

In addition, the equality $\mathbf{a} \cdot \mathbf{b} = |\mathbf{a}||\mathbf{b}|$ implies that $\mathbf{a}$ and $\mathbf{b}$ are parallel by the Cauchy-Schwarz Theorem; but the equality also implies that $\mathbf{a} \cdot \mathbf{b} \geq 0$, so that $\mathbf{a}$ and $\mathbf{b}$ must, in fact, have the same direction. $\square$

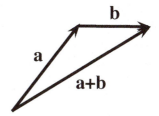

Figure 1.3.7: Geometric representation of the Triangle Inequality.

+ Linear Algebra:

In the + Linear Algebra portion of the previous section, we discussed that a vector $\mathbf{v}$ in $\mathbb{R}^n$ is frequently written as a column vector $[\mathbf{v}]$, a matrix with n rows and 1 column.

There is an operation on matrices called *transpose*. The transpose of an $n \times m$ matrix A is denoted A^{T}, and is the $m \times n$ matrix whose rows are the columns of A, and whose columns are the rows of A.

For instance,

$$\begin{bmatrix} 3 & -5 & 7 \\ -1 & 0 & 2 \end{bmatrix}^{\mathrm{T}} = \begin{bmatrix} 3 & -1 \\ -5 & 0 \\ 7 & 2 \end{bmatrix}.$$

In particular, the transpose of a column vector with n entries is a *row vector* with the same n entries, e.g.,

$$\begin{bmatrix} 2 \\ 5 \\ -1 \\ 7 \end{bmatrix}^{\mathrm{T}} = \begin{bmatrix} 2 & 5 & -1 & 7 \end{bmatrix}.$$

In terms of *matrix multiplication*, the dot product of two vectors $\mathbf{v}$ and $\mathbf{w}$ in $\mathbb{R}^n$, written as column vectors, is defined to be (the unique entry of) the 1×1 matrix given by

$$[\mathbf{v}] \cdot [\mathbf{w}] = [\mathbf{v}]^{\mathrm{T}} [\mathbf{w}].$$

Technically, there is a difference between a 1×1 matrix and the unique entry of the matrix; we shall not distinguish between these two objects.

If $\mathbf{v}$ and $\mathbf{w}$ are vectors in $\mathbb{R}^n$, and A is an $n \times n$ matrix, then the matrix product $A[\mathbf{v}]$ is a column vector in $\mathbb{R}^n$, and $(A[\mathbf{v}])^{\mathrm{T}} = [\mathbf{v}]^{\mathrm{T}} A^{\mathrm{T}}$.

Therefore, we arrive at the important formula for how matrix multiplication interacts with the dot product:

$$A[\mathbf{v}] \cdot [\mathbf{w}] = (A[\mathbf{v}])^{\mathrm{T}} [\mathbf{w}] = [\mathbf{v}]^{\mathrm{T}} \left(A^{\mathrm{T}} [\mathbf{w}] \right) = [\mathbf{v}] \cdot A^{\mathrm{T}} [\mathbf{w}].$$

Online answers to select exercises are here.

1.3.1 Exercises

Basics:

In each of the following exercises, you are given pair of vectors in some $\mathbb{R}^n$; calculate the dot product, and determine if the angle between the vectors is a right, acute, or obtuse angle.

1. $(2, 3)$ and $(-1, 2)$

2. $(\sqrt{2}, 3)$ and $(\sqrt{18}, -2)$

3. $(1, 2, 3)$ and $(3, -2, 1)$

4. $(6, 4, 2)$ and $(-3, -2, -1)$

5. $(6, 4, 2)$ and $(3, 2, 1)$

6. $(2, 0, -1, 3)$ and $(0, 1, 5, 2)$

7. $(\mathbf{i} + 2\mathbf{j}) \cdot (-5\mathbf{i} + 3\mathbf{j})$

8. $(3\mathbf{i} - 2\mathbf{j} + \mathbf{k}) \cdot (5\mathbf{i} - 7\mathbf{k})$

In each of the following exercises, you are given the same pairs of vectors as in Exercises 1-8; find the angles between the vectors. If an angle is not "nice", you may leave your answer with $\cos^{-1}$ in it, or give a decimal approximation from a calculator.

9. $(2, 3)$ and $(-1, 2)$

10. $(\sqrt{2}, 3)$ and $(\sqrt{18}, -2)$

11. $(1, 2, 3)$ and $(3, -2, 1)$

12. $(6, 4, 2)$ and $(-3, -2, -1)$

13. $(6, 4, 2)$ and $(3, 2, 1)$

14. $(2, 0, -1, 3)$ and $(0, 1, 5, 2)$

15. $(\mathbf{i} + 2\mathbf{j}) \cdot (-5\mathbf{i} + 3\mathbf{j})$

16. $(3\mathbf{i} - 2\mathbf{j} + \mathbf{k}) \cdot (5\mathbf{i} - 7\mathbf{k})$

Suppose that a, b, and c are vectors in $\mathbb{R}^n$, such that: $|\mathbf{a}| = 2$, $|\mathbf{b}| = 1$, $|\mathbf{c}| = 5$, $\mathbf{a} \cdot \mathbf{b} = 0$, $\mathbf{a} \cdot \mathbf{c} = 3$, and $\mathbf{b} \cdot \mathbf{c} = -1$. Calculate the following:

17. $\mathbf{c} \cdot (2\mathbf{a} - 3\mathbf{b})$

18. $\mathbf{a} \cdot (2\mathbf{c} - 3\mathbf{b})$

19. $(\mathbf{a} + \mathbf{b}) \cdot (\mathbf{b} - \mathbf{c})$

20. $(\mathbf{b} + \mathbf{c}) \cdot (2\mathbf{a} - 3\mathbf{b} + 5\mathbf{c})$

21. $\dfrac{\mathbf{a}}{|\mathbf{a}|} \cdot \dfrac{\mathbf{b}}{|\mathbf{b}|}$

22. $\dfrac{\mathbf{a}}{|\mathbf{a}|} \cdot \dfrac{\mathbf{c}}{|\mathbf{c}|}$

23. $|\mathbf{a} + 5\mathbf{b}|$ (Hint: Use that $|\mathbf{v}|^2 = \mathbf{v} \cdot \mathbf{v}$.)

24. $|\mathbf{a} - 2\mathbf{b} + 3\mathbf{c}|$

25. Suppose that $\mathbf{a} = (2, 0)$. Show that you **cannot** find $\mathbf{b}$ and $\mathbf{c}$ in $\mathbb{R}^2$, so that $\mathbf{a}$, $\mathbf{b}$, and $\mathbf{c}$ satisfy the conditions for the previous block of exercises.

26. Suppose that $\mathbf{a} = (2, 0, 0)$. Find $\mathbf{b}$ and $\mathbf{c}$ in $\mathbb{R}^3$, so that $\mathbf{a}$, $\mathbf{b}$, and $\mathbf{c}$ satisfy the conditions for the previous block of exercises.

In each of the following exercises, you are given three vectors. In each exercise, determine if any one of the vectors is parallel or perpendicular to either of the other two.

27. $(1, 2)$, $(-1, 3)$, $(6, 2)$

28. $(1, -1)$, $(-3, 3)$, (π, π) ▶

29. $(3, -2, 1)$, $(2, 3, 0)$, $(-4, -6, 0)$

30. $(3, -2, 1)$, $(2, 3, 1)$, $(-4, -6, 2)$

31. $4\mathbf{i} - 2\mathbf{j}$, $\mathbf{i} + \mathbf{j}$, $\mathbf{i} - \mathbf{j}$

32. $4\mathbf{i} - 2\mathbf{j} + \mathbf{k}$, $\mathbf{i} + \mathbf{j} - 2\mathbf{k}$, $\mathbf{i} - \mathbf{j} + 3\mathbf{k}$

In each of the following exercises, you are given a vector $\mathbf{F}$, and a non-zero vector $\mathbf{v}$ in $\mathbb{R}^n$. Find the components of $\mathbf{F}$ parallel and normal to $\mathbf{v}$, i.e., find the orthogonal projection $\mathbf{F_v} = \text{proj}_{\mathbf{v}}(\mathbf{F})$ of $\mathbf{F}$ onto $\mathbf{v}$, and $\mathbf{F}_n = \mathbf{F} - \mathbf{F_v}$. Verify, explicitly, that $\mathbf{F_v}$ and $\mathbf{F}_n$ are orthogonal to each other.

33. $\mathbf{F} = (2, 3)$, $\mathbf{v} = (1, -1)$.

34. $\mathbf{F} = (2, 3)$, $\mathbf{v} = (-2, 2)$.   ▶

35. $\mathbf{F} = (1, 2, 3)$, $\mathbf{v} = (1, -1, 0)$.

36. $\mathbf{F} = (2, 3, 0, 5)$, $\mathbf{v} = (1, -1, 1, -1)$.

37. $\mathbf{F} = \mathbf{i} - \mathbf{j}$, $\mathbf{v} = \mathbf{i} + \mathbf{j}$.

38. $\mathbf{F} = \mathbf{i} - \mathbf{j} + 3\mathbf{k}$, $\mathbf{v} = \mathbf{i} + \mathbf{j} + \mathbf{k}$.

In each of the following exercises, you are given a force **F**, which is acting at one end of a straight rod. Assume that the force and the rod are in the xy-plane, where x and y are measured in meters, and that the rod lies along the positive x-axis, with the origin at the end where **F** acts. All angles are measured counterclockwise from the rod (i.e., from the positive x-axis), with a negative angle indicating the clockwise direction. Find the components of **F** that are parallel and normal to the rod.

39. **F** has magnitude 5 Newtons and acts at an angle of $60°$.

40. **F** has magnitude 5 Newtons and acts at an angle of $-60°$.

41. $\mathbf{F} = (3, 4)$ Newtons.

42. $\mathbf{F} = (-3, 4)$ Newtons.

43. $\mathbf{F} = \mathbf{F}_1 + \mathbf{F}_2$, where $\mathbf{F}_1$ has magnitude 6 Newtons and acts at an angle of $30°$ and $\mathbf{F}_2 = (5, 12)$ Newtons.

44. $\mathbf{F} = \mathbf{F}_1 + \mathbf{F}_2$, where $\mathbf{F}_1$ has magnitude 6 Newtons and acts at an angle of $45°$ and $\mathbf{F}_2 = (-5, 12)$ Newtons.

In each of the following exercises, you are given a constant force **F**, in Newtons, which acts on an object. Also, a displacement vector **d**, in meters, for the object, is given or described. Calculate the work done by the force on the object during the displacement.

45. $\mathbf{F} = (5, -1)$ and $\mathbf{d} = (2, 3)$.

46. $\mathbf{F} = (2, 3, 1)$ and $\mathbf{d} = (-1, 2, -4)$.

47. $\mathbf{F} = (1, 1, 1)$ and **d** goes from the point $(1, 0, -1)$ to the point $(2, 7, 5)$.

48. $\mathbf{F} = (2, 3, 1)$ and **d** goes from the point $(1, 0, -1)$ to the point $(0, 0, 0)$.

49. **F** is a force vector in $\mathbb{R}^2$, which has magnitude 2 and which makes a $30°$ counterclockwise angle with the positive x-axis, and the object moves from $(1, 0)$ to $(0, 1)$.

50. **F** is a force vector in $\mathbb{R}^2$, which has magnitude 6 and which makes a $120°$ counterclockwise angle with the positive x-axis, and the object moves from $(1, 0)$ to $(2, 3)$. ▶

More Depth:

51. Find the angles in the triangle with vertices $(0, 0)$, $(-2, 5)$, and $(4, 0)$. Use a calculator to verify that the angles add up to $180° = \pi$ radians.

52. Find the angles in the quadrilateral with vertices $(0,0)$, $(3,0)$, $(0,4)$, and $(5,5)$. Use a calculator to verify that the angles add up to $360° = 2\pi$ radians.

53. Suppose that $\mathbf{a} = (1,2,-1,2)$ and $\mathbf{b} = (2,4,-2,5)$. Verify the Triangle Inequality, Theorem 1.3.17: $|\mathbf{a}+\mathbf{b}| \leq |\mathbf{a}| + |\mathbf{b}|$.

54. A collection of vectors $\mathbf{v}_1$, $\mathbf{v}_2$, ..., $\mathbf{v}_k$ in $\mathbb{R}^n$ is called an **orthonormal** set of vectors provided that each $\mathbf{v}_i$ is a unit vector, so that $\mathbf{v}_i \cdot \mathbf{v}_i = 1$, and the vectors are pairwise orthogonal, i.e., if $i \neq j$, then $\mathbf{v}_i \cdot \mathbf{v}_j = 0$.

    Suppose that the collection of vectors $\mathbf{v}_1$, $\mathbf{v}_2$, ..., $\mathbf{v}_k$ in $\mathbb{R}^n$ is an orthonormal set. Show that, for all scalars t_1, ..., t_k,

    $$\left| t_1 \mathbf{v}_1 + \cdots + t_k \mathbf{v}_k \right| = \sqrt{t_1^2 + \cdots + t_k^2}.$$

55. Prove Parts 1-4 of Theorem 1.3.2.

+ Linear Algebra:

In each of the following exercises, you are given vectors v and w in $\mathbb{R}^n$, and an $n \times n$ matrix A. Calculate $A[\mathbf{v}]$ and $A^T[\mathbf{w}]$, and then verify that

$$A[\mathbf{v}] \cdot [\mathbf{w}] = [\mathbf{v}] \cdot A^T[\mathbf{w}].$$

56. $\mathbf{v} = (1,2)$, $\mathbf{w} = (3,-1)$, and $A = \begin{bmatrix} 3 & -5 \\ -1 & 2 \end{bmatrix}$.

57. $\mathbf{v} = (2,-7)$, $\mathbf{w} = (0,6)$, and $A = \begin{bmatrix} 4 & 0 \\ 1 & -2 \end{bmatrix}$.

58. $\mathbf{v} = (2,5,3)$, $\mathbf{w} = (1,0,-1)$, and $A = \begin{bmatrix} 0 & 3 & -5 \\ 7 & -1 & 2 \\ 1 & 1 & 1 \end{bmatrix}$.

59. $\mathbf{v} = (0,0,1)$, $\mathbf{w} = (1,1,1)$, and $A = \begin{bmatrix} -1 & 0 & 1 \\ 2 & 4 & 6 \\ 5 & -1 & 2 \end{bmatrix}$.

1.4 Lines, Planes, and Hyperplanes

In this section, we begin by discussing *vector equations* for lines in the xy-plane, and then in xyz-space. These vector equations are equivalent to giving *parameterizations* for the lines.

Then, we discuss planes in $\mathbb{R}^3$. We explain how to describe a plane with a single equation, given a normal vector to the plane and a point on the plane. We also discuss how to parameterize a plane, given a point on the plane and two vectors which are parallel to the plane, but not parallel to each other.

Finally, we look at a generalization of lines in the plane and planes in space. We look at *affine linear subspaces* which have dimension one less than the ambient space; such affine linear subspaces are called *hyperplanes*.

Basics:

Lines in $\mathbb{R}^2$:

You are familiar with lines in the xy-plane given in slope-intercept form by $y = mx + b$ or, more generally, in standard form by $ax + by + c = 0$, where a and b are not both zero. Standard form is more general because it allows for vertical lines, i.e., lines of the form $x = k$, where k is a constant.

Now suppose you are given the pair $(a, b) \neq (0, 0)$ and told that (x_0, y_0) is a point on the line given by $ax + by + c = 0$. Then, c is determined by (a, b) and (x_0, y_0):

$$ax_0 + by_0 + c = 0, \quad \text{and so,} \quad c = -ax_0 - by_0.$$

This means that the points (x, y) on the line are precisely those points such that

$$ax + by - ax_0 - by_0 = 0, \quad \text{that is,} \quad a(x - x_0) + b(y - y_0) = 0.$$

Now we can write this as: $(a, b) \cdot \big[(x, y) - (x_0, y_0)\big] = 0$.

In words, this says that the points (x, y) on the line are precisely those such that the displacement vector from (x_0, y_0) to (x, y) is perpendicular, or normal, to the vector $\mathbf{n} = (a, b)$, i.e., the vector $\mathbf{n} = (a, b)$ is normal to the line.

Thus, we find:

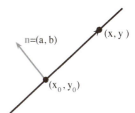

Figure 1.4.1: The vector $\mathbf{n} = (a, b)$ is normal to the line given by $ax + by + c = 0$.

Theorem 1.4.1. *Let $\mathbf{n} = (a, b) \neq \mathbf{0}$. Then, the vector $\mathbf{n}$ is normal to the line in $\mathbb{R}^2$ of points (x, y) such that $ax + by + c = 0$.*

Furthermore, given $\mathbf{n} = (a, b) \neq \mathbf{0}$ and a point (x_0, y_0), the unique line containing (x_0, y_0), and having $\mathbf{n}$ as a normal vector, is the set of points (x, y) such that

$$a(x - x_0) + b(y - y_0) = 0.$$

Example 1.4.8. Find an equation for the plane which contains the point $(3, 1, -2)$ and is normal to the vector $(5, -3, 7)$.

Solution:

This is simple. An equation for the plane is

$$5(x - 3) + (-3)(y - 1) + 7(z - (-2)) = 0.$$

You might simplify this to

$$5(x - 3) - 3(y - 1) + 7(z + 2) = 0;$$

on the other hand, you don't want to expand this into standard form unless you are explicitly instructed to do so.

Example 1.4.9. Consider the plane P given by $6x - 5y + 3z - 7 = 0$. Find a non-zero normal vector to the plane and determine a point on the plane.

Solution:

The components of a non-zero normal vector $\mathbf{n}$ are simply the coefficients of x, y, and z. Thus, $\mathbf{n} = (6, -5, 3)$ is a non-zero normal vector to the plane. Every other non-zero normal vector to the plane is a non-zero scalar multiple of this one specific normal vector.

To find a point on the plane, just pick anything you want for two of the variables and solve for the third. Selecting 0's makes things easy.

Letting $x = 0$ and $y = 0$, we find that $3z - 7 = 0$, so that $z = 7/3$. Thus, $(0, 0, 7/3)$ is a point on the plane. In fact, it's the *z-intercept* of the plane, the point where the plane hits the z-axis.

We've already found **a** point on the plane, but let's go ahead and find the other two intercepts.

The *x-intercept* is where $y = 0$ and $z = 0$. Thus, $6x - 7 = 0$, so that $x = 7/6$. Thus, $(7/6, 0, 0)$ is on the plane.

Finally, the *y-intercept* is where $x = 0$ and $z = 0$. Hence, $-5y - 7 = 0$, and so $y = -7/5$. Therefore, another point on the plane is $(0, -7/5, 0)$.

Example 1.4.10. Consider the plane P given by $6x - 5y + 3z - 7 = 0$. Find a standard equation of the form $3x + by + cz + d = 0$ for the plane Q which contains the point $(2, 1, 3)$ and is parallel to P.

Solution: Two planes are parallel if and only if they have parallel non-zero normal vectors. A non-zero normal vector to P is $\mathbf{n} = (6, -5, 3)$. A non-zero normal vector to Q is $\mathbf{m} = (3, b, c)$. For P and Q to be parallel, $\mathbf{m}$ and $\mathbf{n}$ must be parallel, i.e., there must exist a scalar t such that $(3, b, c) = t(6, -5, 3)$.

Looking at the first component, we conclude that $t = 0.5$. Thus, $b = 0.5(-5) = -2.5$ and $c = 0.5(3) = 1.5$. It follows that the standard equation for Q that we're looking for is of the form

$$3x - 2.5y + 1.5z + d = 0.$$

As we need for $(2, 1, 3)$ to be a point on Q, we must have

$$3(2) - 2.5(1) + 1.5(3) + d = 0,$$

and so, $d = -8$. Therefore, an equation for Q is

$$3x - 2.5y + 1.5z - 8 = 0.$$

Now, how do we specify a line in $\mathbb{R}^3$? Actually, in terms of vector/parametric equations, we do just what we did in $\mathbb{R}^2$.

Theorem 1.4.11. *Let* $\mathbf{v} = (a, b, c) \neq \mathbf{0}$, *and let* (x_0, y_0, z_0) *be a point in* $\mathbb{R}^3$. *Then, a point* (x, y, z) *is on the line which is parallel to* $\mathbf{v}$ *and contains* (x_0, y_0, z_0) *if and only if there exists a scalar* t *such that*

$$(x, y, z) = (x_0, y_0, z_0) + t\mathbf{v} = (x_0, y_0, z_0) + t(a, b, c).$$

This is called a **vector equation for the line**, *and* t *is called a* **parameter**.

Writing out the components separately, this means that (x, y, z) *is on the line which is parallel to the vector* (a, b, c) *and contains the point* (x_0, y_0, z_0) *if and only if there exists a scalar* t *such that*

$$x = x_0 + ta, \qquad y = y_0 + tb, \qquad \text{and} \qquad z = z_0 + tc.$$

These are called (component-wise) **parametric equations for the line**, *and* t *is again called a* **parameter**.

In fact, a vector equation for a line is also frequently referred to as a *parametric equation for a line*. The difference in terminology is whether or not *equation* is used in the singular form or plural. As the vector parametric equation and the component parameterizations are easily interchangeable, there is little harm in this subtle distinction in terminology.

> If a, b, and c are all non-zero, then, eliminating t, we find that a point (x, y, z) is on the line which is parallel to the vector (a, b, c) and contains the point (x_0, y_0, z_0) if and only if
> $$\frac{x - x_0}{a} = \frac{y - y_0}{b} = \frac{z - z_0}{c}.$$
> These are called **symmetric equations for the line** in $\mathbb{R}^3$.

Remark 1.4.12. If one or two of a, b, and c are zero, then you still get equations, without t, that define the line, they're just not so "symmetric". For instance, if $c = 0$, but a and b are non-zero, then, looking back at the parametric equations, we see that the line is given by

$$\frac{x - x_0}{a} = \frac{y - y_0}{b} \qquad \text{and} \qquad z = z_0.$$

If both b and c are zero, but a is non-zero, then the line is given by

$$y = y_0 \qquad \text{and} \qquad z = z_0.$$

Finally, notice that, in every case, the line is described by a pair of equalities, where each equality, by itself, describes a **plane**. Thus, geometrically, the symmetric equations describe a line as the line of intersection of two (non-parallel) planes.

Example 1.4.13. Write a vector equation for the line ℓ in $\mathbb{R}^3$ which contains the point $(-2, 0, 3)$ and is parallel to the vector $(5, 7, 1)$. Also, give equations for a pair of planes whose intersection is the line ℓ.

Solution: The vector equation can be written instantly:

$$(x, y, z) = (-2, 0, 3) + t(5, 7, 1).$$

Symmetric equations would give us equations for the pair of planes that we want. Rather than using "memorized" formulas for the symmetric equations, it's very quick to write out the parametric equations and solve for t in each one:

$$x = -2 + 5t, \qquad y = 0 + 7t, \qquad \text{and} \qquad z = 3 + t.$$

Therefore,

$$t = \frac{x + 2}{5} = \frac{y}{7} = \frac{z - 3}{1}.$$

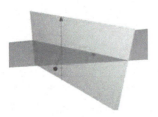

Figure 1.4.4: Two planes intersecting along a line.

Hence, equations for a pair of planes whose intersection is the line ℓ are

$$\frac{x+2}{5} = \frac{y}{7} \quad \text{and} \quad \frac{y}{7} = \frac{z-3}{1},$$

i.e.,

$$7(x+2) - 5y = 0 \quad \text{and} \quad y - 7(z-3) = 0.$$

Example 1.4.14. Let P be the plane given by $3x + 2y - z + 6 = 0$, and let ℓ be the line through the points $(5,3,0)$ and $(6,2,2)$. Determine whether or not P and ℓ intersect and, if they do, determine the point(s) of intersection.

Solution:

We need to produce a non-zero vector $\mathbf{v}$ which is parallel to the line. We get this from our two points by taking the displacement vector

$$\mathbf{v} = (6,2,2) - (5,3,0) = (1,-1,2).$$

Now, we find that line is parameterized by

$$x = 5 + t, \quad y = 3 - t, \quad \text{and} \quad z = 0 + 2t.$$

At a point where the line intersects the plane, we insert these expressions for x, y and z into the equation for the plane to obtain

$$3(5+t) + 2(3-t) - (2t) + 6 = 0. \tag{1.1}$$

We solve this for t to find $t = 27$. Thus, the plane and line intersect at the single point where

$$x = 5 + 27, \quad y = 3 - 27, \quad \text{and} \quad z = 0 + 2(27),$$

i.e., at the point $(32, -24, 54)$.

> You may wonder what would have happened with this method of solution if the line and plane were parallel and non-intersecting, or if the line had been contained in the plane.

> If the line and plane had not intersected, then Formula 1.1 would have reduced to something like $5 = 0$; all the t's would have cancelled out, and you'd be left with something that's not true. This would mean there is no t that makes the line hit the plane, i.e., the line doesn't intersect the plane.

> If the line were contained in the plane, then Formula 1.1 would have reduced to something like $0 = 0$; all the t's would have cancelled out, and you'd be left with something that is true. This would mean that every t that makes the line hit the plane, i.e., the line is contained in the plane.

Example 1.4.15. As we saw in the previous example, if you're given two different points (x_0, y_0, z_0) and (x_1, y_1, z_1), then, the displacement vector $\mathbf{v} = (x_1, y_1, z_1) - (x_0, y_0, z_0)$ is parallel to the line through (x_0, y_0, z_0) and (x_1, y_1, z_1).

Thus,

$$(x, y, z) = (x_0, y_0, z_0) + t\mathbf{v} = (x_0, y_0, z_0) + t\big[(x_1, y_1, z_1) - (x_0, y_0, z_0)\big]$$

is a parameterization of the line through (x_0, y_0, z_0) and (x_1, y_1, z_1).

If you think of t as representing time, then this parameterization is at (x_0, y_0, z_0) at time $t = 0$, and at (x_1, y_1, z_1) at time $t = 1$. Therefore, you can parameterize just the line **segment** between (x_0, y_0, z_0) and (x_1, y_1, z_1) by restricting the value of the parameter to $0 \leq t \leq 1$.

This is called the **standard parameterization of the line segment** from (x_0, y_0, z_0) to the point (x_1, y_1, z_1).

For instance, given the points $(3, 2, 1)$ and $(-1, 0, 5)$, the vector

$$\mathbf{v} \; = \; (-1, 0, 5) - (3, 2, 1) \; = \; (-4, -2, 4)$$

is parallel to the line through the points, and a vector equation for the line is

$$(x, y, z) \; = \; (3, 2, 1) + t(-4, -2, 4).$$

Once you restrict t to $0 \leq t \leq 1$, this equation also gives the standard parameterization of the line segment from $(3, 2, 1)$ to $(-1, 0, 5)$.

More Depth:

In $\mathbb{R}^2$, we used standard equations $ax + by + c = 0$ for lines, and also parameterized lines $(x, y) = (x_0, y_0) + t(a, b)$ (whether written in vector form or not). However, we have not yet discussed *parameterizing planes*.

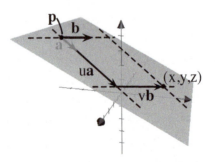

Figure 1.4.5: The plane through $\mathbf{p}$ parallel to the vectors $\mathbf{a}$ and $\mathbf{b}$.

To parameterize a line, we need a point on the line and a non-zero vector which is parallel to the line. To parameterize a plane, we need a point $\mathbf{p}$ on the plane and **two** non-zero vectors $\mathbf{a}$ and $\mathbf{b}$, which are both parallel to the plane, but not parallel to each other. See Figure 1.4.5.

The theorem is:

Theorem 1.4.16. *Let* $\mathbf{p} = (x_0, y_0, z_0)$ *be a point in* $\mathbb{R}^3$. *Let* $\mathbf{a}$ *and* $\mathbf{b}$ *be non-zero vectors in* $\mathbb{R}^3$ *which are not parallel to each other.*

Then, there is a unique plane which contains $\mathbf{p}$ *and is parallel to both* $\mathbf{a}$ *and* $\mathbf{b}$. *A point* (x, y, z) *is on this plane if and only if there exist scalars* u *and* v *such that*

$$(x, y, z) = \mathbf{p} + u\mathbf{a} + v\mathbf{b}.$$

This is a **parameterization** *of the plane.*

We have stated this as a theorem, but we remind you that we are appealing to your intuition before we give a technically accurate definition of what a plane is; we do this in the + Linear Algebra portion of this section.

Example 1.4.17. Give a parameterization of the plane through the point $(1, 2, 3)$ which is parallel to each of the lines parameterized by

$$(x, y, z) = (2, 0, 3) + r(4, -1, 7) \quad \text{and} \quad (x, y, z) = (1, 1, 1) + t(0, 5, -6).$$

Solution:

To parameterize the plane, we want a point on the plane, which is given, and two non-zero vectors, parallel to the plane, but not parallel to each other. But these parallel vectors are easy to produce: the plane is supposed to be parallel to both of the two lines, and you can read off of their parameterizations a non-zero vector parallel to each one. We see that the vectors $(4, -1, 7)$ and $(0, 5, -6)$ are parallel to the plane we're after.

Therefore, a parameterization of the plane is

$$(x, y, z) = (1, 2, 3) + u(4, -1, 7) + v(0, 5, -6).$$

Now we want to give an example in which we use orthogonal projection to find the distance from a point to a plane.

Example 1.4.18. The distance from a point to a plane is the length of the perpendicular line segment from the point to the plane. Let's find a general formula for the distance from an arbitrary point (x_1, y_1, z_1) to the plane in $\mathbb{R}^3$ given by $ax + by + cz + d = 0$, where $(a, b, c) \neq \mathbf{0}$.

For this, we will use orthogonal projection; recall Definition 1.3.9. We pick any point (x_0, y_0, z_0) on the plane, and orthogonally project the vector $\mathbf{d} = \overrightarrow{(x_0, y_0, z_0)(x_1, y_1, z_1)}$ onto the vector $\mathbf{n} = (a, b, c)$, which is a non-zero normal vector to the plane; we then take the length/magnitude of this projection to find the distance. See Figure 1.4.6.

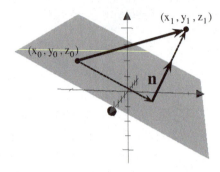

Figure 1.4.6: Orthogonal projection onto $\mathbf{n}$ of a vector from a point on the plane to the point in space.

So, suppose that (x_0, y_0, z_0) is a point on the plane, i.e., suppose that $ax_0 + by_0 + cz_0 + d = 0$. What we shall use below is that this means that $-ax_0 - by_0 - cz_0 = d$.

Now, the orthogonal projection of $\mathbf{d}$ onto $\mathbf{n}$ is

$$\text{proj}_{\mathbf{n}}(x_1 - x_0, y_1 - y_0, z_1 - z_0) = \left(\frac{(x_1 - x_0, y_1 - y_0, z_1 - z_0) \cdot \mathbf{n}}{|\mathbf{n}|^2} \right) \mathbf{n},$$

and the magnitude is

$$\frac{\left| (x_1 - x_0, y_1 - y_0, z_1 - z_0) \cdot \mathbf{n} \right|}{|\mathbf{n}|^2} |\mathbf{n}| = \frac{\left| a(x_1 - x_0) + b(y_1 - y_0) + c(z_1 - z_0) \right|}{\sqrt{a^2 + b^2 + c^2}} =$$

$$\frac{\left| ax_1 + by_1 + cz_1 - ax_0 - by_0 - cz_0 \right|}{\sqrt{a^2 + b^2 + c^2}} = \frac{\left| ax_1 + by_1 + cz_1 + d \right|}{\sqrt{a^2 + b^2 + c^2}}.$$

+ Linear Algebra:

We want to define *affine linear subspaces*, but first, we have to define *linear subspaces* of $\mathbb{R}^n$.

Recall from Definition 1.2.24 that there are axioms that define what a vector space is. Now, suppose that E is a subset of $\mathbb{R}^n$. We want to know when E itself can be considered as a vector space, using the scalar multiplication and vector addition from the (possibly) larger $\mathbb{R}^n$. It is easy to verify that these axioms are satisfied by E, using the restrictions of vector addition and scalar multiplication to the elements of E, provided that just 3 properties are satisfied:

Definition 1.4.19. *A subset E of $\mathbb{R}^n$ is a linear subspace of $\mathbb{R}^n$ if and only if*

1. *the zero vector $\mathbf{0}$ is in E;*

2. *(closure under vector addition) for all $\mathbf{v}$ and $\mathbf{w}$ in E, the vector $\mathbf{v} + \mathbf{w}$ is also in E, and*

3. *(closure under scalar multiplication) for all $\mathbf{v}$ in E and all real numbers t, $t\mathbf{v}$ is also in E.*

Example 1.4.20. The most trivial linear subspace of any $\mathbb{R}^n$ is the **zero subspace**, which is the set containing only the zero vector from $\mathbb{R}^n$. The three properties from Definition 1.4.19 are clearly satisfied.

Example 1.4.21. Let a, b, and c be real numbers. Show that the set E of (x, y) in $\mathbb{R}^2$ such that $ax + by = 0$ is a linear subspace of $\mathbb{R}^2$. Also, show that, if $c \neq 0$, then the set F of (x, y) in $\mathbb{R}^2$ such that $ax + by + c = 0$ is **not** a linear subspace of $\mathbb{R}^2$.

Solution:

1. Certainly, $(0,0)$ is in E since $a \cdot 0 + b \cdot 0 = 0$.

2. Suppose that (x_1, y_1) and (x_2, y_2) are in E. We need to show that $(x_1, y_1) + (x_2, y_2)$ is also in E.

   (x_1, y_1) and (x_2, y_2) being in E means that $ax_1 + by_1 = 0$ and $ax_2 + by_2 = 0$. Adding these two equations, we obtain that

   $$a(x_1 + x_2) + b(y_1 + y_2) = 0,$$

   and this is exactly what it means for $(x_1 + x_2, y_1 + y_2)$ to be in E, i.e., $(x_1, y_1) + (x_2, y_2)$ is in E.

3. Suppose that (x_1, y_1) is in E, and that t is a real number. We need to show that $t(x_1, y_1)$ is also in E.

   (x_1, y_1) being in E means that $ax_1 + by_1 = 0$. Multiplying by t, we find that $a(tx_1) + b(ty_1) = 0$, which means precisely that (tx_1, ty_1) is in E, i.e., $t(x_1, y_1)$ is in E.

To show that F is not a linear subspace, it is enough to show that F fails to have **any** one of the requisite three properties. In fact, just for the practice, we shall show that F satisfies **none** of the properties in Definition 1.4.19. Recall that we are assuming that $c \neq 0$ here.

1. The zero vector is not in F, because $a \cdot 0 + b \cdot 0 + c = c \neq 0$.

2. Suppose that (x_1, y_1) and (x_2, y_2) are in F. Then, $ax_1 + by_1 + c = 0$ and $ax_2 + by_2 + c = 0$. Adding these two equations, we obtain that

$$a(x_1 + x_2) + b(y_1 + y_2) + 2c = 0.$$

But, if $(x_1, y_1) + (x_2, y_2)$ were in F, then we would have to have

$$a(x_1 + x_2) + b(y_1 + y_2) + c = 0.$$

Subtracting one of these last two equations from the other, we find that $c = 0$, which is false.

3. Suppose that (x_1, y_1) is in F, and that t is a real number other than 1. Then, $ax_1 + by_1 + c = 0$, and so $a(tx_1) + b(ty_1) + tc = 0$. If $t(x_1, y_1)$ were in F, then we would have $a(tx_1) + b(ty_1) + c = 0$. Subtracting one of these last two equations from the other, we find that $(t - 1)c = 0$, which means that either $t = 1$ or $c = 0$; a contradiction.

In the + Linear Algebra portion of Section 1.2, we defined a basis for $\mathbb{R}^n$. In fact, every non-zero vector space possesses a basis; in particular, non-zero linear subspaces of $\mathbb{R}^n$ have bases. Suppose that V is a linear subspace of $\mathbb{R}^n$, other than the zero subspace. Then, V has a basis consisting of a finite number of vectors $\mathbf{b}_1$, $\mathbf{b}_2$, ..., $\mathbf{b}_k$ in V such that, for all vectors $\mathbf{v}$ in V, there exist unique scalars a_1, a_2, ... a_k such that

$$\mathbf{v} = a_1\mathbf{b}_1 + a_2\mathbf{b}_2 + \cdots + a_k\mathbf{b}_k.$$

If V is a non-zero linear subspace, then there are an infinite number of choices for a basis for V. However, all bases for a given vector space have the same number of elements; that common number of basis vectors is called the dimension of V. By convention, we also say that dimension of the zero subspace is zero.

One-dimensional linear subspaces of $\mathbb{R}^n$ are lines through the origin. Two-dimensional linear subspaces are planes through the origin.

But, of course, we want to be able to discuss to lines and planes that don't necessarily contain the origin. To accomplish this, we *translate* a linear subspace, in a way that "moves" the origin to some other point $\mathbf{p}$. We do this simply by adding $\mathbf{p}$ to each vector in the linear subspace.

So, we make the following definition.

> **Definition 1.4.22.** *Let V be a linear subspace of $\mathbb{R}^n$, and let $\mathbf{p}$ be a point/vector in $\mathbb{R}^n$. Then, the **translation** of V by $\mathbf{p}$ is the set*
>
> $$\mathbf{p} + V = \{\mathbf{p} + \mathbf{v} \mid \mathbf{v} \in V\}.$$
>
> *An **affine linear subspace** of $\mathbb{R}^n$ is the translation of a linear subspace V, and the **dimension** of an affine linear subspace $\mathbf{p} + V$ is defined to be the dimension of the linear subspace V.*

The adjective *affine* is used to indicate that we are willing to move the origin.

For the dimension of an affine linear subspace to be defined as we have given it, we have to know that, if you write the same affine subspace in different ways, it doesn't change what we're calling the dimension, i.e., you need to know that, if $\mathbf{p} + V = \mathbf{q} + W$, where V and W are linear subspaces, then V and W have the same dimension.

In fact, it is an easy exercise to show that $\mathbf{p} + V = \mathbf{q} + W$ implies that $V = W$. So the dimension of an affine linear subspace is well-defined.

Thus, a 0-dimensional affine linear subspace is just a point. A 1-dimensional affine linear subspace is a line. A 2-dimensional affine linear subspace is a plane.

Now, let's look back at what we saw in Theorem 1.4.1 and Theorem 1.4.6.

Fix a point $\mathbf{p} = (x_0, y_0)$ and a nonzero vector $\mathbf{n}$ is $\mathbb{R}^2$. In Theorem 1.4.1, we saw that you obtain a line by taking the set of points (x, y) in $\mathbb{R}^2$, which includes $\mathbf{p}$, and such that the displacement vector $(x, y) - \mathbf{p}$ is perpendicular to $\mathbf{n}$.

Fix a point $\mathbf{p} = (x_0, y_0, z_0)$ and a nonzero vector $\mathbf{n}$ is $\mathbb{R}^3$. In Theorem 1.4.6, we saw that you obtain a plane by taking the set of points (x, y, z) in $\mathbb{R}^3$, which includes $\mathbf{p}$, and such that the displacement vector $(x, y, z) - \mathbf{p}$ is perpendicular to $\mathbf{n}$.

This may lead you to wonder what happens in higher dimensions. The general theorem is:

> **Theorem 1.4.23.** *Let $\mathbf{a}$ be a non-zero vector in $\mathbb{R}^n$. Then, the set of $\mathbf{x}$ in $\mathbb{R}^n$ such that*
>
> $$\mathbf{a} \cdot \mathbf{x} = a_1 x_1 + a_2 x_2 + \cdots + a_n x_n = 0$$
>
> *is an $(n-1)$-dimensional linear subspace in $\mathbb{R}^n$.*

Now suppose that $\mathbf{a}$ *is a non-zero vector in* $\mathbb{R}^n$ *and* c *is a real number. Let* $\mathbf{p} = (p_1, \ldots, p_n)$ *be any point satisfying the equation*

$$a_1 x_1 + a_2 x_2 + \cdots + a_n x_n + c \;=\; 0.$$

Then,

$$c = -a_1 p_1 - a_2 p_2 - \cdots - a_n p_n$$

and so $a_1 x_1 + a_2 x_2 + \cdots + a_n x_n + c = 0$ *can be rewritten as*

$$\mathbf{a} \cdot (\mathbf{x} - \mathbf{p}) \;=\; a_1(x_1 - p_1) + a_2(x_2 - p_2) + \cdots + a_n(x_n - p_n) \;=\; 0.$$

The points $\mathbf{x}$ *which satisfy this equation are those such that the displacement vector* $\mathbf{x} - \mathbf{p}$ *is perpendicular to the vector* $\mathbf{a}$*; the set of these* $\mathbf{x}$*'s is the translation of* V *by* $\mathbf{p}$*, where* V *is the linear subspace of those* $\mathbf{v}$ *in* $\mathbb{R}^n$ *such that*

$$\mathbf{a} \cdot \mathbf{v} \;=\; 0.$$

Therefore, the equation $\mathbf{a} \cdot (\mathbf{x} - \mathbf{p}) = 0$ *or, equivalently,*

$$a_1 x_1 + a_2 x_2 + \cdots + a_n x_n + c \;=\; 0$$

defines an $(n-1)$*-dimensional affine linear subspace in* $\mathbb{R}^n$*; such an affine linear subspace, one whose dimension is 1 less than the dimension of the surrounding Euclidean space is called a* **hyperplane** *in* $\mathbb{R}^n$*.*

Thus, a hyperplane in $\mathbb{R}^2$ is a line, while a hyperplane in $\mathbb{R}^3$ simply means a plane. A hyperplane in $\mathbb{R}^4$ is a 3-dimensional affine linear subspace.

Online answers to select exercises are here.

Basics:

1.4.1   Exercises

In each of the following problems, you are given two distinct points in $\mathbb{R}^2$ or $\mathbb{R}^3$. First, find a non-zero vector parallel to the line containing the two points. Then, find a vector equation and (component-wise) parametric equations for the line. Finally, if possible, give symmetric equation(s) for the line.

1. $(1, 2)$ and $(-5, 3)$

2. $(1, 0)$ and $(0, 1)$

3. $(1, 2, 3)$ and $(3, 2, 1)$

4. $(1, 0, 0)$ and $(0, -1, -2)$

5. $(\pi, e, \sqrt{2})$ and $(1, 1, 1)$

6. (e, e^2, e^3) and (e^2, e^3, e^4)

In each of the following problems, you are given a point p and a non-zero vector n in $\mathbb{R}^2$ or $\mathbb{R}^3$. Give a standard equation for the line in $\mathbb{R}^2$ or the plane in $\mathbb{R}^3$ which contains the point p and is normal to the vector n.

7. $\mathbf{p} = (2, -1)$ and $\mathbf{n} = (3, 4)$.

8. $\mathbf{p} = (0, 5)$ and $\mathbf{n} = (1, -1)$.

9. $\mathbf{p} = (1, 2, 1)$ and $\mathbf{n} = (7, -6, 2)$.

10. $\mathbf{p} = (\sqrt{5}, 1, -2)$ and $\mathbf{n} = (7, 11, 13)$.

11. $\mathbf{p} = (3, 2, 1)$ and $\mathbf{n} = 5\mathbf{i} - \mathbf{j} + 7\mathbf{k}$.

12. $\mathbf{p} = (5, 0, -3)$ and $\mathbf{n} = 6\mathbf{i} + \sqrt{2}\mathbf{j} - 4\mathbf{k}$.

13. Find a vector equation for the line in $\mathbb{R}^3$ which contains the point $(3, 0, -1)$ and is parallel to the line given by $(x, y, z) = (4, -5, 7) + t(-1, 2, 3)$.

14. Find a vector equation for the line in $\mathbb{R}^3$ which contains the point $(-1, 5, 7)$ and is parallel to the line given by $x = 2 + 3t$, $y = 4 - t$, and $z = 9t$.

15. Determine the points of intersection, if any, of the plane given by $3x - 2y + 5z + 4 = 0$ and the line given by $(x, y, z) = (3, 0, 7) + t(-1, 1, 1)$.

16. Show that the line given by $(x, y, z) = (3, 0, 7) + t(-1, 1, 1)$ is parallel to the plane given by $3x - 2y + 5z + 4 = 0$, **without** explicitly showing that there are no points of intersection.

17. Give a standard equation for a plane that contains the point $(1, -2, 3)$ and is parallel to the plane given by $2x + y - 3z + 7 = 0$.

18. Give a standard equation for a plane that contains the origin and is parallel to the plane given by $x - y + 5z + \sqrt{2} = 0$.

19. Give an equation of the form $ax + 5y + cz + d = 0$ for a plane that contains the point $(1, -2, 3)$ and is parallel to the plane given by $2x + y - 3z + 7 = 0$.

20. Give an equation of the form $ax + by + 4z + d = 0$ for a plane that contains the point $(1, 1, 1)$ and is parallel to the plane given by $2x + y - 3z + 7 = 0$.

21. Determine the points of intersection, if any, of the plane given by $x+2y+5z-8=0$ and the line given by $(x,y,z)=(3,0,7)+t(-1,1,1)$.

22. Determine the points of intersection, if any, of the line given by $(x,y,z)=(-3,1,5)+r(5,0,3)$ and the line given by $(x,y,z)=(3,0,7)+t(-1,1,1)$.

23. Determine a vector equation for a line that contains the point $(7,5,-1)$ and is perpendicular to the plane given by $3x-2y+5z+4=0$.

24. Determine a vector equation for a line that contains the origin and is perpendicular to the plane given by $x+2y-5z+7=0$.

25. Determine a second point on the line which contains the point $(1,1,1)$ and which is parallel to the line given by $(x,y,z)=(2,0,1)+t(4,-5,3)$.

26. Determine a second point on the line which contains the point $(2,4,3)$ and which is parallel to the line given by $(x,y,z)=t(2,-1,2)$.

27. Find the x-, y-, and z-intercepts of the plane given by $3x-2y+5z+4=0$.

28. Find the x-, y-, and z-intercepts of the plane which contains the point $(2,4,-1)$ and which is normal to the vector $(1,1,-1)$.

29. Give the standard parameterization of the line segment from the point $(2,1,3)$ to the point $(5,3,1)$. Explicitly give the restrictions on the parameter.

30. Give the standard parameterization of the line segment from the point $(-1,0,7)$ to the origin. Explicitly give the restrictions on the parameter.

More Depth:

31. Parameterize the plane in $\mathbb{R}^3$ which contains the point $(1,2,3)$ and is parallel to each of the vectors $(5,0,7)$ and $(4,-1,6)$.

32. Parameterize the plane in $\mathbb{R}^3$ which contains the point $(1,2,3)$ and is parallel to the lines given by $(x,y,z)=(3,2,1)+s(1,2,3)$ and $(x,y,z)=(9,1,2)+t(1,-1,1)$.

33. Determine the points of intersection, if any, of the line parameterized by $(x,y,z)=(-3,1,5)+r(5,0,3)$ and the plane parameterized by $(x,y,z)=(3,0,7)+u(-1,1,1)+v(2,5,0)$.

34. Find the distance from the point $(2,0,-5)$ to the plane given by $3x-2y+z+7=0$.

35. Find the distance from the point $(1,1,1)$ to the plane given by $3x-2y+z=7$.

The angle between two planes in $\mathbb{R}^3$, with non-zero normal vectors n_1 and n_2 is the angle between n_1 and n_2, if that angle is $\leq 90°$; otherwise, it is the angle between $-n_1$ and n_2.

36. Find the angle between the planes given by $3x+y-5z+2=0$ and $x+2y+z-4=0$.

37. Find the angle between the planes given by $x+y+z+2=0$ and $x-y+z-2=0$.

The angle between a line, with non-zero parallel vector v, and a plane, with non-zero normal vector n is defined as follows: Take the angle θ between v and n, in degrees. Then, the angle, in degrees, between the line and plane is $|90-\theta|$.

38. Find the angle between the line given by $(x,y,z)=(1,2,0)+t(4,-1,3)$ and the plane given by $-x+3y+z+5=0$.

39. Find the angle between the line given by $(x,y,z)=(-1,5,3)+t(0,-1,2)$ and the plane given by $x+y+z-3=0$.

$\boxed{+ \textbf{ Linear Algebra:}}$

40. Find an equation for the hyperplane in $\mathbb{R}^4$ which contains the point $(1,2,3,4)$ and is normal to the vector $(5,0,-2,3)$.

41. Find an equation for the hyperplane in $\mathbb{R}^5$ which contains the point $(0,1,2,3,4)$ and is normal to the vector $(-1,5,0,-2,3)$.

42. Show that the set E of (x,y) in $\mathbb{R}^2$ such that $xy \geq 0$ contains the zero vector and is closed under scalar multiplication, but is **not** closed under vector addition and, therefore, is not a linear subspace of $\mathbb{R}^2$. $\;\blacktriangleright$

43. Show that the set V of (x,y) in $\mathbb{R}^2$ such that $x^2 + 9y^2 = 6xy$ is a linear subspace of $\mathbb{R}^2$.

44. Prove that, if V and W are linear subspaces of $\mathbb{R}^n$, and $\mathbf{p}$ and $\mathbf{q}$ are in $\mathbb{R}^n$, and $\mathbf{p}+V = \mathbf{q}+W$, then $V=W$.

45. Consider the hyperplane H in $\mathbb{R}^n$ consisting of those $\mathbf{x}$ in $\mathbb{R}^n$ such that $\mathbf{a}\cdot(\mathbf{x}-\mathbf{p})=0$, where $\mathbf{p}$ and $\mathbf{a}$ are in $\mathbb{R}^n$ and $\mathbf{a} \neq 0$. Let $\mathbf{q}$ be a point in $\mathbb{R}^n$. Show that there is a point in H which is closest to $\mathbf{q}$, and that that point is $\mathbf{m} = \mathbf{q} + \mathrm{proj}_{\mathbf{a}}(\mathbf{p}-\mathbf{q})$. Do this by showing two things: 1) That $\mathbf{m}$ is a point in H. 2) That, for all $\mathbf{x}$ in H,

$$|\mathbf{m}-\mathbf{q}| \leq |\mathbf{x}-\mathbf{q}|.$$

1.5   The Cross Product

The dot product of two vectors in $\mathbb{R}^n$ is a scalar, not a vector, and the dot product is defined in all dimensions n.

In this section, we define the *cross product* of two vectors in $\mathbb{R}^3$. This is defined only for vectors in $\mathbb{R}^3$ and the cross product of two vectors is another **vector**, not a scalar. The cross product is **not** commutative or associative, so care must be taken with parentheses and the order in which the vectors are written.

The cross product is slightly difficult to calculate, but contains a great deal of geometric information. It gives a vector that is perpendicular to each of the initial two vectors, and its magnitude is the area of the parallelogram formed from the two vectors.

Basics:

Suppose that $\mathbf{v} = (v_1, v_2, v_3)$ and $\mathbf{w} = (w_1, w_2, w_3)$ are two vectors in $\mathbb{R}^3$. We are first going to algebraically define the *cross product*, $\mathbf{v} \times \mathbf{w}$, of $\mathbf{v}$ and $\mathbf{w}$; after that, we will discuss **why** anyone would want to calculate such a seemingly bizarre thing.

First, we need to define a few concepts from linear algebra, for this will help you calculate the cross product, without having you memorize an awful formula.

An $m \times n$ (read: m by n) *matrix* is a rectangular collection of numbers (or other algebraic objects) with m horizontal rows and n vertical columns. For instance, here's a 2×3 matrix:

$$\begin{bmatrix} 1 & 3 & -7 \\ -5 & 0 & 2 \end{bmatrix}.$$

A matrix with the same number of rows as columns is called a *square matrix*. For every square matrix, there is an important associated number that you can calculate called the *determinant* of the matrix. The determinant is denoted by either writing det in front of the square matrix, or by replacing the brackets around the matrix by vertical lines.

A 1×1 matrix looks like $[a]$, for some real number a. The determinant of this matrix is simply the number a.

So, what's the definition of the determinant of a 2×2 matrix? We define the determinant by

$$\det \begin{bmatrix} a & b \\ c & d \end{bmatrix} = \begin{vmatrix} a & b \\ c & d \end{vmatrix} = ad - bc.$$

This usually thought of as the product of the entries on the diagonal from the upper-left to the lower-right, the **main diagonal**, minus the product of the entries on the other diagonal.

$$\begin{vmatrix} a & b \\ c & d \end{vmatrix} = ad - bc.$$

Figure 1.5.1: The determinant of a 2×2 matrix in terms of the products along diagonals.

While we shall not describe determinants for arbitrarily large square matrices, we need to define the determinant for 3×3 matrices. There are many ways to do this, but we shall use only one, the one that is best for calculating cross products.

Consider a general 3×3 matrix:

$$A = \begin{bmatrix} a & b & c \\ d & e & f \\ g & h & i \end{bmatrix}$$

We shall use what's known as a *cofactor expansion* along the top row to define the determinant of A. This method reduces finding the determinant of a 3×3 matrix to calculating determinants of three 2×2 matrices. We'll first give the formula, and then describe the easy way to remember it.

The determinant of a 3×3 matrix is given by

$$\begin{vmatrix} a & b & c \\ d & e & f \\ g & h & i \end{vmatrix} = \begin{vmatrix} e & f \\ h & i \end{vmatrix} \cdot a - \begin{vmatrix} d & f \\ g & i \end{vmatrix} \cdot b + \begin{vmatrix} d & e \\ g & h \end{vmatrix} \cdot c = \qquad (1.2)$$

$$(ei - fh)a - (di - fg)b + (dh - eg)c.$$

How do you remember this? You do **not** memorize the final result. Instead, you remember how to take determinants of 2×2 matrices, and you remember the formula involving the three 2×2 determinants as follows:

1. You move through each entry in the top row, in order.

2. First, you take the entry a, multiply by $+1$ (we write this because, below, this will alternate to a -1), and you delete the row and column containing a.

$$\begin{vmatrix} \cancel{a} & \cancel{b} & \cancel{c} \\ \cancel{d} & e & f \\ \cancel{g} & h & i \end{vmatrix}$$

Figure 1.5.2: Delete the row and column containing a.

Then, you multiply the determinant of the remaining 2×2 matrix times a, to obtain

$$\begin{vmatrix} e & f \\ h & i \end{vmatrix} \cdot a.$$

3. Then, you move to the second entry in the first row, but you alternate and multiply it by -1, instead of $+1$. You take the entry b, and you delete the row and column containing b.

$$\begin{vmatrix} a & b & c \\ d & e & f \\ g & h & i \end{vmatrix}$$

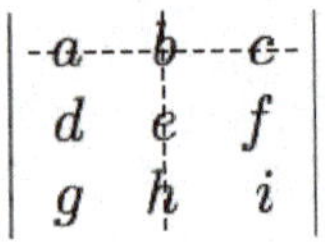

Figure 1.5.3: Delete the row and column containing b.

Then, you multiply the determinant of the remaining 2×2 matrix times $-1 \cdot b$, to obtain

$$-\begin{vmatrix} d & f \\ g & i \end{vmatrix} \cdot b,$$

and you add this to what we already had, to obtain

$$\begin{vmatrix} e & f \\ h & i \end{vmatrix} \cdot a - \begin{vmatrix} d & f \\ g & i \end{vmatrix} \cdot b.$$

4. Finally, you move to the third entry in the first row, but you alternate again and multiply it by $+1$. You take the entry c, and you delete the row and column containing c.

$$\begin{vmatrix} a & b & c \\ d & e & f \\ g & h & i \end{vmatrix}$$

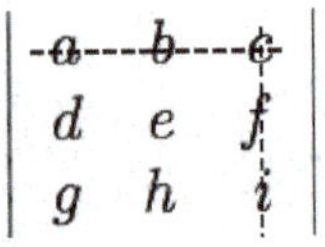

Figure 1.5.4: Delete the row and column containing c.

Then, you multiply the determinant of the remaining 2×2 matrix times $+1 \cdot c$, to obtain

$$\begin{vmatrix} d & e \\ g & h \end{vmatrix} \cdot c,$$

and you add this to what we already had, to obtain the final result that we were after:

$$\begin{vmatrix} e & f \\ h & i \end{vmatrix} \cdot a - \begin{vmatrix} d & f \\ g & i \end{vmatrix} \cdot b + \begin{vmatrix} d & e \\ g & h \end{vmatrix} \cdot c.$$

Example 1.5.1. Calculate the determinant:

$$\begin{vmatrix} 2 & 3 & 5 \\ -1 & 0 & 4 \\ 7 & 1 & 6 \end{vmatrix}.$$

Solution:

We calculate

$$\begin{vmatrix} 2 & 3 & 5 \\ -1 & 0 & 4 \\ 7 & 1 & 6 \end{vmatrix} = \begin{vmatrix} 0 & 4 \\ 1 & 6 \end{vmatrix} \cdot 2 - \begin{vmatrix} -1 & 4 \\ 7 & 6 \end{vmatrix} \cdot 3 + \begin{vmatrix} -1 & 0 \\ 7 & 1 \end{vmatrix} \cdot 5 =$$

$$(0 - 4) \cdot 2 - (-6 - 28) \cdot 3 + (-1 - 0) \cdot 5 = 89.$$

So, that's the determinant of a 3×3 matrix. Now we'll use it to define the cross product of two vectors in $\mathbb{R}^3$. The cross product will be defined in terms of a 3×3 determinant, but the determinant will be of a matrix whose top row consists of the standard basis vectors for $\mathbb{R}^3$, that is, the top row will contain $\mathbf{i}$, $\mathbf{j}$, and $\mathbf{k}$.

Definition 1.5.2. *Suppose that* $\mathbf{v} = (v_1, v_2, v_3)$ *and* $\mathbf{w} = (w_1, w_2, w_3)$ *are vectors in* $\mathbb{R}^3$. *Then, the* **cross product**, $\mathbf{v} \times \mathbf{w}$ *is defined to be the vector given by*

$$\mathbf{v} \times \mathbf{w} = \begin{vmatrix} \mathbf{i} & \mathbf{j} & \mathbf{k} \\ v_1 & v_2 & v_3 \\ w_1 & w_2 & w_3 \end{vmatrix} = \begin{vmatrix} v_2 & v_3 \\ w_2 & w_3 \end{vmatrix} \mathbf{i} - \begin{vmatrix} v_1 & v_3 \\ w_1 & w_3 \end{vmatrix} \mathbf{j} + \begin{vmatrix} v_1 & v_2 \\ w_1 & w_2 \end{vmatrix} \mathbf{k} =$$

$$(v_2 w_3 - v_3 w_2,\ -v_1 w_3 + v_3 w_1,\ v_1 w_2 - v_2 w_1).$$

We want to emphasize that this final result is difficult to memorize; it is much easier to remember and use the determinant formulation.

Example 1.5.3. It's easy to calculate that $\mathbf{i} \times \mathbf{j} = \mathbf{k}$, $\mathbf{j} \times \mathbf{k} = \mathbf{i}$, and $\mathbf{k} \times \mathbf{i} = \mathbf{j}$. We'll demonstrate this first equality, and leave the other two as exercises.

We find

$$\mathbf{i} \times \mathbf{j} = \begin{vmatrix} \mathbf{i} & \mathbf{j} & \mathbf{k} \\ 1 & 0 & 0 \\ 0 & 1 & 0 \end{vmatrix} = \begin{vmatrix} 0 & 0 \\ 1 & 0 \end{vmatrix} \mathbf{i} - \begin{vmatrix} 1 & 0 \\ 0 & 0 \end{vmatrix} \mathbf{j} + \begin{vmatrix} 1 & 0 \\ 0 & 1 \end{vmatrix} \mathbf{k} = 0 \cdot \mathbf{i} - 0 \cdot \mathbf{j} + 1 \cdot \mathbf{k} = \mathbf{k}.$$

Example 1.5.4. Calculate $(3, 1, 2) \times (-5, 0, 4)$.

Solution:

We find

$$(3, 1, 2) \times (-5, 0, 4) = \begin{vmatrix} \mathbf{i} & \mathbf{j} & \mathbf{k} \\ 3 & 1 & 2 \\ -5 & 0 & 4 \end{vmatrix} = \begin{vmatrix} 1 & 2 \\ 0 & 4 \end{vmatrix} \mathbf{i} - \begin{vmatrix} 3 & 2 \\ -5 & 4 \end{vmatrix} \mathbf{j} + \begin{vmatrix} 3 & 1 \\ -5 & 0 \end{vmatrix} \mathbf{k} =$$

$$4\mathbf{i} - 22\mathbf{j} + 5\mathbf{k} = (4, -22, 5).$$

Now that you know **how** to calculate the cross product of two vectors in $\mathbb{R}^3$, you should be asking: "Why would I want to? What is the point of the cross product?"

Actually, both the magnitude and direction of the cross product have very important geometric significance.

Before we state the theorem which explains the importance of the cross product, we first need to discuss some geometry. Suppose that we have two vectors $\mathbf{v}$ and $\mathbf{w}$, which we base at the same point. Then, as we mentioned earlier, these vectors determine a parallelogram, which we say is *spanned by the two vectors*. See Figure 1.5.5.

> Rigorously, the "parallelogram spanned by $\mathbf{v}$ and $\mathbf{w}$" can be taken to the set of points of the form $a\mathbf{v} + b\mathbf{w}$, where $0 \leq a \leq 1$ and $0 \leq b \leq 1$, or any translation of this set obtained by adding a fixed point $\mathbf{p}$ to each point.

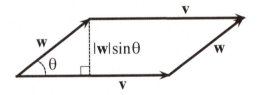

Figure 1.5.5: The area of the parallelogram spanned by $\mathbf{v}$ and $\mathbf{w}$.

The area of the parallelogram is the length of the base times the height; as you can see is Figure 1.5.5, this area is

$$|\mathbf{v}||\mathbf{w}| \sin\theta,$$

where θ is the angle between $\mathbf{v}$ and $\mathbf{w}$. Note that since $0 \leq \theta \leq \pi$ radians, $\sin\theta \geq 0$, and yields the appropriate area of 0 exactly when it should; namely, when $\mathbf{v}$ and $\mathbf{w}$ are parallel, so that the "parallelogram" has no area.

Now we can state the main theorem which explains the geometric significance of the cross product. We will prove this theorem in the More Depth portion of this section.

Theorem 1.5.5. *Suppose that* **v** *and* **w** *are two vectors in* $\mathbb{R}^3$. *Then, the cross product* **v** × **w** *is completely determined by specifying its magnitude and direction as follows:*

1. *The magnitude of* **v** × **w** *is the area of the parallelogram spanned by* **v** *and* **w**, *i.e.,*

$$|\mathbf{v} \times \mathbf{w}| = |\mathbf{v}||\mathbf{w}|\sin\theta,$$

where θ *is the angle between* **v** *and* **w**. *In particular,* **v** × **w** = **0** *if and only if* **v** *and* **w** *are parallel (which allows for* **v** *or* **w** *being* **0***).*

2. *The vector* **v** × **w** *is perpendicular to both* **v** *and* **w**. *Thus, if* **v** *and* **w** *are not parallel, and are drawn based at the same point, then* **v** × **w** *is normal to the plane determined by* **v** *and* **w**.

3. *If* **v** *and* **w** *are not parallel, then the previous property determines the direction of* **v** × **w**, *except that it does not determine which of the two perpendicular directions to the plane* **v** × **w** *points in. This is determined by the* **right-hand rule***: using your right hand, if you point your index finger in the direction of* **v**, *and your middle finger in the direction of* **w**, *then your thumb points in the direction of* **v** × **w**.

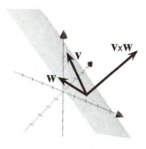

Figure 1.5.6: The direction of **v** × **w** is determined by being normal to the plane and the right-hand rule.

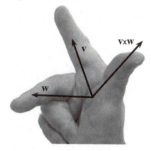

Figure 1.5.7: The direction of **v** × **w** is determined by being normal to the plane and the right-hand rule.

The right-hand rule for the direction of **v** × **w** assumes that you are using a right-handed coordinate system, as we defined in Section 1.1.

Note that Property 1 of Theorem 1.5.5 implies that, for all **v** in $\mathbb{R}^3$, **v** × **v** = **0**.

Example 1.5.6. Let's return to our calculation $(3, 1, 2) \times (-5, 0, 4) = (4, -22, 5)$ from Example 1.5.4. Let's check that our resultant vector is perpendicular to each of the vectors in the cross product; of course, we do this by checking that the dot products are 0.

We find

$$(4, -22, 5) \cdot (3, 1, 2) = 12 - 22 + 10 = 0$$

and

$$(4, -22, 5) \cdot (-5, 0, 4) = -20 + 0 + 20 = 0.$$

Our cross product calculation also tells us that the area of the parallelogram spanned by $(3, 1, 2)$ and $(-5, 0, 4)$ is

$$|(3, 1, 2) \times (-5, 0, 4)| = |(4, -22, 5)| = \sqrt{4^2 + (-22)^2 + 5^2} = \sqrt{525} = 5\sqrt{21}.$$

It follows at once from Property 3 of Theorem 1.5.5, or by direct calculation, that the cross product is **not** commutative. Instead, we have:

Corollary 1.5.7. *The cross product is* **anti-commutative**; *this means that*

$$\mathbf{v} \times \mathbf{w} = -\mathbf{w} \times \mathbf{v}.$$

Example 1.5.8. The cross product is also not associative. For instance, we claim that

$$\mathbf{i} \times (\mathbf{i} \times \mathbf{j}) \neq (\mathbf{i} \times \mathbf{i}) \times \mathbf{j}.$$

To show this, we calculate

$$\mathbf{i} \times (\mathbf{i} \times \mathbf{j}) = \mathbf{i} \times \mathbf{k} = -\mathbf{k} \times \mathbf{i} = -\mathbf{j},$$

while

$$(\mathbf{i} \times \mathbf{i}) \times \mathbf{j} = \mathbf{0} \times \mathbf{j} = \mathbf{0}.$$

However, not all of the familiar algebraic properties of a product fail for the cross product; direct calculation from the definition shows that:

Theorem 1.5.9. *The cross product with a given vector is linear, i.e., it distributes over addition and constants can be "pulled out".*
This means that, for all vectors $\mathbf{p}$, $\mathbf{q}$, *and* $\mathbf{r}$ *in* $\mathbb{R}^3$, *and all scalars* a *and* b,

$$\mathbf{p} \times (a\mathbf{q} + b\mathbf{r}) = a(\mathbf{p} \times \mathbf{q}) + b(\mathbf{p} \times \mathbf{r})$$

and

$$(a\mathbf{q} + b\mathbf{r}) \times \mathbf{p} = a(\mathbf{q} \times \mathbf{p}) + b(\mathbf{r} \times \mathbf{p}).$$

Because the cross product is linear in each "factor", the cross product is frequently said to be *bilinear*.

Now let's look at a series of examples which show the usefulness of the cross product.

Example 1.5.10. Consider the three points $\mathbf{a} = (1, 0, 1)$, $\mathbf{b} = (2, 1, -1)$, and $\mathbf{c} = (0, 3, 0)$. Determine a standard equation for the plane containing these three points, and find the area of the triangle which has $\mathbf{a}$, $\mathbf{b}$, and $\mathbf{c}$ as vertices.

Solution:

Recall from Theorem 1.4.6 that any time you want a standard equation for a plane, you want a point (x_0, y_0, z_0) on the plane and a non-zero normal vector $\mathbf{n}$ to the plane.

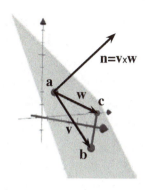

Of course, we have three points on the plane; so we can use any of those as our point on the plane. We'll use $(0, 3, 0)$. But how do we produce a normal vector to the plane?

As you can see in Figure 1.5.8, we let $\mathbf{v}$ be the displacement vector from $\mathbf{a}$ to $\mathbf{b}$, i.e., $\mathbf{v} = (2, 1, -1) - (1, 0, 1) = (1, 1, -2)$, and we let $\mathbf{w}$ be the displacement vector from $\mathbf{a}$ to $\mathbf{c}$, i.e., $\mathbf{w} = (0, 3, 0) - (1, 0, 1) = (-1, 3, -1)$.

Since $\mathbf{a}$, $\mathbf{b}$, and $\mathbf{c}$ are not collinear, $\mathbf{v}$ and $\mathbf{w}$ are not parallel, and so $\mathbf{n} = \mathbf{v} \times \mathbf{w}$ will be a non-zero vector which is normal to the plane.

Figure 1.5.8: Three non-collinear points in $\mathbb{R}^3$ determine a plane.

We calculate

$$\mathbf{n} = \begin{vmatrix} \mathbf{i} & \mathbf{j} & \mathbf{k} \\ 1 & 1 & -2 \\ -1 & 3 & -1 \end{vmatrix} = \begin{vmatrix} 1 & -2 \\ 3 & -1 \end{vmatrix} \mathbf{i} - \begin{vmatrix} 1 & -2 \\ -1 & -1 \end{vmatrix} \mathbf{j} + \begin{vmatrix} 1 & 1 \\ -1 & 3 \end{vmatrix} \mathbf{k} =$$

$$5\mathbf{i} + 3\mathbf{j} + 4\mathbf{k} = (5, 3, 4).$$

Therefore, an equation for the plane is

$$5(x - 0) + 3(y - 3) + 4(z - 0) = 0 \quad \text{or, equivalently,} \quad 5x + 3y + 4z - 9 = 0.$$

You can check that this equation is correct by verifying that the points $\mathbf{a}$, $\mathbf{b}$, and $\mathbf{c}$ satisfy this equation.

The area of triangle with vertices $\mathbf{a}$, $\mathbf{b}$, and $\mathbf{c}$ is half the area of the parallelogram spanned by $\mathbf{v}$ and $\mathbf{w}$. Thus,

$$\text{area of triangle} = \frac{1}{2} |\mathbf{v} \times \mathbf{w}| = \frac{1}{2} \sqrt{5^2 + 3^2 + 4^2} = \frac{5\sqrt{2}}{2}.$$

Example 1.5.11. Consider the two planes given by $(x-1) + 2(y-1) + 2(z-1) = 0$ and $2(x-1) - 3(y-1) + 4(z-1) = 0$. Parameterize, in vector form, the line of intersection of the two planes in (recall Theorem 1.4.11).

Solution:

Note that both planes contain the point $(1, 1, 1)$, and that non-zero normal vectors to the planes are $\mathbf{n}_1 = (1, 2, 2)$ and $\mathbf{n}_2 = (2, -3, 4)$, respectively.

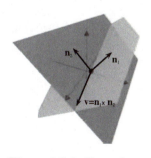

To apply Theorem 1.4.11, we need a point on the line and a non-zero vector parallel to the line. But the line of intersection is in both planes and, hence, is perpendicular to both $\mathbf{n}_1$ and $\mathbf{n}_2$. Thus, $\mathbf{v} = \mathbf{n}_1 \times \mathbf{n}_2$ is a vector parallel to the line.

Figure 1.5.9: Two planes intersecting along a line.

We calculate

$$\mathbf{v} \;=\; \mathbf{n}_1 \times \mathbf{n}_2 \;=\; \begin{vmatrix} \mathbf{i} & \mathbf{j} & \mathbf{k} \\ 1 & 2 & 2 \\ 2 & -3 & 4 \end{vmatrix} \;=\; 14\mathbf{i} - 0\mathbf{j} + (-7)\mathbf{k} \;=\; (14, 0, -7).$$

Therefore, a parameterization of the line of intersection is given by

$$(x, y, z) = (1, 1, 1) + t(14, 0, -7).$$

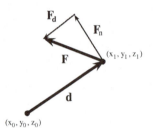

Figure 1.5.10: The torque vector $\mathbf{d} \times \mathbf{F}$ comes out of the page, towards you, by the right-hand rule.

Example 1.5.12. Consider a rigid rod, with one end at the point (x_0, y_0, z_0) and the other end at (x_1, y_1, z_1). Let $\mathbf{d} = (x_1, y_1, z_1) - (x_0, y_0, z_0)$. If a force $\mathbf{F}$ is applied at (x_1, y_1, z_1), then the **torque** produced by $\mathbf{F}$ around (x_0, y_0, z_0) is defined to be

$$\boldsymbol{\tau} \;=\; \mathbf{d} \times \mathbf{F}.$$

The magnitude of the torque is thus $|\mathbf{d}|\,|\mathbf{F}| \sin\theta$, where θ is the angle between $\mathbf{d}$ and $\mathbf{F}$. The magnitude $|\mathbf{d}|$ is referred to as the *length of the lever arm*, and $|\mathbf{F}| \sin\theta$ is the magnitude of the component of the force which is perpendicular to $\mathbf{d}$, that is, the magnitude of $\mathbf{F}_n = \mathbf{F} - \mathbf{F}_\mathbf{d}$ (recall Definition 1.3.9).

Suppose, for instance, that a rigid rod extends from $(1, 2, 3)$ to $(3, 3, 2)$, where all coordinates are measured in meters, and that a force $\mathbf{F} = (2, -1, 0)$ Newtons acts on the rod at $(3, 3, 2)$. Let's determine the torque produced about $(1, 2, 3)$.

We find that $\mathbf{d} = (3, 3, 2) - (1, 2, 3) = (2, 1, -1)$ meters. So, the torque is

$$\boldsymbol{\tau} \;=\; \mathbf{d} \times \mathbf{F} \;=\; \begin{vmatrix} \mathbf{i} & \mathbf{j} & \mathbf{k} \\ 2 & 1 & -1 \\ 2 & -1 & 0 \end{vmatrix} \;=\; -\mathbf{i} - 2\mathbf{j} - 4\mathbf{k} \;=\; (-1, -2, -4) \text{ Newton-meters.}$$

We should remark here that, while a joule is also a Newton-meter, the unit of joules is reserved for work or energy and is not used for torque.

More Depth:

Given three vectors in $\mathbb{R}^3$, we can take what's known as the *triple product*. Before we define and discuss the triple product, it will be helpful to look at the dot product of two vectors in $\mathbb{R}^3$ in a new way.

Suppose that $\mathbf{a} = (a_1, a_2, a_3)$ and $\mathbf{d} = (d_1, d_2, d_3)$ are two vectors in $\mathbb{R}^3$. If we write $\mathbf{d}$ in terms of $\mathbf{i}$, $\mathbf{j}$ and $\mathbf{k}$, then the dot product looks like

$$\mathbf{a} \cdot \mathbf{d} \;=\; (a_1, a_2, a_3) \cdot (d_1 \mathbf{i} + d_2 \mathbf{j} + d_3 \mathbf{k}) \;=\; d_1 a_1 + d_2 a_2 + d_3 a_3;$$

this represents the dot product of $\mathbf{a}$ with $\mathbf{d}$ as replacing the vectors $\mathbf{i}$, $\mathbf{j}$, and $\mathbf{k}$ in the expression for $\mathbf{d}$ with the scalar components of $\mathbf{a}$.

Now, looking back at our definition of the cross product of the vectors $\mathbf{b} = (b_1, b_2, b_3)$ and $\mathbf{c} = (c_1, c_2, c_3)$ in $\mathbb{R}^3$, we have

$$\mathbf{b} \times \mathbf{c} = \begin{vmatrix} \mathbf{i} & \mathbf{j} & \mathbf{k} \\ b_1 & b_2 & b_3 \\ c_1 & c_2 & c_3 \end{vmatrix}.$$

Now, if we dot $\mathbf{a}$ with $\mathbf{b} \times \mathbf{c}$, then, as we saw, we just replace the $\mathbf{i}$, $\mathbf{j}$, and $\mathbf{k}$ in the expression for $\mathbf{b} \times \mathbf{c}$ with the scalar components of $\mathbf{a}$.

Thus we define:

Definition 1.5.13. *Suppose that* $\mathbf{a} = (a_1, a_2, a_3)$, $\mathbf{b} = (b_1, b_2, b_3)$ *and* $\mathbf{c} = (c_1, c_2, c_3)$ *are vectors in* $\mathbb{R}^3$.

    Then, the **triple product** *of* $\mathbf{a}$, $\mathbf{b}$ *and* $\mathbf{c}$ *(in that order) is the real number*

$$\mathbf{a} \cdot (\mathbf{b} \times \mathbf{c}) = \begin{vmatrix} a_1 & a_2 & a_3 \\ b_1 & b_2 & b_3 \\ c_1 & c_2 & c_3 \end{vmatrix}.$$

Either direct calculation or well-known properties of determinants yields that $\mathbf{a} \cdot (\mathbf{b} \times \mathbf{c}) = \mathbf{b} \cdot (\mathbf{c} \times \mathbf{a}) = \mathbf{c} \cdot (\mathbf{a} \times \mathbf{b})$, and that the other three possible orderings negate the triple product.

Example 1.5.14. Let's calculate the triple product of $\mathbf{a} = (2, 3, 5)$, $\mathbf{b} = (-1, 0, 4)$, and $\mathbf{c} = (7, 1, 6)$.

We find

$$\mathbf{a} \cdot (\mathbf{b} \times \mathbf{c}) = \begin{vmatrix} 2 & 3 & 5 \\ -1 & 0 & 4 \\ 7 & 1 & 6 \end{vmatrix} = 89,$$

as we already calculated in Example 1.5.1.

Of course, there remains the question of why you'd **want** to calculate the triple product. We can explain this, after we once again write the absolute value of the dot product in an interesting way.

Suppose that $\mathbf{a}$ and $\mathbf{d} \neq \mathbf{0}$ are vectors in $\mathbb{R}^3$. Recall the definition of the orthogonal projection in Definition 1.3.9. Then, we see that

$$|\mathbf{a} \cdot \mathbf{d}| = |\text{proj}_{\mathbf{d}}\mathbf{a}||\mathbf{d}|.$$

Now, if $\mathbf{b}$ and $\mathbf{c}$ are non-parallel vectors in $\mathbb{R}^3$, then $\mathbf{n} = \mathbf{b} \times \mathbf{c}$ is a non-zero vector which is normal to the parallelogram B spanned by $\mathbf{b}$ and $\mathbf{c}$. Thus,

$$|\mathbf{a} \cdot (\mathbf{b} \times \mathbf{c})| \; = \; |\mathrm{proj}_{\mathbf{n}}\mathbf{a}||\mathbf{b} \times \mathbf{c}|,$$

where $|\mathbf{b} \times \mathbf{c}|$ is the area of the parallelogram B, and $|\mathrm{proj}_{\mathbf{n}}\mathbf{a}|$ is the length of the projection of $\mathbf{a}$ onto a normal vector to the parallelogram, i.e., $|\mathrm{proj}_{\mathbf{n}}\mathbf{a}|$ is the height of the parallelepiped spanned by $\mathbf{a}$, $\mathbf{b}$, and $\mathbf{c}$, with base B.

Thus, we have

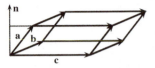

Figure 1.5.11: The parallelepiped spanned by $\mathbf{a}$, $\mathbf{b}$, and $\mathbf{c}$.

Note that we switched to using "det" to denote the determinant to avoid confusion between using vertical lines to denote both the determinant and the absolute value.

Theorem 1.5.15. *Suppose that* $\mathbf{a} = (a_1, a_2, a_3)$, $\mathbf{b} = (b_1, b_2, b_3)$ *and* $\mathbf{c} = (c_1, c_2, c_3)$ *are vectors in* $\mathbb{R}^3$.

Then, the volume of the parallelepiped spanned by $\mathbf{a}$, $\mathbf{b}$, *and* $\mathbf{c}$ *is the absolute value of the triple product, i.e., is*

$$|\mathbf{a} \cdot (\mathbf{b} \times \mathbf{c})| \; = \; \left| \det \begin{bmatrix} a_1 & a_2 & a_3 \\ b_1 & b_2 & b_3 \\ c_1 & c_2 & c_3 \end{bmatrix} \right|,$$

which is independent of the order of the vectors in the triple product.

Example 1.5.16. In Example 1.5.14, we calculated the triple product of $\mathbf{a} = (2, 3, 5)$, $\mathbf{b} = (-1, 0, 4)$, and $\mathbf{c} = (7, 1, 6)$, and found it to be

$$\mathbf{a} \cdot (\mathbf{b} \times \mathbf{c}) \; = \; \det \begin{bmatrix} 2 & 3 & 5 \\ -1 & 0 & 4 \\ 7 & 1 & 6 \end{bmatrix} \; = \; 89.$$

Now, in light of Theorem 1.5.15, we know that that 89 has geometric significance; it's the volume of the parallelepiped spanned by $\mathbf{a}$, $\mathbf{b}$, and $\mathbf{c}$.

Now, we want to discuss the proof of Theorem 1.5.5.

Recall that, if $\mathbf{v} = (v_1, v_2, v_3)$ and $\mathbf{w} = (w_1, w_2, w_3)$ are vectors in $\mathbb{R}^3$, then our definition of the cross product boiled down to

$$\mathbf{v} \times \mathbf{w} \; = \; (v_2w_3 - v_3w_2, \; -v_1w_3 + v_3w_1, \; v_1w_2 - v_2w_1).$$

Property 2 of Theorem 1.5.5 is the easiest to verify; you simply take the dot product of $\mathbf{v} \times \mathbf{w}$ with each of the original vectors and make sure that you get 0 both times.

Thus, we calculate

$$(v_2 w_3 - v_3 w_2, \; -v_1 w_3 + v_3 w_1, \; v_1 w_2 - v_2 w_1) \cdot (v_1, v_2, v_3) =$$

$$v_1 v_2 w_3 - v_1 v_3 w_2 - v_1 v_2 w_3 + v_2 v_3 w_1 + v_1 v_3 w_2 - v_2 v_3 w_1 = 0,$$

and

$$(v_2 w_3 - v_3 w_2, \; -v_1 w_3 + v_3 w_1, \; v_1 w_2 - v_2 w_1) \cdot (w_1, w_2, w_3) =$$

$$v_2 w_1 w_3 - v_3 w_1 w_2 - v_1 w_2 w_3 + v_3 w_1 w_2 + v_1 w_2 w_3 - v_2 w_1 w_3 = 0.$$

Property 1 of Theorem 1.5.5 is a little uglier to verify; we need to show that

$$|(v_2 w_3 - v_3 w_2, \; -v_1 w_3 + v_3 w_1, \; v_1 w_2 - v_2 w_1)| = |\mathbf{v}|\,|\mathbf{w}|\sin\theta.$$

Since $\sin\theta \geq 0$, it is enough to show that the square of both sides of the above equality are equal.

Now,

$$|(v_2 w_3 - v_3 w_2, \; -v_1 w_3 + v_3 w_1, \; v_1 w_2 - v_2 w_1)|^2 =$$

$$(v_2 w_3 - v_3 w_2)^2 + (-v_1 w_3 + v_3 w_1)^2 + (v_1 w_2 - v_2 w_1)^2 =$$

$$v_2^2 w_3^2 - 2 v_2 v_3 w_2 w_3 + v_3^2 w_2^2 + v_1^2 w_3^2 - 2 v_1 v_3 w_1 w_3 + v_3^2 w_1^2 + v_1^2 w_2^2 - 2 v_1 v_2 w_1 w_2 + v_2^2 w_1^2.$$

We also calculate

$$|\mathbf{v}|^2 |\mathbf{w}|^2 \sin^2\theta = |\mathbf{v}|^2 |\mathbf{w}|^2 (1 - \cos^2\theta) = |\mathbf{v}|^2 |\mathbf{w}|^2 \left[1 - \left(\frac{(\mathbf{v} \cdot \mathbf{w})^2}{|\mathbf{v}|^2 |\mathbf{w}|^2} \right) \right] =$$

$$|\mathbf{v}|^2 |\mathbf{w}|^2 - (\mathbf{v} \cdot \mathbf{w})^2 = \left(v_1^2 + v_2^2 + v_3^2 \right) \left(w_1^2 + w_2^2 + w_3^2 \right) - (v_1 w_1 + v_2 w_2 + v_3 w_3)^2.$$

We leave it as an exercise for you to expand this and verify that it matches what we obtained above.

Finally, it remains for us to show that $\mathbf{v} \times \mathbf{w}$ satisfies the right-hand rule. This is actually the most difficult, and we can just give the idea.

As we saw in Example 1.5.3, $\mathbf{i} \times \mathbf{j} = \mathbf{k}$, which satisfies the right-hand rule (assuming that we use a right-handed coordinate system). Suppose you start with two vectors $\mathbf{v}$ and $\mathbf{w}$, which are not parallel, so that $\mathbf{v} \times \mathbf{w} \neq \mathbf{0}$. Then, by rotating all of space, scaling the vectors, and changing the angle between them, all of which can be done in a continuous manner, we can transform the ordered pair $(\mathbf{v}, \mathbf{w})$ into the ordered pair $(\mathbf{i}, \mathbf{j})$, without ever making the pair of vectors parallel, i.e., without ever making the cross product $\mathbf{0}$. Now, the expression for the cross product

$$(v_2 w_3 - v_3 w_2, \; -v_1 w_3 + v_3 w_1, \; v_1 w_2 - v_2 w_1)$$

We won't actually define continuous multivariable functions until Section 1.7.

is a continuous function of the six components. Thus, the cross product changes in a continuous way, and can never suddenly switch directions, from left-handed to right-handed without passing through $\mathbf{0}$, which we avoided.

Therefore, the right-hand rule is always satisfied.

$\boxed{+ \textbf{ Linear Algebra:}}$ Consider two vectors $\mathbf{a} = (a_1, a_2)$ and $\mathbf{b} = (b_1, b_2)$ in $\mathbb{R}^2$. What is the area of the parallelogram spanned by $\mathbf{a}$ and $\mathbf{b}$?

Actually, we answered this question already for vectors in $\mathbb{R}^3$, so we just have to think of $\mathbf{a}$ and $\mathbf{b}$ as being vectors in $\mathbb{R}^3$ by letting $\mathbf{v} = (a_1, a_2, 0)$ and $\mathbf{w} = (b_1, b_2, 0)$.

Then, Theorem 1.5.5 tells us that the area of the parallelogram spanned by $\mathbf{a}$ and $\mathbf{b}$, which is the area of the parallelogram spanned by $\mathbf{v}$ and $\mathbf{w}$, is

$$|\mathbf{v} \times \mathbf{w}| = |(0, 0, a_1 b_2 - a_2 b_1)| = |a_1 b_2 - a_2 b_1| = \left| \det \begin{bmatrix} a_1 & a_2 \\ b_1 & b_2 \end{bmatrix} \right|.$$

Let's denote the matrix formed by putting given vectors $\mathbf{v}_1$, $\mathbf{v}_2$, $\ldots \mathbf{v}_k$ in $\mathbb{R}^n$ in the rows by

$$\begin{bmatrix} - & \mathbf{v}_1 & - \\ - & \mathbf{v}_2 & - \\ & \vdots & \\ - & \mathbf{v}_k & - \end{bmatrix}.$$

Then, we can write that the area of the parallelogram spanned by $\mathbf{a}$ and $\mathbf{b}$ in $\mathbb{R}^2$ is

$$\text{area of parallelogram} = \left| \det \begin{bmatrix} - & \mathbf{a} & - \\ - & \mathbf{b} & - \end{bmatrix} \right|.$$

Now, Theorem 1.5.15 tells us that the volume of the parallelepiped spanned by $\mathbf{a}$, $\mathbf{b}$, and $\mathbf{c}$ in $\mathbb{R}^3$ is

$$\text{volume of parallelepiped} = \left| \det \begin{bmatrix} - & \mathbf{a} & - \\ - & \mathbf{b} & - \\ - & \mathbf{c} & - \end{bmatrix} \right|.$$

This should make you wonder if there's a generalization to higher dimensions. The answer is: yes.

Consider vectors $\mathbf{v}_1$, $\mathbf{v}_2$, $\ldots \mathbf{v}_n$ in $\mathbb{R}^n$. There is a compact n-dimensional object, called a *parallelotope*, which we say is spanned by $\mathbf{v}_1$, $\mathbf{v}_2$, $\ldots \mathbf{v}_n$. The *parallelotope spanned by* $\mathbf{v}_1$, $\mathbf{v}_2$, $\ldots \mathbf{v}_n$ is the set of points

$$t_1 \mathbf{v}_1 + t_2 \mathbf{v}_2 + \cdots + t_n \mathbf{v}_n,$$

where, for each i, $0 \leq t_i \leq 1$. A 2-dimensional parallelotope is a parallelogram. A 3-dimensional parallelotope is a parallelepiped.

The concepts of area and volume are generalized to higher dimensions, in the form of the *n-dimensional volume*; this is slightly confusing in $\mathbb{R}^2$, since it means that the 2-dimensional volume is what is normally called area.

The generalization of Theorem 1.5.15 to all dimensions is:

Theorem 1.5.17. *Suppose that* $\mathbf{v}_1, \mathbf{v}_2, \ldots \mathbf{v}_n$ *are vectors in* $\mathbb{R}^n$.
Then, the n-dimensional volume of the parallelotope spanned by $\mathbf{v}_1, \mathbf{v}_2, \ldots \mathbf{v}_n$
is

$$\left| \det \begin{bmatrix} - & \mathbf{v}_1 & - \\ - & \mathbf{v}_2 & - \\ & \vdots & \\ - & \mathbf{v}_n & - \end{bmatrix} \right|.$$

We assume that you are familiar with determinants of square matrices of arbitrary size.

1.5.1 Exercises

Online answers to select exercises are here.

Basics:

In each of the following exercises, you are given two vectors **v** and **w** in $\mathbb{R}^3$. Calculate the cross product $\mathbf{v} \times \mathbf{w}$.

1. $\mathbf{v} = \mathbf{j}$ and $\mathbf{w} = \mathbf{k}$.

2. $\mathbf{v} = \mathbf{k}$ and $\mathbf{w} = \mathbf{i}$.

3. $\mathbf{v} = (1, 0, 1)$ and $\mathbf{w} = (0, -2, 3)$.

4. $\mathbf{v} = (-1, 2, 3)$ and $\mathbf{w} = (5, 0, 1)$.

5. $\mathbf{v} = (-4, 2, 6)$ and $\mathbf{w} = (2, -1, -3)$.

6. $\mathbf{v} = (7, 2, -1)$ and $\mathbf{w} = (-1, 2, -3)$.

7. $\mathbf{v} = 3\mathbf{i} - 2\mathbf{j} + 5\mathbf{k}$ and $\mathbf{w} = -2\mathbf{j} + \mathbf{k}$.

8. $\mathbf{v} = \mathbf{i} + 4\mathbf{j} - \mathbf{k}$ and $\mathbf{w} = 3\mathbf{i} - 2\mathbf{j} + \mathbf{k}$.

In each of the following exercises, you are given the same pairs of vectors **v** and **w** as you were in the previous exercises. Calculate the area of the parallelogram spanned by **v** and **w** and, if **v** and **w** are not parallel, produce a unit vector which is orthogonal to both **v** and **w**.

9. $\mathbf{v} = \mathbf{j}$ and $\mathbf{w} = \mathbf{k}$.

10. $\mathbf{v} = \mathbf{k}$ and $\mathbf{w} = \mathbf{i}$.

11. $\mathbf{v} = (1, 0, 1)$ and $\mathbf{w} = (0, -2, 3)$.

12. $\mathbf{v} = (-1, 2, 3)$ and $\mathbf{w} = (5, 0, 1)$.

13. $\mathbf{v} = (-4, 2, 6)$ and $\mathbf{w} = (2, -1, -3)$.

14. $\mathbf{v} = (7, 2, -1)$ and $\mathbf{w} = (-1, 2, -3)$.

15. $\mathbf{v} = 3\mathbf{i} - 2\mathbf{j} + 5\mathbf{k}$ and $\mathbf{w} = -2\mathbf{j} + \mathbf{k}$.

16. $\mathbf{v} = \mathbf{i} + 4\mathbf{j} - \mathbf{k}$ and $\mathbf{w} = 3\mathbf{i} - 2\mathbf{j} + \mathbf{k}$.

Suppose that **a**, **b**, and **c** are vectors in $\mathbb{R}^3$ such that $\mathbf{a} \times \mathbf{b} = (2, -4, 2)$, $\mathbf{a} \times \mathbf{c} = (7, 13, -11)$, and $\mathbf{b} \times \mathbf{c} = (1, 7, 1)$. In each of the following exercises, calculate the given expression.

17. $\mathbf{a} \times \mathbf{a}$

18. $\mathbf{b} \times \mathbf{a}$

19. $\mathbf{a} \times (3\mathbf{b} - 2\mathbf{c})$

20. $(2\mathbf{b} - \mathbf{c}) \times (3\mathbf{a} + 5\mathbf{c})$

21. $(\mathbf{c} \times \mathbf{a}) \cdot (1, 1, 1)$

22. $(\mathbf{b} \times \mathbf{a}) \cdot (\mathbf{b} \times \mathbf{c})$

23. $(\mathbf{b} \times \mathbf{a}) \cdot (\mathbf{a} \times \mathbf{b})$

24. $\mathbf{i} \times (\mathbf{c} \times \mathbf{b})$

25. $(\mathbf{b} \times \mathbf{a}) \times (\mathbf{b} \times \mathbf{c})$

26. $(\mathbf{b} \times \mathbf{a}) \times (\mathbf{a} \times \mathbf{b})$

In each of the following exercises, you are given three non-collinear points in $\mathbb{R}^3$. Determine the area of the triangle which has the given points as vertices, and a standard equation for the plane containing the triangle.

27. $(0, 0, 0)$, $(1, 2, 3)$, $(1, 0, 0)$

28. $(0, 0, 0)$, $(-2, -1, -5)$, $(1, 1, 1)$

29. $(1, 0, -1)$, $(2, 5, 3)$, $(3, 2, 1)$

30. $(7, 2, 5)$, $(6, 1, 4)$, $(7, 3, 6)$

In each of the following exercises, you are given a vector equation for a line in space, together with a point which is not on the line. Find a standard equation for the plane containing the line and the point.

31. $(x, y, z) = (1, 0, 1) + t(1, 1, 1)$, and $(3, 2, 1)$

32. $(x, y, z) = t(1, 1, 1)$, and $(3, 2, 1)$

33. $(x, y, z) = (4, -1, 5) + t(2, 0, 7)$, and $(1, 1, 1)$

34. $(x, y, z) = (4, -1, 5) + t(1, 2, 3)$, and $(4, -1, 6)$

In each of the following exercises, consider a rigid rod, with one end at the origin and the other end at $(x_1, y_1, 0)$, where the coordinates are measured in meters. Suppose a force of $\mathbf{F} = (a, b, 0)$ Newtons is applied at $(x_1, y_1, 0)$. Then, calculate the magnitude of the torque produced by $\mathbf{F}$ around the origin, and indicate whether the torque vector comes out the page towards you, or goes into the page (where we take the page to be the xy-plane).

35. $(x_1, y_1) = (0, 1)$, $(a, b) = (3, 4)$.

36. $(x_1, y_1) = (1, 0)$, $(a, b) = (3, 4)$.

37. $(x_1, y_1) = (1, 2)$, $(a, b) = (3, 4)$.

38. $(x_1, y_1) = (1, -2)$, $(a, b) = (-4, 3)$.

More Depth:

39. Find the volume of the parallelepiped spanned by the vectors $(1, 0, 1)$, $(2, -1, 2)$, and $(5, 3, 0)$.

40. Find the volume of the parallelepiped spanned by the vectors $(7, 1, -2)$, $(1, 1, 1)$, and $(3, 2, 1)$.

41. Find the volume of the parallelepiped with one vertex at $(1, 0, 1)$ and adjoining vertices at $(7, 1, 4)$, $(1, 1, 1)$, and $(3, 2, 1)$.

42. Find the volume of the parallelepiped with one vertex at $(-1, 1, 2)$ and adjoining vertices at $(0, 0, 0)$, $(1, 1, 4)$, and $(3, 2, 5)$.

43. Find the volume of the tetrahedron with vertices at $(0, 0, 0)$, $(1, 0, 0)$, $(0, 1, 0)$ and $(0, 0, 1)$.

44. Find the volume of the tetrahedron with vertices at $(1, -1, 0)$, $(2, 0, 0)$, $(0, 2, 1)$ and $(3, 4, 0)$.

$+$ Linear Algebra:

45. Find the 4-dimensional volume of the parallelotope spanned by the vectors $(1, 0, 1, 0)$, $(0, 2, -1, 2)$, $(0, 5, 3, 0)$, and $(1, 2, 3, 4)$.

46. Find the 5-dimensional volume of the parallelotope spanned by the vectors $(1, 0, 1, 0, 1)$, $(0, 2, -1, 2, 0)$, $(0, 5, 3, 0, -1)$, $(1, 2, 3, 4, 5)$, and $(4, 0, 0, -2, 1)$.

1.6 Multi-Component Functions of a Single Variable

Multivariable Calculus is the Calculus of functions of more than one variable. This means that the domains of the functions should be subsets of $\mathbb{R}^n$, where $n \geq 2$.

However, in this section, we look at functions where the domain of the function is a subset of $\mathbb{R}$ and the codomain is a subset of $\mathbb{R}^n$, where $n \geq 2$. This reduces to the study of n single-variable functions, and so is not truly a "multivariable" topic. You may have encountered much of this material in your studies of single-variable Calculus.

A function $f : A \to B$ is specified by giving a domain A, a codomain B, and a rule f that associates, to each element of A, a unique element of B. The codomain may contain more elements than the *range* of f.

Basics:

Suppose that a particle is moving in space in such a way that its x, y, and z coordinates, measured in meters, at time t seconds (measured from some initial starting time), are given by

$$x = x(t) = \cos t, \qquad y = y(t) = \sin t, \qquad \text{and} \qquad z = z(t) = t.$$

Then, the position of the particle is given by the function $\mathbf{p} : \mathbb{R} \to \mathbb{R}^3$ given by

$$\mathbf{p}(t) \; = \; (x(t), y(t), z(t)) \; = \; (\cos t, \sin t, t).$$

More generally, a *single-variable function into* $\mathbb{R}^n$ is any function $\mathbf{p} : A \to B$, where A is a subset of $\mathbb{R}$ and B is a subset of $\mathbb{R}^n$. We also refer to functions into $\mathbb{R}^n$, where $n \geq 2$, as *multi-component functions*. In this context, a function which takes values in $\mathbb{R}$, instead of in a higher-dimensional Euclidean space, is referred to as a *real-valued function*.

A multi-component function is frequently referred to as a *vector-valued function*. We prefer to call it a multi-component function, in order to allow for those cases in which we don't necessarily want to think of the output as a vector.

While it can be very difficult to describe, in any "nice" way, the range of a multi-component function, sometimes it's fairly easy. For instance, looking at the function $\mathbf{p}$ above, you can see that the x- and y-coordinates move around a circle of radius 1, centered at the origin and, as t increases, the z-coordinate increases. Thus, the range of $\mathbf{p}$, which would represent the path of the moving particle, is a spiral; see Figure 1.6.1. It is common to refer to the range of a function as the *image* of the function, particularly when looking at the range/image graphically.

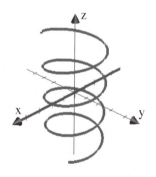

A single-variable function into $\mathbb{R}^n$ is specified by giving the domain and codomain, and the rule which specifies which values that appear in each component of the output. Thus, the rule for such a function $\mathbf{f}$ is given by specifying n real-valued functions, called the *component functions of* $\mathbf{f}$, f_1, f_2, ..., f_n so that

$$\mathbf{f}(t) \; = \; (f_1(t), \ f_2(t), \ \dots, \ f_n(t)).$$

Figure 1.6.1: The range, or image, of $\mathbf{p}(t) = (\cos t, \sin t, t)$.

The domain is frequently assumed to be implicitly given as the largest set of real numbers for which all of the f_i are defined, and the codomain is usually implicitly taken to be all of $\mathbb{R}^n$.

Technically, this means that the domain of the multi-component function is the intersection of the (implicit) domains of all of the component functions.

Definition 1.6.1. *A multi-component function*

$$\mathbf{f}(t) \; = \; (f_1(t), \; f_2(t), \; \ldots, \; f_n(t))$$

is **continuous** *or* **differentiable** *or* **continuously differentiable** *if and only if all of the component functions have the corresponding property.*

     The **derivative***, if it exists is*

$$\mathbf{f}'(t) \; = \; \lim_{h \to 0} \frac{1}{h}\Big(\mathbf{f}(t+h) - \mathbf{f}(t)\Big) \; = \; (f_1'(t), \; f_2'(t), \; \ldots, \; f_n'(t)).$$

When we describe the position of an object as being at a point, we are either thinking of a very small object, and often refer to "a particle", or we are referring to the *center of mass* of the object, which we won't seriously discuss until Section 3.9.

Definition 1.6.2. *Suppose that* $\mathbf{p}(t)$ *is the position of an object at time* t*. Then, provided they exist,* $\mathbf{v}(t) = \mathbf{p}'(t)$ *is the* **velocity** *of the object, the magnitude of the velocity* $|\mathbf{v}(t)|$ *is the* **speed** *of the object, and* $\mathbf{a} = \mathbf{v}'(t) = \mathbf{p}''(t)$ *is the* **acceleration** *of the object.*

Example 1.6.3. Suppose that $\mathbf{q} = (q_1, \ldots, q_n)$ is a point in $\mathbb{R}^n$, and $\mathbf{v} = (v_1, \ldots, v_n)$ is a vector in $\mathbb{R}^n$. Then, as we discussed in Section 1.4, the function

$$\mathbf{p}(t) \; = \; \mathbf{q} + t\mathbf{v} \; = \; (q_1 + tv_1, \, q_2 + tv_2, \, \ldots q_n + tv_n),$$

parameterizes a line in $\mathbb{R}^n$, which means, in particular, that the image of $\mathbf{p}$ is a line.

     If $\mathbf{p}$ is actually specifying the position of an object at time t, then the velocity is

$$\mathbf{v}(t) \; = \; \mathbf{p}'(t) \; = \; (v_1, \ldots, v_n) \; = \; \mathbf{v},$$

i.e., the velocity is constantly $\mathbf{v}$.

Example 1.6.4. Suppose that the position of a particle in $\mathbb{R}^3$, in meters, at time t seconds, is given by

$$\mathbf{p}(t) \; = \; (\cos t, \; \sin t, \; t).$$

We assume here that what we're calling "the particle" doesn't change its mass.

Determine the velocity and speed of the particle, and show that z-component of the net force $\mathbf{F}$ acting on the particle is zero at all times.

Solution:

The velocity is

$$\mathbf{v}(t) \;=\; \mathbf{p}'(t) \;=\; (-\sin t, \, \cos t, \, 1) \;\; \text{m/s}.$$

The speed is the magnitude of the velocity, which is frequently written with a non-boldface v, i.e.,

$$v(t) \;=\; |\mathbf{v}(t)| \;=\; \sqrt{(-\sin t)^2 + (\cos t)^2 + 1^2} \;=\; \sqrt{2} \;\; \text{m/s}.$$

Thus, the velocity is changing, but the speed is constant.

We will need the acceleration of the particle:

$$\mathbf{a}(t) \;=\; \mathbf{v}'(t) \;=\; (-\cos t, \, -\sin t, \, 0) \;\; (\text{m/s})/\text{s}.$$

The question about the force is an application of Newton's 2nd Law of Motion; if we let m be the mass of the particle and let $\mathbf{F} = \mathbf{F}(t)$ be the net force acting on the particle at time t, then

$$\mathbf{F}(t) \;=\; m\mathbf{a}(t) \;=\; m(-\cos t, \, -\sin t, \, 0) \;=\; (-m\cos t, \, -m\sin t, \, 0) \;\; \text{Newtons}.$$

Therefore, the z-component of the force is 0.

Remark 1.6.5. If $\mathbf{p}(t)$ is the position of an object at time t, and $\mathbf{p}$ is continuously differentiable, then, as we discussed in [5], the **distance traveled** by the object between times t_0 and t_1 is the integral of the speed, with respect to time, i.e.,

$$\text{distance traveled} \;=\; \int_{t_0}^{t_1} |\mathbf{p}'(t)| \, dt.$$

If $\mathbf{p}$ is one-to-one, so that the object never backtracks, then this distance traveled is the **length** of the image; in this context, the length is frequently referred to as the **arc length** of the curve which is the image of $\mathbf{p}$.

Recall that a function f is one-to-one if $a \neq b$ implies that $f(a) \neq f(b)$. This is equivalent to: if $f(a) = f(b)$, then $a = b$, i.e., if "two" points in the domain give the same f value, then, in fact, the "two" points were actually the same (one) point.

Consider, for instance, $\mathbf{p} : \mathbb{R} \to \mathbb{R}^2$ given by $\mathbf{p}(t) = (t^2, t^3)$. This function is continuously differentiable and, if it represents the position of an object at time t, the speed of the object is

$$|\mathbf{p}'(t)| \;=\; |(2t, 3t^2)| \;=\; \sqrt{(2t)^2 + (3t^2)^2} \;=\; \sqrt{4t^2 + 9t^4} \;=\; |t|\sqrt{4 + 9t^2}.$$

Hence, the distance traveled between $t = 0$ and $t = 1$ is

$$\int_0^1 t(4 + 9t^2)^{1/2} \, dt \;=\; \frac{1}{27}\left((13)^{3/2} - 8\right).$$

It is easy to verify that $\mathbf{p}$ is one-to-one, so this distance traveled is also the arc length of the image of $\mathbf{p}$ restricted to the closed interval $[0, 1]$.

There are a number of rules for calculating derivatives of single-variable functions. All of these rules follow from straightforward calculations using the component functions; we leave them as exercises.

In these rules, when we add, multiply and compose, we assume that the domains and codomains are what they need to be in order to make the operations defined. Also, when we write "for all t", we mean for all t in the domain.

Theorem 1.6.6. *Suppose that a and b are real numbers, that $\mathbf{f}$ and $\mathbf{g}$ are differentiable single-variable functions into $\mathbb{R}^n$, and that m is a differentiable, real-valued, single-variable function. Then,*

1. *the derivative of a constant function is $\mathbf{0}$, i.e., if $\mathbf{v}$ is a fixed vector such that, for all t, $\mathbf{f}(t) = \mathbf{v}$, then $\mathbf{f}'(t) = \mathbf{0}$ for all t;*

2. *(**linearity**) $a\mathbf{f} + b\mathbf{g}$ is differentiable, and*

$$(a\mathbf{f} + b\mathbf{g})' \;=\; a\mathbf{f}' + b\mathbf{g}';$$

3. *(**Chain Rule**) $\mathbf{f} \circ m$ is differentiable and, for all t,*

$$(\mathbf{f} \circ m)'(t) \;=\; m'(t)\mathbf{f}'(m(t));$$

The dot product is commutative, so it is not important to retain the initial ordering of the functions in its product rule. However, the cross product is **not** commutative; the ordering of the functions in its product rule is crucial.

---

4. *(**Scalar Product Rule**) the scalar product $m\mathbf{f}$ is differentiable and, for all t,*

$$(m\mathbf{f})'(t) \;=\; m(t)\mathbf{f}'(t) + m'(t)\mathbf{f}(t);$$

5. *(**Product Rule for Dot Product**) the dot product $\mathbf{f} \cdot \mathbf{g}$ is differentiable and, for all t,*

$$(\mathbf{f} \cdot \mathbf{g})'(t) \;=\; \mathbf{f}(t) \cdot \mathbf{g}'(t) + \mathbf{f}'(t) \cdot \mathbf{g}(t);$$

6. *(**Product Rule for Cross Product**) if $\mathbf{f}$ and $\mathbf{g}$ are functions into $\mathbb{R}^3$, then the cross product $\mathbf{f} \times \mathbf{g}$ is differentiable and, for all t,*

$$(\mathbf{f} \times \mathbf{g})'(t) \;=\; \mathbf{f}(t) \times \mathbf{g}'(t) \;+\; \mathbf{f}'(t) \times \mathbf{g}(t).$$

Example 1.6.7. Calculate the derivative of $\mathbf{p}(t) = t^2(\cos(7t), \sin(7t))$.

Solution:

We find

$$\mathbf{p}'(t) = \left(t^2(\cos(7t), \sin(7t))\right)' = t^2 \cdot 7(-\sin(7t), \cos(7t)) + 2t(\cos(7t), \sin(7t)) =$$

$$(2t\cos(7t) - 7t^2\sin(7t), \, 2t\sin(7t) + 7t^2\cos(7t).$$

Example 1.6.8. Suppose that we have a differentiable function $\mathbf{a} : \mathbb{R} \to \mathbb{R}^3$, such that $\mathbf{a}(1) = (-2, 0, 1)$ and $\mathbf{a}'(1) = (5, 5, 5)$. Let $\mathbf{b} : \mathbb{R} \to \mathbb{R}^3$ be given by

$$\mathbf{b}(t) = (t^2, t^3, t^4).$$

Calculate $(\mathbf{a} \times \mathbf{b})'(1)$.

Solution:

We first calculate $\mathbf{b}'(t) = (2t, 3t^2, 4t^3)$. Thus, $\mathbf{b}(1) = (1, 1, 1)$ and $\mathbf{b}'(1) = (2, 3, 4)$.

Now, we use the Product Rule for the cross product:

$$(\mathbf{a} \times \mathbf{b})'(1) = \mathbf{a}(1) \times \mathbf{b}'(1) + \mathbf{a}'(1) \times \mathbf{b}(1) =$$

$$(-2, 0, 1) \times (2, 3, 4) + (5, 5, 5) \times (1, 1, 1).$$

Since $(5, 5, 5)$ and $(1, 1, 1)$ are parallel, their cross product is zero. Thus,

$$(\mathbf{a} \times \mathbf{b})'(1) = (-2, 0, 1) \times (2, 3, 4) =$$

$$\begin{vmatrix} \mathbf{i} & \mathbf{j} & \mathbf{k} \\ -2 & 0 & 1 \\ 2 & 3 & 4 \end{vmatrix} = -3\mathbf{i} + 10\mathbf{j} - 6\mathbf{k} = (-3, 10, -6).$$

Example 1.6.9. Suppose that object moves along a circle of radius r, centered at the origin in $\mathbb{R}^2$, in such a way that its position $\mathbf{p}(t)$, at time t, is a differentiable function of t. Show that its position vector $\mathbf{p}(t)$ is always orthogonal to its velocity vector $\mathbf{v}(t) = \mathbf{p}'(t)$.

Solution:

We are given that $|\mathbf{p}(t)| = r$ or, equivalently, $r^2 = |\mathbf{p}(t)|^2 = \mathbf{p}(t) \cdot \mathbf{p}(t)$. Taking derivatives of both sides, using the Product Rule for Dot Products, we obtain

$$0 = \mathbf{p}(t) \cdot \mathbf{p}'(t) + \mathbf{p}'(t) \cdot \mathbf{p}(t) = 2\mathbf{p}(t) \cdot \mathbf{p}'(t).$$

Therefore, $\mathbf{p}(t) \cdot \mathbf{p}'(t) = 0$, and so $\mathbf{p}(t)$ is always orthogonal to the velocity vector $\mathbf{v}(t) = \mathbf{p}'(t)$.

Generally, we think of a single-variable function $\mathbf{p}$ into $\mathbb{R}^n$ as a *parameterization* of the curve which is its image; however, some technical conditions are needed to guarantee that the image of $\mathbf{p}$ is indeed something 1-dimensional, i.e., something that deserves to be called a curve.

We shall give the *regularity* condition below, but, for now, let's assume that the image of $\mathbf{p}$ is a smooth (or, smooth-looking) curve C at a point $\mathbf{p}(t_0)$. How would you define a *tangent line* to C at $\mathbf{p}(t_0)$?

The answer should look familiar; it's what you did when defining tangent lines to the graph of a function of a single variable. You consider *secant lines* of $\mathbf{p}$ at t_0: lines containing the points $\mathbf{p}(t_0)$ and $\mathbf{p}(t_0 + h)$, where $\mathbf{p}(t_0) \neq \mathbf{p}(t_0 + h)$, and we take the limit as $h \to 0$. If the limit of these secant lines exists, we call that line the *tangent line* of $\mathbf{p}$ at t_0.

If the tangent line of $\mathbf{p}$ at t_0 exists, we call any vector which is parallel to that line a *tangent vector* of $\mathbf{p}$ at t_0, and we view all such tangent vectors as being based at $\mathbf{p}(t_0)$. As the vector $\mathbf{p}(t_0 + h) - \mathbf{p}(t_0)$ is parallel to the secant line through $\mathbf{p}(t_0)$ and $\mathbf{p}(t_0 + h)$, and scalar multiplying by $1/h$ yields a parallel vector, we find that

$$\mathbf{p}'(t_0) \;=\; \lim_{h \to 0} \frac{1}{h}\Big(\mathbf{p}(t_0 + h) - \mathbf{p}(t_0)\Big),$$

if it exists, is parallel to the tangent line. See Figure 1.6.2.

The main technical condition that we need is that the tangent vector $\mathbf{p}'(t_0) \neq \mathbf{0}$; thus, we define:

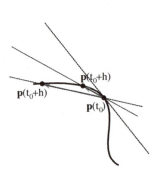

Figure 1.6.2: Secant lines converging to tangent line.

Definition 1.6.10. *A single-variable function* $\mathbf{p}$ *into* $\mathbb{R}^n$ *is* **regular** *at a point* t_0 *in the interior of its domain if and only if* $\mathbf{p}'(t_0)$ *exists and is not the zero vector. We also say that* t_0 *is a* **regular point** *of* $\mathbf{p}$.

The theorem that we need is:

Here, we write "smooth curve" because we assume that you have an intuitive feel for this notion. Later, in Definition 2.6.9, we shall be more technically accurate, and refer to a *1-dimensional* C^1 *submanifold* of Euclidean space. Theorem 1.6.11 will then follow from the Inverse Function Theorem, Theorem 2.6.6.

Theorem 1.6.11. *Suppose that a single-variable function* $\mathbf{p}$ *into* $\mathbb{R}^n$ *is continuously differentiable in an open neighborhood of the point* t_0 *in its domain, and that* $\mathbf{p}$ *is regular at* t_0.

Then, there exists $\epsilon > 0$ *such that the image of the restriction of* $\mathbf{p}$ *to the open interval* $(t_0 - \epsilon, t_0 + \epsilon)$ *is a "smooth curve"* C, *and the vector* $\mathbf{p}'(t_0)$ *is tangent to* C *at the point* $\mathbf{p}(t_0)$.

Thus, there is a well-defined **tangent line** *for* $\mathbf{p}$ *at* t_0, *which is parameterized by*

$$\mathbf{L}(r) \;=\; \mathbf{p}(t_0) + r\mathbf{p}'(t_0).$$

> In particular, if $\mathbf{p}(t)$ is the position of an object at time t, then, at a regular point t_0, the velocity vector $\mathbf{v}(t_0) = \mathbf{p}'(t_0)$ is tangent to the path of the object at the point $\mathbf{p}(t_0)$.

Example 1.6.12. Consider the curve in $\mathbb{R}^2$ parameterized by

$$\mathbf{p}(t) = \left(t^2 - 1,\, t(t^2 - 1)\right).$$

This is frequently referred to as a *node*.

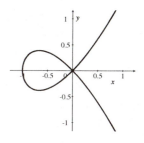

Figure 1.6.3: Image of a regular, but non-simple, parameterization.

This parameterization is not one-to-one: $\mathbf{p}(-1) = \mathbf{p}(1) = (0,0)$. A one-to-one parameterization is usually referred to as a *simple* parameterization. Thus, our current parameterization $\mathbf{p}$ is non-simple, and you see the curve crossing itself when it comes back to the origin. The image is, thus, not smooth at the origin, and it looks like there should be two different tangent lines at the origin.

Nonetheless, this parameterization is regular at each point, for

$$\mathbf{p}'(t) = (2t, 3t^2 - 1),$$

which is never equal to $\mathbf{0}$, for, to make the first component 0, t must be 0, but then the second component is -1, not 0.

But doesn't Theorem 1.6.11 tell us that the image curve should be smooth with a well-defined tangent line? No - the situation is more subtle than that. Theorem 1.6.11 doesn't say that the **entire** image is smooth, but rather that, in some small enough open interval around a regular point, the image of the restriction is smooth, with a well-defined tangent line.

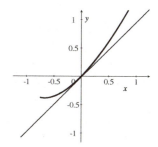

Figure 1.6.4: Image of the restriction of $\mathbf{p}$ to the open interval $(0.5, 1.5)$, and the tangent line at $(0,0)$.

In Figure 1.6.4, you can see the smooth image of the restriction of $\mathbf{p}$ to the open interval $(0.5, 1.5)$. What's a non-zero tangent vector to this fragment of the original image curve? That's easy: $\mathbf{p}'(1) = (2, 2)$. And so, the **unique** tangent line to this portion of the curve at the origin is parameterized by

$$\mathbf{L}(r) = \mathbf{p}(1) + r\mathbf{p}'(1) = (0,0) + r(2,2),$$

i.e., $x(r) = 2r$ and $y(r) = 2r$. This is the line given by $y = x$.

If we restrict $\mathbf{p}$ to the open interval $(-1.5, -0.5)$, we likewise obtain a smooth image with a unique tangent line at $(0,0)$; the tangent line is given by

$$(x(r), y(r)) = \mathbf{L}(r) = \mathbf{p}(-1) + r\mathbf{p}'(-1) = (0,0) + r(-2,2),$$

i.e., $x(r) = -2r$ and $y(r) = 2r$. This is the line given by $y = -x$.

Thus, we see that there **are** two different tangent lines at the origin. This explains why we don't refer to **the** tangent line to the image of a regular curve at a point in the image, but instead refer to the tangent line for $\mathbf{p}$ at particular points t_0 in the domain.

More Depth:

We want to begin with a fairly standard example of an object moving around a circle, but we will be non-standard and will not assume that the speed is constant.

Example 1.6.13. Suppose that an object is moving around a circle of radius $r > 0$ in the xy-plane. Let its position at time t be given by $\mathbf{p}(t)$. We will assume that $\mathbf{p}(t)$ is continuously differentiable, and that the object does not stop, so that $\mathbf{p}'(t)$ is never $\mathbf{0}$. Thus, $\mathbf{p}(t)$ is a regular parameterization, and the non-zero velocity vector $\mathbf{v}(t) = \mathbf{p}'(t)$ is always tangent to the circle. This means that $\mathbf{v}(t)$ is always perpendicular/normal to $\mathbf{p}(t)$, as we saw back in Example 1.6.9.

We would like to look at the acceleration $\mathbf{a}(t)$ of the object, and write it as the sum of two components, $\mathbf{a}_t$ in the tangential direction and $\mathbf{a}_n$ in the normal direction.

As usual, let $\theta(t)$ denote the angle, measured counterclockwise from the positive x-axis to the ray from the origin out to $\mathbf{p}(t)$. We assume that $\theta(t)$ is continuously differentiable. Then,

$$\mathbf{p}(t) \ = \ r\big(\cos\theta(t),\, \sin\theta(t)\big),$$

and

$$\mathbf{v}(t) \ = \ \mathbf{p}'(t) \ = \ r\theta'(t)\big(-\sin\theta(t),\, \cos\theta(t)\big).$$

Note that our assumption that $\mathbf{v}(t) \neq \mathbf{0}$ is equivalent to $\theta'(t) \neq 0$.

We find then that the speed $v = v(t) = |\mathbf{v}(t)|$ is

$$v = v(t) \ = \ |\mathbf{v}(t)| \ = \ r\big|\theta'(t)\big| \cdot \big|\big(-\sin\theta(t),\, \cos\theta(t)\big)\big| \ = \ r\big|\theta'(t)\big|.$$

Now, the acceleration $\mathbf{a}(t)$ is

$$\mathbf{a}(t) \ = \ \mathbf{v}'(t) \ = \ r\theta''(t)\big(-\sin\theta(t),\, \cos\theta(t)\big) \ + \ r\big(\theta'(t)\big)^2\big(-\cos\theta(t),\, -\sin\theta(t)\big).$$

This is already in the form

$$\mathbf{a} \ = \ \mathbf{a}_t \ + \ \mathbf{a}_n,$$

where $\mathbf{a}_t = r\theta''(t)\big(-\sin\theta(t),\, \cos\theta(t)\big)$ and $\mathbf{a}_n = -r\big(\theta'(t)\big)^2\big(\cos\theta(t),\, \sin\theta(t)\big)$.

Therefore, the direction of the normal component of the acceleration is always in the opposite direction of $\mathbf{p}(t)$, i.e., in the direction from the point $\mathbf{p}(t)$ towards the center of the circle. The magnitude of the normal component of the acceleration is

$$|\mathbf{a}_n| \ = \ r\big(\theta'(t)\big)^2 \ = \ \frac{v^2}{r},$$

which is a well-known result in the case when the speed of the object is constant, but is less well known when the speed is not constant.

Finally, if, in fact the speed is constant, then $\left|\theta'(t)\right|$ is constant. Since we are assuming that $\theta'(t)$ is continuously differentiable, this implies that $\theta'(t)$ itself is constant, which, in turn, implies that $\theta''(t) = 0$. And now this implies that the tangential component $\mathbf{a}_t$ of the acceleration is zero. Therefore, for an object moving in a circle, at a constant speed, the acceleration (the entire acceleration vector) points inward towards the center of the circle and has magnitude v^2/r; this is called the *centripetal acceleration*.

In many situations, we need to look at curves with endpoints; this means that we need to consider parameterizations in which the domains are **closed** intervals.

> **Definition 1.6.14.** *A simple regular parameterization of a curve in $\mathbb{R}^n$ consists of:*
>
> 1. *a closed interval $[a, b]$ in $\mathbb{R}^1$, where $a < b$; and*
>
> 2. *a continuously differentiable function $\mathbf{r} : [a, b] \to \mathbb{R}^n$, such that*
>
> 3. $\mathbf{r}$, *restricted to the open interval (a, b), is one-to-one, and*
>
> 4. *for all t in the open interval (a, b), $\mathbf{r}'(t) \neq \mathbf{0}$.*
>
> *The image of a simple regular parameterization is a **regular curve** from $\mathbf{r}(a)$ to $\mathbf{r}(b)$.*

The function $\mathbf{r} : [a, b] \to \mathbb{R}^n$ being continuously differentiable means that it is the restriction of a continuously differentiable function on a larger open set.

The term *simple* refers to Condition 3.

It is important that Definition 1.6.14 allows for the case where $\mathbf{r}(a) = \mathbf{r}(b)$.

Example 1.6.15. If we restrict $\mathbf{p}$ from Remark 1.6.5 to the interval $[0, 1]$, then we obtain the one-to-one parameterization, for $0 \leq t \leq 1$,

$$\mathbf{p}(t) = (t^2, t^3).$$

We find that $\mathbf{p}'(t) = (2t, 3t^2)$, and so the only t at which $\mathbf{p}'(t) = \mathbf{0}$ is when $t = 0$. Thus, this is a simple regular parameterization.

Remark 1.6.16. Suppose that $\mathbf{r} : [a, b] \to \mathbb{R}^n$ is a simple regular parameterization. For each t in the interval $[a, b]$, you can consider the arc length of the image of the restriction of $\mathbf{r}$ to the interval $[a, t]$. This gives the arc length of the image of $\mathbf{r}$ between $\mathbf{r}(a)$ and $\mathbf{r}(t)$, which defines a function $m : [a, b] \to \mathbb{R}$, given by

$$m(t) \;=\; \int_a^t |\mathbf{r}'(u)| \, du.$$

We let $L = m(b)$, which is the arc length of the entire image of $\mathbf{r}$. The image of m is thus the closed interval $[0, L]$.

The function m is continuous and, for all t in the open interval (a, b), the Fundamental Theorem of Calculus tells us that

$$m'(t) \;=\; |\mathbf{r}'(t)| \;>\; 0,$$

where the inequality is a result of $\mathbf{r}$ being regular. Therefore, m is strictly increasing and, hence, when its codomain is restricted to $[0, L]$, we obtain an invertible function, whose inverse $m^{-1} : [0, L] \to [a, b]$ is differentiable for all s in the open interval $(0, L)$. Furthermore, for s in $(0, L)$,

$$(m^{-1})'(s) \;=\; \frac{1}{m'(m^{-1}(s))} \;=\; \frac{1}{|\mathbf{r}'(m^{-1}(s))|}$$

Consider now the function $\hat{\mathbf{r}} : [0, L] \to \mathbb{R}^n$ given by the composition $\hat{\mathbf{r}} = \mathbf{r} \circ m^{-1}$. The image of this function is the same as the image of $\mathbf{r}$. We call this the **reparameterization of r by arc length** and/or say that $\hat{\mathbf{r}}$ is **parameterized by arc length**.

What does this mean? Note that the Chain Rule tells us that

$$\hat{\mathbf{r}}'(s) \;=\; (\mathbf{r} \circ m^{-1})'(s) \;=\; (m^{-1})'(s) \, \mathbf{r}'(m^{-1}(s)) \;=\; \frac{1}{|\mathbf{r}'(m^{-1}(s))|} \, \mathbf{r}'(m^{-1}(s)).$$

Therefore, $|\hat{\mathbf{r}}'(s)| = 1$ and, by our previous discussion, this means that the arc length of the image of $\hat{\mathbf{r}}$ between $\hat{\mathbf{r}}(0) = \mathbf{r}(a)$ and $\hat{\mathbf{r}}(s)$ is

$$\int_0^s |\hat{\mathbf{r}}'(u)| \, du \;=\; \int_0^s 1 \, du \;=\; s,$$

i.e., the parameter of $\hat{\mathbf{r}} = \hat{\mathbf{r}}(s)$ gives you the arc length of the image between $\hat{\mathbf{r}}(0)$ and $\hat{\mathbf{r}}(s)$.

> Note that, if an object has position $\mathbf{p}(t)$ at time t, then $\mathbf{p}$ is parameterized by arc length if and only if the object always has speed 1, or *unit speed*.

Later, especially in Section 4.2, we will need to look at a more-general type of regular curve that also has a direction assigned to it, and so we will need to indicate carefully "which way" the curve goes. This means that we need to discuss what an *orientation* on a curve means, and we need we discuss *piecewise-regular* curves, curves which break up into a finite collection of regular curves. We need to define *oriented, piecewise-regular curves*.

What do we want an "oriented, piecewise-regular curve" to mean? We want the curve to be in one piece (i.e., connected), look smooth, or be able to be chopped up into a finite number number of smooth pieces, and we want to have an indication of which direction on the curve is the positive direction or the direction of motion; we usually indicate this *orientation* with arrows on the curve.

In Figure 1.6.5, we have indicated an oriented, piecewise-regular curve; it is smooth from A to B, and smooth from B to C, even though there's a sharp point at B. We have also indicated the orientation of the curve, the direction of motion, with arrows.

Of course, a picture isn't a mathematical definition. We give the relevant definitions below, but it will help to keep in mind what we're after. Referring back to Figure 1.6.5, we'll call the pieces from A to B and from B to C *oriented regular curves*, and we'll call the entire oriented curve from A to C an *oriented, piecewise-regular curve*.

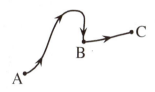

Figure 1.6.5: An oriented, piecewise-regular curve.

Definition 1.6.17. *Suppose that* $\mathbf{r} : [a,b] \to \mathbb{R}^n$ *is a simple regular parameterization of a curve.*

The **orientation** *of the parameterization is given by the continuous collection of unit tangent vectors* $\mathbf{r}'(t)/|\mathbf{r}'(t)|$, *for* $t \neq a, b$.

The image of a simple regular parameterization, together with its orientation, is an **oriented regular curve** *from* $\mathbf{r}(a)$ *to* $\mathbf{r}(b)$.

Remark 1.6.18. If the orientation is specified by the collection of unit vectors $\mathbf{r}'(t)/|\mathbf{r}'(t)|$, then why do we draw arrows which have effectively no length, as in Figure 1.6.5?

The answer is simple really. Vectors of any substantial length would get in the way, as in Figure 1.6.6. So, instead of drawing the unit vectors, we just indicate the direction with little arrowheads on the curve.

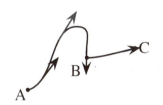

Figure 1.6.6: An oriented, piecewise-regular curve, with unit-length tangent vectors.

As we discussed, we want to allow oriented curves which might have a finite number of sharp corners. This means that we want to chop up a curve into a finite number of oriented regular curves. It is also convenient to allow parameterizations which are constant over time intervals, i.e, which stay at the same point for some interval of time.

There is a subtlety in Definition 1.6.19. We want to allow piecewise-regular parameterizations which double-back over themselves. That is, we want to allow a parameterization that goes from A to B, and then turns around and goes back the other way along the same curve. But you wouldn't see this "doubling-back" in the image of the parameterization. That's why the oriented, piecewise-regular curve consists of the **collection** of images of each of the regular pieces.

Definition 1.6.19. *A piecewise-regular parameterization of a curve in* $\mathbb{R}^n$ *consists of:*

1. *a closed interval* $[a, b]$ *in* $\mathbb{R}^1$, *where* $a < b$; *and*

2. *a continuous function* $\mathbf{r} : [a, b] \to \mathbb{R}^n$, *such that*

3. *there exist a finite number of* t *values,* $t_0, t_1, \ldots, t_k$, *where*

$$a = t_0 \ < \ t_1 \ < \ t_2 \ < \ \cdots \ < \ t_k = b$$

such that $\mathbf{r}$, *restricted to each interval* $[t_{i-1}, t_i]$, *is either a simple, regular parameterization or a constant function.*

Note that the image of a parameterization, restricted to an interval on which the parameterization is constant, is simply a point.

The collection of images of a piecewise-regular parameterization, restricted to each of the intervals $[t_{i-1}, t_i]$, *with their orientations on the non-constant intervals, is an* **oriented, piecewise-regular curve** *from* $\mathbf{r}(a)$ *to* $\mathbf{r}(b)$.

Example 1.6.20. As an example of how we will use oriented, piecewise-regular curves, suppose that an object is moving in $\mathbb{R}^3$, and that its position at time t is given by $\mathbf{r}(t) = (x(t), y(t), z(t))$ meters, where $\mathbf{r} : [a, b] \to \mathbb{R}^3$ is a piecewise-regular parameterization.

Suppose that a constant force $\mathbf{F} = (2, -5, 3)$ Newtons acts on the object as it moves. How much work is done by $\mathbf{F}$?

Over an infinitesimal time interval dt, the displacement of the object is

$$d\mathbf{r} \ = \ \mathbf{r}'(t)\, dt \ = \ (x'(t), y'(t), z'(t))\, dt.$$

Thus, the infinitesimal amount of work done by $\mathbf{F}$ during that infinitesimal displacement is

$$\mathbf{F} \cdot d\mathbf{r} \ = \ (2, -5, 3) \cdot (x'(t), y'(t), z'(t))\, dt \ \text{ joules.}$$

Therefore, the total work done by $\mathbf{F}$ is

$$\int_{t=a}^{t=b} \mathbf{F} \cdot d\mathbf{r} \ = \ \int_a^b (2, -5, 3) \cdot (x'(t), y'(t), z'(t))\, dt \ =$$

$$\int_a^b \big(2x'(t) - 5y'(t) + 3z'(t)\big)\, dt \ = \ 2x(t) - 5y(t) + 3z(t) \, \Big|_a^b \ = \ \mathbf{F} \cdot \mathbf{r}(t) \, \Big|_a^b \ =$$

$$\mathbf{F} \cdot (\mathbf{r}(b) - \mathbf{r}(a)),$$

which is just $\mathbf{F}$ dotted with the net displacement from $A = \mathbf{r}(a)$ to $B = \mathbf{r}(b)$.

You may wonder where we used all of the conditions defining a piecewise-regular curve. In fact, we didn't. However, if you don't want to specify the parameterization itself, and you want to discuss how much work is done by $\mathbf{F}$ as it moves along the oriented, piecewise-regular curve C, that you obtain from $\mathbf{r}$, then you need all of the conditions.

The same argument that we used in the previous in example, for arbitrary constant $\mathbf{F}$, proves:

Theorem 1.6.21. *The work done by a constant force $\mathbf{F}$ along an oriented, piecewise-regular curve C, which starts at a point $\mathbf{A}$ and ends at a point $\mathbf{B}$, is independent of the oriented, piecewise-regular curve and equals simply $\mathbf{F} \cdot \mathbf{d}$, where $\mathbf{d}$ is the displacement vector $\mathbf{d} = \overrightarrow{\mathbf{AB}}$.*

Later, in Section 4.2, we will consider the case of non-constant $\mathbf{F}$.

It will be important for us later to consider curves which end up at the same point at which they started. Typically, we will want to impose conditions on the parameterization of such a *closed curve*, or *loop*, such as requiring it to be piecewise-regular and, possibly, oriented. But, sometimes, one just needs continuity, so we make the following definition.

Definition 1.6.22. *Suppose that $a < b$.* A **continuous parameterization of a closed curve, or loop,** in $\mathbb{R}^n$, *is a continuous function $\mathbf{r} : [a, b] \to \mathbb{R}^n$ such that $\mathbf{r}(a) = \mathbf{r}(b)$.*

When we need more conditions on the parameterization of a closed curve, we will impose them.

Example 1.6.23. Consider the function $\mathbf{r} : [0, 3] \to \mathbb{R}^2$ given by

$$\mathbf{r}(t) \;=\; \begin{cases} (t, 0), & \text{if } 0 \leq t \leq 1; \\ (2 - t, t - 1), & \text{if } 1 \leq t \leq 2; \\ (0, 3 - t), & \text{if } 2 \leq t \leq 3. \end{cases}$$

This is a continuous parameterization of the triangle, a closed curve, with vertices $(0,0)$, $(1,0)$, and $(0,1)$. In fact, it's a piecewise-regular parameterization of the triangle, that starts at the origin, and moves counterclockwise around the triangle, until arriving again at the origin.

Finally, in this section, we want to address the following question:

Suppose that E is a subset of $\mathbb{R}^n$, and that $\mathbf{p}$ is a point in E. What should a tangent vector to E at $\mathbf{p}$ mean?

If you think in terms of a particle, which is moving in such a way that it stays inside of E, then, as we saw in Theorem 1.6.11, the velocity vector of the particle is tangent to the path of the particle. We can use this idea of velocity vectors to paths to define tangent vectors to sets.

> **Definition 1.6.24.** *Let E be a subset of $\mathbb{R}^n$, and let $\mathbf{p}$ be a point in E. A **tangent vector to E at $\mathbf{p}$** is the velocity vector at $\mathbf{p}$ of a continuously differentiable path in E.*
>
> *More precisely, a vector $\mathbf{v}$, based at $\mathbf{p}$, in $\mathbb{R}^n$ is a **tangent vector to E at $\mathbf{p}$** if and only if there exist $\epsilon > 0$, t_0, and a C^1 function*
>
> $$\mathbf{r} : (t_0 - \epsilon, t_0 + \epsilon) \to \mathbb{R}^n$$
>
> *such that, for all t in the interval $(t_0 - \epsilon, t_0 + \epsilon)$, $\mathbf{r}(t)$ is in E, $\mathbf{p} = \mathbf{r}(t_0)$, and $\mathbf{v} = \mathbf{r}'(t_0)$.*

The point of a regular parametrization of curve is that it yields a non-zero tangent vector to the curve at each point.

Online answers to select exercises are here.

Basics:

1.6.1   Exercises

In each of the following exercises, you are given the position vector $\mathbf{p} = \mathbf{p}(t)$, in meters, of a particle at time t seconds. Determine the position, velocity, speed, and acceleration of the particle at the given time t_0 seconds.

1. $\mathbf{p}(t) = (t^2, t^4)$, $t_0 = 1$.

2. $\mathbf{p}(t) = (te^t, e^t \sin t)$, $t_0 = 0$.

3. $\mathbf{p}(t) = (\tan^{-1} t, \ln(1 + t))$, $t_0 = 0$.

4. $\mathbf{p}(t) = (5 \sin t, 5 \cos t)$, $t_0 = \pi/4$.

5. $\mathbf{p}(t) = (1, 2, 3) + t(1, 0, 1)$, $t_0 = 3$.

6. $\mathbf{p}(t) = (5t, 2, e^t)$, $t_0 = 0$.

7. $\mathbf{p}(t) = (\cos(t^2), \sin(t^2), t^2)$, $t_0 = \sqrt{\pi}$.

8. $\mathbf{p}(t) = (e^{-t} \cos t, e^{-t} \sin t, t)$, $t_0 = 100\pi$.

In each of the following exercises, you are given the position vector $\mathbf{p} = \mathbf{p}(t)$ of a particle at time t. (a) Sketch the path of the particle (i.e., the image of $\mathbf{p}$) and indicate the positions at $t = -1$, $t = 0$, $t = 1$, and $t = 2$. (b) On your sketch of the path, sketch the velocity vectors of the particle at $t = -1$, $t = 0$, $t = 1$, and $t = 2$, basing each velocity vector at the position of the particle at that time.

9. $\mathbf{p}(t) = (2t - 1, -3t + 2)$.

10. $\mathbf{p}(t) = (t^2, t)$.

11. $\mathbf{p}(t) = (t^4, t^2)$.

12. $\mathbf{p}(t) = (e^{-2t}, e^{-t})$.

13. $\mathbf{p}(t) = (0, 1, 0) + t(2, 5, -1)$.

14. $\mathbf{p}(t) = (0, 1, 0) + t^2(2, 5, -1)$.

15. $\mathbf{p}(t) = (t, t^2, 3)$.

16. $\mathbf{p}(t) = (t, t^2, t)$.

In the following exercises, you are given the position $\mathbf{p} = \mathbf{p}(t)$ of an object in $\mathbb{R}^n$ at time t. Find the distance traveled by the object between the given times t_0 and t_1.

17. $\mathbf{p}(t) = (\sin(t^2), \cos(t^2))$, $t_0 = 0$, $t_1 = \sqrt{\pi}$.

18. $\mathbf{p}(t) = (t^{3/2}, t)$, $t_0 = 0$, $t_1 = 20/3$.

19. $\mathbf{p}(t) = (1, 0, -2) + t^2(5, 3, -7)$, $t_0 = 1$, $t_1 = 2$.

20. $\mathbf{p}(t) = (2t^{3/2}, \cos t, \sin t)$, $t_0 = 2$, $t_1 = 5$.

Suppose that a and b are differentiable functions from $\mathbb{R}$ into $\mathbb{R}^3$, and that m is a differentiable function from $\mathbb{R}$ to $\mathbb{R}$. Suppose that $\mathbf{a}(0) = (2, -1, 3)$, $\mathbf{b}(0) = (0, 0, 1)$, $m(0) = -5$, $\mathbf{a}'(0) = (1, 0, 0)$, $\mathbf{b}'(0) = (0, 7, 1)$, $m'(0) = 2$ and, finally, that $m(1) = 0$ and $m'(1) = 4$. In each of the following exercises, use the differentiation rules in Theorem 1.6.6 to calculate the indicated derivative.

21. $(3\mathbf{a} - 2\mathbf{b})'(0)$

22. $(\mathbf{a} \cdot \mathbf{b})'(0)$

23. $(\mathbf{a} \times \mathbf{b})'(0)$

24. $(m\mathbf{a})'(0)$

25. $(\mathbf{b} \circ m)'(1)$

26. $(11\mathbf{a} \circ m + 2\mathbf{a} \circ m)'(1)$

27. $[m(\mathbf{a} \cdot \mathbf{b})]'(0)$

28. $[m(\mathbf{a} \times \mathbf{b})]'(0)$

In the following exercises, you given the parameterized curves from Exercises 1-8. Show that each p is regular at the given t_0 and give a parameterization for the tangent line for p at t_0, using r as your parameter.

29. $\mathbf{p}(t) = (t^2, t^4)$, $t_0 = 1$.

30. $\mathbf{p}(t) = (te^t, e^t \sin t)$, $t_0 = 0$.

31. $\mathbf{p}(t) = (\tan^{-1} t, \ln(1 + t))$, $t_0 = 0$.

32. $\mathbf{p}(t) = (5 \sin t, 5 \cos t)$, $t_0 = \pi/4$.

33. $\mathbf{p}(t) = (1, 2, 3) + t(1, 0, 1)$, $t_0 = 3$.

34. $\mathbf{p}(t) = (5t, 2, e^t)$, $t_0 = 0$.

35. $\mathbf{p}(t) = (\cos(t^2), \sin(t^2), t^2)$, $t_0 = \sqrt{\pi}$.

36. $\mathbf{p}(t) = (e^{-t} \cos t, e^{-t} \sin t, t)$, $t_0 = 100\pi$.

More Depth: In each of the following problems, you are given a parameterization $\mathbf{r} = \mathbf{r}(t)$ of a curve in $\mathbb{R}^3$, and you are also given a non-zero vector w. Verify that $\mathbf{r}'(0)$ is perpendicular to w. Let $\mathbf{a} = \mathbf{r}''(0)$, and calculate the orthogonal projection $\mathbf{a_w}$ of a onto w. Also calculate $\mathbf{a}_n$, the component of a normal to w (recall Definition 1.3.9).

37. $\mathbf{r}(t) = \left(e^{3t}, 2t\right)$, $\mathbf{w} = (-2, 3)$.

38. $\mathbf{r}(t) = (4\sin t, 2\sqrt{\cos t})$, $\mathbf{w} = (0, 1)$.

39. $\mathbf{r}(t) = (e^t, e^{2t}, e^{3t})$, $\mathbf{w} = (-1, -1, 1)$.

40. $\mathbf{r}(t) = t(e^t, e^{2t}, e^{3t})$, $\mathbf{w} = (1, -2, 1)$.

41. A constant force of $\mathbf{F} = (1, 3, -7)$ Newtons acts on a particle as it moves with position $\mathbf{p}(t) = (t^2, t^5, t\sqrt{8 + t})$ meters, from time $t = 0$ seconds to time $t = 1$ second. How much work does the force do on the particle?

42. A constant force of $\mathbf{F} = (3, -5, 4)$ Newtons acts on a particle as it moves with position $\mathbf{p}(t) = (\ln(t^2 + 1), \sin(t^3), \tan^{-1} t)$ meters, from time $t = 0$ seconds to time $t = 2$ seconds. How much work does the force do on the particle?

43. Consider the simple regular parameterization of a curve, given by

$$\mathbf{r}(t) = (\sin(t^2), \cos(t^2)),$$

for $0 \le t \le \sqrt{\pi}$. Let $m : [0, \sqrt{\pi}] \to \mathbb{R}$ be the arc length function, measured from $\mathbf{r}(0) = (0, 1)$, so that

$$m(t) = \int_0^t |\mathbf{r}'(u)|\, du$$

   (a) Show that $m(t) = t^2$ and so, in particular, the length of the entire image of $\mathbf{r}$ is π.

   (b) Hence, there is an inverse function $m^{-1} : [0, \pi] \to [0, \sqrt{\pi}]$. What is $m^{-1}(s)$?

   (c) What is the re-parameterization $\hat{\mathbf{r}}$ by arc length?

44. Prove Part (1) of Theorem 1.6.6.

45. Prove Part (2) of Theorem 1.6.6.

46. Prove Part (3) of Theorem 1.6.6.

47. Prove Part (4) of Theorem 1.6.6.

48. Prove Part (5) of Theorem 1.6.6.

49. Prove Part (6) of Theorem 1.6.6.

50. Consider the parameterization $\mathbf{p}(w) = (w^2 - 1, -w(w^2 - 1))$. Let $m(t) = 1 - e^{-t}$, and let $\mathbf{r} = \mathbf{p} \circ m$, i.e.,

$$\mathbf{r}(t) = \left((1 - e^{-t})^2 - 1, \, -(1 - e^{-t})((1 - e^{-t})^2 - 1)\right).$$

(a) Verify that, if $x = x(w) = w^2 - 1$ and $y = y(w) = -w(w^2 - 1)$, then

$$y^2 - x^3 - x^2 \;=\; 0.$$

(b) Sketch, by hand or with the aid of technology, the image of $\mathbf{p} = \mathbf{p}(w)$, for $-\infty < w < \infty$.

(c) Indicate the subset of your image in part (b) that is the image of $\mathbf{r} = \mathbf{r}(t)$, for $-\infty < t < \infty$.

In particular, identity which points on the graph you get as $t \to \infty$, and determine all t values such that $\mathbf{r}(t) = \mathbf{0}$.

(d) If $\mathbf{r}(t)$ gives the position of a particle at time t, what is the speed $s(t)$ of the particle at time t?

(e) What is $\lim_{t \to \infty} s(t)$?

51. Sketch the image of the continuous parameterization $\mathbf{r} : [-1, 2] \to \mathbb{R}^2$ of the closed curve given by $\mathbf{r}(t) = (t, 1 - t^2)$, for $-1 \le t \le 1$, and $\mathbf{r}(t) = (-2t + 3, 0)$, for $1 \le t \le 2$. Show that $\mathbf{r}$ is piecewise-regular, and explicitly give the orientation vectors. Indicate the orientation on your sketch with little arrows.

52. Sketch the image of the continuous parameterization $\mathbf{r} : [0, 2\pi] \to \mathbb{R}^2$ of the closed curve given by $\mathbf{r}(t) = (2 \cos t, 3 \sin t)$. Show that $\mathbf{r}$ is simple regular parameterization, and explicitly give the orientation vectors. Indicate the orientation on your sketch with little arrows.

53. Sketch the image of the continuous parameterization $\mathbf{r} : [0, 2\pi] \to \mathbb{R}^2$ of the closed curve given by $\mathbf{r}(t) = (2 \sin t, 3 \cos t)$. Show that $\mathbf{r}$ is simple regular parameterization, and explicitly give the orientation vectors. Indicate the orientation on your sketch with little arrows.

54. Suppose that you are given a continuously differentiable parameterization $\mathbf{r} = \mathbf{r}(t)$ of a curve in $\mathbb{R}^3$, which is regular at each point. Then, the orientation vectors $\mathbf{T}(t) = \mathbf{r}'(t)/|\mathbf{r}'(t)|$ are well-defined unit tangent vectors.

(a) Show that the vector $\mathbf{w}(t) = \mathbf{T}'(t)$ is orthogonal to $\mathbf{T}$.

(b) Suppose that $\mathbf{w}(t)$, from part (a), is always $\mathbf{0}$. Show that $\mathbf{r}(t)$ must be of the form $\mathbf{r}(t) = f(t)\mathbf{v} + \mathbf{a}$, where $f(t)$ is a scalar function such that $f'(t) > 0$, and $\mathbf{v} \ne \mathbf{0}$ and $\mathbf{a}$ are constant vectors. What can you say about the image of $\mathbf{r}$?

(c) Suppose that $\mathbf{w}(t) \ne \mathbf{0}$. Define the *unit normal vector of* $\mathbf{r}$ to be $\mathbf{N}(t) = \mathbf{w}(t)/|\mathbf{w}(t)|$. Now, define the *unit binormal vector of* $\mathbf{r}$ to be the vector $\mathbf{B}(t) = \mathbf{T}(t) \times \mathbf{N}(t)$. Show that, in fact, $\mathbf{B}(t)$ is a unit vector.

Thus, the triple $\mathbf{T}(t)$, $\mathbf{N}(t)$, and $\mathbf{B}(t)$ is a collection of (possibly) changing unit vectors, which are mutually orthogonal. This triple is extremely important in the advanced study of curves, and is called the Frenet-Serret frame of the parameterization $\mathbf{r}$.

1.7 Multivariable Functions

In this section, we look at true multivariable functions, functions whose domains are subsets of $\mathbb{R}^n$ where $n \geq 2$. We need to discuss basic terminology and concepts for such functions, including continuity.

Differentiability will be discussed in later sections in Chapter 2, and integrability will be covered in Chapter 3.

Consider the function f from $\mathbb{R}^2$ to $\mathbb{R}$ given by

$$f(x,y) \;=\; x^2 - y^2.$$

This is a *multivariable function*, a function of more than one variable. This function is *real-valued*, meaning that its output, for any given input, is a real number, i.e., its codomain is a subset of $\mathbb{R}$.

We can also look at multivariable functions into higher-dimensional Euclidean spaces, such as $\mathbf{m} : \mathbb{R}^3 \to \mathbb{R}^3$ given by

$$\mathbf{m}(r,s,t) \;=\; \big(s\sin t + r^2,\; rst,\; t\ln(s^2+1)\big).$$

This function has three *component functions* $m_1(r,s,t) = s\sin t + r^2$, $m_2(r,s,t) = rst$, and $m_3(r,s,t) = t\ln(s^2+1)$.

More generally, throughout this book, we shall consider functions $\mathbf{f} : E \to F$, where the domain E is a subset of $\mathbb{R}^n$ and the codomain F is a subset of $\mathbb{R}^p$, where n and p are integers that are at least 1. We will frequently refer to such a function simply as a *function of n variables into $\mathbb{R}^p$* or, if $p = 1$, as a *real-valued function of n variables*.

When n and p are large (for instance, ≥ 4), it is common to use x's, as in $\mathbf{x} = (x_1, \ldots, x_n)$, to denote the variables of the domain, and y's, as in $\mathbf{y} = (y_1, \ldots, y_p)$, to denote variables in the codomain. For example, we might write: consider a function

$$\mathbf{y} \;=\; (y_1, y_2, y_3) \;=\; \mathbf{f}(x_1, x_2, x_3, x_4) \;=\; \mathbf{f}(\mathbf{x}),$$

to indicate that we are looking at a function of 4 variables into $\mathbb{R}^3$. Note that this gives variable names to the output of the various component functions of $\mathbf{f}$:

$$y_1 = f_1(\mathbf{x}), \qquad y_2 = f_2(\mathbf{x}), \qquad \text{and} \qquad y_3 = f_3(\mathbf{x}).$$

For real-valued single-variable functions $y = f(x)$, the *graph* is an important object to consider; the graph is the set of points (x, y) in $\mathbb{R}^2$ such that $y = f(x)$. With more variables, things are far more complicated. If we have a function $\mathbf{y} = \mathbf{f}(\mathbf{x})$ of n variables into $\mathbb{R}^p$, then the graph consists of those points

$$(x_1, x_2, \ldots, x_n, y_1, y_2, \ldots, y_p) = (\mathbf{x}, \mathbf{y})$$

in $\mathbb{R}^{n+p}$ such that $\mathbf{y} = \mathbf{f}(\mathbf{x})$.

Of course, we can visualize this graph easily only when $n + p$ equals 2 or 3. When $n = p = 1$, we're in the usual real-valued single-variable function case. When $n = 2$ and $p = 1$, we usually use x and y as the domain variables, and z for the codomain variable, writing

$$z = f(x, y).$$

We will discuss graphs of such functions extensively in the next section.

Given a function $\mathbf{f} : E \to \mathbb{R}^p$ of n variables and a specific point $\mathbf{b}$ in $\mathbb{R}^p$, the *level set where* $\mathbf{f} = \mathbf{b}$ is the set of points $\mathbf{x}$ in E such that $\mathbf{f}(\mathbf{x}) = \mathbf{b}$. Rather than explicitly use the term "level set", we frequently say simply "the set of points such that $\mathbf{f}(\mathbf{x}) = \mathbf{b}$".

For a real-valued function of 2 variables, $z = f(x, y)$, the level set where $f = b$ is frequently called by other names: *level curve*, *contour*, *contour curve*, or *contour line*. We shall explore the Calculus of level sets extensively in Section 2.7.

There are some simple, but, nonetheless, very important types of multivariable functions that we give names to:

Definition 1.7.1. *A function $L : \mathbb{R}^n \to \mathbb{R}$ is* **linear** *if and only if there exists a fixed* $\mathbf{a} = (a_1, a_2, \ldots, a_n)$ *in $\mathbb{R}^n$ such that*

$$L(x_1, x_2, \ldots, x_n) = a_1 x_1 + a_2 x_2 + \cdots + a_n x_n.$$

A function $L : \mathbb{R}^n \to \mathbb{R}$ is **affine linear** *if and only if there exists a fixed* $\mathbf{a} = (a_1, a_2, \ldots, a_n)$ *in $\mathbb{R}^n$ and a real number b such that*

$$L(x_1, x_2, \ldots, x_n) = a_1 x_1 + a_2 x_2 + \cdots + a_n x_n + b.$$

A function $\mathbf{L} : \mathbb{R}^n \to \mathbb{R}^p$ is **linear** *(respectively,* **affine linear***) if and only if each of its component functions is linear (respectively,* **affine linear***).*

Note that this means that a linear function is also affine linear with $b = 0$.

A linear function is also called a *linear transformation*. For more on linear transformations and how they arise in multivariable Calculus, see Theorem 2.2.20.

Example 1.7.2. The function $L : \mathbb{R}^3 \to \mathbb{R}$ given by

$$L(x, y, z) = 3x - 2y + 7z$$

is linear, while the function

$$A(x, y, z) = 3x - 2y + 7z + 5$$

is affine linear, but not linear.

As we saw in Section 1.4, the level set where $L = 0$ is a plane through the origin, while the level set $A = 0$ is a plane which does not contain the origin.

Example 1.7.3. The most basic linear functions from $\mathbb{R}^n$ to $\mathbb{R}$ are the *coordinate functions*. For $1 \le i \le n$, the coordinate function $X_i : \mathbb{R}^n \to \mathbb{R}$ picks out the i-th coordinate. So, for all $\mathbf{p}$ in $\mathbb{R}^n$,

$$X_i(p_1, p_2, \ldots, p_n) = p_i.$$

For example, $X_3(5, 7, -1, 6) = -1$.

If you denote the coordinates of the original point by $\mathbf{x} = (x_1, x_2, \ldots, x_n)$, then

$$X_i(\mathbf{x}) = x_i.$$

This can get very confusing, since many people write x_i for the i-coordinate function. You just need to be careful. For instance, if x_2 is used for the second coordinate function on $\mathbb{R}^3$, then x_2 should not be used to represent the 2nd coordinate of a specific point.

As in single-variable Calculus, *continuity* is an important property for multivariable functions to possess. Suppose that E is a subset of $\mathbb{R}^n$ and F is a subset of $\mathbb{R}^p$. What does it mean for a function $\mathbf{r} : E \to F$ to be continuous at a point $\mathbf{a}$ in E?

Intuitively, it means the same thing that it meant for functions of a single-variable: if $\mathbf{x}$ in E is "close to $\mathbf{a}$", then $\mathbf{r}(\mathbf{x})$ is "close to $\mathbf{r}(\mathbf{a})$". Now, $\mathbf{x}$ being close to $\mathbf{a}$ means that $|\mathbf{x} - \mathbf{a}|$ is close to zero, and $\mathbf{r}(\mathbf{x})$ being close to $\mathbf{r}(\mathbf{a})$ means that $|\mathbf{r}(\mathbf{x}) - \mathbf{r}(\mathbf{a})|$ is close to zero. What continuity at $\mathbf{a}$ should mean is that, if someone tells you how close to zero they want $|\mathbf{r}(\mathbf{x}) - \mathbf{r}(\mathbf{a})|$ to be, you can tell them how close to zero $|\mathbf{x} - \mathbf{a}|$ needs to be to make it happen.

With this preparatory discussion, we can now define continuity.

If $\mathbf{a}$ is an interior point of E (recall Definition 1.1.7), then the *limit of $\mathbf{r}$ as $\mathbf{x}$ approaches* $\mathbf{a}$, $\lim_{\mathbf{x}\to\mathbf{a}} \mathbf{r}(\mathbf{x})$, can be defined, and then continuity at $\mathbf{a}$ is equivalent to $\lim_{\mathbf{x}\to\mathbf{a}} \mathbf{r}(\mathbf{x}) = \mathbf{r}(\mathbf{a})$.

Definition 1.7.4. *Suppose that E is a subset of $\mathbb{R}^n$, F is a subset of $\mathbb{R}^p$, $\mathbf{a}$ is a point in E, and $\mathbf{r}$ is a function from E into F.*

Then, $\mathbf{r}$ is **continuous** *at $\mathbf{a}$ if and only if, for all $\epsilon > 0$, there exists $\delta > 0$ such that, if $\mathbf{x}$ is a point in E such that $|\mathbf{x} - \mathbf{a}| < \delta$, then $|\mathbf{r}(\mathbf{x}) - \mathbf{r}(\mathbf{a})| < \epsilon$.*

We say simply that $\mathbf{r}$ is **continuous** *if and only if it is continuous at each point in its domain E.*

Remark 1.7.5. We should remark that allowing the codomain of $\mathbf{r}$ to be a subset F of $\mathbb{R}^p$, instead of all of $\mathbb{R}^p$, is inconsequential. It is trivial to verify that such a function $\mathbf{r} : E \to F$ is continuous if and only if the extension of $\mathbf{r}$, where you consider the codomain to be all of $\mathbb{R}^p$, is continuous. Thus, henceforth, when discussing continuity, we will assume that the codomain is all of $\mathbb{R}^p$.

It can be very difficult to prove that multivariable functions are continuous, and yet, for many purposes, it's important for us to know that the function that we're dealing with is, in fact, continuous. We typically conclude that a function is continuous by combining two theorems: one which tells us that we just need to look at the component functions, and one which tells us all of our "usual" real-valued functions are continuous.

For the proof of this theorem, see Rudin, [7], Theorem 4.10.

Theorem 1.7.6. *A function $\mathbf{r} : E \to \mathbb{R}^p$, given by $\mathbf{r}(\mathbf{x}) = (r_1(\mathbf{x}), \ldots, r_p(\mathbf{x}))$ is continuous if and only if each of its component functions $r_i : E \to \mathbb{R}$ is continuous.*

Now, recall from single-variable Calculus the notion of an *elementary function*. This same definition applies to multivariable functions.

Definition 1.7.7. *An* **elementary function** *is a function which is a constant function, a power function (with an arbitrary real exponent), a polynomial function, an exponential function, a logarithmic function, a trigonometric function, or inverse trigonometric function, or any finite combination of such functions using addition, subtraction, multiplication, division, or composition.*

Thus,

$$f(x, y, z) = \frac{e^x y \sin z}{7 + y \ln(x + z)}$$

is an elementary function.

As in single-variable Calculus, we have the theorem:

Theorem 1.7.8. *All elementary functions are continuous.*

This theorem follows from the fact that all of the single-variable elementary functions are continuous, and so are the multivariable functions of addition, subtraction, multiplication, and division. See Rudin, [7], Theorem 4.9.

Remark 1.7.9. In no way does Theorem 1.7.8 say anything like "all functions are continuous". You should recall that, even for real-valued single-variable functions, it is easy to give a piecewise definition of a function to produce a discontinuous function. So, certainly, it's easy to give a discontinuous piecewise-defined multivariable function.

Consider, for instance, the function $\mathbf{f} : \mathbb{R}^2 \to \mathbb{R}$ given by

$$f(x, y) = \begin{cases} \dfrac{xy^2}{x^2 + y^4}, & \text{if } (x, y) \neq (0, 0); \\ 0, & \text{if } (x, y) = (0, 0). \end{cases}$$

This function is **not** continuous at $(0, 0)$. We shall have you show in the exercises that the condition of Definition 1.7.4 is not satisfied at $(0, 0)$ for $\epsilon = 1/4$.

We would like to know things like: the set of points (x, y, z) in $\mathbb{R}^3$ which satisfy

$$2 < xe^y + z^2 < 5$$

is an open subset of $\mathbb{R}^3$, while the set of points (x, y, z) which satisfy

$$2 \leq xe^y + z^2 \leq 5$$

is a closed subset of $\mathbb{R}^3$.

For that matter we'd like to know that the level set where

$$xe^y + z^2 = 7$$

is a closed subset of $\mathbb{R}^3$.

First, we need another definition.

There is no claim here that **f** possesses an inverse function, i.e., there may not exist a **function f^{-1}**. The inverse image operation, taking sets to sets, nonetheless makes sense.

We remind you that a subset of $\mathbb{R}^n$ which contains a single point is closed.

In the study of *topology*, Property 2 is how continuity is defined. With our definition of continuity in Definition 1.7.4, the proof is not difficult, and we give it as an exercise. However, you may also consult Theorem and Corollary 4.8 in Rudin, [7].

Definition 1.7.10. *Suppose that we have a function* **f** $: E \to \mathbb{R}^p$, *where E is a subset of $\mathbb{R}^n$. Let B be a subset of $\mathbb{R}^p$. Then, the* **inverse image**, **f**$^{-1}(B)$, *of B is defined to be the set of points* **x** *in E such that* **f**(**x**) *is in B, i.e., the inverse image of B is the set of points which* **f** *takes to points in B.*

If B is a set containing the single point **b**, *then, rather than write* **f**$^{-1}(\{\mathbf{b}\})$, *we write simply* **f**$^{-1}(\mathbf{b})$, *which is precisely the level set of points* **x** *such that* **f**(**x**) = **b**.

The following theorem relates continuity to inverse images of open and closed subsets.

Theorem 1.7.11. *Suppose that we have a function* **f** $: E \to \mathbb{R}^p$, *where E is a subset of $\mathbb{R}^n$. Then, the following are equivalent:*

1. **f** *is continuous.*

2. *For every open subset $\mathcal{U}$ of $\mathbb{R}^p$,* **f**$^{-1}(\mathcal{U})$ *is the intersection of an open subset of $\mathbb{R}^n$ with the domain E.*

3. *For every closed subset C of $\mathbb{R}^p$,* **f**$^{-1}(C)$ *is the intersection of a closed subset of $\mathbb{R}^n$ with the domain E.*

In particular, if $E = \mathbb{R}^n$, and **f** *is continuous, then the inverse images of all open subsets are open, and the inverse images of all closed subsets are closed. Therefore, level sets of continuous functions on $\mathbb{R}^n$ are closed subsets of $\mathbb{R}^n$.*

Example 1.7.12. Consider the function $f : \mathbb{R}^3 \to \mathbb{R}$ given by

$$f(x,y,z) = xe^y + z^2;$$

the domain is all of $\mathbb{R}$, and this is an elementary function, and so it's continuous.

Therefore, the inverse image $f^{-1}\big((2,5)\big)$ of the open interval $(2,5)$ is open. But $f^{-1}\big((2,5)\big)$ is the set of (x,y,z) such that $f(x,y,z)$ is in the interval $(2,5)$, i.e., $f^{-1}\big((2,5)\big)$ is precisely the set of (x,y,z) such that

$$2 < xe^y + z^2 < 5;$$

thus, these strict inequalities, or **any** strict equalities of this sort, on a continuous everywhere-defined function into $\mathbb{R}$, define an open subset of $\mathbb{R}^3$.

Similarly, $f^{-1}\big([2,5]\big)$, the inverse image of the closed interval $[2,5]$ is closed; thus, the inequalities

$$2 \leq xe^y + z^2 \leq 5$$

define a closed set. Moreover, any such inclusive inequalities define a closed subset.

In addition, the level set where

$$xe^y + z^2 = 7$$

is $f^{-1}(7)$, and the set containing just 7 is closed, as it's the closed interval $[7,7]$. Therefore, this level set and, for that matter, **every** level set, of f is closed.

> Recall that the empty set is both open and closed, so that if there are no points which satisfy the relations, the statements are still correct.

In single-variable Calculus, the Extreme Value Theorem is very important; it tells us that a continuous function on a closed, bounded interval $[a,b]$ attains both a minimum value and a maximum value. This result has a generalization to multivariable functions, which will be important to us later when we look at optimization problems in Section 2.10. We remind you of Definition 1.1.14, where we define/characterize the compact subsets of $\mathbb{R}^n$ as the closed and bounded subsets.

> We summarize the content of Theorem 1.7.13 by saying: the continuous image of a compact set is compact.

Theorem 1.7.13. (General Extreme Value Theorem) *Suppose that E is a compact subset of $\mathbb{R}^n$, and $\mathbf{f} : E \to \mathbb{R}^p$ is continuous. Then, the image of E is compact.*

In particular, a continuous function from a compact subset of $\mathbb{R}^n$ into $\mathbb{R}$ attains both a minimum and maximum value.

> This proof is actually quite simple, using the topological definition of compactness; see Rudin, [7], 4.14. Of course, that proof just pushes the real difficulty to proving the Heine-Borel Theorem which identifies the topologically compact subsets of $\mathbb{R}^n$ as the closed, bounded subsets.

> **More Depth:**

There is a another useful theorem involving single-variable functions which generalizes easily: the Intermediate Value Theorem (IVT).

What you probably know as the IVT says something like:

Suppose that a and b are real numbers, with $a < b$, and that $f : [a,b] \to \mathbb{R}$ is a continuous function. If t is between $f(a)$ and $f(b)$, then there exists c in $[a,b]$ such that $f(c) = t$.

A subset I of $\mathbb{R}$ is an interval if and only if, for all x_0 and x_1 in I, if c is a real number such that $x_0 < c < x_1$, then c is also in I. Therefore, what the real-valued single-variable Intermediate Value Theorem says is that the continuous image of an interval is an interval.

Recall now the definition of a connected subset of $\mathbb{R}^n$ from Definition 1.1.16, and also recall that the connected subsets of $\mathbb{R}$ are precisely the intervals. This makes it clear that the following theorem really is a generalization of the usual IVT.

We summarize the content of Theorem 1.7.14 by saying: the continuous image of a connected set is connected.

This proof is actually quite simple, once you first prove some basic properties of how the inverse image behaves with unions and intersections. See Theorem 4.8 of Fleming, [3].

Theorem 1.7.14. (General Intermediate Value Theorem) *Suppose that E is a connected subset of $\mathbb{R}^n$, and $\mathbf{f} : E \to \mathbb{R}^p$ is continuous. Then, the image of E is connected.*

Later, in Chapter 4, we shall need the notion of a *simply-connected subset of* $\mathbb{R}^n$; specifically, we need to understand simply-connected subsets of $\mathbb{R}^2$ and $\mathbb{R}^3$.

The idea is that a simply-connected subset E of $\mathbb{R}^n$ is a subset which is connected (actually, *path-connected*, as we shall define below), and in which any continuous closed path can be continuously deformed, or shrunk, to a point, while keeping the deformed path inside E throughout the deformation. Thus, a region will fail to be simply-connected when there's a continuous closed path which gets "stuck" because it cannot be moved through points in $\mathbb{R}^n$ which are missing from E.

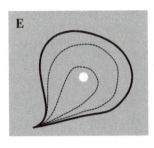

Figure 1.7.1: A hole in $\mathbb{R}^2$ prevents us from collapsing some closed curves.

In $\mathbb{R}^2$, you should think of a simply-connected subset as a connected region with no holes. However, you may remove a finite number of points, or open or closed balls, from $\mathbb{R}^3$ and the resulting space will still be simply-connected.

First, we need to define what it means for a subset of $\mathbb{R}^n$ to be path-connected.

Definition 1.7.15. *Suppose that E is a subset of $\mathbb{R}^n$. Then, E is **path-connected** if and only if, for all $\mathbf{a}$ and $\mathbf{b}$ in E, there exists a continuous function $\mathbf{r} : [0,1] \to E$ (a **continuous path**) such that $\mathbf{r}(0) = \mathbf{a}$ and $\mathbf{r}(1) = \mathbf{b}$, where $[0,1]$ is the closed interval from 0 to 1 in $\mathbb{R}$.*

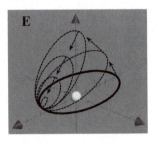

Figure 1.7.2: In $\mathbb{R}^3$, we can move a closed curve around a missing point/ball, and then collapse it to a point.

A path-connected subset of $\mathbb{R}^n$ must be connected, for any separation of E would separate the connected image of any continuous path from points in one piece of the separation to points in the other piece. However, there are exotic subsets of $\mathbb{R}^n$ which are connected, and yet not path-connected. See, for instance, the topologist's sine curve.

Before we define simple-connectedness, it will be helpful to adopt notation for the *unit square* in $\mathbb{R}^2$:

$$[0,1] \times [0,1] \ = \ \{(u,t) \in \mathbb{R}^2 \mid 0 \le u \le 1 \text{ and } 0 \le t \le 1\}.$$

Definition 1.7.16. *Suppose that E is a subset of $\mathbb{R}^n$. Then, E is* **simply-connected** *if and only if, E is path-connected, and every continuous path $\mathbf{r} : [0,1] \to E$ such that $\mathbf{r}(0) = \mathbf{r}(1)$ (every* **closed** *continuous path) can be continuously deformed to a point.*

This second condition means: Let $r : [0,1] \to E$ be any closed continuous path in E. Then, there exists a point $\mathbf{a}$ and a continuous function $H : [0,1] \times [0,1] \to E$ such that, for all u in $[0,1]$, $H(u,0) = \mathbf{r}(u)$ and $H(u,1) = \mathbf{a}$.

Here, you should think of $H(u,t)$ as defining a path $\mathbf{r}_t(u)$ at time t. Then, the conditions say that, at time 0, $\mathbf{r}_0(u) = \mathbf{r}(u)$, so that you start at the path $\mathbf{r}$, and then, as t changes, the path changes in a continuous way, always staying inside E, until, finally, when $t = 1$, $\mathbf{r}_1(u)$ is constantly the single point $\mathbf{a}$. In this respect, you have continuously deformed the closed path $\mathbf{r}$ to a point.

Finally, we wish to mention an important result for continuous functions on compact sets.

Suppose that E is a subset of $\mathbb{R}^n$, F is a subset of $\mathbb{R}^p$, and $\mathbf{r}$ is a function from E into F. Recall the definition of $\mathbf{r}$ being continuous; $\mathbf{r}$ is continuous if and only if, for all $\mathbf{a}$ in E, for all $\epsilon > 0$, there exists $\delta > 0$ such that, if $\mathbf{x}$ is a point in E such that $|\mathbf{x} - \mathbf{a}| < \delta$, then $|\mathbf{r}(\mathbf{x}) - \mathbf{r}(\mathbf{a})| < \epsilon$. This means that, for a continuous function, the choice of δ not only depends on the choice of ϵ, but may also change as the point $\mathbf{a}$ changes. The function $\mathbf{r}$ satisfies a stronger form of continuity if, for all $\epsilon > 0$, there is a single choice of δ that "works" for every $\mathbf{a}$ in E.

Thus, we make the following definition.

Definition 1.7.17. *Suppose that E is a subset of $\mathbb{R}^n$, F is a subset of $\mathbb{R}^p$, and $\mathbf{r}$ is a function from E into F.*

Then, $\mathbf{r}$ is **uniformly continuous** *if and only if, for all $\epsilon > 0$, there exists $\delta > 0$ such that, for all $\mathbf{a}$ in E, if $\mathbf{x}$ is a point in E such that $|\mathbf{x} - \mathbf{a}| < \delta$, then $|\mathbf{r}(\mathbf{x}) - \mathbf{r}(\mathbf{a})| < \epsilon$.*

Theorem 1.7.18. *Suppose that E is a compact subset of $\mathbb{R}^n$, F is a subset of $\mathbb{R}^p$, and $\mathbf{r}$ is a continuous function from E into F. Then, $\mathbf{r}$ is uniformly continuous.*

In other words, continuous functions on compact sets are uniformly continuous.

For the proof, see Theorem 4.19 of Rudin, [7].

Online answers to select exercises are here.

Basics:

1.7.1   Exercises

In each of the following exercises, you are given a function $\mathbf{f} : E \to \mathbb{R}^p$, where E is a subset of some $\mathbb{R}^n$. Determine the values of n and p, give the component functions $f_1 = f_1(x_1, \ldots, x_n)$, $\ldots$, $f_p = f_p(x_1, \ldots, x_n)$ and calculate the value of the function at the given point.

1. $\mathbf{f}(x, y) = (x^2, y^2, x + y, x - y)$; $\mathbf{f}(1, 2)$

2. $\mathbf{f}(a, b, c) = (a^2 + b^2 + c^2, b \sin a)$; $\mathbf{f}(\pi/2, 1, -5)$

3. $\mathbf{f}(t) = (\ln t, \sqrt{t}, t^5)$; $\mathbf{f}(1)$.

4. $f(w, x, y, z) = wx + yz$; $f(1, 2, 3, 4)$. ▶

5. $\mathbf{f}(x_1, x_2, x_3, x_4, x_5) = (y_1, y_2, y_3)$, where $y_1 = x_1 x_2$, $y_2 = x_3 + x_4 x_5$, and $y_3 = x_1 \cos x_2 + x_3 \sin x_4 + e^{x_5}$; $\mathbf{f}(1, -1, 2, -2, 3)$

6. $\mathbf{f}(x_1, x_2, x_3) = (y_1, y_2, y_3, y_4, y_5)$, where $y_1 = x_1 \ln x_3$, $y_2 = x_2 e^{x_3 + x_1}$, $y_3 = x_1$, $y_4 = x_3$, and $y_5 = 0$; $\mathbf{f}(-1, 2, 1)$

In each of the following exercises, decide whether the given function $\mathbf{f}$ (or f) is linear, non-linear but affine linear, or neither. For each function, describe the level set where $\mathbf{f} = \mathbf{0}$ (or $f = 0$).

7. $\mathbf{f}(x, y) = (x^2, y^2, x + y, x - y)$.

8. $\mathbf{f}(x, y, z) = (2x - y, y - z, 5x + y - z + 2)$.

9. $\mathbf{f}(x, y, z) = (2x - y, y - z)$. ▶

10. $f(x, y, z) = x^2 + (y - 1)^2 e^z$.

11. $\mathbf{f}(x, y, z) = (x^2 + (y - 1)^2 e^z, z + 5)$.

12. $\mathbf{f}(x, y) = (3x - y, 0, 12x - 4y, -6x + 2y)$.

13. $\mathbf{f}(x, y) = (3x - y, x + y, 12x - 4y, -6x + 2y)$.

14. $\mathbf{f}(x, y, z) = (x + 2y - z + 1, 3x + 6y - 3z + 3)$.

15. $f(x, y, z) = xyz$.

16. $\mathbf{f}(x, y, z) = (xyz, (x - 1)^4)$.

In each of the following exercises, you are given a elementary real-valued function f. Thus, f is continuous, but its domain may not be all of $\mathbb{R}^n$. Assume that the domain is the largest set for which the formula is defined.

Identify the domain of f, and then decide whether or not the set of those points in the domain which satisfy the given equality or inequality is an open set in $\mathbb{R}^n$, a closed set in $\mathbb{R}^n$, or neither. Recall from the Exercises in Section 1.1 that the intersection of a finite number of open subsets is open, and the intersection of any number of closed subsets is closed.

17. $f(x, y, z) = x^2 - y^2 + z$, and $-3 < f(x, y, z) < 7$.

18. $f(x, y, z) = x^2 - y^2 + z$, and $-3 \leq f(x, y, z) \leq 7$.

19. $f(x, y) = \dfrac{5}{x^2 + y^2}$, and $\dfrac{1}{2} < f(x, y) < 100$.

20. $f(x, y) = \dfrac{5}{x^2 + y^2}$, and $\dfrac{1}{2} \leq f(x, y) \leq 100$.

21. $f(w, x, y, z) = wx^2 + y\sqrt{z}$, and $f(w, x, y, z) = 5$.

22. $f(w, x, y, z) = wx^2 + y\sqrt{z}$, and $1 \leq f(w, x, y, z) \leq 2$.

23. $f(x, y) = x - y^2$, and $0 < f(x, y)$. ▶

24. $f(x, y) = x - y^2$, and $0 < f(x, y) \leq 5$.

25. $f(x, y) = \ln(x - y^2)$, and $-3 < f(x, y) < 8$.

26. $f(x, y, z) = xe^y \tan^{-1} z$, and $f(x, y, z) = 17$.

In each of the following exercises, you are given an elementary real-valued function f which may or may not be defined on all of $\mathbb{R}^n$. You are then given a subset E of the domain of f. Decide if E is compact or not. If E is compact, apply the Extreme Value Theorem, Theorem 1.7.13, to conclude that f attains both a maximum and minimum value when restricted to E. In those exercises in which E is compact, can you determine the extreme values?

27. $f(x, y, z) = x^2 - y^2 + z$, and E is the set of those (x, y, z) such that $x^2 + y^2 + z^2 = 5$.

28. $f(x, y, z) = x^2 - y^2 + z$, and E is the set of those (x, y, z) such that $x + y + z = 5$.

29. $f(x, y) = xy$, and E is the set of those (x, y) such that $0 \leq x \leq 1$. ▶

30. $f(x, y) = xy$, and E is the set of those (x, y) such that $x \geq 0$, $y \geq 0$, and $x + y \leq 5$.

31. $f(x, y) = \dfrac{x \sin y}{x^2 - e^y}$, and E is the set of those (x, y) such that $1 \leq x^2 - e^y \leq 2$.

32. $f(x, y) = \dfrac{x \sin y}{x^2 + e^y}$, and E is the set of those (x, y) such that $1 \leq x^2 + e^y \leq 2$.

More Depth:

In each of the following exercises, you are given an elementary real-valued function f which may or may not be defined on all of $\mathbb{R}^n$. You are then given a subset E of the domain of f. Decide if E is connected or not (by inspection). If E is connected, apply the Intermediate Value Theorem, Theorem 1.7.14, to conclude that the image of f is an interval. If E is connected, can you determine the interval which is the image of f restricted to E?

33. $f(x, y, z) = x^2 - y^2 + z$, and E is the set of those (x, y, z) such that $x^2 + y^2 + z^2 \neq 5$.

34. $f(x, y, z) = x^2 - y^2 + z$, and E is the set of those (x, y, z) such that $x^2 + y^2 + z^2 < 5$.

35. $f(x, y) = xe^y$, and E is the set of those (x, y) such that $|x + y| < 2$.

36. $f(x, y) = xe^y$, and E is the set of those (x, y) such that $|x + y| > 2$.

37. $f(x, y) = xe^y$, and E is the set of those (x, y) such that $x + y \neq 0$.

38. $f(x, y) = xe^y$, and E is the set of those (x, y) such that $x + y = 0$.

39. $f(w, x, y, z) = w \sin x + yz$, and E is the set of those (w, x, y, z) such that $|x| = \pi$.

40. $f(w, x, y, z) = w \sin x + yz$, and E is the set of those (w, x, y, z) such that $|x| \leq \pi$.

41. Consider the triangle T (not filled in) with vertices $(0, 0)$, $(1, 0)$, and $(0, 1)$. Give a continuous function $\mathbf{r} : [0, 1] \to T$ such that $\mathbf{r}(0) = (0.5, 0)$ and $\mathbf{r}(1) = (0.5, 0.5)$.

42. Show that the triangle T from the previous exercise is path-connected.

43. Consider the set E that consists of the two touching circles given by $(x+1)^2 + y^2 = 1$ and $(x - 1)^2 + y^2 = 1$. Give a continuous function $\mathbf{r} : [0, 1] \to E$ such that $\mathbf{r}(0) = (-2, 0)$ and $\mathbf{r}(1) = (2, 0)$.

44. Show that the set E from the previous exercise is path-connected.

Decide, by inspection, if the following subsets of $\mathbb{R}^2$ and $\mathbb{R}^3$ are simply-connected or not.

45. The subset of those (x, y) in $\mathbb{R}^2$ such that $x^2 + y^2 \leq 4$.

46. The subset of those (x, y) in $\mathbb{R}^2$ such that $x^2 + y^2 \geq 4$.

47. The subset of those (x, y) in $\mathbb{R}^2$ such that $y \neq x^2$.

48. The subset of those (x, y) in $\mathbb{R}^2$ such that $y < x^2$.

49. The subset of those (x, y, z) in $\mathbb{R}^3$ such that $3 \leq x^2 + y^2 + z^2$.

50. The subset of those (x, y, z) in $\mathbb{R}^3$ such that $3 \leq x^2 + y^2 + z^2 \leq 7$.

51. The subset of those (x, y, z) in $\mathbb{R}^3$ such that $x^2 + y^2 \neq 0$.

52. The subset of $\mathbb{R}^3$ that consists of all of $\mathbb{R}^3$ except for the points $(-1, 0, 0)$ and $(1, 0, 0)$.

53. The subset of $\mathbb{R}^3$ that consists of all of $\mathbb{R}^3$ except for the circle where $z = 0$ and $x^2 + y^2 = 1$.

54. The subset of $\mathbb{R}^3$ that consists of all of $\mathbb{R}^3$ except for the line segment from $(0, 0, 0)$ to $(1, 0, 0)$.

1.8   Graphing Surfaces

In this section, we want to look at some relatively simple graphs of functions of the form $z = f(x, y)$, and graphs of level surfaces $F(x, y, z) = c$. We shall look at general notions of *cylinders* and what are known as *quadric surfaces*.

This will give us a nice array of basic surfaces to use as examples when we are discussing basic geometric properties and objects, like the *tangent plane*, related to multivariable Calculus.

Basics:

Many of the interesting aspects of multivariable functions arise for real-valued functions of two variables, i.e., for functions from subsets of $\mathbb{R}^2$ into the real numbers, such as

$$f(x, y) = x^2 - y^2.$$

In this case, we typically let z denote the output value, and graph the function $f(x, y)$ by plotting the points (x, y, z) in $\mathbb{R}^3$ such that

$$z = f(x, y).$$

Note that the function $z = f(x, y)$ has exactly one z value for (x, y) in the domain of f, and no z value for (x, y) outside the domain of f. Thus, the graph of a function $z = f(x, y)$ must pass a *vertical line test*: the vertical line at a fixed (x, y) can intersect the graph, at most, once.

We can generalize the situation of the graph of a function $f = f(x, y)$. Writing $z = f(x, y)$ as $z - f(x, y) = 0$, we see that the graph of a function of two variables can be looked at as the level set (or *level surface*) where $F(x, y, z) = 0$ if we let $F(x, y, z) = z - f(x, y)$.

In Section 1.4, we already looked at the most basic type of level surface in $\mathbb{R}^3$; we looked at planes in $\mathbb{R}^3$ as level surfaces of

$$ax + by + cz + d = 0.$$

In this section, we will briefly discuss some basic examples of graphs of functions of two variables and level sets of functions of three variables that have more-interesting features than planes.

Example 1.8.1. Consider the points (x, y, z) in $\mathbb{R}^3$ such that $z = f(x, y) = x^2$.

Some people look at the equation $z = x^2$ and think "Ah. There's no y. So $y = 0$." This is **not** correct. In fact, y gets to be anything; there's no restriction on it.

For instance, $(x, y, z) = (2, 5, 4)$ satisfies $z = x^2$, and so do $(x, y, z) = (2, \pi, 4)$ and $(x, y, z) = (2, \sqrt{7}, 4)$.

So, how do you graph all of the points (x, y, z) such that $z = x^2$?

You graph $z = x^2$ in the xz-plane, and then you extend that parabola parallel to the y-axis to form a surface. This surface has the property that, if you fix the y-coordinate at any value $y = c$, then, in the copy of the xz-plane where $y = c$, you see the graph of $z = x^2$, i.e., you see the "same" parabola each time that you slice the surface with a plane where $y = $ constant. The intersection of the surface with a plane where y equals a constant value is called a y-**cross section** of the surface. All of the y-cross sections of our current surface are parabolas.

> Cross sections are also referred to as *traces*.

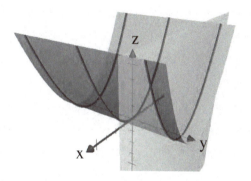

Figure 1.8.1: The parabolic cylinder where $z = x^2$, with y-cross sections.

Any time that you take a plane curve, and extend it parallel to a line, to form a surface, such a surface is called a **cylinder**. Here, because the plane curve that we started with was a parabola, the graph of $z = x^2$ in $\mathbb{R}^3$ is called a **parabolic cylinder**.

What you normally refer to as a cylinder is, more precisely, a **right circular cylinder**; this means that you start with a circle in a plane and extend it parallel to a line which makes a *right* angle with the plane that the figure is in.

Example 1.8.2. For instance, the graph of $x^2 + y^2 = 4$, in $\mathbb{R}^3$, is a right circular cylinder. See Figure 1.8.2. Its z-cross sections are circles of radius 2, centered around the z-axis.

You can also take x- and y-cross sections of this cylinder.

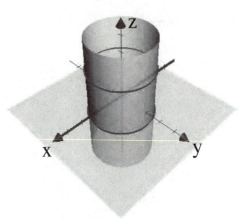

Figure 1.8.2: The right circular cylinder where $x^2 + y^2 = 4$, with z-cross sections.

Suppose you take the cross section where $y = 1$. This means that you're looking at the points (x, y, z) such that $y = 1$ and $x^2 + y^2 = 4$. Thus, $y = 1$ and $x^2 + 1 = 4$, i.e., $x^2 = 3$, so that $x = \pm\sqrt{3}$. Note that z still gets to be anything. Thus, you get a pair of vertical lines in the plane where $y = 1$.

More generally, if $-2 < c < 2$, the cross section where $y = c$ will be the graph $x^2 + c^2 = 4$, i.e., $x = \pm\sqrt{4 - c^2}$ inside the plane where $y = c$. This will be two vertical lines. If $c = \pm 2$, the cross section where $y = c$ will be the single vertical line where $y = c$ and $x = 0$. If $|c| > 2$, then $y = c$ does not intersect the graph of $x^2 + y^2 = 4$ at all, since we cannot have $x^2 = 4 - c^2 < 0$.

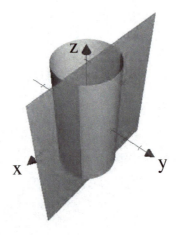

Figure 1.8.3: The right circular cylinder where $x^2 + y^2 = 4$, with the $y = 1$ cross section.

From the symmetry of the equation, you can see that the x-cross sections behave just like the y-cross sections.

One last comment: when one of the coordinate variables is missing from the defining equation, as in $x^2 + y^2 = 4$, then, before you can graph the appropriate set of points, you need to be told **where** your graph lies. If someone says "graph $x^2 + y^2 = 4$", then, by default, you should assume that they mean in the xy-plane, which would give you a circle. If they want the cylinder, then they need to say something like "graph $x^2 + y^2 = 4$ in $\mathbb{R}^3$."

Aside from cylinders, we want to look at graphs of what are known as *quadric surfaces*. These are surfaces which are the graphs of level sets of polynomials with the variables x, y and z of degree 2. Thus, a quadric surface is the graph of something of the form

$$ax^2 + by^2 + cz^2 + dxy + exz + fyz + gx + hy + iz + j = 0,$$

where a, b, c, d, e, f, g, h, i, and j are constants, and not all of a, b, c, d, e, and f are 0.

However, we shall not look at the completely general case, but rather look at quadric surfaces which are oriented in nice ways with respect to the axes.

Example 1.8.3. Consider the function

$$z = f(x, y) = \frac{x^2}{4} + \frac{y^2}{9}.$$

If you fix an x-coordinate, then the corresponding cross section is a parabola, given in terms of y and z. For instance, the $x = 2$ cross section is the graph of

$$z = 1 + \frac{y^2}{9},$$

inside the copy of the yz-plane at $x = 2$.

Similarly, all of the y-cross sections are parabolas.

However, the z-cross sections, which are the level curves of the function f, are:

1. ellipses for $z > 0$,

2. a single point, the origin, for $z = 0$, and

3. the empty set when $z < 0$.

> When we refer to a "z-cross section of $z = f(x, y)$", we are usually picturing the entire surface, together with a z-cross-sectional slice. When we use the term "level curve of $f(x, y)$", we are usually picturing just the graph of the curve in the xy-plane.

Because two out of three of the x-, y-, and z-cross sections are parabolas. This surface is called a **paraboloid**. Because the remaining set of cross sections are (for $z > 0$) ellipses, this surface is called, more specifically, an **elliptic paraboloid**.

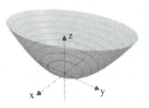

Figure 1.8.4: The elliptic paraboloid given by $z = \frac{x^2}{4} + \frac{y^2}{9}$.

It's easy to move the vertex, or "center", of the paraboloid. The graph of

$$z - 3 = \frac{(x-1)^2}{4} + \frac{(y-2)^2}{9},$$

is the "same" elliptic paraboloid as before, but with the vertex moved to the point $(1, 2, 3)$.

Or, we can turn the elliptic paraboloid upside down, and then raise it 4 units by graphing

$$z = 4 - \frac{x^2}{4} - \frac{y^2}{9}.$$

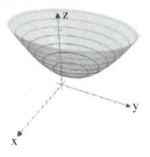

Figure 1.8.5: The elliptic paraboloid given by $z - 3 = \frac{(x-1)^2}{4} + \frac{(y-2)^2}{9}$.

Or, we can turn turn the paraboloid right side up again, put the vertex back at the origin, but change the shapes of the z-cross-sectional ellipses by graphing

$$z = \frac{x^2}{a^2} + \frac{y^2}{b^2},$$

where a and b are any positive constants.

Or, we can have the elliptic paraboloid open up around another axis by interchanging the variables, e.g., by graphing

$$y = \frac{x^2}{a^2} + \frac{z^2}{b^2},$$

which which open around the positive y-axis.

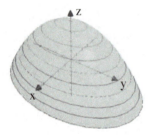

Figure 1.8.6: The elliptic paraboloid given by $z = 4 - \frac{x^2}{4} - \frac{y^2}{9}$.

We could spend a long time discussing the graphs of all of the quadric surfaces, and manipulating the graphs via various *changes of coordinates*; we shall look at aspects of these topics in more depth in Section 2.6 and Section 2.7.

However, for now, we give the quadric surfaces corresponding to some standard normal forms of the defining equations, and also describe the cross sections.

Example 1.8.4. Consider the graph of

$$\frac{x^2}{a^2} + \frac{y^2}{b^2} + \frac{z^2}{c^2} = 1,$$

where a, b and c are positive constants.

When $-a < x < a$, the x-cross sections are ellipses. When $x = \pm a$, the x-cross sections are single points, and when $|x| > a$, the x-cross sections are empty.

The analogous statements are true for y and z:

When $-b < y < b$, the y-cross sections are ellipses. When $y = \pm b$, the y-cross sections are single points, and when $|y| > b$, the y-cross sections are empty.

When $-c < z < c$, the z-cross sections are ellipses. When $z = \pm c$, the z-cross sections are single points, and when $|z| > c$, the z-cross sections are empty.

This quadric surface is called an **ellipsoid**.

Suppose that $a = b = c$, and we call this common value r. Then, our equation can be written

$$x^2 + y^2 + z^2 = r^2,$$

which, as we saw back in Example 1.1.3, describes a sphere of radius 1, centered at the origin.

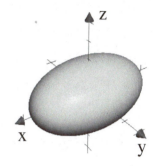

Figure 1.8.7: An ellipsoid: the graph of $\frac{x^2}{a^2} + \frac{y^2}{b^2} + \frac{z^2}{c^2} = 1$.

Example 1.8.5. Consider the graph of

$$z^2 = \frac{x^2}{a^2} + \frac{y^2}{b^2},$$

where a and b are positive constants.

When $z = 0$, you get a single point, the origin. The z-cross sections, for $z \neq 0$, are ellipses.

When $x = 0$, the cross section is the graph of $z^2 = y^2/b^2$, i.e., the pair of lines given by $z = \pm y/b$. The other x-cross sections are hyperbolas.

When $y = 0$, the cross section is the graph of $z^2 = x^2/a^2$, i.e., the pair of lines given by $z = \pm x/a$. The other y-cross sections are hyperbolas.

This is a **cone**, sometimes referred to as a **double cone**. The top half of this cone is the graph of

$$z = \sqrt{\frac{x^2}{a^2} + \frac{y^2}{b^2}},$$

while the bottom half of this cone is the graph of

$$z = -\sqrt{\frac{x^2}{a^2} + \frac{y^2}{b^2}}.$$

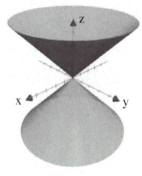

Figure 1.8.8: A cone: the graph of $z^2 = (x^2/a^2) + (y^2/b^2)$.

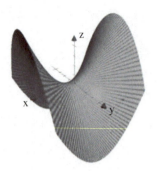

Figure 1.8.9: A hyperbolic paraboloid: the graph of $z = (x^2/a^2) - (y^2/b^2)$.

Example 1.8.6. Consider the graph of

$$z = \frac{x^2}{a^2} - \frac{y^2}{b^2},$$

where a and b are positive constants.

When $z = 0$, the cross section is a pair of lines given by $x/a = \pm y/b$. The z-cross sections, for $z \neq 0$ are hyperbolas.

The x- and y-cross sections are parabolas.

This is called a **hyperbolic paraboloid** or, more simply, a **saddle**.

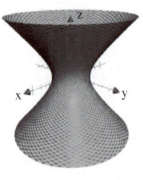

Figure 1.8.10: A hyperboloid of 1 sheet: the graph of $\frac{x^2}{a^2} + \frac{y^2}{b^2} = \frac{z^2}{c^2} + 1$.

Example 1.8.7. Consider the graph of

$$\frac{x^2}{a^2} + \frac{y^2}{b^2} = \frac{z^2}{c^2} + 1,$$

where a, b, and c are positive constants.

The z-cross sections are all ellipses.

If $x = \pm a$, or $y = \pm b$, then the cross sections are pairs of lines. All of the other x- and y-cross sections are hyperbolas.

This graph is called a **hyperboloid, of 1 sheet**, which is to be contrasted with the *hyperboloid of 2 sheets* in the next example.

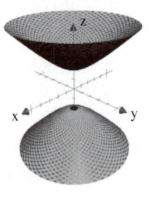

Figure 1.8.11: A hyperboloid of 2 sheets: the graph of $\frac{x^2}{a^2} + \frac{y^2}{b^2} = \frac{z^2}{c^2} - 1$.

Example 1.8.8. Consider the graph of

$$\frac{x^2}{a^2} + \frac{y^2}{b^2} = \frac{z^2}{c^2} - 1,$$

where a, b, and c are positive constants.

When $-c < z < c$, the z-cross sections are all empty. When $z = \pm c$, the cross sections are single points. When $|z| > c$, the cross sections are ellipses.

The x- and y-cross sections are all hyperbolas.

This graph is called a **hyperboloid, of 2 sheets**.

1.8.1   Exercises

Online answers to select exercises are here.

In each of the following exercises, match the equation with its graph in $\mathbb{R}^3$.

Basics:

1. $x^2 = 3z^2 + 2y^2$.

2. $x^2 = 3z^2 - 2y^2$.

3. $x^2 + 1 = 3z^2 + 2y^2$.

4. $z + 1 = 3x^2 + 2y^2$.

5. $z^2 - 1 = 3x^2 + 2y^2$.

6. $-z^2 + 1 = 3x^2 + 2y^2$.

7. $z + 1 = 3x^2 - 2y^2$.

8. $z + 1 = -3x^2 + 2y^2$.

9. $1 = 3z^2 - 2y^2$.

10. $1 = 3z^2 + 2y^2$.

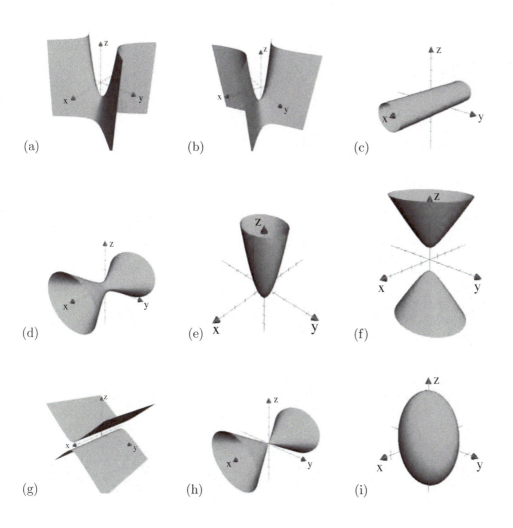

(a)   (b)   (c)

(d)   (e)   (f)

(g)   (h)   (i)

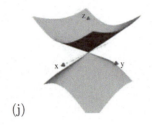

(j)

In each of the following exercises, describe the x-, y-, and z-cross sections, and then sketch the graph of the given equation in $\mathbb{R}^3$.

11. $x^2 - 3y^2 + z = 1$.

12. $x^2 - 3y^2 - z^2 = 1$.

13. $x^2 - 3y^2 + z^2 = 1$.

14. $x^2 + 3y^2 + z^2 = 1$.

15. $x^2 + 3y^2 = (z - 3)^2$.

16. $(x - 1)^2 + 3(y - 2)^2 = (z - 3)^2$.

17. $x^2 + 3y^2 + 1 = z^2$.

18. $x^2 + 3z^2 + 4 = y^2$.

19. $z = e^x$.

20. $z = \sin y$.

In each of the following exercises, the z-cross sections are empty, single points, circles, or ellipses. By analyzing how these cross sections change with z, sketch the corresponding surfaces in $\mathbb{R}^3$.

21. $x^2 + y^2 = e^z$.

22. $x^2 + \dfrac{y^2}{4} = \sin^2 z$.

23. $x^2 + \dfrac{y^2}{4} = \sin z$.

24. $x^2 + y^2 = \dfrac{1}{z^2 + 1}$.

25. $\dfrac{x^2}{9} + \dfrac{y^2}{4} = \ln z$.

26. $\dfrac{x^2}{9} + \dfrac{y^2}{4} = \tan^{-1} z$.

27. A quadric surface M is such that all of its x-cross sections are empty, points, or circles. What are the possible choices for the type of quadric surface that M can be?

28. A quadric surface M is such that none of its x-cross sections are ellipses or circles. What are the possible choices for the type of quadric surface that M can be?

29. A quadric surface M is such that its $z = 0$ cross section is the pair of lines $y = 2x$ and $y = -2x$. What is a possible equation defining M?

30. A quadric surface M is such that its $z = 1$ cross section is the pair of lines $x = 3y$ and $x = -3y$. What is a possible equation defining M?

Suppose that M is a surface defined by $z = f(x, y)$, and suppose that we have a point $(a, b, f(a, b))$ on M.

If we look at the $x = a$ cross section, it is defined inside the plane where $x = a$ by $z = f(a, y)$. Let g_a denote the single-variable function such that $g_a(y) = f(a, y)$. Similarly, if we look at the $y = b$ cross section, it is defined inside the plane where $y = b$ by $z = f(x, b)$. Let h_b denote the single-variable function such that $h_b(x) = f(x, b)$.

Then, if g_a is differentiable at b, we can consider the tangent line to the graph of g_a at $(b, g_a(b))$ to be a line inside the plane where $x = a$; this gives us a line in $\mathbb{R}^3$. Similarly, if h_b is differentiable at a, we can consider the tangent line to the graph of h_b at $(a, h_b(a))$ to be a line inside the plane where $y = b$; this also gives us a line in $\mathbb{R}^3$.

If these two tangent lines are different lines, then, as we shall discuss in Section 2.3, with stronger assumptions on f, the unique plane containing these two lines gives the tangent plane to the graph of f at $(a, b, f(a, b))$.

In the exercises below, assume this tangent plane exists, and is the unique plane containing the two different tangent lines of the x- and y-cross sections at the given point.

31. Show that the tangent plane to the graph of $z = f(x, y)$ at $(a, b, f(a, b))$ is defined by $z - f(a, b) = h_b'(a)(x - a) + g_a'(b)(y - b)$.

32. Find a standard equation for the tangent plane to the graph of $z = f(x, y) = x^2 + 9y^2 + 2$ at $(3, 1, f(3, 1))$. Graph the surface together with the tangent plane.

33. Find a standard equation for the tangent plane to the graph of $z = f(x, y) = x^2$ at $(3, 1, f(3, 1))$. Graph the surface together with the tangent plane.

34. Find a standard equation for the tangent plane to the graph of $z = f(x, y) = -9y^2$ at $(3, 1, f(3, 1))$. Graph the surface together with the tangent plane.

35. Find a standard equation for the tangent plane to the graph of $z = f(x, y) = x^2 - 9y^2$ at $(3, 1, f(3, 1))$. Graph the surface together with the tangent plane.

36. Find a standard equation for the tangent plane to the graph of $z = f(x, y) = x^2 - 9y^2$ at $(0, 0, f(0, 0))$. Graph the surface together with the tangent plane.

Chapter 2

Multivariable Derivatives

For a multivariable function $f = f(x_1, \ldots, x_n)$, the notion of *derivative* becomes more complicated than the situation for a function of a single variable.

There are the *partial derivatives* of f with respect to each of the variables x_i. There is the *total derivative* of f at each point $\mathbf{p}$, where the total derivative of f at $\mathbf{p}$ is, itself, a function – a function which encodes how f changes at $\mathbf{p}$ depending on the direction and the speed.

This leads to discussions of many topics: linear approximation, tangent sets, differentiation rules, methods for obtaining smooth surfaces and other *submanifolds*, local and global extrema, and multivariable Taylor polynomials and series.

2.1 Partial Derivatives

The process of taking the derivative, with respect to a single variable, and holding constant all of the other independent variables, is called finding (or, taking) a *partial derivative*. This is a fundamental mathematical concept that arises in many contexts.

Basics:

If $f(x) = 5x^3$, then $f'(x) = 5 \cdot 3x^2 = 15x^2$. In exactly the same way, if you're given $g(x) = ax^3$, and told that a is a constant, then you find that $g'(x) = a \cdot 3x^2 = 3ax^2$. If you are now told that $a = 5$, you can plug in 5 for a in this latter answer to get what you got before.

Suppose now that you're given the function of two variables $h(x,y) = yx^3$. Since y is one of the independent variables in h, clearly y is not intended to **always** be constant. However, if you're told to assume that, for some physical or mathematical reason, y is held constant at the value $y = 5$, and asked to differentiate h as a function of x, you would look at $h(x,5) = 5x^3$, and differentiate, to once again obtain $15x^2$. If, instead, you're told to assume that y is held constant at the value $y = 7$, and asked to differentiate h as a function of x, you would look at $h(x,7) = 7x^3$, and differentiate to obtain $21x^2$.

More generally, you could just be told to assume that y is held constant, without being told that that constant value is 5 or 7 or anything else specific; then, you can calculate the derivative of yx^3, with respect to x, thinking of y as a constant; you find $y \cdot 3x^2$.

This leads to the following definition:

Definition 2.1.1. *Suppose that we have a real-valued function $z = f(x,y)$ of two real variables. Then, the derivative of f, with respect to x, holding y constant, is called the* **partial derivative of f, with respect to x,** *and is denoted by any of*

$$\frac{\partial z}{\partial x}, \qquad \frac{\partial f}{\partial x}, \qquad f_x(x,y), \qquad \text{or} \qquad f_1(x,y).$$

In the same way, the derivative of f, with respect to y, holding x constant, is called the partial derivative of f, with respect to y, and is denoted by any of

$$\frac{\partial z}{\partial y}, \qquad \frac{\partial f}{\partial y}, \qquad f_y(x,y), \qquad \text{or} \qquad f_2(x,y).$$

> We also use the partial derivative operators:
>
> $$\frac{\partial}{\partial x} \qquad \text{and} \qquad \frac{\partial}{\partial y},$$
>
> *which tell you to take the partial derivatives with respect to x and y, respectively.*

Recall that an **elementary function** of one or more variables is: a function which is a constant function, a power function (with an arbitrary real exponent), a polynomial function, an exponential function, a logarithmic function, a trigonometric function, or inverse trigonometric function, or any finite combination of such functions using addition, subtraction, multiplication, division, or composition.

Since the derivative of a one-variable elementary function is again elementary, taking a partial derivative of an elementary function of more than one variable also yields another elementary function.

We mention the notations $f_1(x, y)$ and $f_2(x, y)$ primarily because you may see them used in other books; there are also technical reasons why these notations are useful in some contexts. However, we shall avoid their use to indicate partial derivatives, since we like to reserve the notations $f_1(x, y)$ and $f_2(x, y)$ for use in denoting the component functions of a multi-component function $\mathbf{f}(x, y) = (f_1(x, y), f_2(x, y))$. In any case, throughout this book, we will explicitly state, or the context will make clear, what we mean by f_1, f_2 or, more generally, f_i.

Example 2.1.2. Consider the fairly simple function

$$z = f(x, y) = x^2 - y^2.$$

We first take the partial derivative, with respect to x, thinking of y as a constant. Let's use all of our various notations just for practice. We find

$$\frac{\partial z}{\partial x} = \frac{\partial f}{\partial x} = f_x(x, y) = f_1(x, y) = \frac{\partial}{\partial x}(x^2 - y^2) = 2x.$$

Now, we take the partial derivative, with respect to y, thinking of x as a constant. We find

$$\frac{\partial z}{\partial y} = \frac{\partial f}{\partial y} = f_y(x, y) = f_2(x, y) = \frac{\partial}{\partial y}(x^2 - y^2) = -2y.$$

Remark 2.1.3. Note that the prime notation for derivatives would be bad when calculating partial derivatives. For instance, $(xy^2 + 5y^3)'$ would be totally ambiguous.

The symbol ∂ looks sort of like a backwards lower-case delta, δ. It is usually read simply as "partial", and $\partial f/\partial x$ is read as "partial f partial x". Occasionally, the symbol ∂ is referred to as a "round d".

Example 2.1.4. Of course, if you have a function, such as $h(t) = 5t + \ln t$, which depends on only one variable, the partial derivative is just the same as the ordinary derivative:

$$\frac{\partial h}{\partial t} = \frac{dh}{dt} = 5 + \frac{1}{t}.$$

It's not **wrong** to write the partial derivative here, but it could be misleading in some cases; it might make someone wonder what the others variables are.

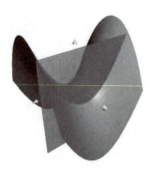

Figure 2.1.1: $y = 0.5$ cross section of $z = x^2 - y^2$.

Example 2.1.5. The derivative of a one-variable function can be interpreted graphically as the slope of the tangent line. Is there also a way to interpret the **partial** derivatives graphically? Yes.

Let's look at the function from Example 2.1.2: $z = f(x, y) = x^2 - y^2$, for which we found that

$$f_x(x, y) = 2x \qquad \text{and} \qquad f_y(x, y) = -2y.$$

From the above, we know that $f_x(0.7, 0.5) = 2(0.7) = 1.4$, but how can we "see" this?

Consider the partial derivative

$$\left.\frac{\partial f}{\partial x}\right|_{(0.7, 0.5)} = f_x(0.7, 0.5).$$

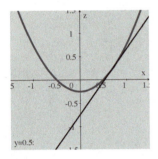

Figure 2.1.2: Slope of tangent line equals $\partial z / \partial x$.

When calculating the partial derivative with respect to x, we hold y constant. This means that, if we want to calculate $f_x(0.7, 0.5)$, we may first fix $y = 0.5$, and then take the one-variable derivative with respect to x. Graphically, this means that we take the $y = 0.5$ cross section of the hyperbolic paraboloid (recall Section 1.8) defined by $z = x^2 - y^2$; this gives us the graph of $z = x^2 - 0.25$ in the copy of the xz-plane where $y = 0.5$.

The partial derivative $f_x(0.7, 0.5)$ is then the slope of the tangent line to the graph of $z = x^2 - 0.25$ at the point where $x = 0.7$; see Figure 2.1.1 and Figure 2.1.2.

Now, consider the partial derivative

$$\left.\frac{\partial f}{\partial y}\right|_{(-1, 0.7)} = f_y(-1, 0.7).$$

Figure 2.1.3: $x = -1$ cross section of $z = x^2 - y^2$.

We know that $f_y(-1, 0.7) = -2(0.7) = -1.4$, and we can see this graphically just as we saw the partial derivative with respect to x.

When calculating the partial derivative with respect to y, x is held constant. Hence, if we want to calculate $f_y(-1, 0.7)$, we may first fix $x = -1$, and then take the one-variable derivative with respect to y. Graphically, this means that we take the $x = -1$

cross section of the hyperbolic paraboloid defined by $z = x^2 - y^2$; this gives us the graph of $z = 1 - y^2$ in the copy of the yz-plane where $x = -1$. The partial derivative $f_y(-1, 0.7)$ is then the slope of the tangent line to the graph of $z = 1 - y^2$ at the point where $y = 0.7$; see Figure 2.1.3 and Figure 2.1.4.

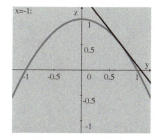

Figure 2.1.4: Slope of tangent line equals $\partial z / \partial y$.

Example 2.1.6. Now we'll calculate partial derivatives that are a bit more difficult. Let's find the partial derivatives of $xy^2 + 5y^3$.

First, calculate the partial derivative with respect to x, by thinking of y as constant; we find

$$\frac{\partial}{\partial x}\left(xy^2 + 5y^3\right) \ = \ 1 \cdot y^2 + 0 \ = \ y^2.$$

Now, calculate the partial derivative with respect to y, by thinking of x as constant; we find

$$\frac{\partial}{\partial y}\left(xy^2 + 5y^3\right) \ = \ x \cdot 2y + 15y^2 \ = \ 2xy + 15y^2.$$

Partial derivatives, by their definition as derivatives, are limits of average rates of change, in which all independent variables, other than one, are held constant. Thus, if $f = f(x, y)$, then

$$f_x(x, y) \ = \ \frac{\partial f}{\partial x} \ = \ \lim_{\Delta x \to 0} \frac{f(x + \Delta x, y) - f(x, y)}{\Delta x}$$

and

$$f_y(x, y) \ = \ \frac{\partial f}{\partial y} \ = \ \lim_{\Delta y \to 0} \frac{f(x, y + \Delta y) - f(x, y)}{\Delta y},$$

provided that these limits exist. Of course, if the limits don't exist, we say that the partial derivatives don't exist.

Note that this means that the units of $\partial f / \partial x$ are the units of f divided by the units of x, and the units of $\partial f / \partial y$ are the units of f divided by the units of y.

These partial rates of change come up often in physical situations. Of course, in lots of problems, the variable names won't be x and y.

Example 2.1.7. The volume, V, of a right circular cylinder is given by

$$V = \pi r^2 h,$$

where r is the radius of the base, and h is the height. Suppose that the cylinder is some sort of container for which the height can vary, such as the interior of a piston.

What is the instantaneous rate of change of the volume, with respect to the height, when the height is 0.3 meters, if the radius is held constant at 0.1 meters?

Solution: We hold r constant and find

$$\frac{\partial V}{\partial h} = \pi r^2,$$

in cubic meters per meter (or, square meters).

Thus, the instantaneous rate of change of the volume, with respect to the height, when the height is 0.3 m, and the radius is held constant at 0.1 m is

$$\frac{\partial V}{\partial h}\bigg|_{(r,h)=(0.1,0.3)} = \pi(0.1)^2 = 0.01\pi \ \text{m}^3/\text{m}.$$

Note that this result is independent of h, so that, in the end, we don't need to use the data that $h = 0.3$ meters.

Example 2.1.8. Suppose that a metal plate occupies the xy-plane, where all distances are measured in meters. Suppose further that this plate is heated in such a way that its temperature at (x, y) meters is given by

$$T(x, y) = 100e^{-0.2x^2 - 0.3y^2} \ ^\circ\text{C}.$$

a. Determine, at an arbitrary point (x, y), the instantaneous rates of change of the temperature with respect to the x- and y-coordinates.

b. What are these instantaneous rates of change at the origin? At the point $(1, -1)$?

Solution:

a. You are being asked to calculate $\dfrac{\partial T}{\partial x}$ and $\dfrac{\partial T}{\partial y}$.

We calculate

$$\frac{\partial T}{\partial x} = 100e^{-0.2x^2 - 0.3y^2}(-0.4x) \ ^\circ\text{C/m}$$

and

$$\frac{\partial T}{\partial y} = 100e^{-0.2x^2 - 0.3y^2}(-0.6y) \quad °C/m.$$

b.   Using our partial derivative formulas from above, we find

$$\left. \frac{\partial T}{\partial x} \right|_{(0,0)} = 0 \quad °C/m,$$

$$\left. \frac{\partial T}{\partial y} \right|_{(0,0)} = 0 \quad °C/m,$$

$$\left. \frac{\partial T}{\partial x} \right|_{(1,-1)} = 100e^{-0.5}(-0.4) \approx -24.2612 \quad °C/m,$$

and

$$\left. \frac{\partial T}{\partial y} \right|_{(1,-1)} = 100e^{-0.5}(0.6) \approx 36.3918 \quad °C/m.$$

The examples so far have all involved functions of two variables. However, of course the idea of a partial derivative makes sense for a real-valued function of any number of variables; when you take a partial derivative with respect to one variable, you treat all of the other independent variables as constants.

For instance, suppose we have a function of three variables $w = f(x, y, z)$. For such a function, there are partial derivatives with respect to x, y, and z. When you take a partial derivative with respect to one of x, y, or z, you assume that the other two independent variables are constant.

Example 2.1.9. Suppose that we have the function

$$w = f(x, y, z) = x\sin(yz) + y^2 e^z + x^3.$$

Then, we find:

$$\frac{\partial w}{\partial x} = \sin(yz) + 0 + 3x^2,$$

$$\frac{\partial w}{\partial y} = x(\cos(yz))z + 2ye^z + 0,$$

and

$$\frac{\partial w}{\partial z} \;=\; x(\cos(yz))y + y^2 e^z + 0.$$

We know from single-variable Calculus that if the derivative of a function exists and is 0 on an open interval, then the function must be constant on the interval. There is a generalization of this result to $\mathbb{R}^n$, but we must replace "open interval" with "connected open subset" (recall Definition 1.1.16).

This follows quickly from Corollary 9.19 and Theorem 9.21 of [7].

Theorem 2.1.10. *If $\mathcal{U}$ is a non-empty connected open subset of $\mathbb{R}^n$, and f is a real-valued function on $\mathcal{U}$ such that all of the partial derivatives of f exist and are 0 at each point in $\mathcal{U}$, then f is constant on $\mathcal{U}$.*

The condition above that all of the partial derivatives are equal to 0 is equivalent to the condition that the vector of partial derivatives is equal to $\mathbf{0}$. For this and other reasons, the vector of partial derivatives will be of extreme importance throughout this book, and so we give it a name.

We have used an arrow on a nabla symbol for the gradient vector. This is standard, though it departs from our usual convention of making vectors boldfaced.

Definition 2.1.11. *The multi-component function*

$$\vec{\nabla} f \;=\; \left(\frac{\partial f}{\partial x_1}, \frac{\partial f}{\partial x_2}, \ldots, \frac{\partial f}{\partial x_n} \right)$$

of partial derivatives of a function $f = f(x_1, \ldots, x_n)$ is called the **gradient vector (function)** *of f.*

Its value at a point $\mathbf{p}$ is denoted either by $\vec{\nabla} f(\mathbf{p})$ or by $\vec{\nabla} f_{|\mathbf{p}}$.

Note that, if the different x_i's have different units, then the separate entries in the gradient vector will have different units. Only if all of the x_i's have the same units can we assign units to the vector itself. For instance, if $f = f(x_1, \ldots, x_n)$ is measured in feet, and all of the x_i's are measured in seconds, then the units on the gradient vector $\vec{\nabla} f$ are feet per second.

Example 2.1.12. Consider $f(x, y) = x^2 - y^2$ from Example 2.1.2. Then,

$$\vec{\nabla} f \;=\; \left(\frac{\partial f}{\partial x}, \frac{\partial f}{\partial y} \right) \;=\; (2x, -2y) \;=\; 2\,(x, -y),$$

and

$$\vec{\nabla}f(3,4) \;=\; (6,-8) \;=\; 2(3,-4).$$

Example 2.1.13. Consider $w = x\sin(yz) + y^2 e^z + x^3$ from Example 2.1.9.

Then, our earlier calculations of the partial derivatives tell us immediately that the gradient vector is given by

$$\vec{\nabla}w \;=\; \left(\sin(yz) + 3x^2,\ xz\cos(yz) + 2ye^z,\ xy\cos(yz) + y^2 e^z\right).$$

Higher-order partial derivatives:

Consider the function

$$f(x,y) = x^2 + 5xy - 4y^2.$$

The partial derivatives are easy to calculate:

$$\frac{\partial f}{\partial x} \;=\; 2x + 5y \qquad \text{and} \qquad \frac{\partial f}{\partial y} \;=\; 5x - 8y.$$

We now want to look at the *second partial derivatives of f*. The first thing to decide is: how many ways can we take a second partial derivative?

A second partial derivative should be a partial derivative of a partial derivative. So, there are four second partial derivatives: you can first take two different first partial derivatives, with respect to x or y and then, for each of those, you can take a partial derivative a second time with respect to x or y.

We introduce new notation, and calculate

$$f_{xx} \;=\; \frac{\partial^2 f}{\partial x^2} \;=\; \frac{\partial}{\partial x}\left(\frac{\partial f}{\partial x}\right) \;=\; \frac{\partial}{\partial x}(2x + 5y) \;=\; 2,$$

$$f_{xy} \;=\; \frac{\partial^2 f}{\partial y \partial x} \;=\; \frac{\partial}{\partial y}\left(\frac{\partial f}{\partial x}\right) \;=\; \frac{\partial}{\partial y}(2x + 5y) \;=\; 5,$$

$$f_{yx} \;=\; \frac{\partial^2 f}{\partial x \partial y} \;=\; \frac{\partial}{\partial x}\left(\frac{\partial f}{\partial y}\right) \;=\; \frac{\partial}{\partial x}(5x - 8y) \;=\; 5,$$

and

$$f_{yy} \;=\; \frac{\partial^2 f}{\partial y^2} \;=\; \frac{\partial}{\partial y}\left(\frac{\partial f}{\partial y}\right) \;=\; \frac{\partial}{\partial y}(5x - 8y) \;=\; -8.$$

The two second partial derivatives f_{xy} and f_{yx} above, the ones with one partial derivative with respect to x and one with respect to y, are called *mixed partial derivatives*.

Note that f_{xy} and f_{yx} are equal in this example. While this is not **always** the case, it's true for "most" of the functions that we deal with. More precisely, we need the continuity condition given in the following theorem.

> **Theorem 2.1.14.** *Suppose that f, f_x, f_y, and f_{xy} exist in an open ball around a point (x_0, y_0), and that f_{xy} is continuous at (x_0, y_0). Then, $f_{yx}(x_0, y_0)$ exists and*
>
> $$f_{yx}(x_0, y_0) \;=\; f_{xy}(x_0, y_0).$$

For the proof, see Theorem 5.3.3 in Trench, [8].

As we usually deal with elementary functions, which are continuous wherever they're defined, and partial derivatives of elementary functions are elementary, the continuity condition in Theorem 2.1.14 typically reduces to the question of whether or not the second partial derivatives are defined.

For an example of a non-elementary function for which the mixed partial derivatives exist, but are not equal, see Example 5.3.4 in Trench, [8].

Example 2.1.15. Just because the order of partial differentiation doesn't (typically) matter as far as the final resulting higher-order partial derivative is concerned, that doesn't mean that calculating the partial derivatives in different orders is equally easy.

Consider

$$h(r, s) \;=\; re^{5s} + \frac{e^r \cos(r + \tan^{-1} r)}{\sqrt{1 - \ln r}}.$$

If you want to calculate the second partial derivative of h, once with respect to r and once with respect to s, it would be a **painful** waste of time to calculate $\partial h / \partial r$ first. If this isn't obvious to you, you should think about it until it's clear.

What you want to do is calculate the partial derivative with respect to s first, since, then, the entire right-hand ugly expression will disappear. Hence, we find that

$$\frac{\partial^2 h}{\partial s \partial r} \;=\; \frac{\partial^2 h}{\partial r \partial s} \;=\; \frac{\partial}{\partial r}\left(\frac{\partial h}{\partial s}\right) \;=\; \frac{\partial}{\partial r}\big(5re^{5s}\big) \;=\; 5e^{5s}.$$

More Depth:

Example 2.1.16. Let's try calculating some more-complicated partial derivatives.

Suppose that

$$z = f(x, y) = x \sin(xy) + 3y^4.$$

We want to calculate the partial derivatives $\partial f / \partial x$ and $\partial f / \partial y$. Note that $x \sin(xy)$ is the product of two functions of x, which will require the Product Rule when differentiating with respect to x. However, when we take the partial derivative with respect to y, we should think of $x\sin(xy)$ as being a "constant" times a function of y; hence, we will not need the Product Rule when applying $\partial/\partial y$.

We find:

$$\frac{\partial f}{\partial x} \; = \; x \cdot \frac{\partial}{\partial x}\big(\sin(xy)\big) \; + \; \sin(xy) \cdot \frac{\partial}{\partial x}(x) \; + \; 0 \; =$$

$$x\cos(xy) \cdot \frac{\partial}{\partial x}(xy) + \sin(xy) \; = \; [x\cos(xy)] \cdot y + \sin(xy) \; = \; xy\cos(xy) + \sin(xy),$$

and

$$\frac{\partial f}{\partial y} \; = \; x \cdot \frac{\partial}{\partial y}\big(\sin(xy)\big) + 12y^3 \; =$$

$$x\cos(xy) \cdot \frac{\partial}{\partial y}(xy) + 12y^3 \; = \; x\cos(xy) \cdot x + 12y^3 \; = \; x^2\cos(xy) + 12y^3.$$

Example 2.1.17. Suppose that we have a fixed number of atoms or molecules of an *ideal gas* in a container, where the container has a volume that can change with time, such as a balloon or the inside of a piston.

The Ideal Gas Law states that there is a relationship between the pressure P, in pascals (Pa, which are N/m^2), that the gas exerts on the container, the temperature T, in kelvins (K), of the gas, and the volume V, in m^3, that the gas occupies; that relationship is

$$PV = kT, \tag{2.1}$$

where k is determined by the number of atoms or molecules of the ideal gas (together with the ideal gas constant). We shall assume a value of $k = 8$ N·m/K in this problem.

If the temperature is held constant at 320 K, what is the instantaneous rate of change of the pressure, with respect to volume, when the volume is 2 m^3? (This is referred to as *isothermal expansion*.)

Solution:

We write P as a function of T and V, and obtain $P = 8TV^{-1}$. What we are being asked for is, precisely, $\partial P/\partial V$, when $(T, V) = (320, 2)$. We find

$$\frac{\partial P}{\partial V} \; = \; -8TV^{-2} \quad \text{Pa/m}^3,$$

and so

$$\frac{\partial P}{\partial V}\bigg|_{(T,V)=(320,2)} \; = \; -8(320)(2^{-2}) \; = \; -640 \quad \text{Pa/m}^3.$$

Example 2.1.18. Suppose that a company makes video game systems and video games, and is willing to sell the systems at below what it costs to make them, knowing that they can make plenty of profit on the games. Their games are backward compatible with older systems, so that, even if they sell no new systems, they will still have some sales of the new games. On the other hand, manufacturing, supply, and shipping constraints imply that, for a large fixed number of systems sold, there is a slight decrease in the amount of profit per game sold.

Suppose that the company has determined that their profit P, in dollars, on sales of the new games and systems, is given by

$$P \;=\; 20g - 100s - 0.01g\sqrt{s},$$

where g is the number of new games sold and s is the number of new game systems sold.

(a) If the company sells no new systems, how much profit do they make per new game sold?

(b) How many new systems would the company have to sell before they'd start losing money regardless of how many new games they sell?

(c) *Marginal profit* means the instantaneous rate of change of the profit. The marginal profit, per new game sold, is the instantaneous rate of change of the profit, with respect to the number of new games sold, holding constant the number of new systems sold.

If the company has fixed sales of $250,000$ new game systems, what is the marginal profit, per new game sold, when the number of new games sold is $1,000,000$?

(d) If the company immediately sells $10,000$ new games, and never sells another one, what is the marginal profit per new system sold, when the number of new systems sold is $10,000$?

Solution:

(a) Selling no new systems means that $s = 0$, and then the profit is simply $P = 20g$. Thus, the company makes a profit of $20 per game, when no systems are sold.

This is so simple that you don't need to think in terms of partial derivatives. However, if we **want** to phrase this in terms of partial derivatives, what we have just found is that

$$\left.\frac{\partial P}{\partial g}\right|_{s=0} = 20$$

dollars per game. This is what you get if you first plug in $s = 0$.

We could, instead, have calculated for arbitrary, but constant, s that

$$\frac{\partial P}{\partial g} = 20 - 0.01\sqrt{s}$$

dollars per game, and then plugged in $s = 0$ to obtain \$20 per game sold.

(b) We rewrite the formula for P as

$$P = \left(20 - 0.01\sqrt{s}\right)g - 100s.$$

If $20 - 0.01\sqrt{s}$ were positive, then, if g were big enough, P would be positive.

For the company to lose money, regardless of the value of g, it would have to be the case that $20 - 0.01\sqrt{s} \le 0$, so that P is negative, no matter how big g is. This is true if and only if $20 \le 0.01\sqrt{s}$, i.e., $2000 \le \sqrt{s}$ or, equivalently, $4,000,000 \le s$.

(c) As we found in (a),

$$\frac{\partial P}{\partial g} = 20 - 0.01\sqrt{s},$$

dollars per game sold. When s is fixed at $250,000$, this means that the marginal profit is

$$\left.\frac{\partial P}{\partial g}\right|_{s=250,000} = 20 - 0.01\sqrt{250,000} = 20 - 5 = 15$$

dollars per new game sold, regardless of the number of games sold.

(d) In this part, g is fixed at $10,000$. The marginal profit, per new game system sold, is

$$\frac{\partial P}{\partial s} = -100 - 0.01g\left(\frac{1}{2}\right)s^{-1/2} = -100 - \frac{0.005g}{\sqrt{s}},$$

dollars per system sold. As $g = 10,000$, when $s = 10,000$, we find that the marginal profit, per new system sold is

$$-100 - \frac{(0.005)(10,000)}{\sqrt{10,000}} = -100.50,$$

dollars per system, i.e., the company **loses** \$100.50 per new system sold.

Example 2.1.19. Suppose that $g(x,y) = xe^y + y^2\tan^{-1}x$. Calculate the dot product

$$\vec{\nabla}g(1,1) \cdot (2,-2).$$

Solution:

We find

$$\vec{\nabla}g = \left(e^y + \frac{y^2}{1+x^2},\ xe^y + 2y\tan^{-1}x\right),$$

and so

$$\vec{\nabla}g(1,1) = \left(e + \frac{1}{2},\ e + 2\cdot\frac{\pi}{4}\right) = \left(e + \frac{1}{2},\ e + \frac{\pi}{2}\right).$$

Therefore,

$$\vec{\nabla}g(1,1)\cdot(2,-2) = \left(e + \frac{1}{2},\ e + \frac{\pi}{2}\right)\cdot(2,-2) =$$

$$2\left(e + \frac{1}{2}\right) + (-2)\left(e + \frac{\pi}{2}\right) = 1 - \pi.$$

In general, partial derivatives make sense for functions of any number of variables, i.e., for $y = f(x_1, x_2, \ldots, x_n)$.

The partial derivative of such an f with respect to x_i, where i is one of the indices from 1 to n, is denoted by

$$\frac{\partial y}{\partial x_i}, \qquad \frac{\partial f}{\partial x_i}, \qquad f_{x_i}, \qquad \text{or} \qquad f_i,$$

and it simply means the derivative of f, with respect to x_i, holding all of the other independent variables constant.

We now give the general, limit-based, definition of the partial derivative; naturally, it amounts to fixing all of the variables except one, and using the limit definition of the derivative in that one changing variable position.

This rigorous definition doesn't make it any easier for us to calculate partial derivatives; we give it for the sake of completeness and because, this definition, written in vector notation, generalizes nicely to allow us to define the derivative at a point, with respect to, or in the "direction of" any vector (which we will look at in Section 2.2).

Definition 2.1.20. *Suppose that $f(x_1, x_2, \ldots, x_n)$ is a real-valued function whose domain is a subset of $\mathbb{R}^n$.*

*Then, we define the **partial derivative of f with respect to x_i** to be*

$$\frac{\partial f}{\partial x_i} = \lim_{h\to 0}\frac{f(x_1,\ldots,x_{i-1},x_i+h,x_{i+1},\ldots,x_n) - f(x_1,\ldots,x_{i-1},x_i,x_{i+1},\ldots,x_n)}{h},$$

*provided that this limit exists. If the limit fails to exist, then we say that the partial derivative is **undefined**.*

Recalling that $\mathbf{e}_i$ denotes the i-th standard basis element (see Section 1.2), and letting $\mathbf{x} = (x_1, x_2, \ldots, x_n)$, then the definition above is equivalent to

$$\frac{\partial f}{\partial x_i} = \lim_{h \to 0} \frac{f(\mathbf{x} + h\mathbf{e}_i) - f(\mathbf{x})}{h}.$$

Example 2.1.21. Resistors, of resistance R_1, R_2, ... , R_n ohms, can be placed in parallel in a circuit to produce a resistor element with a new resistance R_{new} ohms.

The relationship between the resistances is given by

$$\frac{1}{R_{\text{new}}} = \frac{1}{R_1} + \frac{1}{R_2} + \cdots + \frac{1}{R_n},$$

that is,

$$R_{\text{new}} = \left(R_1^{-1} + R_2^{-1} + \cdots + R_n^{-1}\right)^{-1}.$$

Suppose that we have such a parallel resistor element. What is the instantaneous rate of change of the resistance in the new resistor element, with respect to one of the resistances R_i, while holding the other resistances constant?

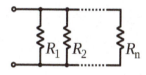

Figure 2.1.5: Resistors in parallel.

Solution:

This isn't too bad. You use the Power Rule and the Chain Rule:

$$\frac{\partial R_{\text{new}}}{\partial R_i} = -\left(R_1^{-1} + R_2^{-1} + \cdots + R_n^{-1}\right)^{-2}\left(-R_i^{-2}\right) = \frac{R_{\text{new}}^2}{R_i^2} = \left(\frac{R_{\text{new}}}{R_i}\right)^2.$$

It is slightly easier, and more elegant, to differentiate implicitly, without ever solving algebraically for R_{new}. That is, you simply take

$$R_{\text{new}}^{-1} = R_1^{-1} + R_2^{-1} + \cdots + R_n^{-1},$$

and take partial derivatives of both sides with respect to R_i, keeping in mind that all of the other **independent** variables are constant as far as R_i is concerned, but the **dependent** variable R_{new} is **not** constant.

We find

$$-R_{\text{new}}^{-2} \frac{\partial R_{\text{new}}}{\partial R_i} = -R_i^{-2},$$

and so recover quickly the result that

$$\frac{\partial R_{\text{new}}}{\partial R_i} = \left(\frac{R_{\text{new}}}{R_i}\right)^2.$$

You can, of course, take second partial derivatives of functions of more than two variables, and you can take partial derivatives of higher order than second order. Assuming existence and continuity of all of the partial derivatives, the order in which you take the partial derivatives doesn't matter; all the matters is how many times you take the partial derivative with respect to each variable.

Definition 2.1.22. *The total number of times that you take individual partial derivatives is called the* **order** *of the resulting partial derivative.*

Before we state the relevant theorem, we'll make another definition.

Definition 2.1.23. *Suppose that $f(x_1, x_2, \ldots, x_n)$ is a real-valued function such that f and all of its partial derivatives of order less than or equal to r are defined and continuous on an open subset $\mathcal{U}$ of $\mathbb{R}^n$. Then, we say that f is r-times* **continuously differentiable** *or, simply,* **of class C^r on $\mathcal{U}$.**

Be aware that some sources use the term *smooth* to mean C^1.

A continuous function is frequently said to be **of class C^0**, *and a function of class C^1 is usually referred to as* **continuously differentiable**. *If f is of class C^r for all r, then we say that f is* **infinitely differentiable**, *or* **of class C^∞**, *or is* **smooth**.

Remark 2.1.24. Having a function be continuously differentiable, i.e., of class C^1, is typically the weakest requirement that we'll need for our functions; being smooth, i.e., of class C^∞, will typically be our strongest requirement. Almost all of our examples will use C^∞ functions.

Occasionally, we'll mention a property that holds for differentiable functions, where we won't even need *continuously* differentiable, but that will be rare. For that matter, there will be a few places where we mention that some property that doesn't hold for every smooth function, but holds for real analytic functions (see Section 2.13).

The term "smooth" for a function, instead of for a geometric object, may seem a bit strange. You might hope that a function has a smooth-looking graph if and only if the function is smooth. However, it is unclear how much differentiability you can actually **see** in a graph; certainly the graphs of C^∞ functions look smooth, but so do graphs of C^1 functions.

So, when we say a function is smooth, we mean that it's infinitely continuously differentiable; when we say that a graph looks smooth, we may be assuming that the function involved is only C^1, and not even C^2. Note that we wrote "looks" smooth in the previous sentence. If we write that a graph "**is** smooth", then we'll mean that the function whose graph we're looking at is smooth.

Try not to worry about these fine distinctions too much; essentially all of the functions that we consider will be C^∞.

Now it is easy for us to state an important theorem on mixed partial derivatives.

Theorem 2.1.25. *Suppose that $f(x_1, x_2, \ldots, x_n)$ is of class C^r on an open subset $\mathcal{U}$ of $\mathbb{R}^n$. Then, at each point $\mathbf{p}$ in $\mathcal{U}$, every partial derivative of order less than, or equal to, r is independent of the order in which the individual partial derivatives are applied.*

For the proof, see Theorem 5.3.4 in Trench, [8].

Example 2.1.26. Suppose that

$$f(x, y, z) = x^5 y^6 z^7 + xe^{yz} + y^2 \sin(3x - 5z).$$

Then, Theorem 2.1.25 implies that

$$\frac{\partial^4 f}{\partial z \partial y \partial x^2} = \frac{\partial^4 f}{\partial z \partial x \partial y \partial x} = \frac{\partial^4 f}{\partial x^2 \partial y \partial z},$$

which are equal to every other 4th order partial derivative that's with respect to x twice, and y and z once each.

They're all equal to

$$840x^3 y^5 z^6 + 90y \cos(3x - 5x).$$

Suppose that we have a multi-component function

$$\mathbf{p}(u, v) = (x(u, v), \ y(u, v), \ z(u, v))$$

from $\mathbb{R}^2$ to $\mathbb{R}^3$. Then, we can take the partial derivatives of $\mathbf{p}$ with respect to u or v; these just yield the vectors of partial derivatives of the component functions, i.e.,

$$\frac{\partial \mathbf{p}}{\partial u} \;=\; \mathbf{p}_u \;=\; (x_u, y_u, z_u),$$

and

$$\frac{\partial \mathbf{p}}{\partial v} \;=\; \mathbf{p}_v \;=\; (x_v, y_v, z_v).$$

Of course, $\mathbf{p}$ could have been a multi-component function from any subset of any $\mathbb{R}^m$ to any subset of any $\mathbb{R}^n$, and we could have done the analogous thing.

In addition, all of our product rules involving multi-component functions from Section 1.6 remain true for partial derivatives.

Example 2.1.27. Let
$$\mathbf{f}(r, \theta) \;=\; (r\cos\theta, r\sin\theta)\,.$$

Then,
$$\mathbf{f}_r \;=\; (\cos\theta, \sin\theta)\,,$$
$$\mathbf{f}_\theta \;=\; (-r\sin\theta, r\cos\theta)\,,$$
$$\mathbf{f}_{rr} \;=\; (0,0)\,,$$
$$\mathbf{f}_{\theta\theta} \;=\; (-r\cos\theta, -r\sin\theta)\,,$$

and
$$\mathbf{f}_{r\theta} \;=\; \mathbf{f}_{\theta r} \;=\; (-\sin\theta, \cos\theta)\,.$$

Example 2.1.28. Suppose that
$$\mathbf{p}(u, v) \;=\; \left(u,\; -2v^3 + uv,\; 3v^4 - uv^2\right).$$

The image of this function is known as the *swallowtail*.

Calculate the cross product $\mathbf{p}_u \times \mathbf{p}_v$, and determine the (u, v) pairs for which the cross product equals $\mathbf{0}$.

Solution:

We quickly calculate
$$\mathbf{p}_u \;=\; (1,\; v,\; -v^2)$$

Figure 2.1.6: The image of $\mathbf{p}$ is the swallowtail.

and

$$\mathbf{p}_v = \big(0, \ -6v^2 + u, \ 12v^3 - 2uv\big) = (-6v^2 + u)\big(0, \ 1, \ -2v\big).$$

Therefore,

$$\mathbf{p}_u \times \mathbf{p}_v = (1, \ v, \ -v^2) \times (-6v^2 + u)\big(0, \ 1, \ -2v\big) =$$

$$(-6v^2 + u)\Big[(1, \ v, \ -v^2) \times \big(0, \ 1, \ -2v\big)\Big] =$$

$$(-6v^2 + u)\begin{vmatrix} \mathbf{i} & \mathbf{j} & \mathbf{k} \\ 1 & v & -v^2 \\ 0 & 1 & -2v \end{vmatrix} = (-6v^2 + u)(-v^2\,\mathbf{i} + 2v\,\mathbf{j} + \mathbf{k}).$$

Thus, the cross product is $\mathbf{0}$ if and only if $-6v^2 + u = 0$, i.e., $u = 6v^2$.

We shall see the relevance of this calculation later, in Section 2.8.

We end this section with a theorem and a definition that we will need later.

Hopefully, you remember from single-variable Calculus that, on an open interval, all of the anti-derivatives of a function look the same, except that they may differ by a constant. A similar result is true for multivariable functions; this easy, but important, result follows quickly from Theorem 2.1.10.

Theorem 2.1.29. *If $\mathcal{U}$ is a non-empty connected open subset of $\mathbb{R}^n$, and f and g are two functions on $\mathcal{U}$ such that all of the partial derivatives of f and g exist and are equal at each point in $\mathcal{U}$, then f and g differ by a constant on $\mathcal{U}$, i.e., there exists a constant C such that, for all $\mathbf{p}$ in $\mathcal{U}$, $f(\mathbf{p}) = g(\mathbf{p}) + C$.*

Proof. Simply apply Theorem 2.1.10 to the function $f - g$, in order to conclude that $f - g$ is constant on $\mathcal{U}$. The desired conclusion follows. $\qquad\square$

Finally, we end with the obvious generalization of the terminology in Definition 2.1.23.

Definition 2.1.30. *Suppose that $\mathcal{U}$ is an open subset of $\mathbb{R}^n$, F is a subset of $\mathbb{R}^k$, and that we have a multi-component function $\mathbf{f} : \mathcal{U} \to F$, where*

$$\mathbf{f}(x_1, \ldots, x_n) = \big(f_1(x_1, \ldots, x_n), \ f_2(x_1, \ldots, x_n), \ \ldots, \ f_k(x_1, \ldots, x_n)\big).$$

> *Then, we say that* $\mathbf{f}$ *is of class* C^r *if and only if each component function is of class* C^r. *In particular,* $\mathbf{f}$ *is of class* C^1 *or* **continuously differentiable** *if and only if each component function is continuously differentiable, and* $\mathbf{f}$ *is* C^∞ *or* **smooth** *if and only if each component function is smooth.*

Online answers to select exercises are here.

2.1.1   Exercises

Basics:

In each of Exercises 1 through 12, find the first-order partial derivatives of the given function with respect to each of the function's variables.

1. $f(x, y) = x^2 y + 2xy + xy^2$

2. $W(p, n) = (pn + 2)^2 - \ln 5$

3. $g(s, t) = s^{2t} - t^{2s}$

4. $z(m, k) = \dfrac{e^{\cos^2(m-k)}}{e^{-\sin^2(m-k)}}$

5. $J(r, \theta) = r \cos \theta - \sin^{-1} r$

6. $s(\alpha, \beta) = 40 - \ln(2\alpha\beta) - \ln(3\alpha) + \ln(5\beta)$

7. $h(x, t) = e^t \sqrt{x^2 - 4t} + 5$

8. $b(\mu, \sigma) = \tan^{-1}(\mu^2 - \sigma) \cdot \log_{10}(\sigma^2 - \mu)$

9. $L(y, \omega, s) = y^5 + y^4\omega^2 s + y^3\omega^4 s^2$

10. $m(a, b, c) = c^{a\sqrt{b}}$

11. $K(x, y, z) = 3x(y + z) - 12z^2(x - y^2) + 4x^3 y^3 z^3$

12. $y(u, v, t) = t^2(u - 5v)^{-t/2}$

13. In Figure 2.1.7, you are given a table of values of a function $f(x, y)$, where, as usual, the x values are listed horizontally, and the y values are given vertically. Assume that the partial derivatives f_x and f_y exist everywhere. Estimate the values of $f_x(-0.8, 1.2)$ and $f_y(-0.8, 1.2)$. Explain why your estimates are reasonable, but also explain why it's possible that your estimates are far from the true values of the partial derivatives.

x y	-1	-0.8	-0.6	-0.4	-0.2	0
2	4.000	3.280	2.720	2.320	2.080	2.000
1.6	3.200	2.624	2.176	1.856	1.664	1.600
1.2	2.400	1.968	1.632	1.392	1.248	1.200
0.8	1.600	1.312	1.088	0.928	0.832	0.800
0.4	0.800	0.656	0.544	0.464	0.416	0.400
0	0.000	0.000	0.000	0.000	0.000	0.000

Figure 2.1.7: The values of $f(x, y)$.

14. In Figure 2.1.8, you are given a table of values of a function $g(x, y)$, where, as usual, the x values are listed horizontally, and the y values are given vertically. Assume that the partial derivatives g_x and g_y exist everywhere. Estimate the values of $g_x(1.8, -1.3)$ and $g_y(1.8, -1.3)$. Explain why your estimates are reasonable, but also explain why it's possible that your estimates are far from the true values of the partial derivatives.

x y	1	1.4	1.8	2.2	2.6	3
-1	-2.000	-2.960	-4.240	-5.840	-7.760	-10.000
-1.1	-2.200	-3.256	-4.664	-6.424	-8.536	-11.000
-1.2	-2.400	-3.552	-5.088	-7.008	-9.312	-12.000
-1.3	-2.600	-3.848	-5.512	-7.592	-10.088	-13.000
-1.4	-2.800	-4.144	-5.936	-8.176	-10.864	-14.000
-1.5	-3.000	-4.440	-6.360	-8.760	-11.640	-15.000

Figure 2.1.8: The values $g(x, y)$.

In Exercises 15 through 21, find the indicated partial derivative at the given point.

15. $f(x, y) = 2(x + y)^2$; $f_x(2, 2)$

16. $g(x, y) = xe^y$; $g_y(3, 1)$

17. $H(u, v) = u^2 v^3 + \sqrt{u + v + 2}$; $H_u(0, 0)$

18. $y(s, \theta) = s \cos(2\theta) + s^2 \sin(\theta)$; $y_\theta(2, -\pi/2)$

19. $T(x_1, x_2) = x_1^2 \ln(x_1 + x_2)$; $T_{x_1}(1, 0)$

20. $m(\delta, y) = (\sec y)^\delta$; $m_y(2,0)$

21. $w(x,t) = x^3 + x^2 t - x^t$; $w_t(4, -4)$

In Exercises 22 through 27, determine the gradient vector function.

22. $q(x,y) = x^3 y - 4x^2 y^2 + xy^3$

23. $r(x,t) = \dfrac{\sqrt{x+3}}{t}$

24. $s(u,v) = v^3 + v^{-2u}$

25. $t(x,y,z) = \dfrac{x^4 y}{4} + \dfrac{x^3 z}{3} + \dfrac{x^2 yz}{2}$

26. $u(r,\theta,\phi) = r\cos^2\theta + r\sin^2\phi$

27. $v(z,s,t) = z\ln s - z^2 \ln t$

In Exercises 28 through 32, calculate the mixed partial derivative, once with respect to each of the function's variables. Note that the order in which you differentiate may make some exercises significantly easier or harder.

28. $f(u,v) = (5u^2 v - 10uv)^2 + 4$

29. $g(x,y) = x^2 y - x + \dfrac{\csc^{-1} y}{\sqrt[5]{\pi y^3 - 1}}$

30. $h(s,t) = \sqrt{s+t}$

31. $P(x,y,z) = 3x^2 yz - 4xy^2 + xz^3$

32. $W(r,\theta,\omega) = r\cos(\theta + \pi\omega)$

33. Suppose that a school of 500 fish is caught in a shark feeding frenzy. If there are s sharks attacking the school, the number of living fish after t minutes can be modeled by $f(s,t) = 400e^{-t} - st + 100$ for $s > 0$ and $t > 0$.

   a. Assuming the number of sharks is constant throughout the frenzy, what is the rate of change of the number of living fish with respect to time?

   b. What is the smallest number of sharks that would be able to devour all of the fish in under 5 minutes?

   c. With this many sharks participating in the feeding frenzy, how quickly are the fish being consumed 1 minute after the frenzy begins?

34. The Ideal Gas Law can be written in the form

$$PV = nRT,$$

where P is the pressure that an ideal gas exerts on the container in pascals, V is the volume of the container in cubic meters, n is the number of moles of the substance present, T is the temperature of the gas in kelvins, and R is a constant approximately equal to 8.314 J/(K · mol).

> A *mole* is approximately 6.022×10^{23} fundamental particles (atoms or molecules) of the gas.

If the pressure is held constant at 2000 pascals and the volume is held constant at 10 cubic meters, what is the instantaneous rate of change of the temperature of the gas, with respect to the number of moles of the substance, when the amount of substance is 4 moles?

35. Suppose that electric charge permeates space in such a way that, at the point (x, y, z), in meters, the charge Q, in coulombs, is given by

$$Q = xy \sin z + yz \cos z + \ln(1 + x^2 + y^2 + z^2).$$

Determine the instantaneous rates of change of the charge, with respect to x, y, and z, at a general point (x, y, z), and also calculate these rates of change at the specific point $(1, 1, 0)$.

36. According to Newton's Law of Universal Gravitation , the magnitude F of the force, in Newtons, of the gravitational attraction between two point-masses M and m, in kilograms, is given by

$$F = \frac{GMm}{r^2},$$

where r is the distance between the point-masses, in meters, and G is the universal gravitational constant, which is approximately 6.67384×10^{-11} m^3/(kg·s^2).

   (a) Calculate $\dfrac{\partial F}{\partial r}$ and $\dfrac{\partial F}{\partial m}$.
   (b) Describe the physical meaning of the partial derivatives which you calculated in part (a).

More Depth:

37. Suppose that you are a park-attendance analyst at Dizzy World, a popular theme park. You and your colleagues have devised the following function to model park attendance throughout the calendar year:

$$D = 1000(100 - p) + 25000 \sin\left(\frac{\pi t}{6}\right)$$

where D is the park attendance in a given month, p is the price of admission, and t is the number of months since January.

a. Assuming that the price of admission is held constant throughout the year, determine a model for the rate of change of the park attendance with respect to time.

b. If the admission price is fixed at $40 per person, what is the rate of change of the park attendance in August?

38. Suppose that Galactic Studios, one of Dizzy World's nearby competitors, announces that, next January, they will be opening a series of new attractions based on a popular movie franchise, *Harvey Putnam and the Wizards of Wargleton.* Your colleague Bob predicts that the popularity of these attractions will inevitably take business away from Dizzy World, but Steve predicts that, in the long run, the influx of tourists will increase attendance at both theme parks. The three of you have modified the existing attendance model as follows:

$$D = 1000(100 - p) + 25000 \sin\left(\frac{\pi t}{6}\right) + \frac{G}{100}\left(q - 20e^{-t/2} - p\right)$$

where D is the attendance at Dizzy World in a given month, p is the price of admission at Dizzy World, G is the park attendance at Galactic Studios, q is the price of admission at Galactic Studios, and t is the number of months since January.

a. If both parks keep their prices fixed at $40, whose prediction is supported by the model?

b. If Dizzy World keeps the price fixed at $40, but Galactic Studios increases their admission price to $50, whose prediction is supported by the model?

b. If Dizzy World keeps the price fixed at $40, but Galactic Studios increases their admission price to $60, whose prediction is supported by the model?

In each of the following exercises, calculate the dot product of the gradient vector of the given function and the given vector.

39. $f(x, y) = x^2 y - 2xy - 5$; $\vec{\nabla} f(2, 4) \cdot (2, 5)$

40. $K(\alpha, \beta) = \alpha \ln(\alpha + \beta)$; $\vec{\nabla} K(1, 1) \cdot (2, 5)$

41. $h(z, t) = z\sqrt{t} + 6z$; $\vec{\nabla} h(0, 3) \cdot (2, 1)$

42. $m(x, y, s) = xy \sin(s) - y \sin(2s)$; $\vec{\nabla} m(1, 2, \pi/4) \cdot (1, 1, \pi/2)$

In the following, determine the given partial derivative of the given multi-component function.

43. $\mathbf{f}(x,y) = \left(x^2y^2, x^2 + y^2, (x+y)^2\right); \mathbf{f}_x$

44. $\mathbf{g}(u,w) = \left(ue^w, u^2e^{-w}, \sqrt{e^w}\right); \mathbf{g}_w$

45. $\mathbf{h}(s,t) = \left(6\ln(st), 3\ln^2(st), \dfrac{s+t}{s-t}\right); \mathbf{h}_{ss}$

46. $\lambda(\kappa, \tau) = \left(e^{\kappa\tau} - 5\kappa\tau, 2\kappa^3 - \tau^3, \dfrac{\kappa}{\tau}\right); \lambda_{\kappa\tau}$

47. $\mathbf{M}(r, \theta, \phi) = \left(r^2 - 1, r\sin\theta\cos\phi, \tan^2(\theta + \phi)\right); \mathbf{M}_{\theta\phi}$

48. $\mathbf{j}(x,y,t) = \left(\dfrac{1}{2}x^4y\sqrt{t}, \dfrac{1}{3}x^6\sqrt{y+t}\right); \mathbf{j}_{xyt}$

49. Consider the swallowtail function $\mathbf{p}(u,v) = \left(u, -2v^3 + uv, 3v^4 - uv^2\right)$, from Example 2.1.28.

    a. Suppose that $\mathbf{p}(u,v) = \mathbf{p}(\hat{u}, \hat{v})$. Show that $u = \hat{u}$ and $v^2 = \hat{v}^2$. Conclude that either

    - $u = \hat{u}$ and $v = \hat{v}$, or
    - $u = \hat{u}$, $v = -\hat{v} \neq 0$, and $u = 2v^2 = 2\hat{v}^2 = \hat{u}$.

    b. Show that the image of the curve $u = 2v^2$ is the curve given by $z = x^2/4$ and $y = 0$, i.e., a parabola in the xz-plane.

    This parabola is the curve along which the swallowtail crosses itself, and is clearly visible in Figure 2.1.6.

50. Why can't there be a real-valued function $f = f(x,y)$, whose domain is $\mathbb{R}^2$, such that, for all x and y,

$$\frac{\partial f}{\partial x} = x^3 + y^2 \quad \text{and} \quad \frac{\partial f}{\partial y} = x - y \, ?$$

Hint: Consider Theorem 2.1.14.

51. Suppose that a and b are constants, and $f = f(x,y)$ is a function on $\mathbb{R}^2$ such that $f_x(x,y) = a$ and $f_y(x,y) = b$. Find such an f, and then use Theorem 2.1.29 to determine what every such possible function f looks like.

2.2   The Total Derivative

In the previous section, we discussed partial derivatives, which represent the instantaneous rates of change of a function, f, with respect to a single variable, while keeping all of the other independent variables constant.

You can think of each partial derivative as the instantaneous rate of change of f, at a point $\mathbf{p}$, as the point "moves" in a direction parallel to the corresponding coordinate axis. Another way to say this is that the partial derivative, with respect to x_i, is the instantaneous rate of change of f, at a point $\mathbf{p}$, as the point moves in the direction of the corresponding standard basis vector, $\mathbf{e}_i$ (recall Section 1.2).

This naturally leads us to look at the instantaneous rates of change of f, at a point $\mathbf{p}$, as the point moves in an arbitrary direction, with an arbitrary speed, i.e., as the point moves with an arbitrary velocity $\mathbf{v}$. Thus, we define the *total derivative* of f, at $\mathbf{p}$, not as a number, but rather as a function which returns a number for each specified velocity vector.

Basics:

If we have a real-valued function $f = f(x_1, \ldots, x_n)$, then, as we saw in Definition 2.1.20, one way to write the definition of the partial derivative with respect to x_i, at a point $\mathbf{p}$, is

$$\frac{\partial f}{\partial x_i}\bigg|_{\mathbf{p}} = \lim_{h \to 0} \frac{f(\mathbf{p} + h\mathbf{e}_i) - f(\mathbf{p})}{h}, \tag{2.2}$$

where $\mathbf{e}_i$ denotes the i-th standard basis element, the vector that's all zeroes, except for a 1 in the i-th component. Written in this form, it is natural to think of the partial derivative $\partial f / \partial x_i$ as the instantaneous rate of change of f with respect to the vector $\mathbf{e}_i$ or in the direction of $\mathbf{e}_i$.

Can we do the same thing with vectors other than the standard basis vectors, that is, can we look at the instantaneous rate of change of f, at $\mathbf{p}$, when we move with arbitrary velocity $\mathbf{v}$? The answer is: "yes".

We will define $d_{\mathbf{p}}f(\mathbf{v})$, the *total derivative of f* at a point $\mathbf{p}$, with respect to an arbitrary vector $\mathbf{v}$. This means that the total derivative of f at $\mathbf{p}$ is actually a **function** from $\mathbb{R}^n$ to $\mathbb{R}$, denoted by $d_{\mathbf{p}}f$. As with single-variable derivatives, the existence of the derivative depends on the existence of a certain limit, which defines when a function is *differentiable at* $\mathbf{p}$.

The total derivative is also referred to as the *differential*. We shall reserve this terminology for the context of approximations or when a specific point $\mathbf{p}$ is not given, as in Section 2.3.

Even though the definitions of the total derivative and differentiability are somewhat technical, and we would like to put them off, we can't really proceed without them. So we will discuss these things first.

Suppose that we have a function $f : E \to \mathbb{R}$, where E is a subset of $\mathbb{R}^n$, and let $\mathbf{p}$ be a point in E. Suppose further that E is a neighborhood of $\mathbf{p}$, i.e., E completely contains an open ball around $\mathbf{p}$ (recall Definition 1.1.4). The easiest example of such a neighborhood would be when E is all of $\mathbb{R}^n$.

The fact that E contains an open ball around $\mathbf{p}$ means precisely that, if a vector $\mathbf{h}$ has sufficiently small magnitude, then $\mathbf{p} + \mathbf{h}$ is in the domain of f. Thus, $f(\mathbf{p} + \mathbf{h})$ is defined for $|\mathbf{h}|$ sufficiently small and, hence, so is the difference $f(\mathbf{p} + \mathbf{h}) - f(\mathbf{p})$.

> Technically, the statement is that, there exists $\epsilon > 0$ (the radius of the open ball) such that, for all vectors $\mathbf{h}$ such that $|\mathbf{h}| < \epsilon$, $\mathbf{p} + \mathbf{h}$ is in E.

Now, you might suspect that we should use a "difference quotient" to define the total derivative, but how do we do this? Dividing by the vector $\mathbf{h}$ is undefined, and dividing by $|\mathbf{h}|$ wouldn't give us the right thing even for functions of a single variable.

Let's revisit the definitions of differentiability and the derivative for a function $f = f(x)$ of a single variable, and try to write things in a way which generalize nicely to the multivariable case.

Suppose that we have a single-variable function $f = f(x)$ and that p is a real number in the domain of f. Then, f is *differentiable* at p if and only if the limit

$$\lim_{h \to 0} \frac{f(p+h) - f(p)}{h}$$

exists; when f is differentiable at p, we call the value of the limit above the *derivative of f at p*.

This means that we say that f is differentiable at p if and only if there exists a number m such that

$$\lim_{h \to 0} \frac{f(p+h) - f(p)}{h} = m.$$

Another way of writing this is that there exists a number m such that

$$\lim_{h \to 0} \frac{f(p+h) - f(p) - mh}{h} = 0.$$

But, the absolute value of 0 is 0, and $h \to 0$ if and only if $|h| \to 0$; so we can say, instead, that f is differentiable at p if and only if there exists a number m such that

$$\lim_{|h| \to 0} \frac{|f(p+h) - f(p) - mh|}{|h|} = 0.$$

Note that the existence of this limit implies, in particular, that f must be defined on a neighborhood of p.

The limit definition above generalizes easily to the multivariable case:

Definition 2.2.1. (Differentiability and the Total Derivative)
Suppose that E is a subset of $\mathbb{R}^n$, that $\mathbf{p}$ is a point in E, and that we have a function $f : E \to \mathbb{R}$.

*Then, f is **differentiable** at $\mathbf{p}$ if and only if there exists a vector $\mathbf{m} = (m_1, \ldots, m_n)$ such that*

$$\lim_{|\mathbf{h}| \to 0} \frac{|f(\mathbf{p} + \mathbf{h}) - f(\mathbf{p}) - \mathbf{m} \cdot \mathbf{h}|}{|\mathbf{h}|} = 0.$$

*If such a vector $\mathbf{m}$ exists, it is unique and we define the **total derivative of** f **at** $\mathbf{p}$ to be the linear function $d_{\mathbf{p}} f : \mathbb{R}^n \to \mathbb{R}$ given by*

$$d_{\mathbf{p}} f(\mathbf{v}) = \mathbf{m} \cdot \mathbf{v} = m_1 v_1 + m_2 v_2 + \cdots + m_n v_n.$$

In order to actually give some examples of using the total derivative, we first need to prove the following relatively easy result.

Theorem 2.2.2. *Suppose that f is a function from a subset of $\mathbb{R}^n$ into $\mathbb{R}$, and that f is differentiable at $\mathbf{p}$.*

Then, all of the partial derivatives of f at $\mathbf{p}$ exist and, for all vectors $\mathbf{v}$ in $\mathbb{R}^n$, the value of $d_{\mathbf{p}} f$ at $\mathbf{v}$ is

$$d_{\mathbf{p}} f(\mathbf{v}) = \lim_{t \to 0} \frac{f(\mathbf{p} + t\mathbf{v}) - f(\mathbf{p})}{t} = \vec{\nabla} f(\mathbf{p}) \cdot \mathbf{v}.$$

In particular, the vector $\mathbf{m}$ in Definition 2.2.1 is equal to $\vec{\nabla} f(\mathbf{p})$.

You may be asking yourself "Why didn't we just define differentiability and the total derivative by the existence and value of the limit in Theorem 2.2.2?". The answer is that we need to consider $f(\mathbf{p} + \mathbf{h}) - f(\mathbf{p})$, where $\mathbf{h}$ approaches $\mathbf{0}$ **in any possible way**, not just by being small scalar multiples of $\mathbf{v}$. This is important for a number of results, including the *Chain Rule*, Theorem 2.4.9.

Proof. If $\mathbf{v} = \mathbf{0}$, the equalities are trivial. So assume that $\mathbf{v} \neq \mathbf{0}$.

Now, since f is differentiable at $\mathbf{p}$, the limit being 0 in Definition 2.2.1 must hold when $\mathbf{h}$ is replaced by $t\mathbf{v}$ as $t \to 0$; that is, there must exist $\mathbf{m}$ in $\mathbb{R}^n$ such that

$$\lim_{t \to 0} \frac{|f(\mathbf{p} + t\mathbf{v}) - f(\mathbf{p}) - \mathbf{m} \cdot t\mathbf{v}|}{|t\mathbf{v}|} = 0.$$

As $\mathbf{v}$ is a constant, we may remove $|\mathbf{v}|$ from the denominator, and also move the magnitude symbols, to obtain

$$\lim_{t \to 0} \left| \frac{f(\mathbf{p} + t\mathbf{v}) - f(\mathbf{p}) - \mathbf{m} \cdot t\mathbf{v}}{t} \right| = \lim_{t \to 0} \left| \frac{f(\mathbf{p} + t\mathbf{v}) - f(\mathbf{p})}{t} - \mathbf{m} \cdot \mathbf{v} \right| = 0.$$

This is equivalent to

$$d_{\mathbf{p}}f(\mathbf{v}) \;=\; \mathbf{m} \cdot \mathbf{v} \;=\; \lim_{t \to 0} \frac{f(\mathbf{p} + t\mathbf{v}) - f(\mathbf{p})}{t}, \tag{2.3}$$

as we wanted.

It remains for us to show that $\mathbf{m} = \vec{\nabla}f(\mathbf{p})$, i.e., for $1 \leq i \leq n$, the i-th component of $\mathbf{m}$, m_i, is equal to the partial derivative $f_{x_i}(\mathbf{p})$. But this is easy; letting the vector $\mathbf{v}$ be $\mathbf{e}_i$ in Formula 2.3, we find

$$m_i \;=\; \mathbf{m} \cdot \mathbf{e}_i \;=\; \lim_{t \to 0} \frac{f(\mathbf{p} + t\mathbf{e}_i) - f(\mathbf{p})}{t} \;=\; f_{x_i}(\mathbf{p}).$$

$\square$

Example 2.2.3. Suppose that $f = f(x, y, z)$ is differentiable at $\mathbf{p} = (3, -1, 2)$, and that $\vec{\nabla}f(\mathbf{p}) = (-5, 7, 11)$. What is the total derivative $d_{\mathbf{p}}f(v_1, v_2, v_3)$?

Solution:

The total derivative is the linear function $d_{\mathbf{p}}f : \mathbb{R}^3 \to \mathbb{R}$ given by

$$d_{\mathbf{p}}f(v_1, v_2, v_3) \;=\; \vec{\nabla}f(\mathbf{p}) \cdot (v_1, v_2, v_3) \;=\; (-5, 7, 11) \cdot (v_1, v_2, v_3) \;=\; -5v_1 + 7v_2 + 11v_3.$$

Remark 2.2.4. According to Theorem 2.2.2, if f is differentiable at $\mathbf{p}$, then the total derivative, $d_{\mathbf{p}}f$, is equal to

$$\lim_{t \to 0} \frac{f(\mathbf{p} + t\mathbf{v}) - f(\mathbf{p})}{t}.$$

If you think of t as representing time, then $\mathbf{p} + t\mathbf{v}$ is the position, at time t, of an object moving with constant velocity $\mathbf{v}$, which is at $\mathbf{p}$ at time 0; we saw this in Example 1.6.3. Without referring to an object, we frequently just talk about $\mathbf{p}$ itself moving.

This means that, intuitively/physically, you should think of $d_{\mathbf{p}}f(\mathbf{v})$ as the instantaneous rate of change of f, at the point $\mathbf{p}$, as $\mathbf{p}$ moves with "velocity" $\mathbf{v}$. We also say "the instantaneous rate of change of f, at the point $\mathbf{p}$, with respect to $\mathbf{v}$".

We shall make the "instantaneous rate of change" interpretation even more precise in Section 2.4.

Example 2.2.5. Since we used partial derivatives to motivate looking at the total derivative, it would be nice to check that the formula in Theorem 2.2.2 yields what we expect.

Assuming that f is differentiable at $\mathbf{p}$, we find easily that

$$d_{\mathbf{p}}f(\mathbf{e}_i) \;=\; \vec{\nabla}f(\mathbf{p}) \cdot \mathbf{e}_i \;=\; \left.\frac{\partial f}{\partial x_i}\right|_{\mathbf{p}},$$

i.e., the instantaneous rate of change of f at $\mathbf{p}$, with respect to the velocity vector $\mathbf{e}_i$ is precisely the partial derivative of f, with respect to x_i, at $\mathbf{p}$.

Before we give any specific examples, we need to deal with the question of how you know when a given function is differentiable, **without** having to prove that some constants exist which make a certain limit equal zero. The following theorem gives the usual criterion that is used to establish differentiability, i.e., to establish the existence of $d_{\mathbf{p}}f$.

For the proof, see Theorem 5.3.10 in Trench, [8].

Recalling Definition 2.1.23, the assumption in Theorem 2.2.6 is that f is of class C^1 on an open neighborhood of $\mathbf{p}$. As we stated earlier, this condition is usually referred to as being *continuously differentiable*. This terminology is used because there is a stronger version of Theorem 2.2.6, which states that the continuity of the partial derivatives of f on an open set $\mathcal{U}$ is equivalent to f being differentiable at each point of $\mathcal{U}$ and for the total derivative $d_{\mathbf{p}}f$ to be a continuous function of $\mathbf{p}$. See [7], Theorem 9.21.

> **Theorem 2.2.6.** *Suppose that all of the partial derivatives of f exist and are continuous on an open neighborhood of $\mathbf{p}$. Then, f is differentiable at $\mathbf{p}$.*

As we typically deal with elementary functions (recall Definition 1.7.7 and Theorem 1.7.8), the question of continuity usually boils down to a question of existence for us, and so, given an elementary function, we proceed with examples and calculations of total derivatives, using the gradient vector formula in Theorem 2.2.2, unless one of the partial derivatives in the gradient vector fails to exist.

Example 2.2.7. Suppose that $f(x,y) = x\cos(\pi y) + y^2 e^x$. Determine the linear function $d_{(-1,1)}f(v_1, v_2)$, and calculate the values of the total derivative of f, at the point $(-1,1)$, with respect to the vectors $(4,-2)$ and $(-3,5)$, i.e., calculate $d_{(-1,1)}f(4,-2)$ and $d_{(-1,1)}f(-3,5)$.

Solution:

Let $\mathbf{v} = (v_1, v_2)$ be a vector in $\mathbb{R}^2$. Then, $d_{(-1,1)}f(\mathbf{v}) = \vec{\nabla}f(-1,1) \cdot \mathbf{v}$; so, we'll first calculate $\vec{\nabla}f(-1,1)$.

We find

$$\vec{\nabla} f \;=\; \left(\frac{\partial f}{\partial x}, \frac{\partial f}{\partial y}\right) \;=\; \left(\cos(\pi y) + y^2 e^x, -\pi x \sin(\pi y) + 2y e^x\right),$$

and so

$$\vec{\nabla} f(-1,1) \;=\; \left(\cos(\pi \cdot 1) + 1^2 e^{-1}, -\pi(-1)\sin(\pi \cdot 1) + 2 \cdot 1 \cdot e^{-1}\right) \;=\; (-1 + e^{-1}, 2e^{-1}).$$

Therefore,

$$d_{(-1,1)} f(\mathbf{v}) \;=\; \vec{\nabla} f(-1,1) \cdot \mathbf{v} \;=\; (-1 + e^{-1}, 2e^{-1}) \cdot (v_1, v_2) \;=\; (-1 + e^{-1})v_1 + 2e^{-1}v_2.$$

Hence,

$$d_{(-1,1)} f(4,-2) \;=\; (-1 + e^{-1}) \cdot 4 \;+\; 2e^{-1} \cdot (-2) \;=\; -4,$$

and

$$d_{(-1,1)} f(-3,5) \;=\; (-1 + e^{-1}) \cdot (-3) \;+\; 2e^{-1} \cdot 5 \;=\; 3 + 7e^{-1}.$$

Example 2.2.8. Suppose that space is being heated in such a way that the temperature, T, in degrees Celsius, is given by $T(x,y,z) = z^2 + 0.5e^{x^2 - y^2}$, where x, y, and z are measured in meters and are all strictly between -3 and 3. If a particle is at the point $(0,1,2)$, and is moving with velocity $(3,-1,0)$ m/s, what instantaneous rate of change in the surrounding temperature does the particle experience?

Solution:

The instantaneous rate of change that we wish to calculate is precisely $d_{(0,1,2)} T(3,-1,0)$. We first calculate

$$\vec{\nabla} T(0,1,2) \;=\; \left(xe^{x^2 - y^2}, -ye^{x^2 - y^2}, 2z\right)\Big|_{(0,1,2)} \;=\; (0, -e^{-1}, 4) \ \ ^\circ\text{C/m}.$$

Thus,

$$d_{(0,1,2)} T(3,-1,0) \;=\; (0, -e^{-1}, 4) \cdot (3,-1,0) \;=\; e^{-1} \ \ ^\circ\text{C/s}.$$

Example 2.2.9. The "velocity" vector in the total derivative doesn't really have to be the *velocity* of some moving object; it needs to be a vector representing the direction and magnitude of the instantaneous rate of change.

Suppose that a company produces x kilograms of chemical A and y kilograms of chemical B, and has determined that its cost, in dollars, of producing these quantities is given by

$$C(x, y) = 5x^2 + 2xy + y^2.$$

If the company is currently producing 10 kilograms of chemical A and 8 kilograms of chemical B, but is changing their production by -1 kilograms per day of chemical A and 2 kilograms per day of chemical B, at what rate is their cost of production changing?

Solution:

You are being asked to determine $d_{(10,8)}C(-1, 2)$, which will have units of dollars per day.

We first find that

$$\vec{\nabla}C(10, 8) = (10x + 2y, 2x + 2y)\Big|_{(10,8)} = (116, 36) \ \$/\text{kg}.$$

Now we find

$$d_{(10,8)}C(-1, 2) = (116, 36) \cdot (-1, 2) = -116 + 72 = -44 \ \$/\text{day};$$

hence, the cost of production is dropping at an instantaneous rate of \$ 44 per day.

It is common to denote the i-th coordinate function simply by the lower-case x_i, instead of X_i. This causes no confusion when discussing the total derivative because, if you see something like $d_{\mathbf{p}}x_i$, it is clear that x_i is being considered as a function and not some particular number.

Example 2.2.10. Recall the coordinate functions from Example 1.7.3 of Section 1.7; for $1 \leq i \leq n$, let $X_i : \mathbb{R}^n \to \mathbb{R}$ be given by $X_i(x_1, \ldots, x_n) = x_i$.

Suppose that $\mathbf{p} = (p_1, \ldots, p_n)$ and $\mathbf{v} = (v_1, \ldots, v_n)$. Then, what is $d_{\mathbf{p}}X_i(\mathbf{v})$?

We find that, regardless of the point at which we calculate the gradient vector, $\vec{\nabla}X_i = (0, \ldots, 0, 1, 0, \ldots 0)$, where the 1 is in the i-th coordinate position.

Hence,

$$d_{\mathbf{p}}X_i(\mathbf{v}) = (0, \ldots, 0, 1, 0, \ldots 0) \cdot (v_1, \ldots, v_n) = v_i.$$

Note that, as functions from $\mathbb{R}^n$ to $\mathbb{R}$, X_i and $d_{\mathbf{p}}X_i$ **are the same function**; they both simply return the i-th coordinate of what they're given.

However, it is frequently a mistake to think of X_i and $d_{\mathbf{p}}X_i$ as really being the same, for the domain of X_i is the set of points in $\mathbb{R}^n$ and the domain of $d_{\mathbf{p}}X_i$ should be thought of as velocity or tangent vectors, based at $\mathbf{p}$.

Example 2.2.11. Suppose that f is a function of a single variable, and that we have points p and v in $\mathbb{R}$. What's the relationship between the total derivative $d_p f(v)$ and the ordinary derivative $f'(p)$?

This is so easy that it may seem a bit confusing. As f is a function of a single variable, its "partial" derivatives are just its single ordinary derivative, and so its gradient vector at p consists solely of $f'(p)$. Thus, $d_p f(v) = f'(p) \cdot v$.

This means that, for a function f of a single variable, $d_p f$ is not the number $f'(p)$, but rather the function given by multiplication by $f'(p)$. If you really want to get $f'(p)$, using $d_p f$, you have to evaluate at 1, that is

$$d_p f(1) = f'(p).$$

> In fact, many references refer to $d_p f$ as simply the *derivative of f at* **p**. One good reason **not** to do this is because, then, the derivative of a function of a single variable would have two different meanings.

Example 2.2.12. Consider now the multi-component function $\mathbf{f} : \mathbb{R}^2 \to \mathbb{R}^3$ given by

$$\mathbf{f}(x, y) = (f_1(x, y), f_2(x, y), f_3(x, y)) = (x^2 y, y \sin x, 5e^{xy}).$$

If $\mathbf{p}$ is a point in $\mathbb{R}^2$, what should/does the total derivative $d_{\mathbf{p}} \mathbf{f}$ mean?

It is defined as you might guess; you take the total derivative of each of the component functions. That is, $d_{\mathbf{p}} \mathbf{f}$ is the function from $\mathbb{R}^2$ to $\mathbb{R}^3$ given by

$$d_{\mathbf{p}} \mathbf{f}(\mathbf{v}) = \left(d_{\mathbf{p}} f_1(\mathbf{v}), d_{\mathbf{p}} f_2(\mathbf{v}), d_{\mathbf{p}} f_3(\mathbf{v}) \right).$$

> See Definition 2.2.17 for a more precise definition.

In this example, $f_1(x, y) = x^2 y$, $f_2(x, y) = y \sin x$, and $f_3(x, y) = 5e^{xy}$. If we let $\mathbf{p} = (\pi, 1)$, then we find that

$$\vec{\nabla} f_1(\pi, 1) = \left. \left(2xy, x^2 \right) \right|_{(\pi, 1)} = (2\pi, \pi^2) = \pi(2, \pi),$$

$$\vec{\nabla} f_2(\pi, 1) = \left. \left(y \cos x, \sin x \right) \right|_{(\pi, 1)} = (-1, 0),$$

and

$$\vec{\nabla} f_3(\pi, 1) = \left. \left(5ye^{xy}, 5xe^{xy} \right) \right|_{(\pi, 1)} = (5e^\pi, 5\pi e^\pi) = 5e^\pi(1, \pi).$$

Now, if we want to calculate $d_{(\pi, 1)} \mathbf{f}(3, 7)$, we need to take three dot products:

$$d_{(\pi, 1)} \mathbf{f}(3, 7) = \left(\vec{\nabla} f_1(\pi, 1) \cdot (3, 7),\ \vec{\nabla} f_2(\pi, 1) \cdot (3, 7),\ \vec{\nabla} f_3(\pi, 1) \cdot (3, 7) \right) =$$

$$\left(\pi(2, \pi) \cdot (3, 7),\ (-1, 0) \cdot (3, 7),\ 5e^\pi(1, \pi) \cdot (3, 7) \right) = \left(\pi(6 + 7\pi),\ -3,\ 5e^\pi(3 + 7\pi) \right).$$

More Depth:

Let's look at a slightly unusual type of problem.

Example 2.2.13. Suppose that $f = f(x, y)$ is differentiable at $\mathbf{p}$, and that $d_{\mathbf{p}}f(3, 2) = 7$ and $d_{\mathbf{p}}f(-1, 5) = -4$. Find the partial derivatives f_x and f_y at $\mathbf{p}$.

Solution:

Since $d_{\mathbf{p}}f(\mathbf{v}) = \overrightarrow{\nabla}f(\mathbf{p}) \cdot \mathbf{v}$, what we are given is that

$$\big(f_x(\mathbf{p}), f_y(\mathbf{p})\big) \cdot (3, 2) \;=\; 7$$

and

$$\big(f_x(\mathbf{p}), f_y(\mathbf{p})\big) \cdot (-1, 5) \;=\; -4.$$

Hence, we simply have to solve two equations with two unknowns:

$$3f_x(\mathbf{p}) + 2f_y(\mathbf{p}) \;=\; 7 \quad \text{and} \quad -1f_x(\mathbf{p}) + 5f_y(\mathbf{p}) \;=\; -4.$$

There are many ways to do this. We'll multiply the second equation by 3, and add the equations; this eliminates the reference to $f_x(\mathbf{p})$. We find $-3f_x(\mathbf{p}) + 15f_y(\mathbf{p}) = -12$, add the equations to obtain

$$17f_y(\mathbf{p}) \;=\; -5, \quad \text{and so} \quad f_y(\mathbf{p}) \;=\; -\frac{5}{17}.$$

We now use either original equation to determine that

$$f_x(\mathbf{p}) \;=\; \frac{43}{17}.$$

As in the case of a single-variable function, differentiability implies continuity.

For the proof, see Theorem 5.3.7 in Trench, [8].

Theorem 2.2.14. *If $f = f(x_1, \ldots, x_n)$ is differentiable at $\mathbf{p}$, then f is continuous at $\mathbf{p}$.*

Example 2.2.15. Of course, continuity does not imply differentiability; you should know that continuity doesn't imply differentiability even for single-variable functions.

A standard two-variable example is

$$f(x,y) \;=\; \sqrt{x^2 + y^2} \;=\; \left(x^2 + y^2\right)^{1/2},$$

which is an elementary function and, hence, is continuous.

Figure 2.2.1: The graph of $z = \sqrt{x^2 + y^2}$ has a sharp point at the origin.

We claim that f is differentiable everywhere, **except** at the origin; in fact, neither f_x nor f_y exists at $(0,0)$, and the lack of existence of either one implies that f is not differentiable at $(0,0)$.

We shall demonstrate that $f_x(0,0)$ does not exist; the proof that $f_y(0,0)$ fails to exist is essentially identical.

By definition,

$$f_x(0,0) \;=\; \lim_{h \to 0} \frac{f(h,0) - f(0,0)}{h} \;=\; \lim_{h \to 0} \frac{\sqrt{h^2} - 0}{h} \;=\; \lim_{h \to 0} \frac{|h|}{h},$$

and this limit does not exist, since $|h|/h$ approaches 1 as h approaches 0 from the right, but approaches -1 as h approaches 0 from the left.

To show that f_x and f_y exist at all points other than $(0,0)$, you simply use the rules of differentiation to obtain

$$f_x \;=\; \frac{x}{\sqrt{x^2 + y^2}} \qquad \text{and} \qquad f_y \;=\; \frac{y}{\sqrt{x^2 + y^2}},$$

which exist, provided that $x^2 + y^2 \neq 0$, i.e., provided that $(x,y) \neq (0,0)$.

We shall return to this example in Example 2.3.9, and look at it graphically, once we discuss the notion of the *tangent plane*.

Remark 2.2.16. Note that the converse of Theorem 2.2.6 is **not** true; it is, in fact, possible for a function to be differentiable at **p** without the partial derivatives being continuous at **p**.

An example of such a point and function are given, in Example 5.3.9 of Trench, [8], by the point $\mathbf{p} = (0,0)$ and the function

$$f(x,y) \;=\; \begin{cases} (x - y)^2 \sin\left(\dfrac{1}{x - y}\right), & \text{if } x \neq y; \\ 0, & \text{if } x = y. \end{cases}$$

You may either consult [8] or verify, as an exercise, that f is, in fact, differentiable at $(0,0)$, $d_{(0,0)}f$ is the function that's constantly zero, but neither f_x nor f_y is continuous at $(0,0)$, since neither $\lim_{(x,y)\to(0,0)} f_x(x,y)$ nor $\lim_{(x,y)\to(0,0)} f_y(x,y)$ exists.

We need to give a rigorous definition of differentiability and the total derivative of a multi-component function, but it really just amounts to what we stated earlier in Example 2.2.12.

Definition 2.2.17. *Suppose that E is a subset of $\mathbb{R}^n$, F is a subset of $\mathbb{R}^k$, and that we have a multi-component function $\mathbf{f} : E \to F$, where*

$$\mathbf{f}(x_1,\ldots,x_n) = \big(f_1(x_1,\ldots,x_n),\ f_2(x_1,\ldots,x_n),\ \ldots,\ f_k(x_1,\ldots,x_n)\big).$$

Then, $\mathbf{f}$ is differentiable at a point $\mathbf{p}$ in E if and only if each f_i is differentiable at $\mathbf{p}$ and, in that case, $d_{\mathbf{p}}\mathbf{f}$ is defined as the function from $\mathbb{R}^n$ to $\mathbb{R}^k$ given by

$$d_{\mathbf{p}}\mathbf{f}(\mathbf{v}) \ = \ \big(d_{\mathbf{p}}f_1(\mathbf{v}),\ldots,d_{\mathbf{p}}f_k(\mathbf{v})\big).$$

Example 2.2.18. Suppose that $\mathbf{f}(x,y,z) = (x^2 - y^3, y^3 - z^4, z^4 - x^2)$. Find all of those vectors $\mathbf{v}$ in $\mathbb{R}^3$ such that $d_{(1,1,1)}\mathbf{f}(\mathbf{v}) = (0,1,0) = \mathbf{j}$.

Solution:

We quickly calculate

$$\vec{\nabla}(x^2 - y^3)\big|_{(1,1,1)} \ = \ (2,-3,0),$$

$$\vec{\nabla}(y^3 - z^4)\big|_{(1,1,1)} \ = \ (0,3,-4),$$

and

$$\vec{\nabla}(z^4 - x^2)\big|_{(1,1,1)} \ = \ (-2,0,4).$$

Therefore, if $\mathbf{v} = (v_1, v_2, v_3)$, we have

$$d_{(1,1,1)}\mathbf{f}(\mathbf{v}) \ = \ \big((2,-3,0)\cdot\mathbf{v},\ (0,3,-4)\cdot\mathbf{v},\ (-2,0,4)\cdot\mathbf{v}\big) \ =$$

$$\big(2v_1 - 3v_2,\ 3v_2 - 4v_3,\ -2v_1 + 4v_3\big).$$

Note that the sum of the three components in this last vector is zero. In other words, the image of $d_{(1,1,1)}\mathbf{f}$ is contained in the subset of $\mathbb{R}^3$ such that the sum of the components is 0.

Therefore, there are **no** vectors $\mathbf{v}$ such that $d_{(1,1,1)}\mathbf{f}(\mathbf{v}) = (0,1,0)$, since the sum of the components of $(0,1,0)$ is 1, not 0.

The following proposition is easy to prove, and will be important to us later; we leave the proof as an exercise.

Proposition 2.2.19. *Suppose that $\mathcal{U}$ is an open subset of $\mathbb{R}^n$ and that $\mathrm{id}_{\mathcal{U}}$ is the identity function on $\mathcal{U}$, i.e., the function from $\mathcal{U}$ to $\mathcal{U}$ such that, for all $\mathbf{x}$ in $\mathcal{U}$, $\mathrm{id}_{\mathcal{U}}(\mathbf{x}) = \mathbf{x}$. Then, for all points $\mathbf{x}$ in $\mathcal{U}$, $\mathrm{id}_{\mathcal{U}}$ is differentiable at $\mathbf{x}$, and*

$$d_{\mathbf{x}}\left(\mathrm{id}_{\mathcal{U}}\right) = \mathrm{id}_{\mathbb{R}^n}.$$

$+$ **Linear Algebra:**

Suppose that E is a subset of $\mathbb{R}^n$, F is a subset of $\mathbb{R}^k$, and that we have a multi-component function $\mathbf{f} : E \to F$, where

$$\mathbf{f}(x_1,\ldots,x_n) = \left(f_1(x_1,\ldots,x_n),\ f_2(x_1,\ldots,x_n),\ \ldots,\ f_k(x_1,\ldots,x_n)\right).$$

Suppose further that $\mathbf{f}$ is differentiable at a point $\mathbf{p}$ in E.

Then, in Definition 2.2.17, we defined the total derivative $d_{\mathbf{p}}\mathbf{f} : \mathbb{R}^n \to \mathbb{R}^k$ by

$$d_{\mathbf{p}}\mathbf{f}(\mathbf{v}) = \left(d_{\mathbf{p}}f_1(\mathbf{v}),\ldots,d_{\mathbf{p}}f_k(\mathbf{v})\right).$$

As $d_{\mathbf{p}}f_i(\mathbf{v}) = \vec{\nabla} f_i(\mathbf{p}) \cdot \mathbf{v}$, for all $\mathbf{v}$, it follows, from properties of the dot product (see Section 1.3) that

$$d_{\mathbf{p}}f_i\left(a\mathbf{v} + b\mathbf{w}\right) = a\,d_{\mathbf{p}}f_i(\mathbf{v}) + b\,d_{\mathbf{p}}f_i(\mathbf{w}).$$

Since we can do this for each component, we conclude:

Theorem 2.2.20. *The function $d_{\mathbf{p}}\mathbf{f} : \mathbb{R}^n \to \mathbb{R}^k$ is a linear transformation, i.e., for all vectors $\mathbf{v}$ and $\mathbf{w}$ in $\mathbb{R}^n$, and for all scalars a and b,*

$$d_{\mathbf{p}}\mathbf{f}(a\mathbf{v} + b\mathbf{w}) = a\,d_{\mathbf{p}}\mathbf{f}(\mathbf{v}) + b\,d_{\mathbf{p}}\mathbf{f}(\mathbf{w}).$$

Example 2.2.21. Suppose that $\mathbf{f} : \mathbb{R}^2 \to \mathbb{R}^4$ is such that

$$d_{(1,3)}\mathbf{f}(1,-5) = (1,2,0,-1) \quad \text{and} \quad d_{(1,3)}\mathbf{f}(2,3) = (3,0,4,7).$$

Using that $2(1,-5) - (2,3) = (0,-13)$, find $d_{(1,3)}\mathbf{f}(0,1)$.

Note that, if $\mathbf{f} = \mathbf{f}(x,y)$, this is equivalent to finding $\mathbf{f}_y(1,3)$, the partial derivative of $\mathbf{f}$ with respect to y at $(1,3)$.

Solution:

As $d_{(1,3)}\mathbf{f}$ is linear, we find

$$-13\, d_{(1,3)}\mathbf{f}(0,1) = d_{(1,3)}\mathbf{f}(0,-13) = d_{(1,3)}\mathbf{f}\big(2(1,-5) - (2,3)\big) =$$

$$2\, d_{(1,3)}\mathbf{f}(1,-5) - d_{(1,3)}\mathbf{f}(2,3) = 2(1,2,0,-1) - (3,0,4,7) = (-1,4,-4,-9).$$

Dividing both sides by -13 (i.e., scalar multiplying both sides by $-1/13$), we find

$$d_{(1,3)}\mathbf{f}(0,1) = -\frac{1}{13}(-1,4,-4,-9) = \frac{1}{13}(1,-4,4,9).$$

All linear transformations are representable by matrix multiplication. Thus, we make the following definition.

Definition 2.2.22. *The* Jacobian matrix *of* $\mathbf{f} : \mathbb{R}^n \to \mathbb{R}^k$ *(or from a subset of* $\mathbb{R}^n$ *to a subset of* $\mathbb{R}^k$*) is the* $k \times n$ *matrix of partial derivatives*

$$\begin{bmatrix} \frac{\partial f_1}{\partial x_1} & \frac{\partial f_1}{\partial x_2} & \cdots & \frac{\partial f_1}{\partial x_n} \\ \frac{\partial f_2}{\partial x_1} & \frac{\partial f_2}{\partial x_2} & \cdots & \frac{\partial f_2}{\partial x_n} \\ \vdots & \vdots & \cdots & \vdots \\ \frac{\partial f_k}{\partial x_1} & \frac{\partial f_k}{\partial x_2} & \cdots & \frac{\partial f_k}{\partial x_n} \end{bmatrix},$$

at any point at which all of the partial derivatives exist.

We write $[d\mathbf{f}]$ *for this matrix, and write* $[d_{\mathbf{p}}\mathbf{f}]$ *for the Jacobian matrix of* $\mathbf{f}$, *evaluated at the point* $\mathbf{p}$.

Remark 2.2.23. Note that the Jacobian matrix of $\mathbf{f}$ has the gradient vector of each component of $\mathbf{f}$ in the corresponding **row**. Therefore, we frequently write

$$[d\mathbf{f}] = \begin{bmatrix} - \vec{\nabla} f_1 - \\ - \vec{\nabla} f_2 - \\ \vdots \\ - \vec{\nabla} f_k - \end{bmatrix},$$

where the horizontal lines are intended to indicate that the gradient vectors are horizontal, i.e., are in the rows.

Now, recall how matrix multiplication is defined and that, when we have a vector $\mathbf{v}$ in $\mathbb{R}^n$, and enclose it in matrix square brackets, $[\mathbf{v}]$, we mean the $n \times 1$ column vector containing the entries of $\mathbf{v}$.

Then, our formula

$$d_{\mathbf{p}}\mathbf{f}(\mathbf{v}) = \big(d_{\mathbf{p}}f_1(\mathbf{v}), \ldots, d_{\mathbf{p}}f_k(\mathbf{v})\big) = \big(\vec{\nabla}f_1(\mathbf{p}) \cdot \mathbf{v}, \; \ldots, \; \vec{\nabla}f_k(\mathbf{p}) \cdot \mathbf{v}\big)$$

becomes, in matrix terms:

Theorem 2.2.24. *The total derivative is calculated via matrix multiplication by the formula*

$$[d_{\mathbf{p}}\mathbf{f}(\mathbf{v})] = [d_{\mathbf{p}}\mathbf{f}][\mathbf{v}];$$

in words, this says that the column matrix of $d_{\mathbf{p}}\mathbf{f}(\mathbf{v})$ equals the Jacobian matrix of $\mathbf{f}$ at $\mathbf{p}$ times the column matrix of $\mathbf{v}$.

Example 2.2.25. Let's look at the function $\mathbf{f} : \mathbb{R}^3 \to \mathbb{R}^3$ from Example 2.2.18:

$$\mathbf{f}(x, y, z) = (x^2 - y^3, y^3 - z^4, z^4 - x^2).$$

Then, the Jacobian matrix of $\mathbf{f}$ is

$$[d\mathbf{f}] = \begin{bmatrix} - & \vec{\nabla}(x^2 - y^3) & - \\ - & \vec{\nabla}(y^3 - z^4) & - \\ - & \vec{\nabla}(z^4 - x^2) & - \end{bmatrix} = \begin{bmatrix} 2x & -3y^2 & 0 \\ 0 & 3y^2 & -4z^3 \\ -2x & 0 & 4z^3 \end{bmatrix}.$$

The Jacobian matrix of $\mathbf{f}$, evaluated at $(1,1,1)$, is

$$
\left[d_{(1,1,1)}\mathbf{f}\right] = \begin{bmatrix} 2x & -3y^2 & 0 \\ 0 & 3y^2 & -4z^3 \\ -2x & 0 & 4z^3 \end{bmatrix}\Bigg|_{(1,1,1)} = \begin{bmatrix} 2 & -3 & 0 \\ 0 & 3 & -4 \\ -2 & 0 & 4 \end{bmatrix}.
$$

Suppose that we wish to calculate $d_{(1,1,1)}\mathbf{f}(7,2,-5)$ using matrices. We find

$$
\left[d_{(1,1,1)}\mathbf{f}(7,2,-5)\right] = \begin{bmatrix} 2 & -3 & 0 \\ 0 & 3 & -4 \\ -2 & 0 & 4 \end{bmatrix}\begin{bmatrix} 7 \\ 2 \\ -5 \end{bmatrix} = \begin{bmatrix} 2\cdot 7 + (-3)\cdot 2 + 0\cdot(-5) \\ 0\cdot 7 + 3\cdot 2 + (-4)\cdot(-5) \\ (-2)\cdot 7 + 0\cdot 2 + 4\cdot(-5) \end{bmatrix}.
$$

Therefore,

$$
\left[d_{(1,1,1)}\mathbf{f}(7,2,-5)\right] = \begin{bmatrix} 8 \\ 26 \\ -34 \end{bmatrix},
$$

i.e., $d_{(1,1,1)}\mathbf{f}(7,2,-5) = (8,26,-34)$.

Online answers to select exercises are here.

Basics:

2.2.1   Exercises

In each of the following exercises, you are given the gradient vector of a differentiable function f at a point $\mathbf{p}$ in $\mathbb{R}^2$ or $\mathbb{R}^3$. Give the total derivative (function) $d_{\mathbf{p}}f(\mathbf{v})$, where $\mathbf{v} = (v_1, v_2)$ or (v_1, v_2, v_3).

1. $\vec{\nabla}f(\mathbf{p}) = (-2,1)$.

2. $\vec{\nabla}f(\mathbf{p}) = (3,5)$.

3. $\vec{\nabla}f(\mathbf{p}) = (-2,1,3)$.

4. $\vec{\nabla}f(\mathbf{p}) = (3,5,-7)$.

In each of the following exercises, determine the total derivative of the given function at the point p with respect to (a) an arbitrary vector (v_1, v_2) or (v_1, v_2, v_3) and (b) the given vector v.

5. $f(x,y) = 12x^2y - 6xy^2 + 12, \mathbf{p} = (2,-2), \mathbf{v} = (1,4)$

6. $g(z,t) = 6^{zt-t^2}, \mathbf{p} = (1,0), \mathbf{v} = (-2,-1)$

7. $h(u,w) = u^2 \sin\left(\dfrac{\pi w}{2}\right), \mathbf{p} = (4,3), \mathbf{v} = (4,3)$

8. $j(\alpha, \beta) = \alpha^2 \beta + \alpha\sqrt{\beta}, \mathbf{p} = (1,9), \mathbf{v} = (1,6)$

9. $\mathcal{P}(x,k) = x^k + 2k^x, \mathbf{p} = (1,e), \mathbf{v} = \left(\sqrt{2}, \sqrt{3}\right)$

10. $z(\omega, s, t) = e^{st} - s\tan^{-1}(\omega t), \mathbf{p} = (\pi, 2, 1), \mathbf{v} = (-1, 0, \pi)$

11. $\mathcal{L}(x,y,z) = x^3 y - x^4 z + 5zy, \mathbf{p} = \left(1, 2, \dfrac{1}{5}\right), \mathbf{v} = (-1, 2, 1)$

12. $x(r,t,\theta) = r^2 t + \cos(\theta + \pi t), \mathbf{p} = \left(2, 1, \dfrac{\pi}{2}\right), \mathbf{v} = (2, 0, \pi)$

13. $T(u,v,w) = 2uv - \log_w(u), \mathbf{p} = (e, 1, e^2), \mathbf{v} = (1, 1, 2)$

14. $y(x_1, x_2, x_3) = \dfrac{x_1}{x_2} - \dfrac{2x_1}{x_3}, \mathbf{p} = (-1, 4, 5), \mathbf{v} = (1, 4, -5)$

15. Suppose that a heated metal plate occupies the portion of the xy-plane where $x^2 + y^2 \leq 4$, where x and y are in meters. The plate is heated in such a way that the temperature, T, in degrees Celsius, is given by

$$T(x,y) = \frac{100x}{x^2 + y^2 + 1}.$$

If an ant is at the point $(-1, 1)$, and is moving with velocity $0.001(3, 4)$ m/s, what instantaneous rate of change in the temperature of the plate does the ant experience?

16. Suppose that the elevation h, above sea level, of a hill, is given by $h = 400 - x^2 - 3y^2$, where x and y are coordinates on the ground, at sea level, and x, y, and h are in feet. If you're at a point on the hill where $(x, y) = (10, 5)$, and moving in such a way that the derivative, with respect to time t, in secs, is

$$\frac{d}{dt}(x,y) = (2, -1) \text{ ft/s},$$

at what rate is your elevation changing?

17. Suppose that a mole is digging underground. The density, in kilograms per cubic meter, of the material through which the mole must dig is given by

$$E(x,y,z) = 1 + z^3 \sqrt{x^2 + y^2},$$

where x indicates how far east the mole is from the entrance point, y indicates how far north the mole is from the entrance point, and z indicates how far below the surface the mole has dug. All distances are measured in meters.

The mole is now 10 meters northwest of the entrance point and 20 meters below the surface, and is moving with a velocity of $(1, -1, 0)$ m/s. What is the instantaneous rate of change of the density of the material through which the mole is digging?

18. Suppose that a company produces w wazbobs and s snargits, and has determined that the cost C and revenue R, in Euros, of manufacturing these items are given by:

$$C(w, s) = 4w - 2ws + 5s \quad \text{and} \quad R(w, s) = 10w + 12s - ws.$$

If the company is currently producing 5 wazbobs and 3 snargits, but is changing production by -1 snargits per day and 2 wazbobs per day, at what rate is their profit changing?

In each of the following exercises, determine the total derivative of the given multi-component function at the point p with respect to (a) the given vector v and (b) an arbitrary vector (v_1, v_2) or (v_1, v_2, v_3).

19. $\mathbf{A}(x, t) = \left(tx^3, x - 5t\right), \mathbf{p} = (3, -2), \mathbf{v} = (1, -2)$

20. $\mathbf{h}(s, t) = \left(s \csc(t), s^2 \csc(t - \pi)\right), \mathbf{p} = \left(2, \dfrac{3\pi}{2}\right), \mathbf{v} = (7, 3)$

21. $\mathbf{g}(x, y) = \left(x^2 - xy, \ln(xy), \ln(x + y)\right), \mathbf{p} = (e, e), \mathbf{v} = (2, 1)$

22. $\mathbf{f}(u, w) = \left(e^{uw}, \sqrt[3]{uw}, 12uw\right), \mathbf{p} = (2, 4), \mathbf{v} = (-3, 1)$

23. $\xi(k, \alpha, t) = \left(\alpha t - ke^t, k \tan^{-1}(\alpha t)\right), \mathbf{p} = (2, 2, 0), \mathbf{v} = (-1, 3, 0)$

24. $\mathbf{W}(x, y, z) = \left(z^2 x, -z^3 y^2 - x\right), \mathbf{p} = \left(\sqrt{2}, 1, -1\right), \mathbf{v} = \left(3, 5, \dfrac{1}{2}\right)$

25. $\mathcal{Z}(r, \omega, x) = \left(r \cos(\omega + \pi x), r \sin(\omega + \pi x), r \tan(\omega + \pi x)\right), \mathbf{p} = \left(2, \dfrac{\pi}{2}, \dfrac{1}{2}\right), \mathbf{v} = (3, -1, 2)$

26. $\mathbf{b}(x_1, x_2, x_3) = \left(\dfrac{3x_2 x_3}{x_1}, \dfrac{2x_1 + x_3}{x_2}, \dfrac{x_1 - x_2}{5x_3}\right), \mathbf{p} = (-1, 2, -1), \mathbf{v} = (0, 0, 0)$

More Depth:

In each of the following exercises, you are given the total derivative of $g(x, y)$ at some point p with respect to two vectors. Use this information to find g_x and g_y at p. Assume that $g(x, y)$ is differentiable at p.

27. $d_{\mathbf{p}}g(6, 10) = \dfrac{26}{3}, d_{\mathbf{p}}g(7, -4) = \dfrac{-1}{3}$

28. $d_{\mathbf{p}}g(1, -2) = 12, d_{\mathbf{p}}g(3, -2) = 16$

29. $d_{\mathbf{p}}g(1, 7) = 11, d_{\mathbf{p}}g(2, 2) = \dfrac{118}{7}$

30. $d_{\mathbf{p}}g\left(3\sqrt{2}, \dfrac{\sqrt{2}}{2}\right) = 8, d_{\mathbf{p}}g(5, 2) = 9\sqrt{2}$

In each of the following exercises, use the given information about the total derivative to solve for the unknown constant A. Assume $A > 0$.

31. $h(s, t) = s^2 t - A s^2 t^2, d_{(1,3)}h(-2, 5) = 11$ ▶

32. $j(u, w) = A\sin(w + \pi u) - A^2 u \cos(w), d_{(1/2, \pi/2)}j(-1, 2\pi) = 20\pi$

33. $k(x, y) = y e^{Ax}, d_{(3,0)}k(4, 7) = 31$

34. $f(x, y, z) = A x^3 \sqrt{y - z}, d_{(2,10,1)}f\left(\dfrac{1}{2}, \dfrac{3}{2}, \dfrac{-3}{2}\right) = 132$

35. Recall Example 2.1.21, in which a new resistor of resistance R_{new} ohms is formed by placing n resistors of resistance $R_1, \ldots, R_n$ ohms in parallel. We will assume that $n = 3$, so that
$$\frac{1}{R_{\text{new}}} = \frac{1}{R_1} + \frac{1}{R_2} + \frac{1}{R_3}.$$
If the triple (R_1, R_2, R_3) is changing, with respect to time, at a rate of $(-2, -1, -3)$ ohms/hour, what is the instantaneous rate of change of R_{new}, with respect to time, at the point where $(R_1, R_2, R_3) = (100, 120, 160)$?

36. Prove Proposition 2.2.19.

$\boxed{+ \textbf{ Linear Algebra:}}$

In each of the following exercises, calculate the specified total derivative using matrices.

37. $\mathbf{A}(x, t) = \left(tx^3, x - 5t\right), d_{(3,-2)}\mathbf{A}(1, -2)$ ▶

38. $\mathbf{h}(s, t) = \left(s\csc(t), s^2 \csc(t - \pi)\right), d_{(2, 3\pi/2)}\mathbf{h}(7, 3)$

39. $\mathbf{g}(x, y) = \left(x^2 - xy, \ln(xy), \ln(x + y)\right), d_{(e,e)}\mathbf{g}(2, 1)$

40. $\mathbf{f}(u, w) = \left(e^{uw}, \sqrt[3]{uw}, 12uw\right), d_{(2,4)}\mathbf{f}(-3, 1)$

41. $\xi(k, \alpha, t) = \left(\alpha t - k e^t, k \tan^{-1}(\alpha t)\right), d_{(2,2,0)}\xi(-1, 3, 0)$

42. $\mathbf{W}(x, y, z) = \left(z^2 x, -z^3 y^2 - x\right), d_{(\sqrt{2}, 1, -1)}\mathbf{W}\left(3, 5, \dfrac{1}{2}\right)$

43. $\mathcal{Z}(r, \omega, x) = \left(r\cos(\omega + \pi x), r\sin(\omega + \pi x), r\tan(\omega + \pi x)\right), d_{(2, \pi/2, 1/2)}\mathcal{Z}(3, -1, 2)$

44. $\mathbf{b}(x_1, x_2, x_3) = \left(\dfrac{3x_2 x_3}{x_1}, \dfrac{2x_1 + x_3}{x_2}, \dfrac{x_1 - x_2}{5x_3}\right), d_{(-1,2,-1)}\mathbf{b}(0, 0, 0)$

2.3 Linear Approximation, Tangent Planes, and the Differential

The definition of the total derivative from the previous section allows us to conclude that, near a given point **p**, a function f can be approximated "well" by an affine linear function, the *linearization of f at* **p**. This approximation is, naturally, referred to as *linear approximation*.

The graph of the linearization is the *tangent plane to the graph of f at* **p**. If you think of linear approximation in terms of changes of the variables and the function value, you are led to approximations in terms of *differentials*.

Basics:

In single-variable Calculus, you should have encountered *linear approximation*: if $f = f(x)$ is differentiable at a, then $f(x)$ is approximately equal to $f(a) + f'(a)(x - a)$, provided that x is "close to a".

Linear approximation is also referred to as the *tangent line approximation*, since the graph of $y = f(a) + f'(a)(x - a)$ is the tangent line to the graph of $y = f(x)$ at the point $(a, f(a))$. Graphically, linear/tangent line approximation means simply that the line which best approximates the graph of f at the point $(a, f(a))$ is the line given by $y = f(a) + f'(a)(x - a)$.

We can look at linear approximation from the point of view of the changes in x and f. The linear approximation, above, can be rewritten as

$$\Delta f = f(x) - f(a) \approx f'(a)(x - a) = f'(a)\Delta x,$$

when Δx is close to 0.

Looking at this last form, it is common to formally define new "variables" df and dx, which are related by $df = f'(a)dx$, and then to say that, near a, if dx is a small Δx (small, meaning close to 0), then the *differential df* is approximately the change in f, i.e., if $|dx|$ is small, then $\Delta f \approx df$. In this guise, the approximation is normally referred to as *differential approximation*.

We wish to look at all of this again — linear approximation, tangent sets, and differential approximations — but, now, for multivariable functions.

Throughout this section, we assume that f is a real-valued function, whose domain is a neighborhood of a point **p** in $\mathbb{R}^n$, and that f is differentiable at **p**.

The definition of the total derivative in Definition 2.2.1 tells us that

$$\lim_{|\mathbf{h}| \to 0} \frac{\left| f(\mathbf{p} + \mathbf{h}) - f(\mathbf{p}) \; - \; d_{\mathbf{p}} f(\mathbf{h}) \right|}{|\mathbf{h}|} \;=\; 0. \tag{2.4}$$

The limit in Formula 2.4 implies that, when $|\mathbf{h}|$ is close to 0, the quantity

$$\left| f(\mathbf{p} + \mathbf{h}) - f(\mathbf{p}) \; - \; d_{\mathbf{p}} f(\mathbf{h}) \right|$$

must also be close to 0 or, equivalently,

$$f(\mathbf{p} + \mathbf{h}) - f(\mathbf{p}) \; - \; d_{\mathbf{p}} f(\mathbf{h})$$

must be close to the zero vector, $\mathbf{0}$, i.e.,

$$f(\mathbf{p} + \mathbf{h}) \;\approx\; f(\mathbf{p}) \; + \; d_{\mathbf{p}} f(\mathbf{h}).$$

> In fact, as we shall make precise in Theorem 2.3.15, $f(\mathbf{p} + \mathbf{h}) - f(\mathbf{p}) \; - \; d_{\mathbf{p}}(\mathbf{h})$ must approach $\mathbf{0}$ **faster** than $|\mathbf{h}|$ approaches 0.

Writing $\mathbf{x}$ for $\mathbf{p} + \mathbf{h}$, so that $\mathbf{h} = \mathbf{x} - \mathbf{p}$, we immediately conclude:

Proposition 2.3.1. (Linear Approximation) *If $\mathbf{x}$ is close to $\mathbf{p}$, i.e., if $|\mathbf{x} - \mathbf{p}|$ is close to 0, then*

$$f(\mathbf{x}) \;\approx\; f(\mathbf{p}) \; + \; d_{\mathbf{p}} f(\mathbf{x} - \mathbf{p}) \;=\; f(\mathbf{p}) \; + \; \vec{\nabla} f(\mathbf{p}) \cdot (\mathbf{x} - \mathbf{p}).$$

In particular, suppose $f = f(x, y)$, and $\mathbf{p} = (a, b)$. In this case, if (x, y) is close to (a, b), then

$$f(x, y) \;\approx\; f(a, b) \; + \; f_x(a, b)(x - a) \; + \; f_y(a, b)(y - b).$$

The point of multivariable linear approximation is the same as it was in the single variable case; we want to approximate a complicated function by a simple function, an affine linear function, for values of the independent variable(s) near some given point.

> Recall the definition of an affine linear function from Definition 1.7.1.

Example 2.3.2. Consider the function

$$f(x, y) \;=\; 4 + x - x^2 - y^3.$$

Let's use linear approximation at $(1, 1)$ to approximate $f(0.9, 1.2)$ and $f(1.01, 1.05)$.

We apply the formula

$$f(x, y) \;\approx\; f(a, b) \; + \; f_x(a, b)(x - a) \; + \; f_y(a, b)(y - b),$$

where $(a, b) = (1, 1)$, and first using $(x, y) = (0.9, 1.2)$, and then using $(x, y) = (1.01, 1.05)$.

We find $f(1, 1) = 3$. Now we calculate

$$f_x(x, y) \;=\; 1 - 2x, \qquad \text{and} \qquad f_y(x, y) \;=\; -3y^2,$$

and so

$$f_x(1, 1) \;=\; -1, \qquad \text{and} \qquad f_y(1, 1) = -3.$$

Therefore, for each of our (x, y) pairs, which are both near $(1, 1)$, we will use that

$$f(x, y) \;\approx\; 3 - 1(x - 1) - 3(y - 1).$$

Now, we easily find

$$f(0.9, 1.2) \;\approx\; 3 - 1(0.9 - 1) - 3(1.2 - 1) \;=\; 2.5$$

and

$$f(1.01, 1.05) \;\approx\; 3 - 1(1.01 - 1) - 3(1.05 - 1) \;=\; 2.84.$$

It is a **HUGE** mistake to use the functions f_x and f_y, in terms of x and y, in the linear approximation. You are supposed to be approximating a complicated function by a nice, simple, affine linear function. Hence, you must calculate the partial derivatives at the point (a, b), so that you end up with **constants** for coefficients.

While it's not a huge mistake, it is a serious waste of time to "simplify" this affine linear function by distributing the coefficients over $(x - a)$ and $(y - b)$ (here, $(x - 1)$ and $(y - 1)$). The intention is to use linear approximation when (x, y) is close to (a, b), i.e., when both $(x - a)$ and $(y - b)$ are close to 0. Hence, it is typically best to leave these close-to-0 quantities as they are.

The actual values of $f(0.9, 1.2)$ and $f(1.01, 1.05)$ are 2.362 and 2.832275, respectively. Not surprisingly, the second approximation is significantly better; after all, $(1.01, 1.05)$ is significantly closer to $(1, 1)$ than $(0.9, 1.2)$ is.

Example 2.3.3. We don't have to be given much information to use linear approximation.

Suppose that $f = f(x, y)$ is differentiable at $(3, 7)$, and that $f(3, 7) = 4$ and $\vec{\nabla} f(3, 7) = (-1, 5)$. Can we reasonably approximate the value of $f(2.9, 7.2)$?

Sure. We just apply linear approximation, using that $(2.9, 7.2)$ is "close" to $(3, 7)$:

$$f(2.9, 7.2) \;\approx\; f(3, 7) \;+\; \vec{\nabla} f(3, 7) \cdot \big((2.9, 7.2) - (3, 7)\big) \;=\; 4 + (-1, 5) \cdot (-0.1, 0.2).$$

Thus, we obtain that $f(2.9, 7.2) \;\approx\; 4 + 0.1 + 1 \;= 5.1$.

We give a name to the affine linear function on the right in the formulas of Proposition 2.3.1.

Definition 2.3.4. *The affine linear function on $\mathbb{R}^n$ given by*

$$L_f(\mathbf{x}; \mathbf{p}) = f(\mathbf{p}) + \vec{\nabla} f(\mathbf{p}) \cdot (\mathbf{x} - \mathbf{p})$$

is called the **linearization of f at $\mathbf{p}$.**

If it is clear what f and $\mathbf{p}$ are, we sometimes write simply $L(\mathbf{x})$ for the linearization.

> The linearization is the *1st-order Taylor polynomial of f*. As is the case for single-variable functions, multivariable functions can have Taylor polynomials of all orders. We shall discuss this in Section 2.13.

Using our new notation and terminology, we can rephrase Proposition 2.3.1:

Proposition 2.3.5. (**Linear Approximation**) *If $\mathbf{x}$ is close to $\mathbf{p}$, i.e., if $|\mathbf{x} - \mathbf{p}|$ is close to 0, then*

$$f(\mathbf{x}) \approx L_f(\mathbf{x}; \mathbf{p}).$$

Let's look again at Example 2.3.2, but now using our new terminology.

Example 2.3.6. Consider the function

$$f(x, y) = 4 + x - x^2 - y^3.$$

Find the linearization of f at $(1, 1)$, and use it to approximate the values of $f(0.9, 1.2)$ and $f(1.01, 1.05)$.

Solution:

We did all of the work for this in Example 2.3.2. We found that $f(1, 1) = 3$ and $\vec{\nabla} f(1, 1) = (f_x(1, 1),\ f_y(1, 1)) = (-1, -3)$.

Thus,

$$L(x, y) = f(1, 1) + \vec{\nabla} f(1, 1) \cdot \big((x, y) - (1, 1)\big) = 3 + (-1, -3) \cdot (x - 1, y - 1),$$

and so

$$L(x, y) = 3 - (x - 1) - 3(y - 1).$$

> We repeat our earlier warning: It is a **HUGE** mistake to use $\vec{\nabla} f$, in terms of x and y, in the linearization. You are supposed to be approximating a complicated function by a nice, simple affine linear function. Hence, you must calculate $\vec{\nabla} f$ at $\mathbf{p}$, so that you end up with **constants** for coefficients.

Of course, this gives us the same approximations that we had in Example 2.3.2:

$$f(0.9, 1.2) \approx L_f\big((0.9, 1.2); (1, 1)\big) = 3 - 1(0.9 - 1) - 3(1.2 - 1) = 2.5$$

and

$$f(1.01, 1.05) \approx L_f\big((1.01, 1.05); (1, 1)\big) = 3 - 1(1.01 - 1) - 3(1.05 - 1) = 2.84.$$

Example 2.3.7. Suppose that $g : \mathbb{R}^4 \to \mathbb{R}$ is given by

$$g(x_1, x_2, x_3, x_4) \;=\; x_2 e^{x_1} + x_3 \cos(2\pi x_4).$$

Find the linearization of g at $(0, 1, -2, 1)$, and use it to approximate $g(0.01, 1.02, -2.005, 0.99)$.

Solution: Writing $\mathbf{x}$ for (x_1, x_2, x_3, x_4), the linearization of g at $(0, 1, -2, 1)$ is

$$L_g\big(\mathbf{x}; (0, 1, -2, 1)\big) \;=\; g(0, 1, -2, 1) + \vec{\nabla} g(0, 1, -2, 1) \cdot \big(\mathbf{x} - (0, 1, -2, 1)\big).$$

We first find $g(0, 1, -2, 1) = 1 \cdot e^0 - 2 \cos(2\pi \cdot 1) = -1$. Now, we calculate

$$\frac{\partial g}{\partial x_1} \;=\; x_2 e^{x_1}, \qquad \frac{\partial g}{\partial x_2} \;=\; e^{x_1}, \qquad \frac{\partial g}{\partial x_3} \;=\; \cos(2\pi x_4),$$

and

$$\frac{\partial g}{\partial x_4} \;=\; -2\pi x_3 \sin(2\pi x_4).$$

Therefore,

$$\vec{\nabla} g(0, 1, -2, 1) \;=\; \big(1 \cdot e^0, \; e^0, \; \cos(2\pi \cdot 1), \; -2\pi(-2)\sin(2\pi \cdot 1)\big) \;=\; (1, \; 1, \; 1, \; 0).$$

Thus,

$$L_g\big(\mathbf{x}; (0, 1, -2, 1)\big) \;=\; -1 \;+\; (1,1,1,0) \cdot \big(\mathbf{x} - (0, 1, -2, 1)\big) \;=$$

$$-1 + 1(x_1 - 0) + 1(x_2 - 1) + 1\big(x_3 - (-2)\big) + 0(x_4 - 1),$$

that is

$$L(\mathbf{x}) \;=\; L_g\big(\mathbf{x}; (0, 1, -2, 1)\big) \;=\; -1 + x_1 + (x_2 - 1) + (x_3 + 2),$$

where we have left parentheses around quantities that will be close to 0.

Finally, we look at the linear approximation

$$g(0.01, 1.02, -2.005, 0.99) \;\approx\; L(0.01, 1.02, -2.005, 0.99) \;=$$

$$-1 + 0.01 + (1.02 - 1) + (-2.005 + 2) \;=\; -1 + 0.01 + 0.02 - 0.005 \;=\; -0.975.$$

You may check that the actual answer, to 8 decimal places, is -0.97079242, so our estimate is reasonably close.

Recall that, if f is a function of one variable, then the graph of the linearization of f, at p, is the tangent line to the graph of f at $(p, f(p))$; that is, the tangent line is the graph of $y = f(p) + f'(p)(x - p)$. Why should this be true? Because the linearization is the affine linear function that best approximates f at points near p; so the line you get from graphing the linearization should be the line which best approximates the graph of f near points with x-coordinate close to p, and that line is the tangent line to the graph.

In the same way, if we have $z = f(x, y)$, then the graph of the linearization $z = L_f\big((x, y); (a, b)\big)$ is the tangent **plane** to the graph of $z = f(x, y)$ at the point $(a, b, f(a, b))$. More generally, in any number of dimensions, we can consider the *tangent set* to the graph of f:

Definition 2.3.8. *Suppose that f is a real-valued function on a subset of $\mathbb{R}^n$, and that f is differentiable at* **p**. *Then, the graph of the linearization of f at* **p** *is called the* **tangent set to the graph of f at the point $(\mathbf{p}, f(\mathbf{p}))$.**

In other words, the tangent set to the graph of f at the point $(\mathbf{p}, f(\mathbf{p}))$ is the set of points $(\mathbf{x}, z)$ in $\mathbb{R}^{n+1}$ such that

$$z = f(\mathbf{p}) + \vec{\nabla} f(\mathbf{p}) \cdot (\mathbf{x} - \mathbf{p}).$$

Frequently, it is convenient to refer to the tangent set as being "at the point **p**", instead of "at the point $(\mathbf{p}, f(\mathbf{p}))$"; it is implicit that the z-coordinate is the value of f at **p**.

In particular, suppose that $f = f(x, y)$ is a real-valued function on a subset of $\mathbb{R}^2$, and that f is differentiable at (a, b). Then, the **tangent plane to the graph of f at the point $(a, b, f(a, b))$ is the set of points (x, y, z) in $\mathbb{R}^3$ such that**

$$z = f(a, b) + f_x(a, b)(x - a) + f_y(a, b)(y - b).$$

In terms of tangent vectors, defined in Definition 1.6.24, the tangent set is characterized by containing the point $(\mathbf{p}, f(\mathbf{p}))$ and being parallel to every tangent vector to the graph of f at $(\mathbf{p}, f(\mathbf{p}))$.

Example 2.3.9. Consider again the function $z = g(x, y) = 4 + x - x^2 - y^3$ from Example 2.3.2 and Example 2.3.6, but now let's look at the tangent plane to its graph at the point where $(x, y) = (1, 1)$.

We already produced the linearization of g at $(1, 1)$ in Example 2.3.6; the linearization of g at $(1, 1)$ is

$$L(x, y) = 3 - (x - 1) - 3(y - 1),$$

where we remind you that you do **not** want to simplify this; you want to have the quantities $x - 1$ and $y - 1$ in parentheses.

Figure 2.3.1: The tangent plane to $z = 4 + x - x^2 - y^3$ at $(1, 1, 3)$.

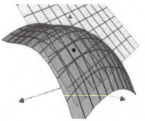

Therefore, the tangent plane to the graph of g at the point where $(x, y) = (1, 1)$ is the graph of

$$z = 3 - (x - 1) - 3(y - 1).$$

We give you two views of the graph of g together with the graph of L, the tangent plane, in Figure 2.3.1 and Figure 2.3.2. The bold black dot is at the point $(1, 1, 3)$.

Figure 2.3.2: Another view of the same tangent plane.

Remark 2.3.10. It is important that the tangent plane to the graph of function f, at a point where $(x, y) = (a, b)$, contains the tangent lines to the corresponding cross sections of the graph of f.

How do you see this? Let's consider the $x = a$ cross section. This cross section of the graph of f is just the graph of the function obtained from f by setting x equal to a; let's call this cross section function $h(y)$. In other words, we let $h(y) = f(a, y)$. We would like to see that the tangent line (inside the copy of the yz-plane given by $x = a$) to the graph of h, at the point where $y = b$, is contained in the tangent plane to the graph of f at the point where $(x, y) = (a, b)$.

In fact, the tangent line to the graph of h isn't just contained in the tangent plane, it's the $x = a$ cross section of the tangent plane. Why? Because the tangent line to the graph of h, where $y = b$, is given by

$$z = h(b) + h'(b)(y - b),$$

inside the plane $x = a$.

In terms of f, the two equations above for the tangent line become

$$x = a \quad \text{and} \quad z = f(a, b) + f_y(a, b)(y - b).$$

On the other hand, the tangent plane to the graph of f is the graph of

$$z = f(a, b) + f_x(a, b)(x - a) + f_y(a, b)(y - b).$$

Figure 2.3.3: A cross section of the tangent plane gives the tangent line to the cross section of the graph of f.

The $x = a$ cross section of the tangent plane is thus given by $x = a$ and

$$z = f(a, b) + f_x(a, b)(a - a) + f_y(a, b)(y - b) = f(a, b) + f_y(a, b)(y - b);$$

the same equations that we had before, which shows that the tangent line to the $x = a$ cross section of the graph of f is the $x = a$ cross section of the tangent plane. The same reasoning applies to the cross section where $y = b$.

Consider, for instance, the function $z = g(x, y) = 4 + x - x^2 - y^3$ from the previous example, Example 2.3.9. We found that the tangent plane was the graph of the linearization, i.e., the graph of $z = 3 - (x - 1) - 3(y - 1)$.

If we look inside the cross section where $x = 1$, then we find that the graph of g becomes the graph of $z = 4 - y^3$ and the tangent line, at $y = 1$, is given by $z = 3 - 3(y - 1)$, inside the plane where $x = 1$. This cross section and corresponding tangent line can easily be seen in Figure 2.3.3.

Example 2.3.11. Let us return to Example 2.2.15, in which we considered

$$z = f(x, y) = \sqrt{x^2 + y^2}.$$

We mentioned at the time that f is not differentiable at the origin, and that the graph has a sharp point there. Thus, the tangent plane to the graph at the point $(0, 0, 0)$ is **undefined**.

Thinking about Remark 2.3.10, it is instructive to look at the $x = 0$ and $y = 0$ cross sections. The $x = 0$ cross section is the graph of $z = \sqrt{0^2 + y^2} = |y|$. Similarly, the $y = 0$ cross section is the graph of $z = |x|$. As you should recall from single-variable Calculus, there is no tangent line to the graph of the absolute value function at the origin.

Figure 2.3.4: The graph of $z = \sqrt{x^2 + y^2}$ has no tangent plane at the origin.

Linear approximation tells us that, if $\mathbf{x}$ is close to $\mathbf{p}$, then

$$f(\mathbf{x}) \approx f(\mathbf{p}) + d_{\mathbf{p}}f(\mathbf{x} - \mathbf{p}),$$

or, equivalently,

$$f(\mathbf{x}) - f(\mathbf{p}) \approx d_{\mathbf{p}}f(\mathbf{x} - \mathbf{p}).$$

Writing $\Delta \mathbf{x}$ in place of $\mathbf{x} - \mathbf{p}$, and writing Δf in place of $f(\mathbf{x}) - f(\mathbf{p})$, we arrive at:

> **Proposition 2.3.12. (Differential Approximation)** *If $\mathbf{x}$ is close to $\mathbf{p}$, so that $\Delta \mathbf{x}$ is close to 0, then the corresponding change in f, Δf, is approximated by*
>
> $$\Delta f \approx d_{\mathbf{p}}f(\Delta \mathbf{x}) = \vec{\nabla} f(\mathbf{p}) \cdot \Delta \mathbf{x}.$$

Of course, this is just a different version of linear approximation, but one that is used when what you're worried about is the change in various quantities, not the actual values themselves.

Example 2.3.13. Suppose that $h(x, y, z) = 2z \ln(e + xy)$. Then, $h(0, -2, 5) = 10$. Use differential approximation to estimate the change in h, if $\Delta(x, y, z) = (0.01, 0.05, -0.2)$.

Solution:

We use that

$$\Delta h \;\approx\; \vec{\nabla} h(0, -2, 5) \cdot (0.01, 0.05, -0.2),$$

and so we need to calculate the gradient vector.

We find

$$\frac{\partial h}{\partial x} \;=\; \frac{2zy}{e + xy}, \qquad \frac{\partial h}{\partial y} \;=\; \frac{2zx}{e + xy}, \qquad \text{and} \qquad \frac{\partial h}{\partial z} \;=\; 2\ln(e + xy),$$

and so

$$\vec{\nabla} h(0, -2, 5) \;=\; \left(-\frac{20}{e}, \; 0, \; 2 \right).$$

Therefore,

$$\Delta h \;\approx\; \left(-\frac{20}{e}, \; 0, \; 2 \right) \cdot (0.01, 0.05, -0.2) \;=\; -\frac{0.2}{e} + 0 - 0.4 \;\approx\; -0.473575888.$$

Example 2.3.14. Suppose that we have a right circular cylinder of radius 0.5 feet and height 1 foot. If we increase the radius by 0.1 feet and decrease the height by 0.1 feet, then differential approximation tells us that the change in the volume is approximately what?

Solution:

The volume of the cylinder is given by $V = \pi r^2 h$. We will use that

$$\Delta V \;\approx\; \vec{\nabla} V(\mathbf{p}) \cdot \Delta(r, h) \;=\; \left(\frac{\partial V}{\partial r}, \frac{\partial V}{\partial h} \right)\Bigg|_{\mathbf{p}} \cdot (\Delta r, \Delta h),$$

where $\mathbf{p} = $ (original radius, original height) $= (0.5, 1)$, $\Delta r = 0.1$, and $\Delta h = -0.1$.

We find

$$\Delta V \;\approx\; \left(2\pi rh, \; \pi r^2 \right)\Big|_{(r,h)=(0.5,1)} \cdot (0.1, \; -0.1) \;=\; (\pi, \; 0.25\pi) \cdot (0.1, \; -0.1) \;=\; 0.075\pi \text{ ft}^3.$$

Before we give more examples, we first want to make more precise the way in which $f(\mathbf{p} + \mathbf{h})$ gets "close to" $f(\mathbf{p}) + d_{\mathbf{p}}f(\mathbf{h})$ as $\mathbf{h}$ gets "close to" $\mathbf{0}$. We will show that the difference between $f(\mathbf{p} + \mathbf{h})$ and $f(\mathbf{p}) + d_{\mathbf{p}}f(\mathbf{h})$ is the product of $|\mathbf{h}|$ with a function that approaches 0 as $\mathbf{h}$ approaches $\mathbf{0}$; hence, the difference approaches 0 faster than just $|\mathbf{h}|$ does.

More Depth:

Theorem 2.3.15. *Suppose that E is a subset of $\mathbb{R}^n$, that $\mathbf{p}$ is a point in E, and that $f : E \to \mathbb{R}$ is differentiable at $\mathbf{p}$.*

Then, there exists a real-valued function $\mathcal{E}$ on a neighborhood $\mathcal{W}$ of $\mathbf{0}$ in $\mathbb{R}^n$, which is continuous at $\mathbf{0}$, such that $\mathcal{E}(\mathbf{0}) = 0$, and such that, for all $\mathbf{h}$ in $\mathcal{W}$, $\mathbf{p} + \mathbf{h}$ is in the domain of f, and

$$f(\mathbf{p} + \mathbf{h}) \;=\; f(\mathbf{p}) \;+\; d_{\mathbf{p}}f(\mathbf{h}) + |\mathbf{h}|\mathcal{E}(\mathbf{h}).$$

Proof. Actually, this is quite easy. Define

$$\mathcal{E}(\mathbf{h}) \;=\; \begin{cases} \dfrac{f(\mathbf{p} + \mathbf{h}) - f(\mathbf{p}) - d_{\mathbf{p}}f(\mathbf{h})}{|\mathbf{h}|}, & \text{if } \mathbf{h} \neq \mathbf{0}; \\[2ex] 0, & \text{if } \mathbf{h} = \mathbf{0}. \end{cases}$$

Then the continuity of $\mathcal{E}$ is immediate from the fact that f is differentiable at $\mathbf{p}$. The theorem follows immediately. $\qquad\square$

Now let's consider a slightly complicated tangent plane question.

Example 2.3.16. Determine all of the points on the graph of

$$z \;=\; f(x, y) \;=\; \frac{5x}{x^2 + y^2 + 1}$$

at which the tangent planes are horizontal.

Solution:

A horizontal plane has an equation of the form $z = c$, where c is a constant. In order for $z = f(\mathbf{p}) + \vec{\nabla}f(\mathbf{p}) \cdot (\mathbf{x} - \mathbf{p})$ to be of this form, $\vec{\nabla}f(\mathbf{p})$ would have to be $\mathbf{0}$. Thus, we need to find those points where both partial derivatives of f are 0, i.e., we need to simultaneously solve

$$\frac{\partial f}{\partial x} = 0 \quad \text{and} \quad \frac{\partial f}{\partial y} = 0.$$

We calculate

$$\frac{\partial f}{\partial x} = 5 \cdot \frac{(x^2 + y^2 + 1)(1) - x(2x)}{\left(x^2 + y^2 + 1\right)^2} = \frac{5(y^2 - x^2 + 1)}{\left(x^2 + y^2 + 1\right)^2}$$

and

$$\frac{\partial f}{\partial y} = 5x \cdot \frac{\partial}{\partial y}\left[\left(x^2 + y^2 + 1\right)^{-1}\right] = \frac{-10xy}{\left(x^2 + y^2 + 1\right)^2}.$$

Thus, both partial derivatives are 0 if and only if

$$y^2 - x^2 + 1 = 0 \quad \text{and} \quad xy = 0.$$

Hence, we must have $x = 0$ or $y = 0$. But if $x = 0$, then $y^2 + 1 = 0$, which is impossible (in the real numbers, where we're working). Therefore, we must have $y = 0$, and then $-x^2 + 1 = 0$, so that $x = \pm 1$.

Finally, we conclude that the tangent plane to the graph is horizontal precisely when $(x, y) = (1, 0)$ and $(x, y) = (-1, 0)$. The points in space are thus $(1, 0, f(1, 0)) = (1, 0, 5/2)$ and $(-1, 0, f(-1, 0)) = (-1, 0, -5/2)$.

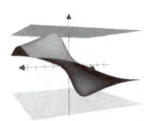

Figure 2.3.5: The graph of $z = f(x, y)$ with horizontal tangent planes.

Example 2.3.17. Let's look at the tangent set in a higher-dimensional example. Consider the function from Example 2.3.7:

$$g(x_1, x_2, x_3, x_4) = x_2 e^{x_1} + x_3 \cos(2\pi x_4).$$

The graph of $z = g(x_1, x_2, x_3, x_4)$ is the set of points $\left(x_1, x_2, x_3, x_4, g(x_1, x_2, x_3, x_4)\right)$ in $\mathbb{R}^5$. So, obviously, we're not going to try to picture this!

We found the linearization of g at $(0, 1, -2, 1)$ to be

$$L(\mathbf{x}) = -1 + x_1 + (x_2 - 1) + (x_3 + 2).$$

So the tangent set to the graph of g at $(0, 1, -2, 1, -1)$ is the set of points in $\mathbb{R}^5$ of the form

$$\left(x_1, \ x_2, \ x_3, \ x_4, \ -1 + x_1 + (x_2 - 1) + (x_3 + 2)\right).$$

We want to give another example of differential approximation, but first, we need to have a serious discussion of why the notation will look the way that it will look.

Recall Example 2.2.10, in which we saw that, if x_i (which we'll write instead of X_i) is the i-th coordinate function on $\mathbb{R}^n$, then $d_{\mathbf{p}}x_i(v_1, v_2, \ldots, v_n) = v_i$. Since this result does not depend on the point $\mathbf{p}$, it is standard to write simply dx_i in place of $d_{\mathbf{p}}x_i$ (though we shall include the $\mathbf{p}$ subscript initially below).

Now, if f is differentiable at $\mathbf{p}$, then

$$d_{\mathbf{p}}f(v_1, v_2, \ldots, v_n) = \vec{\nabla} f(p) \cdot \mathbf{v} = \left.\frac{\partial f}{\partial x_1}\right|_{\mathbf{p}} \cdot v_1 + \left.\frac{\partial f}{\partial x_2}\right|_{\mathbf{p}} \cdot v_2 + \cdots + \left.\frac{\partial f}{\partial x_n}\right|_{\mathbf{p}} \cdot v_n.$$

Combining this with the $d_{\mathbf{p}}x_i$'s, and writing $\mathbf{v}$ for $(v_1, v_2, \ldots, v_n)$, we obtain

$$d_{\mathbf{p}}f(\mathbf{v}) = \left.\frac{\partial f}{\partial x_1}\right|_{\mathbf{p}} \cdot d_{\mathbf{p}}x_1(\mathbf{v}) + \left.\frac{\partial f}{\partial x_2}\right|_{\mathbf{p}} \cdot d_{\mathbf{p}}x_2(\mathbf{v}) + \cdots + \left.\frac{\partial f}{\partial x_n}\right|_{\mathbf{p}} \cdot d_{\mathbf{p}}x_n(\mathbf{v}).$$

Since we now have functions applied to $\mathbf{v}$ on both sides, we can suppress the $\mathbf{v}$, we just write an equality of functions, from $\mathbb{R}^n$ to $\mathbb{R}$:

$$d_{\mathbf{p}}f = \left.\frac{\partial f}{\partial x_1}\right|_{\mathbf{p}} \cdot d_{\mathbf{p}}x_1 + \left.\frac{\partial f}{\partial x_2}\right|_{\mathbf{p}} \cdot d_{\mathbf{p}}x_2 + \cdots + \left.\frac{\partial f}{\partial x_n}\right|_{\mathbf{p}} \cdot d_{\mathbf{p}}x_n.$$

Finally, since everything is now calculated at an arbitrary point $\mathbf{p}$, the explicit reference to $\mathbf{p}$ is usually omitted. We arrive at:

Definition 2.3.18. *The **differential** of f is*

$$df = \frac{\partial f}{\partial x_1} dx_1 + \frac{\partial f}{\partial x_2} dx_2 + \cdots + \frac{\partial f}{\partial x_n} dx_n = \vec{\nabla} f \cdot d\mathbf{x}.$$

Thus, df gives you $d_{\mathbf{p}}f$ for each point $\mathbf{p}$ at which f is differentiable.

> In advanced Calculus books or books on differential geometry, the notation df is used in two different ways, and confusion is avoided by either the context or an explicit statement of what is meant.

> If you want to impress, or horrify, your friends, you can tell them that we have defined the differential as a section of the bundle of vector bundle homomorphisms from the tangent bundle of $\mathbb{R}^n$ to the pull-back of the tangent bundle of $\mathbb{R}$, instead of being a map between total spaces of the tangent bundles.

When you use the differential in approximation problems, what you do is "replace" each dx_i by the actual small change Δx_i, and then the differential yields the approximate change in Δf. Of course, what we think of as "replacing" is actually evaluating both sides at the vector $\Delta \mathbf{x}$, and then using differential approximation at the point $\mathbf{p}$ in question, i.e., using $\Delta f \approx d_{\mathbf{p}}f(\Delta \mathbf{x}) = \vec{\nabla} f(\mathbf{p}) \cdot \Delta \mathbf{x}$.

Example 2.3.19. Suppose that

$$\omega(x, y, z) \ = \ e^{x^2 y} \sin\left(\frac{\pi z}{2}\right).$$

Calculate the general differential $d\omega$, then calculate, $d_{(3,2,1)}\omega$ and $d_{(3,2,1)}\omega(-2, 5, 7)$. Finally, if (x, y, z) is initially $(3, 2, 1)$ and changes by $\Delta(x, y, z) = (-0.02, 0.05, 0.07)$, then approximate the corresponding change in ω.

Solution:

We first calculate

$$d\omega \ = \ \frac{\partial \omega}{\partial x}\, dx \ + \ \frac{\partial \omega}{\partial y}\, dy \ + \ \frac{\partial \omega}{\partial z}\, dz \ =$$

$$2xy e^{x^2 y} \sin\left(\frac{\pi z}{2}\right)\, dx \ + \ x^2 e^{x^2 y} \sin\left(\frac{\pi z}{2}\right)\, dy \ + \ \frac{\pi e^{x^2 y}}{2} \cos\left(\frac{\pi z}{2}\right)\, dz.$$

This is the general differential.

To find $d_{(3,2,1)}\omega$, we now plug $(x, y, z) = (3, 2, 1)$ into the coefficients of dx, dy, and dz. We obtain:

> Recall that it is standard to **not** subscript dx, dy, and dz with the point $\mathbf{p}$, even when calculating the differential at a specific point $\mathbf{p}$.

$$d_{(3,2,1)}\omega \ = \ 12e^{18}\, dx \ + \ 9e^{18}\, dy \ + \ 0\, dz \ = \ 3e^{18}\left(4\, dx + 3\, dy\right).$$

Now we quickly obtain

$$d_{(3,2,1)}\omega(-2, 5, 7) \ = \ 3e^{18}\big(4 \cdot (-2) + 3 \cdot 5\big) \ = \ 21e^{18}.$$

The approximate change in f is given by differential approximation

$$\Delta f \ \approx \ d_{(3,2,1)}\omega(-0.02, 0.05, 0.07) \ = \ d_{(3,2,1)}\omega\Big(0.01(-2, 5, 7)\Big) \ =$$

$$0.01\, d_{(3,2,1)}\omega(-2, 5, 7) \ = \ 0.01 \cdot 21e^{18} \ = \ 0.21e^{18}.$$

Of course, this result is precisely what you get from taking

$$d_{(3,2,1)}\omega \ = \ 12e^{18}\, dx \ + \ 9e^{18}\, dy \ + \ 0\, dz,$$

and replacing (dx, dy, dz) by the actual changes $(\Delta x, \Delta y, \Delta z) = (-0.02, 0.05, 0.07)$.

Note that, in the previous example, our approximate change in ω was $0.21e^{18}$; this number is **huge**. Consequently, it would be reasonable to expect that the error using linear approximation would be large, in an absolute sense, but might be small relative to the size of ω itself. To make the point more clear, consider a specific example: a change in length of 1 foot for something that's 0.5 feet long seems like a big change, while a change of 1 foot for something that's 10,000 feet long doesn't seem like much of a change.

It is frequently the case that what you care about isn't just the change in a function f, but its change, **relative to its initial size or value**; that is, rather than calculating just Δf, you want to calculate $\frac{\Delta f}{f}$. Of course, this is approximated by $\frac{df}{f}$, the *relative differential*.

Example 2.3.20. The power loss across a resistor is given by

$$P = I^2 R,$$

where P is the power loss in watts, I is the current, in amps, and R is the resistance, in ohms.

Express the relative differential of P in terms of the relative differentials of I and R, and use this to approximate the relative change in P, if I goes up by 10%, but R goes down by 5%.

Solution:

We find

$$dP = \frac{\partial P}{\partial I}\, dI + \frac{\partial P}{\partial R}\, dR = 2IR\, dI + I^2\, dR.$$

Thus,

$$\frac{dP}{P} = \frac{2IR\, dI + I^2\, dR}{I^2 R} = 2 \cdot \frac{dI}{I} + \frac{dR}{R},$$

which expresses the relative differential of P in terms of the relative differentials of I and R.

We now replace dI with the actual ΔI and dR with the actual ΔR or, here, dI/I with $\Delta I/I$ and dR/R with $\Delta R/R$. We are given that

$$\frac{\Delta I}{I} = 0.1 \quad \text{and} \quad \frac{\Delta R}{R} = -0.05.$$

Therefore, we find that

$$\frac{\Delta P}{P} \approx \frac{dP}{P} = 2(0.1) + (-0.05) = 0.15,$$

i.e., the relative change in P is approximately 15%.

Online answers to select exercises are here.

Basics:

2.3.1   Exercises

In each of the following exercises, you are given the values of $f(\mathbf{p})$ and $\vec{\nabla} f(\mathbf{p})$ for a differentiable function f. (a) Give a formula for the linear approximation of $f(x, y)$ or $f(x, y, z)$ at an arbitrary point (x, y) or (x, y, z) that's close to $\mathbf{p}$, and (b) use your answer to part (a) to approximate the value of f at the given point q.

1. $f(1, 2) = 7$, $\vec{\nabla} f(1, 2) = (-3, 4)$, $\mathbf{q} = (1.001, 2.003)$

2. $f(200, -100) = 1$, $\vec{\nabla} f(200, -100) = (0.01, 0.03)$, $\mathbf{q} = (202, -101)$

3. $f(3, 2, 1) = -4$, $\vec{\nabla} f(3, 2, 1) = (-2, 0, 1)$, $\mathbf{q} = (3.1, 2.2, 1.3)$

4. $f(0, 0, 0) = 1000$, $\vec{\nabla} f(0, 0, 0) = (10, -12, 11)$, $\mathbf{q} = (0.01, 0.02, -0.01)$

In each of the following exercises, you are given a function and two points, p and a. Use linear approximation at p to approximate the value of the function at a.

5. $f(x, y) = 5x^3 y - \sqrt{xy}$, $\mathbf{p} = (3, 3)$, $\mathbf{a} = (2.995, 3.003)$ ▶

6. $g(s, t) = e^{s^2 - t}$, $\mathbf{p} = (1, 0)$, $\mathbf{a} = (1.01, -0.04)$

7. $H(u, w) = 2^u \cos(\pi w)$, $\mathbf{p} = (2, 1)$, $\mathbf{a} = (1.99, 1.01)$

8. $L(x, y, z) = (x + y - 2zy)^2$, $\mathbf{p} = (-1, 2, -2)$, $\mathbf{a} = (-1.02, 2.009, -1.992)$

9. $z(x_1, x_2, x_3) = x_2 e^{x_3} - 3x_1^2 x_3$, $\mathbf{p} = (-4, 0.5, 0)$, $\mathbf{a} = (-3.98, 0.51, 0.003)$

10. $\kappa(x, t, \omega, \mu) = \dfrac{\mu \tan(\pi t)}{\sqrt{x - \omega}}$, $\mathbf{p} = (5, -1, 1, 2)$, $\mathbf{a} = (4.999, -1.001, 0.999, 2.002)$

In each of the following exercises, (a) find the linearization of f at p, (b) give an equation for the tangent set to the graph of f at $(\mathbf{p}, f(\mathbf{p}))$, and (c) use the linearization to estimate the value of f at a.

11. $f(x, y) = x^3 y - 5x^2 y^2 + 2xy^3$, $\mathbf{p} = (-1, 3)$, $\mathbf{a} = (-0.999, 2.997)$ ▶

12. $f(x, y) = x^2 \sec(\pi y)$, $\mathbf{p} = (3, 2)$, $\mathbf{a} = (3.1, 2.08)$

13. $f(x, y) = \dfrac{e^{3y - 2x}}{e^{y - 4x}}$, $\mathbf{p} = (2, -2)$, $\mathbf{a} = (2.06, -2.06)$

14. $f(x, M, \delta) = Mx^2 - \frac{2}{3} M^2 \delta^2 + x^5$, $\mathbf{p} = (2, -3, 7)$, $\mathbf{a} = (2.005, -2.995, 6.995)$

15. $f(x, y, z) = \dfrac{z}{y\sqrt{x}}$, $\mathbf{p} = (100, 4, 50)$, $\mathbf{a} = (102, 3.9, 50)$

16. $f(x, y, t, s) = (3x + y)\cot(\pi t + 2\pi s^2), \mathbf{p} = (0.5, 1, 0.5, 2), \mathbf{a} = (0.51, 0.98, 0.51, 2.03)$

In each of the following exercises, you are given a function g and a starting point p. Use differential approximation with the given Δ to estimate the change in g.

17. $g(x, y) = 7x(y^2 - 5), \mathbf{p} = (3, 6), \Delta(x, y) = (0.01, 0.005)$

18. $g(x, y) = e^{x^y}, \mathbf{p} = (1, 27), \Delta(x, y) = (0.3, 0.3)$

19. $g(s, t, \omega) = \frac{1}{3}s^2 t^2 - \sqrt{t}\cos(\omega), \mathbf{p} = (-2, 4, 0), \Delta(s, t, \omega) = (0.004, -0.007, 0.2)$

20. $g(a, b, u, v) = a^b - 3au + 3bv + u^v, \mathbf{p} = (2, 2, 3, 2), \Delta(a, b, u, v) = (0.02, 0, 0.03, 0)$

21. $g(k_1, k_2, k_3, k_4) = \sqrt[3]{k_1 + 2k_2} - (k_3 + 2k_4)^3, \mathbf{p} = (4, 30, 1, 2), \Delta(k_1, k_2, k_3, k_4) = (-1, 1, -0.05, 0.05)$

22. A cardboard box is measured to have length, width, and height of 2, 3, and 1 feet, respectively, to enclose a volume of 6 cubic feet. However, more-careful measurements show that the box is really 2.01 by 2.98 by 1.03 feet. (a) Use linear approximation to estimate the revised (measured) volume of the box. (b) Use differential approximation to estimate the change in the (measured) volume of the box. (c) Show that your answers to part (a) and (b) agree.

23. Recall Example 2.1.17, in which we had a fixed number of atoms or molecules of an ideal gas in a container, where the container had a volume that can change with time, such as a balloon or the inside of a piston.

    The Ideal Gas Law can be written as $P = \dfrac{kT}{V}$, where P is the pressure, in pascals (Pa, which are N/m^2), T is the temperature T, in kelvins (K), V is the volume V, in m^3, and k is a constant, which we shall assume has a value of $k = 8$ N·m/K in this problem.

    Suppose that $T = 100.32$ and $V = 10.17$. Use linear approximation to estimate P. What is the actual value of P?

24. An online store sets up a script which automatically adjusts the price of a collectible snargit over time. The price function is given by $S = 1000 + n\sqrt{2t}$, where S is the price (in Swiss francs) of the snargit, $t \geq 0$ is the number of days since the sales began, and $n \geq 0$ is the number of snargits that have already been sold.

    On the 30th day of sales, a total of 647 snargits have been sold. In the next 24 hours, 33 more sales are completed. Approximately how much has the price increased in this 24-hour period?

| More Depth: | In each of the following exercises, determine all of the points on the graph of the given function at which the tangent plane is horizontal. |

25. $f(x, y) = \dfrac{x^2 y}{xy + 3}$ ▶

26. $g(s, t) = \dfrac{3s^2}{t^2 - s^2 - 5}$

27. $j(\sigma, \omega) = (\sigma + 7)^2 \sin^2(\omega)$

28. $k(x_1, x_2) = \dfrac{2x_1}{3x_1^2 + x_2^2 - 13}$

29. $N(x, t) = \dfrac{t - 5}{(x + 1)^2 + (t - 1)^2}$

30. $P(u, v) = \cos(u - v)\cos(u + v)$

In each of the following exercises, find an equation for the tangent set of the given function at the point p.

31. $f(x, y, z) = (x + 3z)(y^2 - 3yz), \mathbf{p} = (5, -2, 3)$

32. $S(u, v, w) = ue^{v^2 - 2w^2}, \mathbf{p} = (-7, 3, 2)$

33. $H(x_1, x_2, x_3, x_4) = x_1^{x_2} \sin(x_3 + x_4), \mathbf{p} = (1, 2, 3\pi, 4\pi)$

34. $\xi(x, y, \mu, \omega) = x^2 \mu^3 - 4y\mu + 4x\omega - y^2\omega^3, \mathbf{p} = (-2, 3, 1, -2)$

In each of the following exercises, find (a) the general differential dg and (b) approximate the change in g, if the point at which g is evaluated is initially p and then changes by Δ.

35. $g(u, v, w) = u^2 v - e^{-vw^2}, \mathbf{p} = (6, 1, 0), \Delta(u, v, w) = (0.03, -0.01, 0.01)$ ▶

36. $g(s, t, \omega) = \dfrac{t^s \cos(\pi\omega)}{s}, \mathbf{p} = (3, 1, 3), \Delta(s, t, \omega) = (-0.02, 0.05, -0.02)$

37. $g(x, y, z) = \dfrac{x^5 y^3 z}{2} - \dfrac{x^4 y^2 z^2}{3}, \mathbf{p} = (2, -3, 8), \Delta(x, y, z) = (-0.001, 0.001, 0.002)$

38. $g(p, t, \mu) = p\ln(t - 2\mu), \mathbf{p} = (-7, 3, 1), \Delta(p, t, \mu) = (0.3, 0.1, 0.3)$

39. The Ideal Gas Law, which we looked at earlier, can be written in the form

$$PV = NkT,$$

where P is the pressure of the gas measured in atmospheres, N is the number of gas particles in the system, T is the temperature in kelvins, and V is the volume of the gas in cubic meters. k is Boltzmann's constant, which has a value of approximately $1.38 \times 10^{-23} J \cdot K^{-1}$.

    a. Approximate the relative change in pressure if the temperature increases by 7%, the volume increases by 5%, and the number of particles remains unchanged.

    b. Approximate the relative change in temperature if the number of particles increases by 12%, the volume decreases by 3%, and the pressure increases by 5%.

    c. Imagine an alternate universe in which Boltzmann's constant is not a constant at all, but a variable. Approximate the relative change in k if the temperature decreases by 20%, the pressure decreases by 7%, the number of particles increases by 10%, and the volume increases by 14%.

40. Recall Example 2.1.21, in which a new resistor of resistance R_{new} ohms is formed by placing n resistors of resistance $R_1, \ldots, R_n$ ohms in parallel. We will assume that $n = 3$, so that

$$\frac{1}{R_{\text{new}}} = \frac{1}{R_1} + \frac{1}{R_2} + \frac{1}{R_3}.$$

Suppose that, initially, $(R_1, R_2, R_3) = (100, 120, 160)$, but then R_1 increases by 3%, R_2 decreases by 1%, and R_3 increases by 2%.

    (a) Use relative differentials to approximate the relative change in R_{new}.

    (b) What is the exact relative change in R_{new}?

2.4   Differentiation Rules

In single-variable Calculus, there are various rules for calculating derivatives: the Product Rule, the Quotient Rule, and Chain Rule, among others. We wish to have similar rules for derivatives of multivariable functions, rules which apply to gradient vectors, total derivatives, or differentials.

As we shall see in this section, most of the rules that we want follow easily from their single-variable counterparts. However, the multivariable Chain Rule does not follow from the single-variable Chain Rule; it's really something new, and we will spend the majority of this section discussing it.

In the Basics portion of this section, we shall deal with the differentiation rules in terms of partial derivatives and gradient vectors. In the More Depth portion, we shall give the results in terms of total derivatives and differentials.

Basics:

In single-variable Calculus, if you want to calculate the derivative $(3x^5 - 7\sin x)'$, you can quickly compute $3 \cdot 5x^4 - 7\cos x$. This requires you to know the derivatives of x^5 and $\sin x$, and also requires you to use something that you might not even think about anymore: differentiation is a *linear operation*, i.e., if a and b are constants, and $f = f(x)$ and $g = g(x)$ are functions, then $(af + bg)' = af' + bg'$. That's why you can split up sums and differences and pull out constants.

We would like to have such linearity for a multivariable function f. Of course, partial differentiation is just differentiation with respect to a single variable, and so partial differentiation is linear. Applying this to each component of the gradient vector of f immediately yields

---

Theorem 2.4.1. (Linearity of Differentiation) *Suppose that f and g are real-valued functions on a subset of $\mathbb{R}^n$, and that a and b are constants. Then, at any point at which $\vec{\nabla}f$ and $\vec{\nabla}g$ exist, $\vec{\nabla}(af + bg)$ exists and*

$$\vec{\nabla}(af + bg) \;=\; a\,\vec{\nabla}f \;+\; b\,\vec{\nabla}g.$$

Example 2.4.2. Suppose that we have functions $f = f(x, y, z)$ and $g = g(x, y, z)$, and we know that $\vec{\nabla}f(3, 5, -7) = (2, 9, 1)$ and $\vec{\nabla}g(3, 5, -7) = (-1, 0, 4)$. Calculate $\vec{\nabla}(3f - 2g)(3, 5, -7)$.

Solution:

We find

$$\vec{\nabla}(3f - 2g)(3, 5, -7) \;=\; 3\vec{\nabla}f(3, 5, -7) - 2\vec{\nabla}g(3, 5, -7) \;=\; 3(2, 9, 1) - 2(-1, 0, 4) \;=$$

$$(6, 27, 3) + (2, 0, -8) \;=\; (8, 27, -5).$$

The Product Rule for multivariable functions is easy to derive from the single-variable Product Rule. If we have $f = f(\mathbf{x})$ and $g = g(\mathbf{x})$, then, for each variable x_i, the single-variable Product Rule gives us

$$\frac{\partial (fg)}{\partial x_i} \;=\; f \frac{\partial g}{\partial x_i} \;+\; g \frac{\partial f}{\partial x_i}.$$

From this it follows quickly that:

> **Theorem 2.4.3. (Product Rule)** *Suppose that f and g are real-valued functions on a subset of $\mathbb{R}^n$. Then, at any point $\mathbf{p}$ at which $\vec{\nabla}f$ and $\vec{\nabla}g$ exist, $\vec{\nabla}(fg)$ exists and*
> $$\vec{\nabla}(fg)(\mathbf{p}) \;=\; f(\mathbf{p})\,\vec{\nabla}g(\mathbf{p}) \;+\; g(\mathbf{p})\,\vec{\nabla}f(\mathbf{p}).$$
> *We usually write simply*
> $$\vec{\nabla}(fg) \;=\; f\,\vec{\nabla}g \;+\; g\,\vec{\nabla}f.$$

Example 2.4.4. Calculate the gradient vector $\vec{\nabla}\left[\left(x^2 y + \cos z\right)\left(z \ln y + \sqrt{x}\right)\right]$.

Solution:

We find

$$\vec{\nabla}\left[\left(x^2 y + \cos z\right)\left(z \ln y + \sqrt{x}\right)\right] \;=$$

$$\left(x^2 y + \cos z\right)\vec{\nabla}\left(z \ln y + \sqrt{x}\right) \;+\; \left(z \ln y + \sqrt{x}\right)\vec{\nabla}\left(x^2 y + \cos z\right) \;=$$

$$\left(x^2 y + \cos z\right)\left(\frac{1}{2}x^{-1/2},\; \frac{z}{y},\; \ln y\right) \;+\; \left(z \ln y + \sqrt{x}\right)\left(2xy,\; x^2,\; -\sin z\right).$$

There is no good reason to try to "simplify" this gradient vector answer by multiplying through by the scalars and adding/combining the vectors. Leave the expression as it is; if you have to evaluate the gradient vector at a point, plug the point into this expanded form, and combine the vectors at the end, if it's desirable.

The Quotient Rule for multivariable functions is also easy to derive from the single-variable Quotient Rule. If we have $f = f(\mathbf{x})$ and $g = g(\mathbf{x})$, then, for each variable x_i, the single-variable Quotient Rule gives us

$$\frac{\partial}{\partial x_i}\left(\frac{f}{g}\right) = \frac{g\dfrac{\partial f}{\partial x_i} - f\dfrac{\partial g}{\partial x_i}}{g^2},$$

provided that g is not 0.

From this it follows quickly that:

Theorem 2.4.5. (Quotient Rule) *Suppose that f and g are real-valued functions on a subset of $\mathbb{R}^n$. Then, at any point $\mathbf{p}$ at which $\vec{\nabla}f$ and $\vec{\nabla}g$ exist, and at which $g \neq 0$, $\vec{\nabla}(f/g)$ exists and*

$$\vec{\nabla}\left(\frac{f}{g}\right)\bigg|_{\mathbf{p}} = \frac{g(\mathbf{p})\,\vec{\nabla}f(\mathbf{p}) - f(\mathbf{p})\,\vec{\nabla}g(\mathbf{p})}{\left(g(\mathbf{p})\right)^2}.$$

We usually write simply

$$\vec{\nabla}\left(\frac{f}{g}\right) = \frac{g\,\vec{\nabla}f - f\,\vec{\nabla}g}{g^2}.$$

Example 2.4.6. Calculate $\vec{\nabla}\left(\dfrac{y^2\tan^{-1}x}{y - x^3}\right)$ at the point $(1, 5)$.

Solution: We find

$$\vec{\nabla}\left(\frac{y^2\tan^{-1}x}{y - x^3}\right) = \frac{(y - x^3)\,\vec{\nabla}(y^2\tan^{-1}x) - (y^2\tan^{-1}x)\,\vec{\nabla}(y - x^3)}{(y - x^3)^2} =$$

$$\frac{(y - x^3)\left(\dfrac{y^2}{1 + x^2},\ 2y\tan^{-1}x\right) - (y^2\tan^{-1}x)(-3x^2,\ 1)}{(y - x^3)^2}.$$

Therefore, at $(1, 5)$, we find

$$\vec{\nabla}\left(\frac{y^2\tan^{-1}x}{y - x^3}\right)(1, 5) =$$

$$\frac{(5 - 1^3)\left(\dfrac{5^2}{1 + 1^2},\ 2 \cdot 5 \tan^{-1} 1\right) - (5^2 \tan^{-1} 1)(-3 \cdot 1^2,\ 1)}{(5 - 1^3)^2} =$$

$$\frac{4\left(\dfrac{25}{2},\ 10 \cdot \dfrac{\pi}{4}\right) - \left(25 \cdot \dfrac{\pi}{4}\right)(-3,\ 1)}{16} =$$

$$\frac{1}{16}\left(\frac{200 + 75\pi}{4},\ \frac{15\pi}{4}\right) = \frac{5}{64}\left(40 + 15\pi,\ 3\pi\right).$$

The *Power Rule* for multivariable functions follows immediately from the single-variable Power Rule together with the single-variable Chain Rule. Consider the single-variable power function $g = g(z) = z^\alpha$, where α is an arbitrary real number, and a multivariable function $f = f(\mathbf{x})$. Then, for each x_i,

$$\frac{\partial}{\partial x_i}(f^\alpha) = \alpha f^{\alpha - 1} \frac{\partial f}{\partial x_i},$$

provided that $f^{\alpha - 1}$ and $\partial f / \partial x_i$ exist.

Theorem 2.4.7. (Power Rule) *Suppose $f^{\alpha - 1}$ and $\vec{\nabla} f$ exist. Then, $\vec{\nabla}(f^\alpha)$ exists and*

$$\vec{\nabla}(f^\alpha) = \alpha f^{\alpha - 1} \vec{\nabla} f.$$

Example 2.4.8. Suppose that $h(3, 7, 1) = 4$ and $\vec{\nabla} h(3, 7, 1) = (-8, 0, 4)$. Calculate $\vec{\nabla}\sqrt{h}$ at $(3, 7, 1)$.

Solution:

From the Power Rule, we know that

$$\vec{\nabla}\sqrt{h} = \vec{\nabla}(h^{1/2}) = \frac{1}{2}h^{-1/2}\vec{\nabla}h = \frac{1}{2\sqrt{h}}\vec{\nabla}h.$$

Therefore,

$$\vec{\nabla}\sqrt{h}\,\Big|_{(3,7,1)} = \frac{1}{2\sqrt{h(3, 7, 1)}}\vec{\nabla}h(3, 7, 1) = \frac{1}{4}(-8, 0, 4) = (-2, 0, 1).$$

Now we come to the Chain Rule. This result does **not** follow quickly from the single-variable Chain Rule. Before we look at the multivariable case, let's recall how the single-variable Chain Rule looks in Leibniz's notation (the notation with the d's); the Chain Rule says that, if f is a function of x, and x is itself a function of another variable t, then, if you consider f as a function of t, the derivatives are related by

$$\frac{df}{dt} = \frac{df}{dx} \cdot \frac{dx}{dt},$$

i.e., it looks like algebraic cancellation of the dx's occurs.

It is important to realize that "consider f as a function of t" means, technically, that you look at the two functions $f(x)$ and $x(t)$ and take the derivative, with respect to t, of the composition $f(x(t)) = (f \circ x)(t)$. In this form, the single-variable Chain Rule is

$$(f \circ x)'(t) = f'\big(x(t)\big) \cdot x'(t).$$

Now let's look at the multivariable case.

Suppose that f is a function of x, y, and z, and that x, y, and z are themselves functions of two other variables s and t. We usually write $f = f(x, y, z)$, and $x = x(s, t)$, $y = y(s, t)$ and $z = z(s, t)$. Then, we can think of f as a function of s and t, and the question is:

Can we express $\partial f / \partial s$ and $\partial f / \partial t$ in terms of the partial derivatives of f, with respect to x, y, and z, and the partial derivatives of x, y, and z, with respect to s and t?

The answer is "yes", and one form of the *multivariable Chain Rule* gives us that

$$\frac{\partial f}{\partial s} = \frac{\partial f}{\partial x} \cdot \frac{\partial x}{\partial s} + \frac{\partial f}{\partial y} \cdot \frac{\partial y}{\partial s} + \frac{\partial f}{\partial z} \cdot \frac{\partial z}{\partial s} \tag{2.5}$$

and

$$\frac{\partial f}{\partial t} = \frac{\partial f}{\partial x} \cdot \frac{\partial x}{\partial t} + \frac{\partial f}{\partial y} \cdot \frac{\partial y}{\partial t} + \frac{\partial f}{\partial z} \cdot \frac{\partial z}{\partial t}. \tag{2.6}$$

How can you possibly remember these formulas??? Actually, they're not too bad.

If you want to calculate $\partial f / \partial s$, you take the partial derivatives of f with respect to each of the variables that it explicitly depends on — here, x, y, and z — and then you multiply each of those partial derivatives by what you would need to multiply by so that, if algebraic cancellation were occurring, you'd get $\partial f / \partial s$ each time, and then you add.

So, you take f. Then, you think

$$\frac{\partial f}{\partial s} = \frac{\partial f}{\partial x} \cdot ? + \frac{\partial f}{\partial y} \cdot ? + \frac{\partial f}{\partial z} \cdot ?,$$

where you have to fill in the question marks with the appropriate quantities. What are the appropriate quantities?

You multiply $\partial f / \partial x$ by whatever looks like the right "fraction" so that, if algebraic cancellation were occurring, you'd get the $\partial f / \partial s$ that you're after. Thus, you multiply $\partial f / \partial x$ by $\partial x / \partial s$. Similarly, you multiply $\partial f / \partial y$ by $\partial y / \partial s$ and you multiply $\partial f / \partial z$ by $\partial z / \partial s$.

Of course, when calculating $\partial f / \partial t$, you proceed along the exact same lines; you just replace all of the s's with t's.

Note that Formula 2.5 and Formula 2.6 can also be written in terms of the dot product:

$$\frac{\partial f}{\partial s} = \vec{\nabla} f(x, y, z) \cdot \frac{\partial}{\partial s}(x, y, z),$$

and

$$\frac{\partial f}{\partial t} = \vec{\nabla} f(x, y, z) \cdot \frac{\partial}{\partial t}(x, y, z).$$

> Algebraic cancellation in fraction multiplication is definitely **not** what's going on in the Chain Rule. If it were, we would get 3 times $\partial f / \partial s$, not just $\partial f / \partial s$. Thinking this way is just a helpful way of remembering the Chain Rule.

The general theorem is a bit long to state, but keep our discussion above in mind as you read it. In Remark 2.4.20, we will give an idea of why the formulas look like they do; for the actual proof, see Theorem 5.4.3 in Trench, [8].

Theorem 2.4.9. (Chain Rule for Partial Derivatives) *Suppose that* $\mathbf{t} = (t_1, \ldots, t_m)$, *that* $\mathbf{x} = (x_1, \ldots, x_n)$, *and that* $f(\mathbf{x})$ *is a real-valued function of* $\mathbf{x}$, *while* $\mathbf{x}$ *is itself a multi-component function* $\mathbf{x} = \mathbf{x}(\mathbf{t})$. *Then, by composing, we may consider* f *as a function of* $\mathbf{t}$.

Let $\mathbf{a}$ *be a specific value of* $\mathbf{t}$, *i.e., a specific point in* $\mathbb{R}^m$, *and fix a specific coordinate function* t_i. *Suppose that the vector of partial derivatives*

$$\frac{\partial \mathbf{x}}{\partial t_i} = \left(\frac{\partial x_1}{\partial t_i}, \frac{\partial x_2}{\partial t_i}, \ldots, \frac{\partial x_m}{\partial t_i} \right)$$

exists at $\mathbf{a}$, *and that* f *is differentiable at* $\mathbf{x}(\mathbf{a})$.

Then, $\partial f / \partial t_i$ *exists at* $\mathbf{a}$, *and*

$$\left. \frac{\partial f}{\partial t_i} \right|_{\mathbf{a}} = \vec{\nabla} f(\mathbf{x}(\mathbf{a})) \cdot \left. \frac{\partial \mathbf{x}}{\partial t_i} \right|_{\mathbf{a}}.$$

If $m = 1$, so that $\mathbf{x} = \mathbf{x}(t)$ depends on a single variable, then $f(\mathbf{x}(t))$ is also a function of a single variable, and the partial derivatives with respect to t_i, in the Chain Rule, are usually replaced by ordinary derivatives with respect to t. This does have the unfortunate effect of destroying the "algebraic cancellation look" of the Chain Rule.

The references to a specific $\mathbf{a}$ are usually omitted, and we write simply

$$\frac{\partial f}{\partial t_i} = \vec{\nabla} f(\mathbf{x}) \cdot \frac{\partial \mathbf{x}}{\partial t_i}$$

or

$$\frac{\partial f}{\partial t_i} = \frac{\partial f}{\partial x_1} \cdot \frac{\partial x_1}{\partial t_i} + \frac{\partial f}{\partial x_2} \cdot \frac{\partial x_2}{\partial t_i} + \cdots + \frac{\partial f}{\partial x_n} \cdot \frac{\partial x_n}{\partial t_i}.$$

Here, where we write "so that g is explicitly a function of u and v", we are technically dealing with the composition of $g = g(x, y)$ with the function given by $(x, y) = (u - v, u + v)$.

Example 2.4.10. We'll start with something simple. Let $g(x, y) = xy$, $x = u - v$, and $y = u + v$.

Let's calculate the partial derivatives $\partial g / \partial u$ and $\partial g / \partial v$ two different ways: first, by using the Chain Rule and, second, by plugging the x and y expressions into g, so that g is explicitly a function of u and v, and then calculating the partial derivatives.

• Using the Chain Rule, we find

$$\frac{\partial g}{\partial u} = \frac{\partial g}{\partial x}\frac{\partial x}{\partial u} + \frac{\partial g}{\partial y}\frac{\partial y}{\partial u} = y \cdot 1 + x \cdot 1 = (u + v) + (u - v) = 2u,$$

and

$$\frac{\partial g}{\partial v} = \frac{\partial g}{\partial x}\frac{\partial x}{\partial v} + \frac{\partial g}{\partial y}\frac{\partial y}{\partial v} = y \cdot (-1) + x \cdot 1 = (-u - v) + (u - v) = -2v.$$

• Plugging $x = u - v$ and $y = u + v$ into $g(x, y)$, we obtain $g = (u - v)(u + v) = u^2 - v^2$. So, we easily calculate that

$$\frac{\partial g}{\partial u} = 2u \quad \text{and} \quad \frac{\partial g}{\partial v} = -2v,$$

which, of course, agrees with what we obtained from the Chain Rule.

Example 2.4.11. Suppose that $T = T(u, v, w)$, and that $u = r \cos t$, $v = 5r - 3t$, and $w = r^2 e^t$. Also, suppose that $\vec{\nabla} T(2, 10, 4) = (-1, 3, 2)$. Calculate

$$\left. \frac{\partial T}{\partial r} \right|_{(r,t)=(2,0)} \quad \text{and} \quad \left. \frac{\partial T}{\partial t} \right|_{(r,t)=(2,0)}.$$

Solution:

First, note that, when $(r,t) = (2,0)$, $(u,v,w) = (2,10,4)$, the point at which you're given the value of the gradient vector; this had better be the case, or you wouldn't have the data that you need to solve the problem.

Now we will use the Chain Rule twice, in its gradient form:

$$\frac{\partial T}{\partial r} = \vec{\nabla}T(u,v,w) \cdot \frac{\partial}{\partial r}(u,v,w) \quad \text{and} \quad \frac{\partial T}{\partial t} = \vec{\nabla}T(u,v,w) \cdot \frac{\partial}{\partial t}(u,v,w), \quad (2.7)$$

where we will, in the end, apply these at $(r,t) = (2,0)$ and $(u,v,w) = (2,10,4)$.

As we are given the gradient vector at the point that we need, what we still need to do is calculate the partial derivatives of $(u,v,w) = (r\cos t, \, 5r - 3t, \, r^2 e^t)$ at $(r,t) = (2,0)$. We find

$$\frac{\partial}{\partial r}(u,v,w)\Big|_{(r,t)=(2,0)} = \frac{\partial}{\partial r}(r\cos t, \, 5r - 3t, \, r^2 e^t)\Big|_{(r,t)=(2,0)} =$$

$$(\cos t, \, 5, \, 2re^t)\Big|_{(r,t)=(2,0)} = (1,5,4)$$

and

$$\frac{\partial}{\partial t}(u,v,w)\Big|_{(r,t)=(2,0)} = \frac{\partial}{\partial t}(r\cos t, \, 5r - 3t, \, r^2 e^t)\Big|_{(r,t)=(2,0)} =$$

$$(-r\sin t, \, -3, \, r^2 e^t)\Big|_{(r,t)=(2,0)} = (0,-3,4).$$

Now, applying the Chain Rule, as given in Formula 2.7, we obtain

$$\frac{\partial T}{\partial r}\Big|_{(r,t)=(2,0)} = \vec{\nabla}T(2,10,4) \cdot (1,5,4) = (-1,3,2) \cdot (1,5,4) = -1 + 15 + 8 = 22$$

and

$$\frac{\partial T}{\partial t}\Big|_{(r,t)=(2,0)} = \vec{\nabla}T(2,10,4) \cdot (0,-3,4) = (-1,3,2) \cdot (0,-3,4) = 0 - 9 + 8 = -1.$$

Example 2.4.12. Suppose that

$$f(x,y) = x^2 y + xy^3, \qquad x = 5s^2 + 7st, \qquad \text{and} \qquad y = 3st + 4t^2.$$

We'll calculate the partial derivatives $\partial f/\partial s$ and $\partial f/\partial t$ by using the Chain Rule.

We quickly find

$$\frac{\partial f}{\partial s} = \frac{\partial f}{\partial x}\frac{\partial x}{\partial s} + \frac{\partial f}{\partial y}\frac{\partial y}{\partial s} = \left(2xy + y^3\right)\left(10s + 7t\right) + \left(x^2 + 3xy^2\right)\left(3t\right)$$

and

$$\frac{\partial f}{\partial t} = \frac{\partial f}{\partial x}\frac{\partial x}{\partial t} + \frac{\partial f}{\partial y}\frac{\partial y}{\partial t} = \left(2xy + y^3\right)\left(7s\right) + \left(x^2 + 3xy^2\right)\left(3s + 8t\right).$$

Should you leave your answer in this form? That depends.

If you are given a specific (s,t) pair, then you can calculate the corresponding (x,y) pair, using that $x = 5s^2 + 7st$ and $y = 3st + 4t^2$, and then plug those two pairs of numbers into the calculated partial derivatives.

If you want your final answers in terms of just s and t, then, at this point, you should replace each occurrence of x by $5s^2 + 7st$ and each occurrence of y by $3st + 4t^2$. We shall do that here, in order to make a point.

We obtain

$$\frac{\partial f}{\partial s} = \left[2(5s^2 + 7st)(3st + 4t^2) + (3st + 4t^2)^3\right]\left(10s + 7t\right) +$$
$$\left[(5s^2 + 7st)^2 + 3(5s^2 + 7st)(3st + 4t^2)^2\right]\left(3t\right)$$

and

$$\frac{\partial f}{\partial t} = \left[2(5s^2 + 7st)(3st + 4t^2) + (3st + 4t^2)^3\right]\left(7s\right) +$$
$$\left[(5s^2 + 7st)^2 + 3(5s^2 + 7st)(3st + 4t^2)^2\right]\left(3s + 8t\right).$$

Notice how much more manageable the partial derivatives looked in terms of x, y, s, and t, even though we had to keep around the two original equations $x = 5s^2 + 7st$ and $y = 3st + 4t^2$. The Chain Rule "automatically" breaks up the partial derivatives into more manageable pieces.

Don't panic if you don't know what electric potential means; knowing what it means is actually irrelevant to solving the problem. However, if you're interested: an electric potential function is a function ϕ such that $\vec{\nabla}\phi$ (or, $-\vec{\nabla}\phi$ for physicists), is the associated *electric field*. Hence, electric potential is defined only up to the addition of an arbitrary constant. Equivalently, you can take the potential to be zero at a specific point.

We shall look at *potential functions* and *vector fields* in depth in Chapter 4.

Example 2.4.13. Suppose that a particle is moving through space, and that, at time $t = 2$ seconds, the particle is at the point $(3, 4, 7)$, in meters, and is moving with an instantaneous velocity of $(-2, 1, 5)$ m/s. Suppose there is an electric potential in space, given at each point by $\phi(x, y, z) = xy - z^2$ volts, where x, y and z are in meters.

Determine the instantaneous rate of change, with respect to time, of the electric potential at the particle's position, at time $t = 2$ seconds.

Solution:

Let the position of the particle, in meters, at time t seconds, be given by $\mathbf{p}(t) = (x(t), y(t), z(t))$, so that the velocity is given by $\mathbf{v}(t) = \mathbf{p}'(t)$. Then, the given data about the position and velocity is that $\mathbf{p}(2) = (3, 4, 7)$ meters and $\mathbf{v}(2) = \mathbf{p}'(2) = (-2, 1, 5)$ m/s. At any time t, the potential at the particle's location is $\phi(\mathbf{p}(t)) = (\phi \circ \mathbf{p})(t)$. Hence, what we want to calculate is $(\phi \circ \mathbf{p})'(2)$.

Before we apply the Chain Rule, recall that, for a function of a single variable, the ordinary derivative and its only partial derivative are the same. Now, the Chain Rule tells us that

$$(\phi \circ \mathbf{p})'(t) = \frac{d}{dt}(\phi \circ \mathbf{p})(t) = \vec{\nabla}\phi(\mathbf{p}(t)) \cdot \frac{d\mathbf{p}}{dt} = \vec{\nabla}\phi(\mathbf{p}(t)) \cdot \mathbf{v}(t) \ \text{volts/s}.$$

Therefore, at time $t = 2$ seconds, we have

$$(\phi \circ \mathbf{p})'(2) = \vec{\nabla}\phi(\mathbf{p}(2)) \cdot \mathbf{v}(2) = \vec{\nabla}\phi(3, 4, 7) \cdot (-2, 1, 5) \ \text{volts/s},$$

and we see that we need to calculate $\vec{\nabla}\phi(3, 4, 7)$ before we can finish the problem.

We find

$$\vec{\nabla}\phi = \vec{\nabla}(xy - z^2) = (y, x, -2z) \ \text{volts/m}$$

and so

$$\vec{\nabla}\phi(3, 4, 7) = (4, 3, -14) \ \text{volts/m}.$$

Finally, combining our calculations above, we find

$$(\phi \circ \mathbf{p})'(2) = (4, 3, -14) \cdot (-2, 1, 5) = -8 + 3 - 70 = -75 \ \text{volts/s}.$$

More Depth:

We are going to essentially restate the differentiation rules that we have already established, but now we will give them in terms of total derivatives and differentials. We will also repeat the setup from most of our earlier examples, but now ask for results in terms of total derivatives and differentials. The formulas that we derive here follow immediately from the earlier formulas involving partial derivatives and gradient vectors, **after** we know that the functions involved are, in fact, differentiable. We omit the proofs of differentiability, and refer you to Theorems 5.3.9 and 5.4.3 of Trench, [8].

> Note that this is the linearity of the "differentiation at $\mathbf{p}$" operation, which is applied to a linear combination of functions.

> We have already seen another type of linearity for the total derivative: for a fixed function f, the function $d_{\mathbf{p}}f$ is linear. This linearity means that $d_{\mathbf{p}}f(a\mathbf{v} + b\mathbf{w}) = a\,d_{\mathbf{p}}f(\mathbf{v}) + b\,d_{\mathbf{p}}f(\mathbf{w})$.

Theorem 2.4.14. (Linearity of Differentiation) *If f and g are differentiable at $\mathbf{p}$, then $af + bg$ is differentiable at $\mathbf{p}$, and*

$$d_{\mathbf{p}}(af + bg) \;=\; a\,d_{\mathbf{p}}f + b\,d_{\mathbf{p}}g.$$

In differential notation, we omit the references to $\mathbf{p}$, and write simply

$$d(af + bg) \;=\; a\,df + b\,dg.$$

Example 2.4.15. Suppose that we have functions $f = f(x, y, z)$ and $g = g(x, y, z)$, and we know that $\vec{\nabla} f(3, 5, -7) = (2, 9, 1)$ and $\vec{\nabla} g(3, 5, -7) = (-1, 0, 4)$. Calculate $d_{(3,5,-7)}(3f - 2g)(6, 4, 2)$.

Solution:

From Example 2.4.2, we know that

$$\vec{\nabla}(3f - 2g)(3, 5, -7) \;=\; (8, 27, -5).$$

Therefore,

$$d_{(3,5,-7)}(3f - 2g)(6, 4, 2) \;=\; \left[\vec{\nabla}(3f - 2g)(3, 5, -7)\right] \cdot (6, 4, 2) \;=\;$$

$$(8, 27, -5) \cdot (6, 4, 2) \;=\; 48 + 108 - 10 \;=\; 146.$$

Theorem 2.4.16. (Product Rule) *Suppose that f and g are real-valued functions on a subset of $\mathbb{R}^n$. If f and g are differentiable at $\mathbf{p}$, then fg is differentiable at $\mathbf{p}$ and*

$$d_{\mathbf{p}}(fg) \;=\; f(\mathbf{p})\,d_{\mathbf{p}}g \;+\; g(\mathbf{p})\,d_{\mathbf{p}}f.$$

In differential notation, we omit the references to $\mathbf{p}$, and write simply

$$d(fg) \;=\; f\,dg \;+\; g\,df.$$

Example 2.4.17. Calculate the differential $d\left[\left(x^2y + \cos z\right)\left(z\ln y + \sqrt{x}\right)\right]$.

Solution:

We can read the differential calculation directly from the gradient vector calculation in Example 2.4.4. However, we can just recalculate everything, always using differentials:

$$d\left[\left(x^2y + \cos z\right)\left(z\ln y + \sqrt{x}\right)\right] =$$

$$\left(x^2y + \cos z\right)d\left(z\ln y + \sqrt{x}\right) + \left(z\ln y + \sqrt{x}\right)d\left(x^2y + \cos z\right) =$$

$$\left(x^2y + \cos z\right)\left(\tfrac{1}{2}x^{-1/2}\,dx + \frac{z}{y}\,dy + \ln y\,dz\right) +$$

$$\left(z\ln y + \sqrt{x}\right)\left(2xy\,dx + x^2\,dy - \sin z\,dz\right).$$

Theorem 2.4.18. (Quotient Rule) *Suppose that f and g are real-valued functions on a subset of $\mathbb{R}^n$. If f and g are differentiable at $\mathbf{p}$, and $g(\mathbf{p}) \neq 0$, then f/g is differentiable at $\mathbf{p}$ and*

$$d_{\mathbf{p}}\left(\frac{f}{g}\right) = \frac{g(\mathbf{p})\,d_{\mathbf{p}}f - f(\mathbf{p})\,d_{\mathbf{p}}g}{\left(g(\mathbf{p})\right)^2}.$$

In differential notation, we omit the references to $\mathbf{p}$, and write simply

$$d\left(\frac{f}{g}\right) = \frac{g\,df - f\,dg}{g^2}.$$

Example 2.4.19. Calculate $d_{(1,5)}\left(\dfrac{y^2\tan^{-1}x}{y - x^3}\right)(-4, 3)$, and interpret your result in terms of the rate of change of the function.

Solution: We will not use our calculation of the gradient vector from Example 2.4.6, but instead use differential notation throughout. You should compare the two calculations.

We find

$$d\left(\frac{y^2\tan^{-1}x}{y - x^3}\right) = \frac{(y - x^3)\,d(y^2\tan^{-1}x) - (y^2\tan^{-1}x)\,d(y - x^3)}{(y - x^3)^2} =$$

$$\frac{(y - x^3)\left(\dfrac{y^2}{1 + x^2}\,dx + 2y\tan^{-1}x\,dy\right) - (y^2\tan^{-1}x)(-3x^2\,dx + dy)}{(y - x^3)^2}.$$

Therefore, at $(1, 5)$, we find

$$d_{(1,5)}\left(\frac{y^2\tan^{-1}x}{y - x^3}\right) =$$

$$\frac{(5 - 1^3)\left(\dfrac{5^2}{1 + 1^2}\,dx + 2\cdot 5\tan^{-1}1\,dy\right) - (5^2\tan^{-1}1)(-3\cdot 1^2\,dx + dy)}{(5 - 1^3)^2} =$$

$$\frac{4\left(\dfrac{25}{2}\,dx + 10\cdot\dfrac{\pi}{4}\,dy\right) - \left(25\cdot\dfrac{\pi}{4}\right)(-3\,dx + dy)}{16} =$$

$$\frac{1}{16}\left[\left(\frac{200 + 75\pi}{4}\right)dx + \frac{15\pi}{4}\,dy\right].$$

Hence,

$$d_{(1,5)}\left(\frac{y^2\tan^{-1}x}{y - x^3}\right)(-4, 3) =$$

$$\frac{1}{16}\left[\left(\frac{200 + 75\pi}{4}\right)(-4) + \frac{15\pi}{4}(3)\right] = -\frac{1}{64}(800 + 255\pi).$$

This tells us that, when we're at the point $(1, 5)$, and moving with velocity $(-4, 3)$, the function $\dfrac{y^2\tan^{-1}x}{y - x^3}$ is changing at an instantaneous rate of $-\frac{1}{64}(800 + 255\pi)$; in particular, the negative sign tells us that the function is decreasing.

Remark 2.4.20. Before we give the total derivative version of the Chain Rule, we would like to give you some idea of why the Chain Rule for Partial Derivatives, Theorem 2.4.9, looks the way it does. This is just intended to give you the general idea; there are subtle points to the actual proof. See the proof of Theorem 5.4.3 in Trench, [8], for the details.

Suppose that we have functions $f = f(x, y)$, $x = x(s, t)$, and $y = y(s, t)$. We are going to look at $\partial f/\partial s$, and see where the Chain Rule comes from. We will leave t fixed, and assume that s changes by some (small) amount Δs. Then, there are corresponding changes in x, y, and f; we define

$$\Delta x = x(s + \Delta s, t) - x(s, t), \qquad \Delta y = y(s + \Delta s, t) - y(s, t),$$

and

$$\Delta f \;=\; f\big(x(s+\Delta s,t),y(s+\Delta s,t)\big) - f\big(x(s,t),y(s,t)\big).$$

Thus,

$$\frac{\Delta f}{\Delta s} \;=\; \frac{f\big(x(s+\Delta s,t),y(s+\Delta s,t)\big) - f\big(x(s,t),y(s,t)\big)}{\Delta s} \;=\;$$

$$\frac{f\big(x(s,t)+\Delta x,y(s,t)+\Delta y\big) - f\big(x(s,t),y(s,t)\big)}{\Delta s},$$

which, suppressing the (s,t) reference, gives us

$$\frac{\Delta f}{\Delta s} \;=\; \frac{f\big(x+\Delta x,y+\Delta y\big) - f\big(x,y\big)}{\Delta s}.$$

We now use "mathematician's basic trick #1"; we add 0 in a clever way. Our clever way will be to add $0 = -f(x,y+\Delta y) + f(x,y+\Delta y)$ to the numerator to obtain

$$\frac{\Delta f}{\Delta s} \;=\; \frac{f\big(x+\Delta x,y+\Delta y\big) - f\big(x,y+\Delta y\big) + f\big(x,y+\Delta y\big) - f\big(x,y\big)}{\Delta s} \;=\;$$

$$\frac{f\big(x+\Delta x,y+\Delta y\big) - f\big(x,y+\Delta y\big)}{\Delta s} \;+\; \frac{f\big(x,y+\Delta y\big) - f\big(x,y\big)}{\Delta s},$$

where the point is that, in the numerator of the first fraction, the second coordinate doesn't change and, in the numerator of the second fraction, the first coordinate doesn't change; thus, we are set up for partial derivatives.

Now, we use "mathematician's basic trick #2" twice; we multiply by 1 in a clever way to obtain

$$\frac{\Delta f}{\Delta s} \;=\; \frac{f\big(x+\Delta x,y+\Delta y\big) - f\big(x,y+\Delta y\big)}{\Delta x}\cdot\frac{\Delta x}{\Delta s} \;+\; \frac{f\big(x,y+\Delta y\big) - f\big(x,y\big)}{\Delta y}\cdot\frac{\Delta y}{\Delta s}.$$

It may look like the first factor on the right should approach $\partial f/\partial x$ at $(x,y+\Delta y)$, not at (x,y). But keep in mind that Δy is approaching 0.

As Δs approaches 0, continuity implies that Δx, Δy, and Δf also approach 0, and the previous formula yields

$$\frac{\partial f}{\partial s} \;=\; \frac{\partial f}{\partial x}\cdot\frac{\partial x}{\partial s} \;+\; \frac{\partial f}{\partial y}\cdot\frac{\partial y}{\partial s},$$

which is what the Chain Rule says.

Of course, the same type of argument applies to $\partial f/\partial t$.

We now wish to look at what the Chain Rule for Partial Derivatives tells us about total derivatives. Recall that, for a differentiable real-valued function $h = h(\mathbf{z})$, there is an equality

$$d_{\mathbf{p}} h(\mathbf{e}_i) \;=\; \left. \frac{\partial h}{\partial z_i} \right|_{\mathbf{p}},$$

where $\mathbf{e}_i$ is the i-th standard basis vector.

Now, suppose that $\mathbf{t} = (t_1, \ldots, t_m)$, that $\mathbf{x} = (x_1, \ldots, x_n)$, and that $f(\mathbf{x})$ is a real-valued function of $\mathbf{x}$, while $\mathbf{x}$ is itself a multi-component function $\mathbf{x} = \mathbf{x}(\mathbf{t})$. Then, by composing, we may consider f as a function of $\mathbf{t}$. Fix a point $\mathbf{a}$, at which $\mathbf{x}$ is differentiable (see Definition 2.2.17), and suppose that f is differentiable at $\mathbf{x}(\mathbf{a})$. Then, using the Chain Rule for Partial Derivatives, we find

$$d_{\mathbf{a}}(f \circ \mathbf{x})(\mathbf{e}_i) \;=\; \left. \frac{\partial}{\partial t_i}(f \circ \mathbf{x}) \right|_{\mathbf{a}} \;=\; \vec{\nabla} f(\mathbf{x}(\mathbf{a})) \cdot \left. \frac{\partial \mathbf{x}}{\partial t_i} \right|_{\mathbf{a}} \;=\; d_{\mathbf{x}(\mathbf{a})} f \left(\left. \frac{\partial \mathbf{x}}{\partial t_i} \right|_{\mathbf{a}} \right) \;=$$

$$d_{\mathbf{x}(\mathbf{a})} f \left(\left. \frac{\partial \mathbf{x}}{\partial t_i} \right|_{\mathbf{a}} \right) \;=\; d_{\mathbf{x}(\mathbf{a})} f \big(d_{\mathbf{a}} x_1(\mathbf{e}_i), \; d_{\mathbf{a}} x_2(\mathbf{e}_i), \; \ldots, \; d_{\mathbf{a}} x_n(\mathbf{e}_i) \big) \;=\; d_{\mathbf{x}(\mathbf{a})} f \big(d_{\mathbf{a}} \mathbf{x}(\mathbf{e}_i) \big).$$

Hence,

$$d_{\mathbf{a}}(f \circ \mathbf{x})(\mathbf{e}_i) \;=\; \big(d_{\mathbf{x}(\mathbf{a})} f \circ d_{\mathbf{a}} \mathbf{x} \big)(\mathbf{e}_i).$$

As this is true for all of the standard basis vectors, and the total derivative is a linear function, it follows that, for all vectors $\mathbf{v}$ in $\mathbb{R}^m$,

$$d_{\mathbf{a}}(f \circ \mathbf{x})(\mathbf{v}) \;=\; \big(d_{\mathbf{x}(\mathbf{a})} f \circ d_{\mathbf{a}} \mathbf{x} \big)(\mathbf{v}),$$

i.e., that there is an equality of functions

$$d_{\mathbf{a}}(f \circ \mathbf{x}) \;=\; d_{\mathbf{x}(\mathbf{a})} f \circ d_{\mathbf{a}} \mathbf{x}.$$

If f were, in fact, a multi-component function $\mathbf{f}$, applying the above argument to each component yields the following, extremely elegant, version of the Chain Rule:

Theorem 2.4.21. (Chain Rule for Total Derivatives) *Suppose that A is a subset of $\mathbb{R}^m$, B is a subset of $\mathbb{R}^n$, C is a subset of $\mathbb{R}^k$, and that we have functions $\mathbf{x} : A \to B$ and $\mathbf{f} : B \to C$. Furthermore, suppose that $\mathbf{x}$ is differentiable at $\mathbf{a}$, and that $\mathbf{f}$ is differentiable at $\mathbf{x}(\mathbf{a})$.*

Then, $\mathbf{f} \circ \mathbf{x}$ is differentiable at $\mathbf{a}$, and

$$d_{\mathbf{a}}(\mathbf{f} \circ \mathbf{x}) \;=\; d_{\mathbf{x}(\mathbf{a})} \mathbf{f} \circ d_{\mathbf{a}} \mathbf{x}.$$

For the proof, see Theorem 5.4.3 in Trench, [8].

Example 2.4.22. Recall Example 2.4.11. We had $T = T(u, v, w)$, $u = r \cos t$, $v = 5r - 3t$, and $w = r^2 e^t$. We were also given that $\vec{\nabla}T(2, 10, 4) = (-1, 3, 2)$.

Now, let $\mathbf{m}(r, t) = (u, v, w) = (r \cos t, 5r - 3t, r^2 e^t)$, and note that $\mathbf{m}(2, 0) = (2, 10, 4)$. Calculate $d_{(2,0)}(T \circ \mathbf{m})(-1, 7)$.

Solution:

The Chain Rule gives us

$$d_{(2,0)}(T \circ \mathbf{m})(-1, 7) = d_{\mathbf{m}(2,0)}T\big(d_{(2,0)}\mathbf{m}(-1, 7)\big) =$$

$$\vec{\nabla}T(2, 10, 4) \cdot d_{(2,0)}\mathbf{m}(-1, 7) =$$

$$\vec{\nabla}T(2, 10, 4) \cdot \Big(d_{(2,0)}u(-1, 7), d_{(2,0)}v(-1, 7), d_{(2,0)}w(-1, 7)\Big) =$$

$$\vec{\nabla}T(2, 10, 4) \cdot \Big(\vec{\nabla}u(2, 0) \cdot (-1, 7), \vec{\nabla}v(2, 0) \cdot (-1, 7), \vec{\nabla}w(2, 0) \cdot (-1, 7)\Big). \qquad (2.8)$$

We were explicitly given that $\vec{\nabla}T(2, 10, 4) = (-1, 3, 2)$, and the gradient vectors of u, v, and w at $(2, 0)$ can easily be read from the partial derivative calculations in Example 2.4.11.

We found

$$\frac{\partial}{\partial r}(u, v, w)\Big|_{(r,t)=(2,0)} = (1, 5, 4)$$

and

$$\frac{\partial}{\partial t}(u, v, w)\Big|_{(r,t)=(2,0)} = (0, -3, 4).$$

From this, we conclude

$$\vec{\nabla}u(2, 0) = (1, 0), \quad \vec{\nabla}v(2, 0) = (5, -3), \quad \text{and} \quad \vec{\nabla}w(2, 0) = (4, 4).$$

Now, plugging everything into Formula 2.8, we find

$$d_{(2,0)}(T \circ \mathbf{m})(-1, 7) = (-1, 3, 2) \cdot \Big((1, 0) \cdot (-1, 7), (5, -3) \cdot (-1, 7), (4, 4) \cdot (-1, 7)\Big) =$$

$$(-1, 3, 2) \cdot (-1, -26, 24) = -29.$$

We shall return to this example later, in the $+$ *Linear Algebra* subsection, and look at how matrix multiplication encodes the calculations in a nice way.

If $\mathbf{x} = \mathbf{x}(t)$ is a function of a single variable, then the Chain Rule for $\mathbf{f} \circ \mathbf{x}$ can be written in a simpler form. We leave the proof of this as an exercise for you.

Corollary 2.4.23. *Suppose that* $\mathbf{x} = \mathbf{x}(t)$ *is a single-variable function into* $\mathbb{R}^n$, $\mathbf{f}$ *is a function from a subset of* $\mathbb{R}^n$ *into* $\mathbb{R}^k$. *Suppose that* $\mathbf{x}$ *is differentiable at* t *and that* $\mathbf{f}$ *is differentiable at* $\mathbf{x}(t)$.

Then, $\mathbf{f} \circ \mathbf{x}$ *is differentiable at* t, *and*

$$(\mathbf{f} \circ \mathbf{x})'(t) = d_{\mathbf{x}(t)}\mathbf{f}(\mathbf{x}'(t)).$$

Remark 2.4.24. Let's look back at Example 2.4.13. We had a particle which was moving through space and, at time $t = 2$ seconds, the particle was at the point $(3, 4, 7)$, in meters, and was moving with an instantaneous velocity of $(-2, 1, 5)$ m/s. We assumed that there was an electric potential in space, given at each point by $\phi(x, y, z) = xy - z^2$ volts, where x, y and z were in meters. The problem was to determine the instantaneous rate of change, with respect to time, of the electric potential at the particle's position, at time $t = 2$ seconds.

As we discussed back in Section 2.2, the total derivative $d_{\mathbf{p}}\phi(\mathbf{v})$ is the instantaneous rate of change of ϕ, when you're at the point $\mathbf{p}$, and moving with velocity $\mathbf{v}$. Thus, in terms of total derivatives, precisely what we needed to calculate was $d_{(3,4,7)}\phi(-2, 1, 5)$, which we know equals $\vec{\nabla}\phi(3, 4, 7) \cdot (-2, 1, 5)$ volts/s, which, of course, is exactly what we calculated in Example 2.4.13.

What we want to emphasize in this remark is that the Chain Rule actually gives us **more** justification for interpreting the total derivative $d_{\mathbf{p}}\phi(\mathbf{v})$ is the instantaneous rate of change of ϕ, when you're at the point $\mathbf{p}$, and moving with velocity $\mathbf{v}$.

Back in Theorem 2.2.2, we stated and proved that, if f is differentiable at $\mathbf{p}$, then

$$d_{\mathbf{p}}f(\mathbf{v}) = \lim_{t \to 0} \frac{f(\mathbf{p} + t\mathbf{v}) - f(\mathbf{p})}{t}.$$

Define $\mathbf{x}(t) = \mathbf{p} + t\mathbf{v}$, so that $\mathbf{x}(0) = \mathbf{p}$ and $\mathbf{x}'(t) = \mathbf{v}$. Then, $\mathbf{x}(t)$ would give the position of a moving object which is at $\mathbf{p}$ at time 0, and is moving in a straight line with constant velocity $\mathbf{v}$.

Our justification for referring to $d_{\mathbf{p}}f(\mathbf{v})$ as "the instantaneous rate of change of f, when you're at the point $\mathbf{p}$, and moving with velocity $\mathbf{v}$" was that this instantaneous rate of change is given by $(f \circ \mathbf{x})'(0)$ and, using the definition of the derivative and the equality from above, we have

$$(f \circ \mathbf{x})'(0) \ = \ \lim_{t \to 0} \frac{(f \circ \mathbf{x})(t) - (f \circ \mathbf{x})(0)}{t} \ = \ \lim_{t \to 0} \frac{f(\mathbf{p} + t\mathbf{v}) - f(\mathbf{p})}{t} \ = \ d_{\mathbf{p}} f(\mathbf{v}).$$

Now, however, we can use the Chain Rule to see that it was irrelevant that the object was moving with **constant** velocity $\mathbf{v}$; we just need that the instantaneous velocity is $\mathbf{v}$, when the object is at $\mathbf{p}$, in order to conclude that $(f \circ \mathbf{x})'(0) = d_{\mathbf{p}} f(\mathbf{v})$. Why?

Suppose that all we know about $\mathbf{x}(t)$ is that $\mathbf{p} = \mathbf{x}(0)$ and $\mathbf{v} = \mathbf{x}'(0)$. Then, using the Chain Rule or, more specifically, Corollary 2.4.23, we find

$$(f \circ \mathbf{x})'(0) \ = \ d_{\mathbf{x}(0)} f\big(\mathbf{x}'(0)\big) \ = \ d_{\mathbf{p}} f(\mathbf{v}).$$

+ **Linear Algebra:**

Recall the definition of the Jacobian matrix from Definition 2.2.22, and how you calculate total derivatives via Jacobian matrices and matrix multiplication from Theorem 2.2.24. Composition of linear transformations corresponds to matrix multiplication. So, in matrix terms, the Chain Rule becomes:

Theorem 2.4.25. (Matrix form of the Chain Rule) *Suppose that A is a subset of $\mathbb{R}^m$, B is a subset of $\mathbb{R}^n$, C is a subset of $\mathbb{R}^k$, and that we have functions $\mathbf{x} : A \to B$ and $\mathbf{f} : B \to C$. Furthermore, suppose that $\mathbf{x}$ is differentiable at $\mathbf{a}$, and that $\mathbf{f}$ is differentiable at $\mathbf{x}(\mathbf{a})$.*

Then, $\mathbf{f} \circ \mathbf{x}$ is differentiable at $\mathbf{a}$, and we have an equality of matrices

$$\big[d_{\mathbf{a}}(\mathbf{f} \circ \mathbf{x})\big] \ = \ \big[d_{\mathbf{x}(\mathbf{a})}\mathbf{f}\big]\big[d_{\mathbf{a}}\mathbf{x}\big].$$

Example 2.4.26. Let's look again at Example 2.4.22. We had $T = T(u, v, w)$, $u = r \cos t$, $v = 5r - 3t$, and $w = r^2 e^t$. We were also given that $\vec{\nabla} T(2, 10, 4) = (-1, 3, 2)$. We let $\mathbf{m}(r, t) = (u, v, w) = (r \cos t, 5r - 3t, r^2 e^t)$, and noted that $\mathbf{m}(2, 0) = (2, 10, 4)$. The problem was to calculate $d_{(2,0)}(T \circ \mathbf{m})(-1, 7)$.

Let's see how you would approach this with Jacobian matrices.

We ignore the $(-1, 7)$ for the time being. We first want to calculate the Jacobian matrix $\big[d_{(2,0)}(T \circ \mathbf{m})\big]$. The Chain Rule tells us that

$$\big[d_{(2,0)}(T \circ \mathbf{m})\big] \ = \ \big[d_{\mathbf{m}(2,0)} T\big]\big[d_{(2,0)}\mathbf{m}\big].$$

We calculate

$$[d_{\mathbf{m}(2,0)}T] = [d_{(2,10,4)}T] = \begin{bmatrix} \dfrac{\partial T}{\partial u} & \dfrac{\partial T}{\partial v} & \dfrac{\partial T}{\partial w} \end{bmatrix}\bigg|_{(2,10,4)} = \begin{bmatrix} -1 & 3 & 2 \end{bmatrix}$$

and

$$[d_{(2,0)}\mathbf{m}] = \begin{bmatrix} - & \vec{\nabla} u & - \\ - & \vec{\nabla} v & - \\ - & \vec{\nabla} w & - \end{bmatrix}\bigg|_{(2,0)} = \begin{bmatrix} - & \vec{\nabla}(r\cos t) & - \\ - & \vec{\nabla}(5r - 3t) & - \\ - & \vec{\nabla}(r^2 e^t) & - \end{bmatrix}\bigg|_{(2,0)} =$$

$$\begin{bmatrix} \cos t & -r\sin t \\ 5 & -3 \\ 2re^t & r^2 e^t \end{bmatrix}\bigg|_{(2,0)} = \begin{bmatrix} 1 & 0 \\ 5 & -3 \\ 4 & 4 \end{bmatrix}.$$

Thus,

$$[d_{(2,0)}(T \circ \mathbf{m})] = [d_{\mathbf{m}(2,0)}T][d_{(2,0)}\mathbf{m}] =$$

$$\begin{bmatrix} -1 & 3 & 2 \end{bmatrix}\begin{bmatrix} 1 & 0 \\ 5 & -3 \\ 4 & 4 \end{bmatrix} = \begin{bmatrix} 22 & -1 \end{bmatrix}.$$

Finally, we conclude that

$$[d_{(2,0)}(T \circ \mathbf{m})(-1,7)] = [d_{(2,0)}(T \circ \mathbf{m})]\begin{bmatrix} -1 \\ 7 \end{bmatrix} = \begin{bmatrix} 22 & -1 \end{bmatrix}\begin{bmatrix} -1 \\ 7 \end{bmatrix} = [-29],$$

i.e., $d_{(2,0)}(T \circ \mathbf{m})(-1,7) = -29$, as we found in Example 2.4.22.

One advantage to our current approach is that, if we now want to calculate the total derivative $d_{(2,0)}(T \circ \mathbf{m})$ with respect to some other velocity vector, then we have to redo very little work. For instance, we find

$$[d_{(2,0)}(T \circ \mathbf{m})(5,4)] = [d_{(2,0)}(T \circ \mathbf{m})]\begin{bmatrix} 5 \\ 4 \end{bmatrix} = \begin{bmatrix} 22 & -1 \end{bmatrix}\begin{bmatrix} 5 \\ 4 \end{bmatrix} = [106],$$

that is, $d_{(2,0)}(T \circ \mathbf{m})(5,4) = 106$.

2.4.1 Exercises

Online answers to select exercises are here.

Basics:

Suppose that $f(\mathbf{p}) = 3$, $g(\mathbf{p}) = 5$, $\vec{\nabla} f(\mathbf{p}) = (2, 7, 0)$, and $\vec{\nabla} g(\mathbf{p}) = (1, -7, 8)$. Determine the following gradient vectors at the point p.

1. $\vec{\nabla}(f + 4g)$

2. $\vec{\nabla}(fg)$

3. $3\vec{\nabla} f - \vec{\nabla}(2g)$

4. $\vec{\nabla}\left(\dfrac{f}{g}\right)$ ▶

5. $\vec{\nabla}\left(\dfrac{g}{f}\right)$

6. $\vec{\nabla}\left(\dfrac{fg}{f + g}\right)$

Suppose that $k(\mathbf{p}) = 8$ and $\vec{\nabla} k(\mathbf{p}) = (-1, 16, 5)$. Determine each of the following gradient vectors at the point p.

7. $\vec{\nabla} 4k^2$

8. $\vec{\nabla} k^3$

9. $\vec{\nabla} \sqrt[3]{k}$ ▶

10. $\vec{\nabla}\left(\dfrac{3}{k}\right)$

11. $\vec{\nabla}(k - k^3)$

12. $\dfrac{-1}{3}\vec{\nabla}\sqrt{k^3}$

Calculate the following gradient vectors.

13. $\vec{\nabla}\left[(x^y - z^3 y)(e^y - z^2 \tan y)\right]$

14. $\vec{\nabla}\left[(u^2 v - \sqrt{w^2 + u})(w^3 u + \ln v)\right]$ at the point $(u, v, w) = (5, e, 2)$

15. $\vec{\nabla}\left[\dfrac{x + \sin(t + \pi\omega)}{x + \cos(t + \pi\omega)}\right]$ at the point $(x, t, \omega) = \left(-2, \frac{\pi}{2}, 1\right)$ ▶

16. $\vec{\nabla}\left[\dfrac{x_1 x_2 x_3}{x_1 + x_2^2 + x_3^3}\right]$

17. $\vec{\nabla}\left[\sqrt{e^x \tan y - e^z \cos y}\right]$

18. $\vec{\nabla}\left[(q^2\bar{p}\hat{q} - p\bar{q}^2\hat{p}^3)^3\right]$ at the point $(q, \bar{q}, \hat{q}, p, \bar{p}, \hat{p}) = (1, 3, -1, -3, 1, 3)$

Suppose that the real-valued function $f = f(x, y)$ is differentiable at $(1, 2)$, and that $\vec{\nabla}f(1, 2) = (6, -3)$. In each of the following exercises, you are given $x = x(t)$ and $y = y(t)$. Verify that $(x(0), y(0)) = (1, 2)$, and then calculate $\partial f/\partial t = df/dt$ at $t = 0$.

19. $x = 1 + 3\sin t$, and $y = 2\cos t$.

20. $x = e^{2t}$, and $y = 2e^t$.

21. $x = 1$, and $y = 2 + \tan^{-1} t$.

22. $x = t^9 + 1$, and $y = t^5 + 2$.

In each of the following exercises, calculate $\partial f/\partial s$, $\partial f/\partial t$, and $\partial f/\partial w$, where applicable.

23. $f(x, y) = x^2 y^2, x = 5s - t, y = 3s^2 - t$

24. $f(u, v) = u\sin(2v), u = e^{st}, v = st$

25. $f(x, y, z) = z^2 \ln(x - y), x = s + t, y = \dfrac{s}{t}, z = st$

26. $f(x_1, x_2, x_3) = \dfrac{2x_1}{x_2 - x_3}, x_1 = e^s + t^2, x_2 = s^2 + t^2, x_3 = s^2 t^2$

27. $f(x, y) = x^3 - y, x = s + 3w, y = t - 2w$

28. $f(u, v) = \sqrt{u - v}, u = s^2 t - w, v = st^2 + w$

29. $f(x, y, z) = x^2 y - yz + 3zx, x = s - t, y = 2t + w, z = w - 3s$

30. $f(a, b, c) = \dfrac{a + b}{2a - c}, a = sw, b = st^2, c = stw$

31. Suppose that a heated metal plate occupies the portion of the xy-plane where $x^2 + y^2 \leq 25$, where x and y are in meters. The plate is heated in such a way that, at the point $(x, y) = (-2\sqrt{2}, 2\sqrt{2})$, the partial derivatives of the temperature, T, in degrees Celsius, are given by

$$\frac{\partial T}{\partial x} = 3°\text{C/m} \quad \text{and} \quad \frac{\partial T}{\partial y} = -1°\text{C/m}.$$

Recall that, in polar coordinates, $x = r\cos\theta$ and $y = r\sin\theta$. Calculate $\partial T/\partial r$ and $\partial T/\partial \theta$ at $(r, \theta) = (4, 3\pi/4)$, where r is in meters and θ is in radians. Physically, what do $\partial T/\partial r$ and $\partial T/\partial \theta$ measure?

32. Gravitational and electromagnetic forces are acting on a charged particle in space. Let $F = F(x, y, z)$ be the magnitude of the net force, in Newtons, which acts on the particle at the position (x, y, z) meters. Suppose that, at $(x, y, z) = (1, 3, 2)$, the partial derivatives of F, in N/m, are

$$F_x = 4.5, \quad F_y = 10, \quad \text{and} \quad F_z = -2.$$

If the position of the particle, in meters, at time t seconds, is given by

$$\mathbf{p}(t) = (t^2, t^3 + 2, t + 1),$$

what is the rate of change of the magnitude of the net force on the particle, with respect to time, at time $t = 1$?

33. A company produces wazbobs, which have three main components: cranks, gizmos, and beepers. The cost of producing a wazbob is given by the following, in dollars:

$$W(c, g, b) = c + 2g + \frac{b^2}{2}$$

Where c, g, and b represent the cost of each crank, gizmo, and beeper, respectively. These quantities depend on the market prices of the raw materials: $c = 0.08s + 0.02w$, $g = 8 + 0.05s + 0.25i$, and $b = 2 + 0.03s + 0.1w$, where s is the cost of silicon in dollars per kilogram, w is the cost of copper wiring in dollars per meter, and i is the cost of Irukandji jellyfish venom in dollars per liter.

a. Use the multivariable Chain Rule to find $\partial W / \partial s$.

b. Use substitution to write W explicitly as a function of s, w, and i. What is the derivative of this function with respect to s?

c. Are the answers to part (a) and (b) different? Why or why not?

| More Depth: |

Suppose that $\vec{\nabla} f(\mathbf{p}) = (2, 7, 0)$ and $\vec{\nabla} g(\mathbf{p}) = (1, -7, 8)$. **Calculate the following total derivatives, possibly leaving $f(\mathbf{p})$ and $g(\mathbf{p})$ in your answer.**

34. $d_{\mathbf{p}}(f + 4g)(-1, -3, -5)$

35. $d_{\mathbf{p}}(fg)(2, 0, 0)$

36. $3d_{\mathbf{p}}f(1, 4, -7) - d_{\mathbf{p}}(2g)(1, 4, -7)$

37. $d_{\mathbf{p}}\left(\dfrac{f}{g}\right)(1, \pi, \pi^2)$

38. $d_{\mathbf{p}}\left(\dfrac{g}{f}\right)(\ln(2), \ln(3), \ln(5))$

39. $d_{\mathbf{p}}\left(\dfrac{fg}{f+g}\right)(-1,0,1)$

Calculate the following differentials and total derivatives.

40. $d\left[(3x^2z - 2y)(x^3z - y^2)\right]$

41. $d\left[(e^x y + \sqrt{x^2 + z})(e^{z-y} + \sqrt{x+y})\right]$

42. $d\left[\dfrac{xy - e^z}{x^2y + e^z}\right]$

43. $d\left[\dfrac{x}{e^y} - \dfrac{2x}{e^z}\right]$

44. $d_{(3,3)}\left[(2e^{xy} - x^2y)(\sqrt{xy} - y^2x)\right](-2,1)$

45. $d_{(5,1/5)}\left[(x^2 - 3y)(xy^2 - xy + x^2y)\right](25,-5)$

46. $d_{(2,1,-\sqrt{2})}\left[\dfrac{xy}{z}\right](-\sqrt{2},\sqrt{2},1)$

47. $d_{(-1,2)}\left[\dfrac{x^2 + y^2}{x^2y^2}\right](3,5)$

48. Suppose that $S = S(r,\phi,\theta), r = s+t, \phi = \sqrt{s}+t$, and $\theta = s+\sqrt{t}$. Suppose that $\vec{\nabla}S(5,5,3) = \left(2,\dfrac{1}{2},-9\right)$. Let $\mathbf{x}(s,t) = (r,\phi,\theta) = (s+t,\sqrt{s}+t,s+\sqrt{t})$.

    Calculate $d_{(1,4)}(S \circ \mathbf{x})(-3,2)$.

49. Suppose that $\lambda = \lambda(x,y), x = c^2v + \mu$, and $y = v - c\sin\mu$. Suppose that $\vec{\nabla}\lambda(3 + \pi, 3) = (-3,\pi)$. Let $\mathbf{m}(c,v,\mu) = (x,y) = (c^2v + \mu, v - c\sin\mu)$.

    Calculate $d_{(1,3,\pi)}(\lambda \circ \mathbf{m})(10,10,10)$.

50. Prove that, if f is differentiable at $\mathbf{p}$ and a is a constant, then af is differentiable at $\mathbf{p}$.

51. Prove Corollary 2.4.23. (Hint: Recall Example 2.2.11.)

$+$ Linear Algebra:

52. Formulate a Product Rule for Dot Product in terms of Jacobian matrices.

53. Formulate a Product Rule for Cross Product in terms of Jacobian matrices.

54. Suppose that $L = L(x,y,z), x = 2uv, y = u^2v$, and $z = e^{uv}$. Suppose further that $\vec{\nabla}L(2,2,e) = (4,-2,e)$. Let $\mathbf{g}(u,v) = (x,y,z) = (2uv, u^2v, e^{uv})$. Calculate each of the following total derivatives using matrices:

    a. $d_{(2,1/2)}(L \circ \mathbf{g})(4,1)$

    b. $d_{(2,1/2)}(L \circ \mathbf{g})(-2,e)$

    c. $d_{(2,1/2)}(L \circ \mathbf{g})(0,0)$

55. Consider Exercise 48. Calculate $d_{(1,4)}(S \circ \mathbf{x})(-3,2)$ using matrices.

56. Consider Exercise 49. Calculate $d_{(1,3,\pi)}(\lambda \circ \mathbf{m})(10,10,10)$ using matrices.

2.5   The Directional Derivative

In Section 2.2, we defined the total derivative $d_{\mathbf{p}}f(\mathbf{v})$ of a real-valued function f, at a point $\mathbf{p}$, with respect to any vector $\mathbf{v}$. If f is a function of position and $\mathbf{v}$ is a velocity vector, then $d_{\mathbf{p}}f(\mathbf{v})$ represents the instantaneous rate of change of f, with respect to time, as you move from $\mathbf{p}$ with velocity $\mathbf{v}$.

However, if we want to analyze how f changes as we change direction, but not speed, then we replace the arbitrary vector $\mathbf{v}$ by a vector which specifies only direction, i.e., by a unit vector, a vector of magnitude 1; we usually denote unit vectors by $\mathbf{u}$.

It is classical to introduce new notation for $d_{\mathbf{p}}f(\mathbf{u})$; it is usually denoted by $D_{\mathbf{u}}f(\mathbf{p})$, and is called the *directional derivative of f, at $\mathbf{p}$, in the direction $\mathbf{u}$.*

If f is a function of position and $\mathbf{u}$ is velocity, then $\mathbf{u}$ being a unit vector means that the speed is 1, and $D_{\mathbf{u}}f(\mathbf{p})$ is the rate of change of f, at $\mathbf{p}$, with respect to time, as we move in the direction of $\mathbf{u}$ with speed 1. If $\mathbf{u}$ is a unitless direction, then $D_{\mathbf{u}}f(\mathbf{p})$ can be interpreted as a rate of change with respect to distance.

We want to fix the point $\mathbf{p}$ and address questions such as "In what direction does f increase as rapidly as possible, and what is the rate of change of f in that direction?" and "In what directions does f have no rate of change?". These are questions about $D_{\mathbf{u}}f(\mathbf{p})$ and, naturally, the gradient of f at $\mathbf{p}$ is of fundamental importance.

Basics:

Suppose that $f = f(\mathbf{x})$ is real-valued and differentiable at $\mathbf{p}$. Then, as we discussed in Section 2.2, we have the notion of the total derivative $d_{\mathbf{p}}f(\mathbf{v})$ of f, at $\mathbf{p}$, with respect to the vector $\mathbf{v}$. If f is a function of position and $\mathbf{v}$ is a velocity vector, then $d_{\mathbf{p}}f(\mathbf{v})$ represents the instantaneous rate of change of f, with respect to time, as you move from $\mathbf{p}$ with velocity $\mathbf{v}$.

From Theorem 2.2.2, we know that the total derivative can be calculated by

$$d_{\mathbf{p}}f(\mathbf{v}) \; = \; \vec{\nabla}f(\mathbf{p}) \cdot \mathbf{v}.$$

It follows from this formula that, for any constant c, $d_{\mathbf{p}}f(c\mathbf{v}) = c\, d_{\mathbf{p}}f(\mathbf{v})$. For $c > 0$, this means, intuitively, that, if you multiply your speed by c, then the rate of change of f will also be multiplied by c.

Recall from Section 1.2 that a *direction* actually **means** a unit vector; though sometimes, instead of saying "the direction $\mathbf{u}$", we say the more common "the direction of $\mathbf{u}$".

Hence, if we're trying to decide what direction to move in to maximize or minimize the value of $d_{\mathbf{p}}f(\mathbf{v})$, it is important that we use the same speed in each direction, since multiplying the speed would multiply the value of $d_{\mathbf{p}}f(\mathbf{v})$. In theory, we could use any fixed positive speed, but the simplest/nicest one to pick is 1, i.e., we require that $|\mathbf{v}| = 1$. Thus, we look at $d_{\mathbf{p}}f(\mathbf{u})$, where $\mathbf{u}$ is a unit vector.

If $\mathbf{u}$ is a velocity vector with magnitude 1, then $d_{\mathbf{p}}f(\mathbf{u})$ is an instantaneous rate of change of f with respect to time. But this is the same as the instantaneous rate of change of f with respect to distance multiplied by the rate of change of the distance with respect to time, i.e., times the speed, which is 1. Therefore, the number that we obtain from $d_\mathbf{p}f(\mathbf{u})$ can also be interpreted as the rate of change of f with respect to distance.

In fact, if $f = f(\mathbf{x})$ is a function of position, then $\vec{\nabla}f(\mathbf{p})$ has units which are the units of f divided by length units. Thus, if you are given $\mathbf{u}$ as a unitless unit vector, rather than as a velocity vector, then

$$d_\mathbf{p}f(\mathbf{u}) \; = \; \vec{\nabla}f(\mathbf{p}) \cdot \mathbf{u}$$

has units which are the units of f divided by length units, and $d_\mathbf{p}f(\mathbf{u})$ is the instantaneous rate of change of f, at $\mathbf{p}$, with respect to distance in $\mathbb{R}^n$, as you move in the direction $\mathbf{u}$.

On the other hand, if the units of the components of $\mathbf{x}$ are mixed, then $d_\mathbf{p}f(\mathbf{u})$ may be considered as a pure number, without units. The point is: the proper interpretation of $d_\mathbf{p}f(\mathbf{u})$ depends on the context.

In any case, we make the following definition.

Definition 2.5.1. (Directional Derivative) *Suppose that f is a differentiable function of n variables at a point $\mathbf{p}$ in $\mathbb{R}^n$, and suppose that $\mathbf{u}$ is a unit vector in $\mathbb{R}^n$. Then, the* **directional derivative of f, at $\mathbf{p}$, in the direction of $\mathbf{u}$** *is*

$$D_\mathbf{u}f(\mathbf{p}) \; = \; d_\mathbf{p}f(\mathbf{u}) \; = \; \lim_{t \to 0^+} \frac{f(\mathbf{p}+t\mathbf{u}) - f(\mathbf{p})}{t}$$

and represents the instantaneous rate of change of f, at $\mathbf{p}$, with respect to distance in $\mathbb{R}^n$, in the direction of $\mathbf{u}$.

As a number, this is equal to the instantaneous rate of change of f, at $\mathbf{p}$, with respect to time, in the direction of $\mathbf{u}$, as $\mathbf{p}$ moves with speed 1.

In Theorem 2.2.2, we stated that the total derivative equals the two-sided limit; so, of course, it also equals the limit from the right. The reason for taking the limit from the right, here, is that, if $t > 0$, then $t\mathbf{u}$ has the same direction as $\mathbf{u}$, and so seems more natural when discussing moving in the direction of $\mathbf{u}$.

We have not stated here that $f = f(\mathbf{x})$ is a function of position, or that the components of $\mathbf{x}$ have length units. We are using, abstractly, that $\mathbf{x}$ specifies a "position" in $\mathbb{R}^n$, and we are referring to the distance in $\mathbb{R}^n$, without units. We do this even though the variables $x_1, \ldots, x_n$, in a given problem, may have physical meaning other than position.

Remark 2.5.2. The notation $D_\mathbf{u}f(\mathbf{p})$ is standard for the directional derivative, but is a bit misleading; it makes it look like we fix $\mathbf{u}$ and change the point $\mathbf{p}$, i.e., look at $D_\mathbf{u}f$ as a function of $\mathbf{p}$. This is typically **not** the case. In standard applications, the point $\mathbf{p}$ is fixed, and we want to look at what happens as we change $\mathbf{u}$. Nonetheless, we shall adhere to the standard convention and denote the directional derivative by $D_\mathbf{u}f(\mathbf{p})$.

There is, in fact, one good reason to fix $\mathbf{u}$ and look at the function $D_\mathbf{u}f$; it generalizes the notion of a partial derivative. For instance, if $f = f(x,y,z)$ and we look at the standard basis $\mathbf{i}, \mathbf{j}, \mathbf{k}$ (of unit vectors) for $\mathbb{R}^3$, then it's easy to see that

$$D_{\mathbf{i}}f = \frac{\partial f}{\partial x}, \qquad D_{\mathbf{j}}f = \frac{\partial f}{\partial y}, \qquad \text{and} \qquad D_{\mathbf{k}}f = \frac{\partial f}{\partial z}.$$

More generally, if $f = f(x_1, \ldots, x_n)$, and $\mathbf{e}_i$ denotes the i-th standard basis element (which is a unit vector) for $\mathbb{R}^n$, then

$$D_{\mathbf{e}_i}f \;=\; \frac{\partial f}{\partial x_i}.$$

Suppose now that you're given a function f and a point $\mathbf{p}$, and you want to know in what direction $\mathbf{p}$ should move (or, you should move in from $\mathbf{p}$) so that f increases as rapidly as possible, or decreases as rapidly as possible. That is, you want to determine in what directions $\mathbf{u}$ the directional derivative $D_{\mathbf{u}}f(\mathbf{p})$ attains a maximum or minimum value (presuming, for the moment, that such directions exist). This follows easily from our geometric interpretation of the dot product from Section 1.3, and the fact that the magnitude of a unit vector is 1.

Theorem 2.5.3. *Suppose that f is differentiable at $\mathbf{p}$ in $\mathbb{R}^n$, and suppose that $\mathbf{u}$ is a unit vector. Then, the directional derivative of f, at $\mathbf{p}$, in the direction of $\mathbf{u}$ is given by*

$$D_{\mathbf{u}}f(\mathbf{p}) \;=\; d_{\mathbf{p}}f(\mathbf{u}) \;=\; \vec{\nabla}f(\mathbf{p}) \cdot \mathbf{u} \;=\; |\vec{\nabla}f(\mathbf{p})|\cos\theta,$$

where θ is the angle between $\vec{\nabla}f(\mathbf{p})$ and $\mathbf{u}$ (assuming that $\vec{\nabla}f(\mathbf{p}) \neq \mathbf{0}$).

Therefore, either $\vec{\nabla}f(\mathbf{p}) = \mathbf{0}$, and $D_{\mathbf{u}}f(\mathbf{p}) = 0$ for all $\mathbf{u}$, or $\vec{\nabla}f(\mathbf{p}) \neq 0$ and

  a. *the direction $\mathbf{u}$ in which $D_{\mathbf{u}}f(\mathbf{p})$ attains its maximum value is the direction of $\vec{\nabla}f(\mathbf{p})$, i.e., $\mathbf{u} = \vec{\nabla}f(\mathbf{p})/|\vec{\nabla}f(\mathbf{p})|$, and that maximum value of $D_{\mathbf{u}}f(\mathbf{p})$ is $|\vec{\nabla}f(\mathbf{p})|$;*

  b. *the direction $\mathbf{u}$ in which $D_{\mathbf{u}}f(\mathbf{p})$ attains its minimum value is the direction of $-\vec{\nabla}f(\mathbf{p})$, i.e., $\mathbf{u} = -\vec{\nabla}f(\mathbf{p})/|\vec{\nabla}f(\mathbf{p})|$, and that minimum value of $D_{\mathbf{u}}f(\mathbf{p})$ is $-|\vec{\nabla}f(\mathbf{p})|$; and*

  c. *$D_{\mathbf{u}}f(\mathbf{p}) = 0$ if and only if $\mathbf{u}$ is perpendicular to $\vec{\nabla}f(\mathbf{p})$.*

Proof. If $\vec{\nabla}f(\mathbf{p}) = \mathbf{0}$, then $D_{\mathbf{u}}f(\mathbf{p}) = \vec{\nabla}f(\mathbf{p}) \cdot \mathbf{u} = 0$, regardless of what $\mathbf{u}$ is. Assume now that $\vec{\nabla}f(\mathbf{p}) \neq \mathbf{0}$.

Then, by Theorem 2.2.2 and Theorem 1.3.2

$$D_{\mathbf{u}}f(\mathbf{p}) \;=\; \vec{\nabla}f(\mathbf{p}) \cdot \mathbf{u} \;=\; \left|\vec{\nabla}f(\mathbf{p})\right| \cos\theta,$$

where θ is the angle between $\vec{\nabla}f(\mathbf{p})$ and $\mathbf{u}$.

Part (c) follows immediately from this formula.

To see parts (a) and (b), notice that $\left|\vec{\nabla}f(\mathbf{p})\right|$ does not change as $\mathbf{u}$ changes, only θ does. Now, the maximum (resp., minimum) value of $\cos\theta$ is 1 (resp., -1), which occurs when $\theta = 0$ (resp., $\theta = \pi$ radians). Parts (a) and (b) follow at once. $\qquad\square$

Example 2.5.4. Let

$$f(x,y) \;=\; xe^y + x^3 + \sin y,$$

and suppose we're at the point $(1,0)$.

    a. If we move towards the point $(2,3)$, what is the rate of change of f with respect to distance?

    b. In what direction should we head to maximize the rate of change of f, and what is the rate of change if we move in that direction?

    c. In what direction should we head to minimize the rate of change of f, and what is the rate of change if we move in that direction?

Solution:

Of course, all of these are directional derivative questions, and we will need $\vec{\nabla}f(1,0)$, so we'll calculate that first.

$$\vec{\nabla}f(1,0) \;=\; \left(\frac{\partial f}{\partial x}, \frac{\partial f}{\partial y}\right)\Bigg|_{(1,0)} \;=\; \left(e^y + 3x^2,\; xe^y + \cos y\right)\Big|_{(1,0)} \;=\; (4,2).$$

a.   This part of the question is just asking us to calculate $D_{\mathbf{u}}f(1,0)$, but there is a mildly "tricky" part in how $\mathbf{u}$ is specified. We aren't moving in the direction of the **vector** $(2,3)$, but rather moving from the point $(1,0)$ towards the **point** $(2,3)$, i.e., in the direction of the vector $\overrightarrow{(1,0)(2,3)} = (2,3) - (1,0) = (1,3)$.

Therefore,

$$\mathbf{u} \;=\; \frac{(1,3)}{|(1,3)|} \;=\; \frac{1}{\sqrt{1^2+3^2}}\,(1,3) \;=\; \frac{1}{\sqrt{10}}\,(1,3),$$

and

$$D_{\mathbf{u}}f(1,0) \;=\; \vec{\nabla}f(1,0)\cdot\mathbf{u} \;=\; (4,2)\cdot\frac{1}{\sqrt{10}}\,(1,3) \;=\; \frac{1}{\sqrt{10}}\,(4,2)\cdot(1,3) \;=\; \frac{1}{\sqrt{10}}\,10 \;=\; \sqrt{10}.$$

b. You should immediately know the answer to part b. You want to pick $\mathbf{u}$ to maximize $D_{\mathbf{u}}f(1,0)$; so, $\mathbf{u}$ needs to be in the direction of $\vec{\nabla}f(1,0)$ and the maximum value is $|\vec{\nabla}f(1,0)|$. Since we know that $\vec{\nabla}f(1,0) = (4,2) = 2(2,1)$, we want

$$\mathbf{u} \;=\; \frac{(4,2)}{|(4,2)|} \;=\; \frac{2(2,1)}{|2(2,1)|} \;=\; \frac{2(2,1)}{2\sqrt{5}} \;=\; \frac{1}{\sqrt{5}}\,(2,1),$$

and, in this direction, $D_{\mathbf{u}}f(1,0) = |(4,2)| = 2\sqrt{5}$.

c. There is essentially no extra work to do here; the answers are the negatives of the answers in part b: the minimum value of $D_{\mathbf{u}}f(1,0)$ occurs when

$$\mathbf{u} \;=\; \frac{-(4,2)}{|(4,2)|} \;=\; -\frac{1}{\sqrt{5}}\,(2,1),$$

and, in this direction, $D_{\mathbf{u}}f(1,0) = -|(4,2)| = -2\sqrt{5}$.

Example 2.5.5. In the previous example, you should have noticed that all that was used about the function f was its gradient vector at the given point. Thus, instead of giving a formula for f at arbitrary x and y, we could simply give that f is differentiable at $\mathbf{p}$ and give $\vec{\nabla}f(\mathbf{p})$, and you could still answer all of the directional derivative questions.

Suppose, for instance, that g is differentiable at $(-1,3)$ and that $\vec{\nabla}g(-1,3) = (2,7)$.

a. If we move in the direction of the vector $(-12,5)$, what is the rate of change of g with respect to distance?

b. In what direction should we head to maximize the rate of change of g, and what is the rate of change if we move in that direction?

Solution:

a. The unit vector in the direction of $(-12,5)$ is

$$\mathbf{u} \;=\; \frac{1}{|(-12,5)|}(-12,5) \;=\; \frac{1}{13}(-12,5).$$

The rate of change that we're asked for is:

$$D_{\mathbf{u}}g(-1,3) \; = \; \vec{\nabla}g(-1,3) \cdot \mathbf{u} \; = \; (2,7) \cdot \frac{1}{13}(-12,5) \; = \; \frac{11}{13}.$$

b.   The direction in which the change in g is a maximum is in the direction of the gradient vector; the unit vector is this direction is

$$\mathbf{u} \; = \; \frac{1}{|(2,7)|}(2,7) \; = \; \frac{1}{\sqrt{53}}(2,7).$$

Furthermore, with this $\mathbf{u}$, the value of $D_{\mathbf{u}}g(-1,3)$ is $|\vec{\nabla}g(-1,3)| = \sqrt{53}$.

Example 2.5.6. Suppose that the elevation z of a hill is given in terms of x and y coordinates (at sea level, approximating the Earth as being flat) by

$$z \; = \; 400 - 5x^2 - 3y^2,$$

where x, y, and z are in feet, and $5x^2 + 3y^2 \leq 400$. Note that directions in the xy-plane would correspond to what you would read on a compass; this is to be contrasted with directions in xyz-space.

If we're at the point on the surface of the hill where $(x,y) = (2,10)$, in what xy-direction should we head to ascend the hill as rapidly as possible?

Solution:

We want the direction $\mathbf{u}$ in $\mathbb{R}^2$ such that the rate of change of the height is as big as possible, i.e., so that $D_{\mathbf{u}}z(2,10)$ attains its maximum value. We know from Theorem 2.5.3 that $\mathbf{u}$ should be in the direction of $\vec{\nabla}z(2,10)$.

We calculate

$$\vec{\nabla}z(2,10) \; = \; (-10x, -6y)\Big|_{(2,10)} \; = \; (-20,-60) \; = \; -20(1,3),$$

and so, the direction that we're after is

$$\mathbf{u} \; = \; \frac{-20(1,3)}{|-20(1,3)|} \; = \; \frac{-20(1,3)}{20\sqrt{1^2+3^2}} \; = \; -\frac{1}{\sqrt{10}}(1,3).$$

Example 2.5.7. Recall Example 2.2.8, in which space was being heated in such a way that the temperature, T, in degrees Celsius, was given by

$$T(x,y,z) = z^2 + 0.5e^{x^2 - y^2},$$

where x, y, and z are measured in meters and are all strictly between -3 and 3. As before, suppose a particle is at the point $(0, 1, 2)$.

    a. What is the instantaneous rate of change of the temperature, with respect to distance, at the particle's location, if the particle moves in the direction of the vector $(3, -1, 0)$?

    b. What is the instantaneous rate of change of the temperature, with respect to time, at the particle's location, if the particle moves in the direction of the vector $(3, -1, 0)$ at a speed of 7 m/s?

    c. In what direction would the particle need to move to cool off as rapidly as possible, and what would be the rate of change of the temperature, with respect to distance, if the particle moved in this direction?

    d. In what directions would the particle need to move so that there is no instantaneous change in the temperature?

Solution:

    Parts (a), (c), and (d) are questions about directional derivatives. Part (b) can be answered easily after answering Part (a).

    We first calculate the gradient vector, or look up our previous calculation in Example 2.2.8:

$$\vec{\nabla}T(0,1,2) = \left. \left(xe^{x^2-y^2},\ -ye^{x^2-y^2},\ 2z \right) \right|_{(0,1,2)} = (0,\ -e^{-1},\ 4)\ \,^\circ\mathrm{C/m}.$$

    a. This question is just asking for $D_{\mathbf{u}}T(0,1,2)$, where $\mathbf{u}$ is the unit vector in the direction of $(3, -1, 0)$, i.e.,

$$\mathbf{u} = \frac{1}{\sqrt{10}}(3, -1, 0).$$

Thus,

$$D_{\mathbf{u}}T(0,1,2) = \vec{\nabla}T(0,1,2) \cdot \mathbf{u} = (0, -e^{-1}, 4) \cdot \frac{1}{\sqrt{10}}(3, -1, 0) = \frac{1}{e\sqrt{10}}\ \,^\circ\mathrm{C/m}.$$

    b. The quick answer is that you multiply the rate of change that we found in Part (a) by 7 m/s to obtain

$$\left(\frac{1}{e\sqrt{10}}\ \,^\circ\mathrm{C/m} \right) \cdot (7\ \mathrm{m/s}) = \frac{7}{e\sqrt{10}}\ \,^\circ\mathrm{C/s}.$$

If you want some mathematical notation for what we calculated, remember that the total derivative $d_{\mathbf{p}}T(\mathbf{v})$ is the instantaneous rate of change of T, at the point $\mathbf{p}$, when you move with arbitrary velocity $\mathbf{v}$. Therefore, what we were asked for was $d_{(0,1,2)}T(\mathbf{v})$, where $\mathbf{v}$ is the unique vector with the same direction as $(3,-1,0)$ and magnitude 7; this is 7 times the unit vector in the direction of $(3,-1,0)$, i.e.,

$$\mathbf{v} \;=\; 7 \cdot \frac{1}{\sqrt{10}}\,(3,-1,0)$$

We can check that this yields what we found above:

$$d_{(0,1,2)}T(\mathbf{v}) \;=\; \vec{\nabla}T(0,1,2) \cdot \mathbf{v} \;=\; (0,\,-e^{-1},\,4) \cdot \frac{7}{\sqrt{10}}\,(3,-1,0) \;=\; \frac{7}{e\sqrt{10}} \quad {}^{\circ}\mathrm{C/s}.$$

c. Cooling off as rapidly as possible means that the rate of change of the temperature is as negative as possible. This occurs in the direction opposite the gradient, i.e., the particle would need to move in the direction

$$\mathbf{u} \;=\; -\frac{\vec{\nabla}T(0,1,2)}{\left|\vec{\nabla}T(0,1,2)\right|} \;=\; -\frac{(0,\,-e^{-1},\,4)}{\left|(0,\,-e^{-1},\,4)\right|} \;=\; -\frac{1}{\sqrt{16+e^{-2}}}\,(0,\,-e^{-1},\,4).$$

And, when the particle moves in this direction, the rate of change of T is

$$-\left|\vec{\nabla}T(0,1,2)\right| \;=\; -\sqrt{16+e^{-2}} \quad {}^{\circ}\mathrm{C/m}.$$

d. In order to have 0 instantaneous rate of change, we would need

$$D_{\mathbf{u}}T(0,1,2) \;=\; \vec{\nabla}T(0,1,2) \cdot \mathbf{u} \;=\; 0,$$

i.e., for $\mathbf{u}$ to be perpendicular to $\vec{\nabla}T(0,1,2) = (0,\,-e^{-1},\,4)$. A vector (a,b,c) is perpendicular to $(0,\,-e^{-1},\,4)$ if and only if

$$(0,\,-e^{-1},\,4) \cdot (a,b,c) \;=\; -be^{-1} + 4c \;=\; 0.$$

This means that we need $b = 4ec$, so that our vector is $(a, 4ec, c)$, and we want to make it a unit vector; so, we want $\mathbf{u}$ to be any vector of the form

$$\frac{(a, 4ec, c)}{\left|(a, 4ec, c)\right|},$$

for all values of a and c, other than $a = c = 0$.

More Depth:

Example 2.5.8. Let's return to Example 2.5.6, in which the elevation z of a hill was given in terms of x and y coordinates by

$$z = 400 - 5x^2 - 3y^2,$$

where x, y, and z are in feet, and $5x^2 + 3y^2 \leq 400$.

We found that, if we are at $(x, y) = (2, 10)$, then, to ascend as rapidly as possible, we should move in the xy-direction that's in the same direction as the gradient vector

$$\vec{\nabla} z(2, 10) = -20(1, 3),$$

i.e., in the direction

$$\mathbf{u} = -\frac{1}{\sqrt{10}}(1, 3) = \left(-\frac{1}{\sqrt{10}}, -\frac{3}{\sqrt{10}}\right).$$

But, clearly, moving in an xy-direction, and staying on the surface of the hill, must imply that we move in a corresponding z-direction as well. So, our question now is: in what **xyz**-direction should we head to ascend the hill as rapidly as possible? In other words, we now want to know the instantaneous rate of change of the z-coordinate as well.

But this is easy.

We know that, if $\mathbf{u}$ is in the xy-direction of $\vec{\nabla} z(2, 10)$, then $D_{\mathbf{u}} z(2, 10) = \left|\vec{\nabla} z(2, 10)\right| = 20\sqrt{10}$. But, the physical meaning of $D_{\mathbf{u}} z(2, 10)$ is precisely the instantaneous rate of change of z, as (x, y) moves in the direction $\mathbf{u}$. So, what's the instantaneous rate of change of (x, y, z) as (x, y) moves in the direction $\mathbf{u}$? It's the vector whose first two components are the components of $\mathbf{u}$, and whose third component is $D_{\mathbf{u}} z(2, 10)$, i.e., the xyz-direction in which you would need to move to ascend as rapidly as possible is in the direction of

$$\left(-\frac{1}{\sqrt{10}}, -\frac{3}{\sqrt{10}}, 20\sqrt{10}\right) = \frac{1}{\sqrt{10}}(-1, -3, 200),$$

i.e., in the direction of the unit vector

$$\frac{1}{\sqrt{40,010}}(-1, -3, 200).$$

Example 2.5.9. Suppose you have a function $w = w(x, y)$ which is differentiable at a point $\mathbf{p}$. Suppose further that, at $\mathbf{p}$, the directional derivative of w in the direction of the vector $(2, 3)$ equals 7, and the directional derivative in the direction of the vector $(1, -2)$ equals 5. What is the directional derivative of w, at $\mathbf{p}$, in the direction of the vector $(-1, 4)$?

Solution: How can you possibly do this? Well...we have only one formula for calculating $D_{\mathbf{u}}w(\mathbf{p})$:

$$D_{\mathbf{u}}w(\mathbf{p}) = \vec{\nabla}w(\mathbf{p}) \cdot \mathbf{u}.$$

The $\mathbf{u}$ that we need to use $(-1,4)/|(-1,4)|$, so the question is: how do we find $\vec{\nabla}w(\mathbf{p})$? The answer is that we use the given directional derivative data to come up with two linear equations and two unknowns that we need to solve to find the partial derivatives of w at $\mathbf{p}$.

Suppose that $\vec{\nabla}w(\mathbf{p}) = (a,b)$. Let $\mathbf{u}_1 = (2,3)/|(2,3)|$ and $\mathbf{u}_2 = (1,-2)/|(1,-2)|$.

Then, we are told that

$$7 = D_{\mathbf{u}_1}w(\mathbf{p}) = \vec{\nabla}w(\mathbf{p}) \cdot \mathbf{u}_1 = (a,b) \cdot \frac{(2,3)}{|(2,3)|} = \frac{1}{\sqrt{13}}(2a + 3b)$$

and

$$5 = D_{\mathbf{u}_2}w(\mathbf{p}) = \vec{\nabla}w(\mathbf{p}) \cdot \mathbf{u}_2 = (a,b) \cdot \frac{(1,-2)}{|(1,-2)|} = \frac{1}{\sqrt{5}}(a - 2b).$$

We want to solve for a and b.

We leave it as an exercise for you to solve

$$2a + 3b = 7\sqrt{13}, \qquad a - 2b = 5\sqrt{5}$$

and show that

$$a = \frac{14\sqrt{13} + 15\sqrt{5}}{7} \qquad \text{and} \qquad b = \frac{7\sqrt{13} - 10\sqrt{5}}{7}.$$

Now, we can calculate what we want

$$D_{\mathbf{u}}w(\mathbf{p}) = \vec{\nabla}w(\mathbf{p}) \cdot \mathbf{u} = \left(\frac{14\sqrt{13} + 15\sqrt{5}}{7}, \frac{7\sqrt{13} - 10\sqrt{5}}{7}\right) \cdot \frac{(-1,4)}{|(-1,4)|} =$$

$$\frac{1}{\sqrt{17}} \cdot \frac{-14\sqrt{13} - 15\sqrt{5} + 28\sqrt{13} - 40\sqrt{5}}{7} = \frac{14\sqrt{13} - 55\sqrt{5}}{7\sqrt{17}} \approx -2.5.$$

No one said that the answer would be pretty.

2.5.1　Exercises

Online answers to select exercises are here.

Basics:

In each of the following exercises, you are given the value of $\vec{\nabla}f(\mathbf{p})$ for a differentiable function f. You are also given a vector **v**. (a) Calculate the rate of change of f, with respect to distance, at **p**, in the direction of **v**. (b) Determine the direction in which one should head from **p** in order to make f decrease as rapidly as possible, with respect to distance, and give the value of this minimum rate of change. (c) Describe the directions, at **p**, in which the rate of change of f is 0.

1. $\vec{\nabla}f(1,2) = (-3,4)$, $\mathbf{v} = (5,-1)$

2. $\vec{\nabla}f(200,-100) = (0.01, 0.03)$, $\mathbf{v} = (1,1)$

3. $\vec{\nabla}f(3,2,1) = (-2,0,1)$, $\mathbf{v} = (0,5,12)$

4. $\vec{\nabla}f(0,0,0) = (10,-12,11)$, $\mathbf{v} = (-1,-1,-1)$

In each of the following exercises, you are given a differentiable real-valued function f, a point **p** and a vector **v**. Determine the unit vector **u** in the direction of **v**, and then calculate $D_{\mathbf{u}}f(\mathbf{p})$, i.e., calculate the instantaneous rate of change of f, at **p**, with respect to distance.

5. $f(x,y) = x^2 + y^3$, $\mathbf{p} = (1,1)$, $\mathbf{v} = (-3,4)$.

6. $f(x,y) = xe^y - y\sin(xy)$, $\mathbf{p} = (0.5, \pi)$, $\mathbf{v} = (1,-1)$.

7. $f(x,y) = \dfrac{x+y}{x-y}$, $\mathbf{p} = (2,1)$, $\mathbf{v} = (\sqrt{3}, \sqrt{13})$.

8. $f(x,y) = \dfrac{x\ln(x+y)}{y^2+1}$, $\mathbf{p} = (0.8, 0.2)$, $\mathbf{v} = (1,1)$.

9. $f(x,y) = 5xy^2 + x^3y$, $\mathbf{p} = (-1,1)$, **v** is the vector from $(-1,1)$ to $(3,4)$.

10. $f(x,y) = e^{-x^2-y^2} + 2xy$, $\mathbf{p} = (0,1)$, **v** is the vector from $(0,1)$ to $(-5,2)$.

11. $f(x,y,z) = x^2 + y^3 + z^5$, $\mathbf{p} = (1,-1,1)$, $\mathbf{v} = (1,0,1)$.

12. $f(x,y,z) = x^2 + y^3 + z^5$, $\mathbf{p} = (-1,1,-1)$, $\mathbf{v} = (1,1,1)$.

13. $f(x,y,z) = xe^y + 5y\tan^{-1}z$, $\mathbf{p} = (-1,0,1)$, $\mathbf{v} = (3,2,\sqrt{3})$.

14. $f(x,y,z) = xe^y + 5y\tan^{-1}z$, $\mathbf{p} = (0,1,0)$, $\mathbf{v} = (3,2,\sqrt{3})$.

15. $f(x,y,z) = \sqrt{x} + \sqrt[3]{y} + \sqrt[5]{z}$, $\mathbf{p} = (1,1,1)$, **v** is the vector from $(1,1,1)$ to $(2,-3,4)$.

16. $f(x, y, z) = \sqrt{x} + \sqrt[3]{y} + \sqrt[5]{z}$, $\mathbf{p} = (4, 8, 32)$, $\mathbf{v}$ is the vector from $(1, 1, 1)$ to $(0, 0, 0)$.

17. $f(w, x, y, z) = wx + yz$, $\mathbf{p} = (1, 2, 3, 4)$, $\mathbf{v} = (1, 0, -1, 0)$.

18. $f(w, x, y, z) = xe^{2w} + ze^{3y}$, $\mathbf{p} = (\ln 3, 1, \ln 2, 1)$, $\mathbf{v} = (1, 1, -1, -1)$.

In each of the following exercises, you are give a real-valued differentiable function f and a point p. Determine (a) the direction (as a unit vector) in which f increases most rapidly at p, (b) the maximum value of the rate of change of f at p, with respect to distance, (c) the direction (as a unit vector) in which f decreases most rapidly at p, (d) the minimum value of the rate of change of f at p, with respect to distance, and (e) a description of the directions from p in which f remains constant. Note that the functions and points are those from the previous exercises.

19. $f(x, y) = x^2 + y^3$, $\mathbf{p} = (1, 1)$.

20. $f(x, y) = xe^y - y\sin(xy)$, $\mathbf{p} = (0.5, \pi)$.

21. $f(x, y) = \dfrac{x + y}{x - y}$, $\mathbf{p} = (2, 1)$.

22. $f(x, y) = \dfrac{x\ln(x + y)}{y^2 + 1}$, $\mathbf{p} = (0.8, 0.2)$.

23. $f(x, y) = 5xy^2 + x^3 y$, $\mathbf{p} = (-1, 1)$.

24. $f(x, y) = e^{-x^2 - y^2} + 2xy$, $\mathbf{p} = (0, 1)$.

25. $f(x, y, z) = x^2 + y^3 + z^5$, $\mathbf{p} = (1, -1, 1)$.

26. $f(x, y, z) = x^2 + y^3 + z^5$, $\mathbf{p} = (-1, 1, -1)$.

27. $f(x, y, z) = xe^y + 5y\tan^{-1} z$, $\mathbf{p} = (-1, 0, 1)$.

28. $f(x, y, z) = xe^y + 5y\tan^{-1} z$, $\mathbf{p} = (0, 1, 0)$.

29. $f(x, y, z) = \sqrt{x} + \sqrt[3]{y} + \sqrt[5]{z}$, $\mathbf{p} = (1, 1, 1)$.

30. $f(x, y, z) = \sqrt{x} + \sqrt[3]{y} + \sqrt[5]{z}$, $\mathbf{p} = (4, 8, 32)$.

31. $f(w, x, y, z) = wx + yz$, $\mathbf{p} = (1, 2, 3, 4)$.

32. $f(w, x, y, z) = xe^{2w} + ze^{3y}$, $\mathbf{p} = (\ln 3, 1, \ln 2, 1)$.

33. Suppose that the xy-plane has an electric charge, in coulombs, given by

$$Q = xy + x^2 \sin(\pi y),$$

where x and y are in meters. Also, suppose that a particle is at the point $(1, 1)$.

(a) If the particle moves in the direction in which the charge increases most rapidly, in what direction does it move?

(b) If the particle moves in the direction given in part (a), with speed 1 m/s, at what rate is the charge changing, in coulombs per second, when the particle is at $(1, 1)$?

(c) If the particle moves in the direction given in part (a), with speed 7 m/s, at what rate is the charge changing, in coulombs per second, when the particle is at $(1, 1)$?

34. The elevation above sea level, in meters, of a valley is given by $z = -5x^2 - 3y^2$, where x and y are also in meters, and the minus signs indicate that the valley is below sea level. Suppose that a woman is at a point in the valley where $(x, y) = (1, 2)$.

    (a) In what direction (in the xy-plane) should the woman head to ascend as rapidly as possible?

    (b) At what rate, in meters of elevation per meter of distance in the xy-plane, will the woman ascend if she moves in the direction from (a)?

35. Suppose that space is being heated in such a way that the temperature, T, in degrees Celsius, is given by $T(x, y, z) = x/(y^2 + z^2 + 0.01)$, where x, y, and z are measured in meters, and are all of absolute value less than 10. Suppose that a particle is at the point $(1, -1, 2)$.

    a. What is the instantaneous rate of change of the temperature, with respect to distance, at $(1, -1, 2)$, if the particle moves toward the origin?

    b. What is the instantaneous rate of change of the temperature, with respect to time, at $(1, -1, 2)$ if the particle moves toward the origin at a speed of 7 m/s?

    c. In what direction would the particle need to move to cool off as rapidly as possible, and what would be the rate of change of the temperature, with respect to distance, if the particle moved in this direction?

    d. In what directions would the particle need to move so that there is no instantaneous change in the temperature?

36. A lab produces a quantity of a liters of a chemical solution A, b liters of solution B, and c liters of solution C. These chemicals are used in making p liters of a solution P, and it has been found that

$$p = 100ab^{-1}c^{-1} + 50a^{-3}bc^{-2} + 25a^{-0.5}b^{-0.5}c.$$

The triple (a, b, c) is referred to as the *production level*, and the distance between two production levels (a_1, b_1, c_1) and (a_2, b_2, c_2) is given by the Euclidean distance,

$$\sqrt{(a_2 - a_1)^2 + (b_2 - b_1)^2 + (c_2 - c_1)^2},$$

because this distance is relevant to the cost of production.

Suppose that the lab currently produces $(a, b, c) = (3, 2, 1)$ liters.

(a) In what direction should the lab change its production level in order to make p increase as rapidly as possible, with respect to distance?

(b) If the lab changes its production level as in part (a), what will the rate of change of p, with respect to distance be (initially)?

(c) If the lab wants to change its production level, but keep p constant, in what directions can the lab change the production level?

37. Suppose that we have a fixed number of atoms or molecules of an ideal gas in a container, where the container has a volume that can change with time.

Recall the Ideal Gas Law, as it was stated in Example 2.1.17: $PV = kT$, where P, is the pressure, in pascals (Pa, which are N/m^2), that the gas exerts on the container, T, is the temperature, in kelvins (K), of the gas, V, is the volume, in m^3, that the gas occupies, and the constant k is determined by the number of atoms or molecules of the ideal gas (together with the ideal gas constant).

Suppose that $k = 8$ N·m/K is fixed, and that the temperature and volume are currently given by $(T, V) = (320 \text{ K}, 2 \text{ m}^3)$.

If (T, V) changes in the direction of $(-1, 2)$, what is the rate of change of pressure, with respect to distance in the (T, V)-plane? In particular, does P increase or decrease?

> **More Depth:**

38. As we shall discuss later, in Example 4.3.15, the gravitational potential and the electric potential from two masses, or charges, located at $(0, 0, 0)$ and $(1, 0, 0)$ are both functions of the form

$$f(x, y, z) = \frac{A}{\sqrt{x^2 + y^2 + z^2}} + \frac{B}{\sqrt{(x-1)^2 + y^2 + z^2}},$$

where A and B are constants. In the following questions, your answers will depend on A and B.

(a) Calculate $\vec{\nabla} f(2, 0, 0)$.

(b) If an object is at $(2, 0, 0)$, and moves along a straight line to the point $(1, 1, 1)$, what is the instantaneous rate of change of the potential, with respect to distance, when the particle is at $(2, 0, 0)$?

(c) If an object is at $(2, 0, 0)$, in what direction should it move to make the potential drop as rapidly as possible?

(d) If the object moves in the direction from part (b), what is the instantaneous rate of change of the potential, with respect to distance, when the particle is at $(2, 0, 0)$?

(e) If an object is at $(2, 0, 0)$, describe the directions in which it can move in order to keep the potential constant, i.e., equal to $f(2, 0, 0)$.

39. The elevation z of a rolling hill, in feet, above sea level, is given in terms of x- and y-coordinates at sea level by $z = 400 - x^2(1 + \sin y) - y^2(1 + \cos x)$, where x and y are also in feet, and each has absolute value less than 10. A man is on the hill at the point where $(x, y) = (\pi, -\pi)$. In what direction in $\mathbb{R}^3$ should the man move in order to ascend as rapidly as possible?

40. In Exercise 34, above, answer part (a) again, but, this time, give the direction in xyz-space, i.e., give a unit vector in $\mathbb{R}^3$ for the direction.

41. Suppose you have a function $f = f(x, y)$ which is differentiable at a point $\mathbf{p}$. Suppose further that, at $\mathbf{p}$, the directional derivative of f in the direction of the vector $(1, 1)$ equals -2, and the directional derivative in the direction of the vector $(1, -2)$ equals 5. What is the directional derivative of f, at $\mathbf{p}$, in the direction of the vector $(-1, 6)$?

42. Suppose you have a function $f = f(x, y)$ which is differentiable at a point $\mathbf{p}$. Suppose further that, at $\mathbf{p}$, the directional derivative of f in the direction of the vector $(-1, 1)$ equals 3, and the directional derivative in the direction of the vector $(2, -1)$ equals 5. What are the directional derivatives of f, at $\mathbf{p}$, in the directions of the vectors $(2, 0)$ and $(0, 3)$?

43. The elevation z of a hill, in feet, above sea level, is given in terms of x- and y-coordinates at sea level by $z = f(x, y)$, where x and y are also in feet, and f is a differentiable function. A man is on the hill at the point where $(x, y) = (a, b)$. The man measures that, as he moves northwest, i.e., in the direction of $(-1, 1)$, he initially ascends at a rate of 2 feet of elevation for every foot in the xy-plane. He also measures that, as he moves southwest, i.e., in the direction of $(-1, -1)$, he initially descends at a rate of 0.5 feet of elevation for every foot in the xy-plane. In what direction in the xy-plane should the man move in order to descend as rapidly as possible?

44. The elevation z of a hill, in feet, above sea level, is given in terms of x- and y-coordinates at sea level by $z = f(x, y)$, where x and y are also in feet, and f is a differentiable function. A woman is on the hill at the point where $(x, y) = (a, b)$. The woman measures that, as she moves east, i.e., in the direction of $(1, 0)$, she initially ascends at a rate of 2 feet of elevation for every foot in the xy-plane. She also measures that, as she moves south, i.e., in the direction of $(0, -1)$, she initially descends at a rate of 0.5 feet of elevation for every foot in the xy-plane. At what rate will the woman's elevation change if she moves northwest?

2.6 Change of Coordinates

A *change of coordinates* is the mathematically precise way of describing what's going on when two people are discussing the same function or geometric object, even though they're describing the situation using different variables. We need to have a one-to-one correspondence, a *bijective* function, that takes us from one set of coordinates to the other. If we want the change of coordinates to "preserve" all of our derivative calculations, and take smooth graphs to other smooth graphs, then we need for the change of coordinates to have continuous partial derivatives of all orders, and we need for the inverse of the change of coordinates (the "change of coordinates in the opposite direction") to also have continuous partial derivatives of all orders.

As we shall see in many places throughout this book, changes of coordinates allow us to change complicated-looking problems into much easier ones.

Basics:

Changing coordinates is a fundamental process in mathematics. Changing coordinates can take a difficult problem and put it into a simple form. *Smooth changes of coordinates* allow us to simplify problems involving differentiation and integration, and let us give a rigorous definition of what it means for a graph, or any subset of $\mathbb{R}^n$ to be *smooth*, where we want the rigorous definition of "smooth" to agree with what you intuitively see as "smooth".

Example 2.6.1. Perhaps the most common type of change of coordinates, one which you probably do without really thinking about it, is a *scaling*. For example, in some physical situation, one person measures in terms of x grams and y meters and another person, who's looking at the same situation, measures in terms of $\hat{x}$ kilograms and $\hat{y}$ centimeters. Then, when the second person has $\hat{x} = 1$ and $\hat{y} = 2$, the first person has $x = 1000$ and $y = 0.02$.

In this example, to go from one set of coordinates to the other, you use $x = 1000\hat{x}$ and $y = \hat{y}/100$ or, equivalently, $\hat{x} = x/1000$ and $\hat{y} = 100y$. A scaling just means that the new coordinates are obtained from the old ones by multiplying by non-zero constants (which are usually positive for a scaling).

As you know, whenever you, or a computer/calculator, draw(s) graphs, you pick scales for the various axes; this scaling of the axes pretty much guarantees that you won't see any effect on the graphs from scaling the coordinates. For instance, suppose that x, y, $\hat{x}$, and $\hat{y}$ are as we described above, and consider the function $z = f(x, y) = x^2 y$. In terms of $\hat{x}$ and $\hat{y}$, $z = (1000\hat{x})^2(\hat{y}/100) = 10,000\hat{x}^2\hat{y}$. If you use technology to graph these, and let it automatically choose the ranges of the variables and scales, you are likely to see no difference in the graphs. But, if you do see a difference, it will just be that the graph is stretched or compressed in certain directions, i.e., the graph has been scaled.

Example 2.6.2. But what about other coordinate changes? What if we let $u = 2x$ and $v = x + y$? Is it reasonable to think of (u, v) as a "change of coordinates" from (x, y)?

Well...if you're given (x, y), it determines a unique pair (u, v). Conversely, given (u, v), you can solve to determine a unique pair (x, y), namely $x = u/2$ and $y = v - x = v - u/2$. Therefore, there is a one-to-one correspondence, or a *bijection*, between (x, y) pairs and (u, v) pairs. Another way to say this is that the function

$$(u, v) = \mathbf{\Phi}(x, y) = (2x, x + y)$$

has an *inverse function*

$$(x, y) = \mathbf{\Phi}^{-1}(u, v) = (u/2,\, v - u/2).$$

Why do we care about having a one-to-one correspondence? Suppose that two people are discussing a mathematical situation, problem, or function, and Person A uses coordinates x and y, while Person B uses coordinates u and v. If Person A is looking at a specific point $(x, y) = (3, -1)$, then that point should correspond to a unique point (u, v) for Person B, and vice-versa.

But is bijectivity all that we want to require of a coordinate change?

No. We want Person A to be able to calculate partial and total derivatives in his/her coordinates, and translate that into partial and total derivative calculations in Person B's coordinates, and vice-versa. In particular, if Person A has a function f which has continuous partial derivatives of any order r in his/her coordinates, then Person B should also find that f has continuous partial derivatives of order r in his/her coordinates, and vice-versa. Note that, technically, "the function f" in different coordinates means the function f composed with the appropriate coordinate change function.

Thus, recalling the terminology from Definition 2.1.30, what we want from a change of coordinates $\mathbf{\Phi} : \mathbb{R}^n \to \mathbb{R}^n$ is that it's a bijection and that, for all r-times continuously differentiable (i.e., C^r) functions $\mathbf{f}$ or smooth (i.e., C^∞) functions $\mathbf{f}$, from $\mathbb{R}^n$ to $\mathbb{R}^k$, the functions $\mathbf{f} \circ \mathbf{\Phi}$ and $\mathbf{f} \circ \mathbf{\Phi}^{-1}$ are also C^r or C^∞. The Chain Rule tells us that what we need is for $\mathbf{\Phi}$ and $\mathbf{\Phi}^{-1}$ themselves to be of class C^r or C^∞. Frequently, however, we don't need for $\mathbf{\Phi}$ to be defined on **all** of $\mathbb{R}^n$, but rather merely on a open subset.

Hence, we make the following definition:

Definition 2.6.3. (Change of Coordinates) *A C^r **change of coordinates** (on an open subset of $\mathbb{R}^n$) is a bijection $\Phi : \mathcal{A} \to \mathcal{B}$, where $\mathcal{A}$ and $\mathcal{B}$ are open subsets of $\mathbb{R}^n$, such that Φ and Φ^{-1} are of class C^r. A C^∞ change of coordinates is called a* **smooth change of coordinates**.

A function Φ from a subset of $\mathbb{R}^n$ to a subset of $\mathbb{R}^n$ is called a C^r (resp., smooth) **local change of coordinates at p** *provided that there exist open subsets $\mathcal{A}$ and $\mathcal{B}$ of the domain and codomain, respectively, of Φ such that* **p** *is in $\mathcal{A}$ and the restriction $\Phi : \mathcal{A} \to \mathcal{B}$ is a C^r (resp., smooth) change of coordinates.*

> Another term for a change of coordinates is a *diffeomorphism*, provided that $1 \le r \le \infty$.

We will use smooth changes of coordinates almost exclusively, since we are interested in partial derivatives of all orders.

Example 2.6.4. Let's look back at our function Φ from Example 2.6.2:

$$(u, v) = \Phi(x, y) = (2x, x + y), \qquad (x, y) = \Phi^{-1}(u, v) = (u/2, \, v - u/2).$$

This Φ is a coordinate change from all of $\mathbb{R}^2$ to $\mathbb{R}^2$, and is, in fact, a **smooth** coordinate change, since the component functions of both Φ and Φ^{-1} are clearly infinitely continuously differentiable. In fact, since all of the component functions are linear, all of the partial derivatives of all of the component functions of order 2 or more are zero. Such a change of coordinates is called a *linear change of coordinates*.

Let's look at what happens when we compose the "change of coordinates" Φ or, rather, its inverse Φ^{-1} with another function.

Consider the function

$$z = f(x, y) = \frac{5x}{x^2 + y^2 + 1}.$$

In terms of u and v, this corresponds to the function

$$z = \left(f \circ \Phi^{-1}\right)(u, v) = \frac{5u/2}{(u/2)^2 + (v - u/2)^2 + 1}.$$

The graphs of these two functions appear in Figure 2.6.1 and Figure 2.6.2.

Note that the graphs appear qualitatively similar; the graph in Figure 2.6.2 looks like the graph in Figure 2.6.1, just stretched a little and turned a bit. The fact that the graphs appear similar is not just a result of Φ being a bijection, but also because Φ and Φ^{-1} are smooth.

For instance, since f is differentiable, its graph is smooth. Now, since Φ^{-1} is also differentiable, the Chain Rule tells us that $f \circ \Phi^{-1}$ is differentiable, and so $f \circ \Phi^{-1}$

Figure 2.6.1: Graph of $z = 5x/(x^2 + y^2 + 1)$.

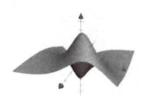

Figure 2.6.2: Graph of $z = (5u/2)/[(u/2)^2 + (v - u/2)^2 + 1]$.

has a smooth graph. In addition, the Chain Rule tells us how to calculate the partial derivatives and total derivative of $f \circ \mathbf{\Phi}^{-1}$ in terms of the partial or total derivatives of f and $\mathbf{\Phi}^{-1}$.

Note that we have $z = f(x, y)$ and $z = (f \circ \mathbf{\Phi}^{-1})(u, v)$. The partial derivatives $\partial z / \partial x$ and $\partial z / \partial y$ refer to the partial derivatives of this first function, while $\partial z / \partial u$ and $\partial z / \partial v$ refer to the partial derivatives of this second function. Let's use the multivariable Chain Rule to calculate $\partial z / \partial u$ and $\partial z / \partial v$ from $\partial z / \partial x$ and $\partial z / \partial y$.

We found in Example 2.3.16 that

$$\frac{\partial z}{\partial x} = \frac{5(y^2 - x^2 + 1)}{\left(x^2 + y^2 + 1\right)^2} \quad \text{and} \quad \frac{\partial z}{\partial y} = \frac{-10xy}{\left(x^2 + y^2 + 1\right)^2}.$$

Now, if we want $\partial z / \partial u$ and $\partial z / \partial v$, we use the Chain Rule, recalling that $x = u/2$ and $y = v - u/2$:

$$\frac{\partial z}{\partial u} = \frac{\partial z}{\partial x} \frac{\partial x}{\partial u} + \frac{\partial z}{\partial y} \frac{\partial y}{\partial u} = \frac{5(y^2 - x^2 + 1)}{\left(x^2 + y^2 + 1\right)^2} \cdot \frac{1}{2} + \frac{-10xy}{\left(x^2 + y^2 + 1\right)^2} \cdot -\frac{1}{2} =$$

$$\frac{5(y^2 - x^2 + 1) + 10xy}{2\left(x^2 + y^2 + 1\right)^2}$$

and

$$\frac{\partial z}{\partial v} = \frac{\partial z}{\partial x} \frac{\partial x}{\partial v} + \frac{\partial z}{\partial y} \frac{\partial y}{\partial v} = \frac{5(y^2 - x^2 + 1)}{\left(x^2 + y^2 + 1\right)^2} \cdot 0 + \frac{-10xy}{\left(x^2 + y^2 + 1\right)^2} \cdot 1 =$$

$$\frac{-10xy}{\left(x^2 + y^2 + 1\right)^2}.$$

If you want these partial derivatives in terms of u and v, you must now plug in $x = u/2$ and $y = v - u/2$ into these results.

Example 2.6.5. It's not difficult to come up with examples of smooth changes of coordinates which are not smooth **everywhere**, i.e., local smooth changes of coordinates.

Consider, for instance,

$$(u, v) = \mathbf{\Phi}(x, y) = (y + e^x - 1, \ x^3).$$

It's not difficult to solve $u = y + e^x - 1$, $v = x^3$ for x and y. You should find

$$x = v^{1/3} \quad \text{and} \quad y = u - e^{(v^{1/3})} + 1.$$

Figure 2.6.3: Graph of $z = 4 - x^2 - y^2$.

Since we can explicitly solve for x and y, we see that $\boldsymbol{\Phi}$ is a bijection, and it's clear that $\boldsymbol{\Phi}$ is smooth. However, since

$$\frac{\partial \left(v^{1/3}\right)}{\partial v} \;=\; \frac{1}{3}\, v^{-2/3},$$

$\boldsymbol{\Phi}^{-1}$ is **not** differentiable anywhere that $v = 0$. As $v = 0$ means that $x = 0$, we see that $\boldsymbol{\Phi}$ is a smooth change of coordinates, locally, at any point at which $x \neq 0$.

You may wonder what effect $\boldsymbol{\Phi}$ or $\boldsymbol{\Phi}^{-1}$ has on graphs near/at points where $x = 0$ versus points where $x \neq 0$. First, note that $\boldsymbol{\Phi}$ takes the origin to the origin, i.e., $\boldsymbol{\Phi}(0,0) = (0,0)$ and so $\boldsymbol{\Phi}^{-1}(0,0) = (0,0)$. This will be important below when we look at graphs to see the effect of applying $\boldsymbol{\Phi}^{-1}$; it means that we want to compare the graph of some $f = f(x,y)$, when (x,y) is close to the origin, with the graph of $f(\boldsymbol{\Phi}^{-1}(u,v))$, when (u,v) is also close to the origin.

Now, consider the function $z = f(x,y) = 4 - x^2 - y^2$. The graph of this function is a circular paraboloid, which peaks at $(x,y,z) = (0,0,4)$; see Figure 2.6.3.

In Figure 2.6.4, we have graphed $z = f(\boldsymbol{\Phi}^{-1}(u,v))$. As you can see, the lack of differentiability of $\boldsymbol{\Phi}^{-1}$, where $v = 0$, has produced a sharp edge at points where $v = 0$, i.e., at points over/under the u-axis. Note, however, that at points where $v \neq 0$, the graph remains smooth.

Figure 2.6.4: Graph of $z = 4 - (v^{1/3})^2 - (u - e^{v^{1/3}} + 1)^2$.

More Depth:

Suppose that $\boldsymbol{\Phi}$ is a function of class C^r, where $1 \leq r \leq \infty$, from a subset of $\mathbb{R}^n$ to a subset of $\mathbb{R}^n$. If $\mathbf{p}$ is a point in the domain of $\boldsymbol{\Phi}$, we would like to know if $\boldsymbol{\Phi}$ is a C^r smooth local change of coordinates at $\mathbf{p}$.

Given $\mathbf{u} = \boldsymbol{\Phi}(\mathbf{x})$, you can, of course, try to explicitly solve for $\mathbf{x}$ as a function of $\mathbf{u}$ and, if you succeed, then check whether or not the $\boldsymbol{\Phi}^{-1}$ that you produce is of class C^r. However, it's possible that an inverse function of class C^r exists, even if you can't produce it algebraically, and just knowing that the inverse **exists** is extremely important.

How can you tell if a function $\boldsymbol{\Phi}$, of class C^r, is a local C^r change of coordinates at a point $\mathbf{p}$ if you can't actually produce a formula for $\boldsymbol{\Phi}^{-1}$? The *Inverse Function Theorem* tells us that it all depends on whether or not the linear function $d_{\mathbf{p}}\boldsymbol{\Phi}$ has an inverse or not, and **that** condition is easy to check.

Before we state the theorem, let's quickly prove the easy part of it. Suppose that $\boldsymbol{\Phi}$ is a local C^r change of coordinates at the point $\mathbf{p}$, where $1 \leq r \leq \infty$. We want to see that $d_{\mathbf{p}}\boldsymbol{\Phi}$ is a bijection, i.e., that it possesses an inverse function (which is then necessarily linear).

As $\boldsymbol{\Phi}$ is a local C^r change of coordinates, there exist open sets $\mathcal{A}$ and $\mathcal{B}$ such that $\mathbf{p}$ is in $\mathcal{A}$, and so that the restriction $\boldsymbol{\Phi} : \mathcal{A} \to \mathcal{B}$ is of class C^r and possesses an inverse function $\mathbf{\Phi}^{-1} : \mathcal{B} \to \mathcal{A}$ which is also of class C^r. Now, recalling that $\mathrm{id}_\mathcal{A}$ and $\mathrm{id}_\mathcal{B}$ denote the identity functions on $\mathcal{A}$ and $\mathcal{B}$, we have, by definition of the inverse function, that

$$\mathrm{id}_\mathcal{A} = \mathbf{\Phi}^{-1} \circ \mathbf{\Phi} \qquad \text{and} \qquad \mathrm{id}_\mathcal{B} = \mathbf{\Phi} \circ \mathbf{\Phi}^{-1}.$$

Now, recalling from Proposition 2.2.19 that $d_\mathbf{p}(\mathrm{id}_\mathcal{A})$ and $d_{\mathbf{\Phi}(\mathbf{p})}(\mathrm{id}_\mathcal{B})$ are both the identity function $\mathrm{id}_{\mathbb{R}^n}$ on $\mathbb{R}^n$, and applying the Chain Rule at $\mathbf{p}$ and at $\mathbf{\Phi}(\mathbf{p})$, we find

$$\mathrm{id}_{\mathbb{R}^n} = d_\mathbf{p}(\mathbf{\Phi}^{-1} \circ \mathbf{\Phi}) = d_{\mathbf{\Phi}(\mathbf{p})}\mathbf{\Phi}^{-1} \circ d_\mathbf{p}\mathbf{\Phi}$$

and

$$\mathrm{id}_{\mathbb{R}^n} = d_{\mathbf{\Phi}(\mathbf{p})}(\mathbf{\Phi} \circ \mathbf{\Phi}^{-1}) = d_{\mathbf{\Phi}(\mathbf{\Phi}^{-1}(\mathbf{p}))}\mathbf{\Phi} \circ d_{\mathbf{\Phi}(\mathbf{p})}\mathbf{\Phi}^{-1} = d_\mathbf{p}\mathbf{\Phi} \circ d_{\mathbf{\Phi}(\mathbf{p})}\mathbf{\Phi}^{-1}.$$

This tells us that $d_\mathbf{p}\mathbf{\Phi}$ has an inverse function, and that inverse function is precisely $d_{\mathbf{\Phi}(\mathbf{p})}\mathbf{\Phi}^{-1}$; in particular, $d_\mathbf{p}\mathbf{\Phi}$ is a bijection.

The Inverse Function Theorem tells us, somewhat surprisingly, that the converse of this statement is also true.

Theorem 2.6.6. (Inverse Function Theorem) *Suppose that $1 \leq r \leq \infty$, and that $\mathbf{\Phi}$ is a function from a subset of $\mathbb{R}^n$ to a subset of $\mathbb{R}^n$, which is defined and of class C^r on an open neighborhood $\mathcal{A}$ of a point $\mathbf{p}$. Then, $\mathbf{\Phi}$ is a local C^r change of coordinates at $\mathbf{p}$ if and only if $d_\mathbf{p}\mathbf{\Phi}$ is a bijection and, when these equivalent conditions hold,*

$$d_{\mathbf{\Phi}(\mathbf{p})}(\mathbf{\Phi}^{-1}) = (d_\mathbf{p}\mathbf{\Phi})^{-1}.$$

For the proof in the C^1 case, see Theorem 6.3.4 in Trench, [8]. For the case where $r > 1$, see Section 4.5 of [3].

It is also true that, if $\mathbf{\Phi}$ is real-analytic, i.e., has a convergent power series in an open neighborhood of $\mathbf{p}$ and $d_\mathbf{p}\mathbf{\Phi}$ is a bijection, then, locally, there is a real-analytic inverse function of Φ. Real analytic functions are also said to be **of class** C^ω. See Section 2.13.

Remark 2.6.7. We should compare the multivariable Inverse Function Theorem with the single variable Inverse Function Theorem; see, for instance, [4].

Suppose that Φ is a function from a subset of $\mathbb{R}$ to a subset of $\mathbb{R}$. Recall that, in this setting, $d_p\Phi(v) = \Phi'(p) \cdot v$. Thus, $d_p\Phi$ is a bijection if and only if $\Phi'(p) \neq 0$ and, in this case, $(d_p\Phi)^{-1}(w) = w/\Phi'(p)$.

Then, Theorem 2.6.6 tells us that, if $r \geq 1$ and the derivative $\Phi^{(r)}$ is continuous on an open neighborhood of p, then, at p, Φ has a local inverse function Φ^{-1}, of class C^r, if and only if $\Phi'(p) \neq 0$ and, when these equivalent conditions hold,

$$(d_p\Phi)^{-1} = d_{\Phi(p)}(\Phi^{-1}).$$

That is, for all w,

$$\frac{w}{\Phi'(p)} = (d_p\Phi)^{-1}(w) = d_{\Phi(p)}\Phi^{-1}(w) = (\Phi^{-1})'(\Phi(p)) \cdot w,$$

or, equivalently,

$$\frac{1}{\Phi'(p)} = \left(\Phi^{-1}\right)'\left(\Phi(p)\right).$$

For C^1 functions, this is the usual single-variable Inverse Function Theorem.

> The single-variable Inverse Function Theorem does not require that the function be **continuously** differentiable, but it also doesn't allow you to conclude that the inverse is continuously differentiable.

Example 2.6.8. Consider the function

$$(u, v) = \mathbf{\Phi}(x, y) = (x\sin(xy),\ y^5 + 2y^2 + x).$$

We claim that this is a smooth local change of coordinates at $(1, 0)$.

Certainly, $\mathbf{\Phi}$ is a smooth function, as all of the partial derivatives, of all orders, exist and are continuous. We want to show that $d_{(1,0)}\mathbf{\Phi} : \mathbb{R}^2 \to \mathbb{R}^2$ is a bijection; we'll do this by producing the inverse function. We need the first partial derivatives of u and v:

$$\frac{\partial u}{\partial x} = xy\cos(xy) + \sin(xy), \quad \frac{\partial u}{\partial y} = x^2\cos(xy), \quad \frac{\partial v}{\partial x} = 1, \quad \text{and} \quad \frac{\partial v}{\partial y} = 5y^4 + 4y.$$

Thus,

$$\frac{\partial u}{\partial x}\bigg|_{(1,0)} = 0, \quad \frac{\partial u}{\partial y}\bigg|_{(1,0)} = 1, \quad \frac{\partial v}{\partial x}\bigg|_{(1,0)} = 1, \quad \text{and} \quad \frac{\partial v}{\partial y}\bigg|_{(1,0)} = 0.$$

Hence, $d_{(1,0)}u(a, b) = (0, 1) \cdot (a, b) = b$ and $d_{(1,0)}v(a, b) = (1, 0) \cdot (a, b) = a$, and so

$$d_{(1,0)}\mathbf{\Phi}(a, b) = \left(d_{(1,0)}u(a, b),\ d_{(1,0)}v(a, b)\right) = (b, a).$$

Therefore, $d_{(1,0)}\mathbf{\Phi}$ is clearly a bijection, where the inverse function is $\left(d_{(1,0)}\mathbf{\Phi}\right)^{-1}(b, a) = (a, b)$.

The Inverse Function Theorem tells us then that $\mathbf{\Phi}$ possesses a smooth local inverse function $\mathbf{\Phi}^{-1}$, and that

$$\left(d_{(1,0)}\mathbf{\Phi}\right)^{-1} = d_{\mathbf{\Phi}(1,0)}\left(\mathbf{\Phi}^{-1}\right) = d_{(0,1)}\left(\mathbf{\Phi}^{-1}\right).$$

Thus, without explicitly solving algebraically for $\mathbf{\Phi}^{-1}$ (which we can't do anyway), we still know that, in an open neighborhood of $\mathbf{\Phi}(1, 0) = (0, 1)$, the inverse of $\mathbf{\Phi}$ exists, is smooth, and

$$d_{(0,1)}\left(\mathbf{\Phi}^{-1}\right)(b, a) = (a, b).$$

We now want to give a rigorous definition of what it means for a subset of Euclidean space to be "smooth". Of course, our rigorous definition is supposed to correspond to what you see as "smooth".

Our approach to this is to describe basic subsets that we clearly want to call "smooth", and then obtain all other smooth subsets by applying smooth changes of coordinates to these basic smooth subsets. So...what basic subsets do we definitely want to call "smooth"?

Well, the x-axis definitely looks smooth (and flat) in the xy-plane. The x-axis and the xy-plane definitely look smooth in xyz-space, i.e., in $\mathbb{R}^3$. In general, we want to say that the subset of $\mathbb{R}^n$ in which we set any number of the last coordinates equal to 0 is **smooth** inside of $\mathbb{R}^n$, and then we want to be able to alter these via smooth changes of coordinates. Thus, we make the following definition:

Definition 2.6.9. (Submanifolds and Hypersurfaces) *For $0 \leq k \leq n$, we let $\mathbb{R}^k \times \{0\}$ denote the set of n-tuples of real numbers of the form $(x_1, x_2, \ldots, x_k, 0, 0, \ldots, 0)$. We refer to this set as the **basic k-dimensional smooth submanifold of $\mathbb{R}^n$**.*

---

*Suppose that a point $\mathbf{p}$ is in a subset E of $\mathbb{R}^n$. Then, E **is a C^r k-dimensional submanifold of $\mathbb{R}^n$ at $\mathbf{p}$** if and only if there exists a local C^r change of coordinates $\Phi : \mathcal{A} \to \mathcal{B}$ at $\mathbf{p}$, where $\mathcal{A}$ is an open neighborhood of $\mathbf{p}$ in $\mathbb{R}^n$ and $\mathcal{B}$ is an open neighborhood of $\mathbf{0}$ in $\mathbb{R}^n$, such that $\Phi(\mathbf{p}) = \mathbf{0}$ and such that, for all $\mathbf{x}$ in $\mathcal{A} \cap E$, $\Phi(\mathbf{x})$ is in $\mathbb{R}^k \times \{0\}$ and, for all $\mathbf{y}$ in $\mathcal{B} \cap (\mathbb{R}^k \times \{0\})$, $\Phi^{-1}(\mathbf{y})$ is in E.*

In words, this says that E is a C^r k-dimensional submanifold of $\mathbb{R}^n$ at $\mathbf{p}$ if and only if there exists a local C^r change of coordinates on $\mathbb{R}^n$ at $\mathbf{p}$ which, locally, takes E to $\mathbb{R}^k \times \{0\}$.

---

*When we say that a subset E of $\mathbb{R}^n$ is **smooth** at a point $\mathbf{p}$, we mean that E is a C^∞ submanifold, of some dimension, at $\mathbf{p}$. The subset E is a **smooth hypersurface** at $\mathbf{p}$ if E is a smooth $(n-1)$-dimensional submanifold of $\mathbb{R}^n$ at $\mathbf{p}$, i.e., E is smooth at $\mathbf{p}$ and of one dimension less than the surrounding space.*

---

*We say that the entire subset E of $\mathbb{R}^n$ is a C^r **k-dimensional submanifold of $\mathbb{R}^n$**, or a **smooth k-dimensional submanifold**, or a **smooth hypersurface** if and only if E has the respective property at each point $\mathbf{p}$ in E.*

Thus, a smooth hypersurface in $\mathbb{R}^2$ is a smooth curve, and a smooth hypersurface in $\mathbb{R}^3$ is just a smooth surface (without the "hyper").

Remark 2.6.10. Understand what's going on here. We want a rigorous definition of smooth. We decided that certain basic sets were smooth submanifolds, and we want

C^r changes of coordinates to preserve whether or not a submanifold is C^r. Thus, a C^r submanifold is a set which, locally, results from applying a C^r change of coordinates to a basic smooth submanifold.

Be aware that, when a subset of $\mathbb{R}^n$ **looks** smooth, it may be merely a C^1 submanifold, not a C^∞ submanifold. It is debatable how much differentiability you can actually see.

Example 2.6.11. Prove that the sphere in $\mathbb{R}^3$, given by $x^2 + y^2 + z^2 = 1$, is a smooth hypersurface at $(0, 0, 1)$.

Solution:

Define the function

We shall look at problems like this in more detail and generality in Section 2.7.

$$\mathbf{\Phi}(x, y, z) \;=\; (x, \; y, \; x^2 + y^2 + z^2 - 1).$$

Then $\mathbf{\Phi}$ is certainly a smooth function. We claim that $\mathbf{\Phi}$ is a smooth local change of coordinates at $(0, 0, 1)$ and takes the sphere to the xy-plane, i.e., to $\mathbb{R}^2 \times \{0\}$. This would prove what we want.

It's easy to calculate that

$$d_{(0,0,1)}\mathbf{\Phi}(a, b, c) \;=\; (a, b, 2c),$$

and so $d_{(0,0,1)}\mathbf{\Phi}$ is a bijection with $\left(d_{(0,0,1)}\mathbf{\Phi}\right)^{-1}(\alpha, \beta, \gamma) = (\alpha, \beta, \gamma/2)$. Therefore, by the Inverse Function Theorem, $\mathbf{\Phi}$ is a smooth local change of coordinates at $(0, 0, 1)$, and you can easily check that $\mathbf{\Phi}(0, 0, 1) = (0, 0, 0)$. Furthermore, the sphere that we're looking at is precisely where $x^2 + y^2 + z^2 - 1$ equals 0, so that $\mathbf{\Phi}(x, y, z) = (x, y, 0)$ if and only if (x, y, z) is on the sphere. This is what we wanted to show; so the sphere is a smooth hypersurface at $(0, 0, 1)$.

You can use a similar argument at each point on the sphere to show that the entire sphere is a smooth hypersurface in $\mathbb{R}^3$.

+ Linear Algebra:

Linear algebra and matrices are extremely useful when dealing with changes of coordinates and the Inverse Function Theorem. Recall from Definition 2.2.22 that the matrix of partial derivatives of a multi-component function $\mathbf{\Phi}$ is called the Jacobian matrix of $\mathbf{\Phi}$. For a function, such as our $\mathbf{\Phi}$'s, from a subset of $\mathbb{R}^n$ to a subset of $\mathbb{R}^n$, the Jacobian matrix will be a square, $n \times n$, matrix, and the linear function $d_{\mathbf{p}}\Phi$ is a bijection if and only if the Jacobian matrix $[d_{\mathbf{p}}\Phi]$ is invertible.

To decide if a matrix is invertible, we may check that the determinant is non-zero, which works well for small matrices, or we can perform Gauss-Jordan elimination. If you want to actually calculate the inverse map $(d_{\mathbf{p}}\Phi)^{-1} = d_{\Phi(\mathbf{p})}(\Phi^{-1})$, then you want to calculate the inverse matrix $[d_{\mathbf{p}}\Phi]^{-1}$, and so using row operations would be the way to go. Of course, if you have technology at your disposal, you may use a calculator of computer to calculate determinants and/or inverse matrices.

The matrix form of the Inverse Function Theorem is:

> **Theorem 2.6.12. (Inverse Function Theorem, matrix form)** *Suppose that* $1 \leq r \leq \infty$, *and that* Φ *is a function from a subset of* $\mathbb{R}^n$ *to a subset of* $\mathbb{R}^n$, *which is defined and of class* C^r *on an open neighborhood* $\mathcal{A}$ *of a point* $\mathbf{p}$. *Then,* Φ *is a local* C^r *change of coordinates at* $\mathbf{p}$ *if and only if* $[d_{\mathbf{p}}\Phi]$ *is an invertible matrix and, when these equivalent conditions hold,*
> $$\left[d_{\Phi(\mathbf{p})}\left(\Phi^{-1}\right) \right] \;=\; [d_{\mathbf{p}}\Phi]^{-1}.$$

Example 2.6.13. Find all of those points at which the smooth function

$$\Phi(x,y) = (x^2 + y^2,\ x^3 + y^3)$$

is a local smooth change of coordinates.

Solution:

The determinant will work well for us here. The Jacobian matrix of $d\Phi$ is easy to calculate:

$$[d\Phi] \;=\; \begin{bmatrix} 2x & 2y \\ 3x^2 & 3y^2 \end{bmatrix}.$$

This matrix is invertible precisely where its determinant is non-zero.

We calculate

$$\det \begin{bmatrix} 2x & 2y \\ 3x^2 & 3y^2 \end{bmatrix} \;=\; (2x)(3y^2) - (3x^2)(2y) \;=\; 6xy^2 - 6x^2 y \;=\; 6xy(y - x).$$

Thus, Φ is a local smooth change of coordinates at any point where $xy(x - y) \neq 0$, i.e., everywhere except along the x-axis, the y-axis, and the line where $y = x$.

Example 2.6.14. Verify that, at $(\ln 3, 1, 0)$, the function

$$\Phi(x, y, z) = \left(xyz, \ z^2 + e^{xy}, \ x + \frac{z}{y} \right)$$

is a local smooth change of coordinates on $\mathbb{R}^3$, which takes the point $(\ln 3, 1, 0)$ to the point $(0, 3, \ln 3)$. Determine the total derivative of Φ^{-1} at $(0, 3, \ln 3)$.

Solution:

First, it's easy to calculate that $\Phi(\ln 3, 1, 0) = (0, 3, \ln 3)$.

We now calculate the Jacobian matrix

$$[d\Phi] = \begin{bmatrix} yz & xz & xy \\ ye^{xy} & xe^{xy} & 2z \\ 1 & -\dfrac{z}{y^2} & \dfrac{1}{y} \end{bmatrix},$$

and so

$$\left[d_{(\ln 3, 1, 0)} \Phi \right] = \begin{bmatrix} 0 & 0 & \ln 3 \\ 3 & 3\ln 3 & 0 \\ 1 & 0 & 1 \end{bmatrix}.$$

Using technology or some easy row operations, you find that $\left[d_{(\ln 3, 1, 0)} \Phi \right]$ is invertible and its inverse is

$$\left[d_{(\ln 3, 1, 0)} \Phi \right]^{-1} = \begin{bmatrix} -\dfrac{1}{\ln 3} & 0 & 1 \\[2mm] \dfrac{1}{(\ln 3)^2} & \dfrac{1}{3\ln 3} & -\dfrac{1}{\ln 3} \\[2mm] \dfrac{1}{\ln 3} & 0 & 0 \end{bmatrix}.$$

Thus, near $(\ln 3, 1, 0)$, Φ has a local inverse Φ^{-1}, near $(0, 3, \ln 3)$, and

$$d_{(0, 3, \ln 3)} \left(\Phi^{-1} \right) = \left(d_{(\ln 3, 1, 0)} \Phi \right)^{-1}.$$

and so

$$\left[d_{(0, 3, \ln 3)} \left(\Phi^{-1} \right) \right] = \left[\left(d_{(\ln 3, 1, 0)} \Phi \right)^{-1} \right] = \left[d_{(\ln 3, 1, 0)} \Phi \right]^{-1}.$$

Therefore,

$$d_{(0, 3, \ln 3)} \left(\Phi^{-1} \right)(a, b, c) = \left[d_{(\ln 3, 1, 0)} \Phi \right]^{-1} \begin{bmatrix} a \\ b \\ c \end{bmatrix} =$$

$$\left(-\frac{a}{\ln 3} + c, \ \frac{a}{(\ln 3)^2} + \frac{b}{3\ln 3} - \frac{c}{\ln 3}, \ \frac{a}{\ln 3} \right).$$

Finally, we wish to discuss **tangent spaces** to submanifolds of $\mathbb{R}^n$ which are (at least) C^1.

Recall that we defined tangent vectors to a subset of $\mathbb{R}^n$ back in Definition 1.6.24; we defined them to be derivatives of parameterized curves in the subset.

The following theorem tells us that, at each point $\mathbf{p}$ in a C^1 k-dimensional submanifold E of $\mathbb{R}^n$, the set of tangent vectors is actually a k-dimensional linear subspace of $\mathbb{R}^n$.

Theorem 2.6.15. *Suppose that E is a C^1 k-dimensional submanifold of $\mathbb{R}^n$ at a point $\mathbf{p}$ in E. Let $\boldsymbol{\Phi} : \mathcal{A} \to \mathcal{B}$ be a local C^1 change of coordinates at $\mathbf{p}$, where $\mathcal{A}$ is an open neighborhood of $\mathbf{p}$ in $\mathbb{R}^n$ and $\mathcal{B}$ is an open neighborhood of $\mathbf{0}$ in $\mathbb{R}^n$, such that $\boldsymbol{\Phi}(\mathbf{p}) = \mathbf{0}$ and such that, for all $\mathbf{x}$ in $\mathcal{A} \cap E$, $\boldsymbol{\Phi}(\mathbf{x})$ is in $\mathbb{R}^k \times \{\mathbf{0}\}$ and, for all $\mathbf{y}$ in $\mathcal{B} \cap (\mathbb{R}^k \times \{\mathbf{0}\})$, $\boldsymbol{\Phi}^{-1}(\mathbf{y})$ is in E.*

Then, the set of tangent vectors to E at $\mathbf{p}$ is precisely the image of the restriction of $(d_{\mathbf{p}}\boldsymbol{\Phi})^{-1} = d_{\mathbf{0}}\left(\boldsymbol{\Phi}^{-1}\right)$ to $\mathbb{R}^k \times \{\mathbf{0}\}$. In particular, the set of tangent vectors to E at $\mathbf{p}$ form a k-dimensional linear subspace of $\mathbb{R}^n$.

Proof. This is actually easy. Suppose that $\mathbf{v}$ is a tangent vector to E at $\mathbf{p}$. So there exist $\epsilon > 0$, t_0, and a C^1 function $\mathbf{r} : (t_0 - \epsilon, t_0 + \epsilon) \to \mathbb{R}^n$ such that, for all t in the interval $(t_0 - \epsilon, t_0 + \epsilon)$, $\mathbf{r}(t)$ is in E, $\mathbf{p} = \mathbf{r}(t_0)$, and $\mathbf{v} = \mathbf{r}'(t_0)$.

Then, $\boldsymbol{\Phi} \circ \mathbf{r}$ is a C^1 function into $\mathbb{R}^k \times \{\mathbf{0}\}$, and so $\mathbf{w} := (\boldsymbol{\Phi} \circ \mathbf{r})'(t_0)$ is in $\mathbb{R}^k \times \{\mathbf{0}\}$. Now, the Chain Rule, as given in Corollary 2.4.23, tells us that

$$(\boldsymbol{\Phi} \circ \mathbf{r})'(t_0) \;=\; d_{\mathbf{p}}\boldsymbol{\Phi}(\mathbf{v}).$$

Hence, $\mathbf{v} = (d_{\mathbf{p}}\boldsymbol{\Phi})^{-1}(\mathbf{w})$, as desired.

Suppose, conversely, that $\mathbf{w}$ is in $\mathbb{R}^k \times \{\mathbf{0}\}$, and let $\mathbf{v} = (d_{\mathbf{p}}\boldsymbol{\Phi})^{-1}(\mathbf{w})$. Define $\mathbf{r} : (-\epsilon, \epsilon) \to \mathbb{R}^n$ by $\mathbf{r}(t) = \boldsymbol{\Phi}^{-1}(t\mathbf{w})$, where $\epsilon > 0$ is chosen small enough so that $t\mathbf{w}$ is always in $\mathcal{B}$. Then, $\mathbf{r}(t)$ is always in E, $\mathbf{r}(0) = \mathbf{p}$, and

$$\mathbf{r}'(0) = d_{\mathbf{0}}\left(\boldsymbol{\Phi}^{-1}\right)(\mathbf{w}) = (d_{\mathbf{p}}\boldsymbol{\Phi})^{-1}(\mathbf{w}) = \mathbf{v},$$

i.e., $\mathbf{v}$ is a tangent vector to E at $\mathbf{p}$. $\qquad\square$

In light of the above theorem, we make a definition:

Definition 2.6.16. *The set of tangent vectors to a C^1 submanifold E of $\mathbb{R}^n$ at a point $\mathbf{p}$ is called the **tangent space to E at $\mathbf{p}$**, and is denoted by $T_{\mathbf{p}}E$.*

Remark 2.6.17. Note that $\mathbb{R}^n$ is a submanifold of itself and that $T_{\mathbf{p}}\mathbb{R}^n$ is all of $\mathbb{R}^n$, because we can use the parameterization $\mathbf{r}(t) = \mathbf{p} + t\mathbf{v}$, no matter what $\mathbf{v}$ is. In fact, for the same reason, if $\mathcal{U}$ is an open subset of $\mathbb{R}^n$ and $\mathbf{p}$ is in $\mathcal{U}$, then $T_{\mathbf{p}}\mathcal{U}$ is all of R^n.

2.6.1   Exercises

Online answers to select exercises are here.

Basics:

In each of the following exercises, you are given open subsets $\mathcal{A}$ and $\mathcal{B}$ of some $\mathbb{R}^n$, and an infinitely differentiable function $\Phi : \mathcal{A} \to \mathcal{B}$. Show that each Φ is a smooth change of coordinates by explicitly producing Φ^{-1}, and noting that the component functions of Φ^{-1} are infinitely differentiable.

1. $\mathcal{A} = \mathbb{R}^2$, $\mathcal{B} = \mathbb{R}^2$, $\boldsymbol{\Phi}(x, y) = (3x + y, x - 5y)$.

2. $\mathcal{A} = \mathbb{R}^2$, $\mathcal{B} = \{(u, v) \in \mathbb{R}^2 \mid v > 0\}$, $\boldsymbol{\Phi}(x, y) = (7y, 5e^x)$.

3. $\mathcal{A} = \mathbb{R}^2$, $\mathcal{B} = \{(u, v) \in \mathbb{R}^2 \mid 3u - v > 0\}$, $\boldsymbol{\Phi}(x, y) = (x + e^y, 3x - e^y)$.

4. $\mathcal{A} = \{(x, y) \in \mathbb{R}^2 \mid x > 0 \text{ and } y \neq 0\}$, $\mathcal{B} = \{(u, v) \in \mathbb{R}^2 \mid u \neq 0\}$, $\boldsymbol{\Phi}(x, y) = (y^3, \ln x)$.

5. $\mathcal{A} = \mathbb{R}^3$, $\mathcal{B} = \mathbb{R}^3$, $\boldsymbol{\Phi}(x, y, z) = (1 + y + 2z, 3x + 4z, 5 - x + y)$.

6. $\mathcal{A} = \mathbb{R}^3$, $\mathcal{B} = \{(u, v, w) \in \mathbb{R}^3 \mid u > 0, -\pi/2 < w < \pi/2\}$, $\boldsymbol{\Phi}(x, y, z) = \left(e^{x-y}, x + z, \tan^{-1} y\right)$.

In each exercise below, use technology to graph the hyperbolic paraboloid given by $z = f(x, y) = (x - 1)^2 - (y - 1)^2$, and then take the given $\boldsymbol{\Phi}$, from Exercises 1 through 4, and compare the hyperbolic paraboloid with the graph of $z = (f \circ \boldsymbol{\Phi}^{-1})(u, v)$. In particular, compare the graph of f near where $(x, y) = (1, 1)$ with the graph of $f \circ \boldsymbol{\Phi}^{-1}$ near where $(u, v) = \boldsymbol{\Phi}(1, 1)$, and describe what similarities and differences you see. When graphing, note the given restrictions on the domain and codomain of each $\boldsymbol{\Phi}$.

7. $\mathcal{A} = \mathbb{R}^2$, $\mathcal{B} = \mathbb{R}^2$, $\boldsymbol{\Phi}(x, y) = (3x + y, x - 5y)$.

8. $\mathcal{A} = \mathbb{R}^2$, $\mathcal{B} = \{(u, v) \in \mathbb{R}^2 \mid v > 0\}$, $\boldsymbol{\Phi}(x, y) = (7y, 5e^x)$.

9. $\mathcal{A} = \mathbb{R}^2$, $\mathcal{B} = \{(u, v) \in \mathbb{R}^2 \mid 3u - v > 0\}$, $\boldsymbol{\Phi}(x, y) = (x + e^y, 3x - e^y)$.

10. $\mathcal{A} = \{(x, y) \in \mathbb{R}^2 \mid x > 0 \text{ and } y \neq 0\}$, $\mathcal{B} = \{(u, v) \in \mathbb{R}^2 \mid u \neq 0\}$, $\boldsymbol{\Phi}(x, y) = (y^3, \ln x)$.

In each of the following exercises, you are given open subsets $\mathcal{A}$ and $\mathcal{B}$ of some $\mathbb{R}^n$, and an infinitely differentiable bijection $\Phi : \mathcal{A} \to \mathcal{B}$. Show, however, that Φ is NOT a smooth change of coordinates by explicitly producing Φ^{-1}, and determining points in $\mathcal{B}$ where Φ^{-1} is not differentiable.

11. $\mathcal{A} = \mathbb{R}^2$, $\mathcal{B} = \mathbb{R}^2$, $\Phi(x, y) = (y^3, x - y)$.

12. $\mathcal{A} = \mathbb{R}^2$, $\mathcal{B} = \{(u, v) \in \mathbb{R}^2 \mid v > -1\}$, $\Phi(x, y) = (y, e^{(x^5)} - 1)$.

13. $\mathcal{A} = \{(x, y) \in \mathbb{R}^2 \mid y \neq -1\}$, $\mathcal{B} = \{(u, v) \in \mathbb{R}^2 \mid u > -1, v \neq -1\}$, $\Phi(x, y) = (e^x - 1, (y^3 + 1)^{-1} - 1)$.

14. $\mathcal{A} = \{(x, y) \in \mathbb{R}^2 \mid x > -1, y > -1\}$, $\mathcal{B} = \{(u, v) \in \mathbb{R}^2 \mid u < 1, v < 1\}$, $\Phi(x, y) = \left(\dfrac{x}{x + 1}, \left(\dfrac{y}{y + 1} \right)^3 \right)$.

15. $\mathcal{A} = \mathbb{R}^3$, $\mathcal{B} = \mathbb{R}^3$, $\Phi(x, y, z) = (x^5 + 3y, z, y)$.

16. $\mathcal{A} = \mathbb{R}^3$, $\mathcal{B} = \mathbb{R}^3$, $\Phi(x, y, z) = (x^5 + y^3, z, y^3 + z)$.

17. $\mathcal{A} = \{(x, y, z) \in \mathbb{R}^3 \mid z > -1\}$, $\mathcal{B} = \{(u, v, w) \in \mathbb{R}^3 \mid w > -1\}$, $\Phi(x, y, z) = (x^5 + 3e^y - 3, \ln(z + 1), e^y - 1)$.

18. $\mathcal{A} = \{(x, y, z) \in \mathbb{R}^3 \mid x > -1, y > -1, z > -1\}$, $\mathcal{B} = \{(u, v, w) \in \mathbb{R}^3 \mid u < 1, v < 1, w < 1\}$, $\Phi(x, y, z) = \left(\dfrac{x}{x + 1}, \left(\dfrac{y}{y + 1} \right)^5, \dfrac{z}{z + 1} \right)$.

Using technology, graph the hyperbolic paraboloid $z = f(x, y) = x^2 - y^2$. Then, using technology, take the given Φ, from Exercises 11 through 14, and compare the hyperbolic paraboloid with the graph of $z = (f \circ \Phi^{-1})(u, v)$. Identify the places on the graph where you see the effects of the lack of differentiability of Φ^{-1}. Note that each $\Phi(x, y)$ takes the origin to the origin, i.e., $\Phi(0) = 0$. Also, note the given restrictions on the domain and codomain of each Φ.

19. $\mathcal{A} = \mathbb{R}^2$, $\mathcal{B} = \mathbb{R}^2$, $\Phi(x, y) = (y^3, x - y)$.

20. $\mathcal{A} = \mathbb{R}^2$, $\mathcal{B} = \{(u, v) \in \mathbb{R}^2 \mid v > -1\}$, $\Phi(x, y) = (y, e^{(x^5)} - 1)$.

21. $\mathcal{A} = \{(x, y) \in \mathbb{R}^2 \mid y \neq -1\}$, $\mathcal{B} = \{(u, v) \in \mathbb{R}^2 \mid u > -1, v \neq -1\}$, $\Phi(x, y) = (e^x - 1, (y^3 + 1)^{-1} - 1)$.

22. $\mathcal{A} = \{(x, y) \in \mathbb{R}^2 \mid x > -1, y > -1\}$, $\mathcal{B} = \{(u, v) \in \mathbb{R}^2 \mid u < 1, v < 1\}$, $\Phi(x, y) = \left(\dfrac{x}{x + 1}, \left(\dfrac{y}{y + 1} \right)^3 \right)$.

In each of the following exercises, you are given a (local) change of coordinates $(u, v) = \Phi(x, y)$ and the partial derivatives of a differentiable function $z = f(x, y)$. Determine, in terms of u and v, the partial derivatives, $\partial z / \partial u$ and $\partial z / \partial v$, of $z = (f \circ \Phi^{-1})(u, v)$.

23. $\partial z / \partial x = y$, $\partial z / \partial y = x$, $(u, v) = \Phi(x, y) = (5x - y, x + y)$.   ▶

24. $\partial z / \partial x = xy$, $\partial z / \partial y = e^y + (x^2/2)$, $(u, v) = \Phi(x, y) = (5x - y, x + y)$.

25. $\partial z / \partial x = y$, $\partial z / \partial y = x$, $(u, v) = \Phi(x, y) = (7y, 5e^x)$.

26. $\partial z / \partial x = xy$, $\partial z / \partial y = e^y + (x^2/2)$, $(u, v) = \Phi(x, y) = (7y, 5e^x)$.

In each of the following exercises, you are given a (local) change of coordinates $(u, v, w) = \Phi(x, y, z)$ and the partial derivatives of a differentiable function $p = f(x, y, z)$. Determine, in terms of u, v, and w, the partial derivatives, $\partial p / \partial u$, $\partial p / \partial v$, and $\partial p / \partial w$, of $p = (f \circ \Phi^{-1})(u, v, w)$.

27. $\partial p / \partial x = yz + 2x$, $\partial p / \partial y = xz - 2y$, $\partial p / \partial z = xy$, $(u, v, w) = \Phi(x, y, z) = (1 + y + 2z, 3x + 4z, 5 - x + y)$.

28. $\partial p / \partial x = e^y$, $\partial p / \partial y = xe^y - \tan^{-1} z$, $\partial p / \partial z = -\dfrac{y}{1 + z^2}$, $(u, v, w) = \Phi(x, y, z) = (1 + y + 2z, 3x + 4z, 5 - x + y)$.

29. $\partial p / \partial x = yz + 2x$, $\partial p / \partial y = xz - 2y$, $\partial p / \partial z = xy$, $(u, v, w) = \Phi(x, y, z) = \left(e^{x-y}, x + z, \tan^{-1} y\right)$.   ▶

30. $\partial p / \partial x = e^y$, $\partial p / \partial y = xe^y - \tan^{-1} z$, $\partial p / \partial z = -\dfrac{y}{1 + z^2}$, $(u, v, w) = \Phi(x, y, z) = \left(e^{x-y}, x + z, \tan^{-1} y\right)$.

> **More Depth:**

In each of the following exercises, you are given a function Φ from a subset of $\mathbb{R}^2$ into $\mathbb{R}^2$, and a point **p**. Verify that Φ is a smooth change of coordinates at **p**, and find an explicit formula for $d_{\Phi(\mathbf{p})} \left(\Phi^{-1}\right)(r, s)$. (Note that this does NOT require you find an explicit formula for Φ^{-1} itself.)

31. $\Phi(x, y) = (xy, x^2 + y^2)$, $\mathbf{p} = (1, 2)$.

32. $\Phi(x, y) = (x \cos y, x \sin y)$, $\mathbf{p} = (1, 0)$.   ▶

33. $\Phi(x, y) = \left(\dfrac{e^x}{y}, \dfrac{y^2}{x^3 + 1}\right)$, $\mathbf{p} = (0, 1)$.

34. $\Phi(x, y) = (\tan^{-1}(xy), y \ln x)$, $\mathbf{p} = (1, 2)$.

In each of the following exercises, verify that the given equation defines a smooth hypersurface in the appropriate $\mathbb{R}^n$ at the given point p.

35. $xy = 0$, $\mathbf{p} = (0,1)$

36. $y^2 - x^3 - x^2 = 0$, $\mathbf{p} = (3,6)$.

37. $x^2 + y^2 - z^2 = 0$, $\mathbf{p} = (3,4,5)$.

38. $x\tan^{-1}y + z^3 = 1$, $\mathbf{p} = (1,0,1)$.

39. $wx - yz = 0$, $\mathbf{p} = (w,x,y,z) = (1,0,3,0)$.

40. $y\ln(x^2 - z^2) + \sqrt{w} = 2$, $\mathbf{p} = (w,x,y,z) = (4,1,1,0)$.

+ **Linear Algebra:**

In each of the following exercises, use matrices to verify that the given Φ is a local change of coordinates at the given point p, and find an explicit formula for $d_{\Phi(\mathbf{p})}(\Phi^{-1})(\mathbf{v})$.

41. $\Phi(x,y) = (5x - y, x + y)$, $\mathbf{p} = (1,3)$.

42. $\Phi(x,y) = (7y, 5e^x)$, $\mathbf{p} = (0,0)$.

43. $\Phi(x,y,z) = (y^2, x^3, z^5)$, $\mathbf{p} = (3,2,1)$.

44. $\Phi(x,y,z) = \left(\dfrac{y}{z}, \dfrac{z}{x}, x\right)$, $\mathbf{p} = (1,2,4)$.

45. $\Phi(x,y,z) = (y\cos x, y\sin x, z)$, $\mathbf{p} = (\pi,1,2)$.

46. $\Phi(w,x,y,z) = (x,w,y,z)$, $\mathbf{p} = (-2,-1,0,1)$.

If Φ is a smooth local change of coordinates at p, then the determinant of the total derivative matrix, $\det[d_{\mathbf{p}}\Phi]$, is non-zero. If $\det[d_{\mathbf{p}}\Phi]$ is positive, then Φ is said to be *orientation-preserving* at p; if $\det[d_{\mathbf{p}}\Phi]$ is negative, then Φ is said to be *orientation-reversing* at p.

In each of the following exercises, use the determinant to verify that the given smooth Φ is a smooth local change of coordinates at the given point, and decide whether the change of coordinates is orientation-preserving or orientation-reversing.

47. $\Phi(x,y) = (5x - y, x + y)$, $\mathbf{p} = (1,3)$.

48. $\Phi(x,y) = (7y, 5e^x)$, $\mathbf{p} = (0,0)$.

49. $\Phi(x,y,z) = (y^2, x^3, z^5)$, $\mathbf{p} = (3,2,1)$.

50. $\boldsymbol{\Phi}(x, y, z) = \left(\dfrac{y}{z}, \dfrac{z}{x}, x \right)$, $\mathbf{p} = (1, 2, 4)$.

51. $\boldsymbol{\Phi}(x, y, z) = (y \cos x, y \sin x, z)$, $\mathbf{p} = (\pi, 1, 2)$.

52. $\boldsymbol{\Phi}(w, x, y, z) = (x, w, y, z)$, $\mathbf{p} = (-2, -1, 0, 1)$.

2.7   Level Sets and Gradient Vectors

In this section, we will look at the set of points $\mathbf{x}$ where a real-valued function F is equal to a given constant; such a set of points is called a *level set of f*.

If F is continuously differentiable and $\mathbf{p}$ is a point at which $\vec{\nabla}F(\mathbf{p}) \neq \mathbf{0}$, then the level set of points $\mathbf{x}$ such that $F(\mathbf{x}) = F(\mathbf{p})$ looks *smooth at* $\mathbf{p}$ and has a well-defined *tangent set*, which is the best affine linear approximation to the level set at $\mathbf{p}$. The gradient vector $\vec{\nabla}F(\mathbf{p})$ is perpendicular to the tangent set and, thus, we say that $\vec{\nabla}F(\mathbf{p})$ is perpendicular to the level set itself at $\mathbf{p}$.

For level curves in $\mathbb{R}^2$, the tangent set is the familiar tangent line and, for level surfaces in $\mathbb{R}^3$, the tangent set is the tangent plane; this definition agrees with our earlier definition of the tangent plane, in Definition 2.3.8, for graphs of functions $z = f(x, y)$.

In the More Depth portion of this section, we will discuss the tangent set in terms of our definition of *tangent vectors* in Definition 1.6.24.

Basics:

If you think about all of the curves that you've ever looked at in the xy-plane, they typically fall into two categories: graphs of functions $y = f(x)$, such as $y = 3x + 7$, $y = x^2$, $y = x + e^x$, and graphs of sets of points (x, y) such that some function $F(x, y)$ equals a given constant value, such as $x^2 + y^2 = 4$, $x^2 - y^2 = 1$, $x + \sin(xy) = 5$.

However, we don't have to think of these as **different** categories of graphs; given $f(x)$, we can define $F(x, y) = y - f(x)$, and then the set of points (x, y) where $y = f(x)$ is the same as the set of points (x, y) where $F(x, y) = 0$. Thus, looking at sets of points where $F(x, y)$ equals a constant is the more-general manner of looking at curves in the plane.

Curves in the xy-plane are 1-dimensional, but sit inside the 2-dimensional xy-plane. Surfaces in XYZ-space are 2-dimensional objects which sit inside 3-dimensional space.

In the same way, we can look at surfaces in $\mathbb{R}^3$ by considering graphs of functions $z = f(x, y) = x^2 - y^2$ or by looking at where a function $F(x, y, z)$ equals a constant, like the sphere given by $F(x, y, z) = x^2 + y^2 + z^2 = 4$. Again, this latter way of looking at surfaces is more general, since $z = f(x, y)$ can be rewritten as $z - f(x, y) = 0$, e.g., $z = x^2 - y^2$ can rewritten as $z - x^2 + y^2 = 0$.

Hence, we make the following definition, which we looked at briefly in Section 1.7:

If the level set where $F(x, y) = c$ is 1-dimensional, then, naturally, the level set is referred to as the **level curve** (even if it's a straight line). Similarly, if the level set where $F(x, y, z) = c$ is a surface, then the level set is referred to as the **level surface**.

> **Definition 2.7.1.** *Suppose that $F = F(x_1, \ldots, x_n) = F(\mathbf{x})$ is a real-valued function. Let c be a constant. Then, the set of points $\mathbf{x}$ such that $F(\mathbf{x}) = c$ is called the* **level set where** $F = c$.
>
> *If $\mathbf{p}$ is a point in the domain of F, then the set of $\mathbf{x}$ such that $F(\mathbf{x}) = F(\mathbf{p})$ is the* **level set of F containing $\mathbf{p}$**.

Example 2.7.2. Let $F(x, y) = x^2 - y^2$.

The level curve where $F = 1$ is given by $x^2 - y^2 = 1$, and is a hyperbola which has its vertices on the x-axis at $(x, y) = (\pm 1, 0)$, and has the lines $y = x$ and $y = -x$ as asymptotes.

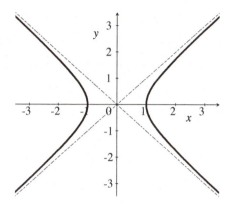

Figure 2.7.1: The level curve where $F(x, y) = x^2 - y^2 = 1$.

We would like to consider other level curves of F and compare them. Thus, we graph them all in the same xy-plane. See Figure 2.7.2.

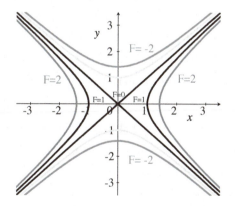

Figure 2.7.2: The level curves where $F = x^2 - y^2$ equals -2, -1, 0, 1, and 2.

When $c > 0$, the level curve where $F = x^2 - y^2 = c$ is a hyperbola with it vertices on the x-axis at $(x, y) = (\pm\sqrt{c}, 0)$, and asymptotes given by $y = x$ and $y = -x$. When $c < 0$, the level curve where $F = x^2 - y^2 = c$ is where $y^2 - x^2 = -c$, and now $-c > 0$; this describes a hyperbola with vertices on the y-axis, at $(x, y) = (0, \pm\sqrt{-c})$, and the same asymptotes given by $y = \pm x$.

Note that the level "curve" where $F = 0$ is the set of points where $x^2 - y^2 = (x - y)(x + y) = 0$, i.e., the two lines $x - y = 0$ and $x + y = 0$. These, of course, are

Perhaps the most famous type of singularity is the non-smooth point which a black hole is thought to produce in the space-time continuum.

precisely the asymptotes of all of the other level curves of F.   These two lines cross each other at $(0,0)$, producing a point at which the level curve is not smooth; such a point is called a *singularity*.

Remark 2.7.3. In Example 2.7.2, we looked at the level curves of $F(x,y) = x^2 - y^2$. These level curves are precisely the z-cross sections, or contours, of the graph of $z = x^2 - y^2$, which we know, from Section 1.8, is a hyperbolic paraboloid. The only difference

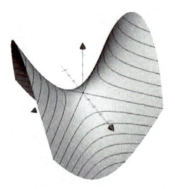

Figure 2.7.3: The hyperbolic paraboloid $z = x^2 - y^2$, with level curves.

between the z-cross sections and the level curves is that we think of the cross sections as being in different copies of the xy-plane, corresponding to the given value of z, while we typically think of the level curves as all being in one xy-plane. Why do we want to think of these things differently? When we look at the graph of $z = x^2 - y^2$ and take z-cross sections, it is the surface given by $z = x^2 - y^2$ that we are really interested in. When we look at the level curves of a function $F(x,y)$, it is often the level curves themselves, as graphs in the xy-plane, that we care about.

In a sense, the only difference between thinking of z-cross sections and level curves is psychological.

Example 2.7.4. While some collections of level sets can be sketched easily by hand, like those in Example 2.7.2, others are far more complicated. However, graphing software can easily produce graphs of more-complicated level sets.

Consider the function

$$F(x,y) \;=\; x^3 + y^2 - x + y.$$

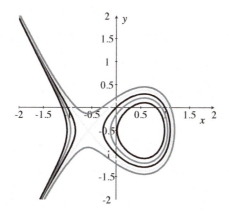

Figure 2.7.4: Some level curves of $F = x^3 + y^2 - x + y$.

Here, we have had software sketch the level curves where F equals -0.27, -0.135, 0, 0.135, and 0.27 in black, red, blue, green, and purple, respectively.

Note that each of the level curves where F equals -0.27, -0.135, and 0 is in two pieces, i.e., is not connected. Also note that the green level curve where $F = 0.135$ appears to cross itself and produce a singularity, a non-smooth point, on the level curve near the point $(x, y) = (-0.6, -0.5)$.

Example 2.7.5. Let's look at the level surfaces of the function $S(x, y, z) = x^2 + y^2 + z^2$. Suppose that $S(x, y, z) = c$.

- If $c < 0$, then there are no (real) points which satisfy $x^2 + y^2 + z^2 = c$, so the graph is empty.

- If $c = 0$, then there is exactly one point which satisfies $x^2 + y^2 + z^2 = c$, namely, the origin. Note that the graph here is **not** 2-dimensional; it's a point, which is 0-dimensional.

- If $c > 0$, then the graph of $x^2 + y^2 + z^2 = c$ is a sphere, centered at the origin, of radius $\sqrt{c}$ (recall Section 1.8).

Note that we really shouldn't refer to the $c < 0$ and $c = 0$ cases as level "surfaces", since the graphs are not 2-dimensional. "Level sets" would be more technically correct.

Figure 2.7.5: The level "surface" $x^2 + y^2 + z^2 = c < 0$ is empty.

Figure 2.7.6: The level "surface" $x^2 + y^2 + z^2 = c = 0$ is a single point.

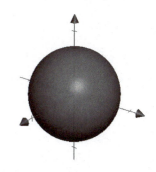

Figure 2.7.7: The level surface $x^2 + y^2 + z^2 = c > 0$ is a sphere.

Figure 2.7.8: The level surface $z^2 - 2z - x^2 - y^2 = 0$.

Example 2.7.6. Let's look at another example of level surfaces. Let $G(x, y, z) = z^2 - 2z - x^2 - y^2$.

- Consider the graph of the level surface where $G = 0$. Note that

$$G \; = \; z^2 - 2z - x^2 - y^2 \; = \; (z-1)^2 - x^2 - y^2 - 1.$$

Thus, as we discussed in Section 1.8, the graph of G is a hyperboloid of two sheets, just shifted up one unit from a standard one, which is centered at $\mathbf{0}$. See Figure 2.7.8. Note that the hyperboloid looks smooth at each point, and is 2-dimensional.

- Consider the graph of the level surface where $G = -1$. This is where

$$G \; = \; z^2 - 2z - x^2 - y^2 \; = \; (z-1)^2 - x^2 - y^2 - 1 \; = \; -1,$$

i.e., where $(z-1)^2 \; = \; x^2 + y^2$.

As we discussed in Section 1.8, the graph of G is a (double) cone, just shifted up one unit from a standard one, which is centered at $\mathbf{0}$. See Figure 2.7.9. Note that the level surface has a sharp point at $(0, 0, 1)$. This point, where the graph is not smooth, is again called a singularity.

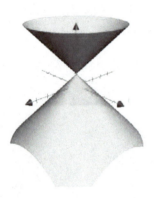

Figure 2.7.9: The level surface $z^2 - 2z - x^2 - y^2 = -1$ has a singularity.

Now that we've looked at a few examples of level curves and surfaces, there are questions that you may be asking: Why are level sets smooth at some points, but not at others? At the places where the level sets are smooth, is there a nice way to obtain a *tangent set*, like a tangent line or tangent plane? And, recalling the title of this section, what do gradient vectors have to do with level sets?

To answer the smoothness question, we need to appeal to the Implicit Function Theorem, which we will look at and discuss more thoroughly in Theorem 2.7.16 and Theorem 2.12.2. One way to address the tangent set question is to define *tangent vectors* in terms of parameterized curves; we shall take this approach in the More Depth portion of this section. But the quickest way for us to see the relationship between level sets and the gradient vector is to use our earlier results on the directional derivative from Section 2.5.

From Theorem 2.5.3, we know that $D_{\mathbf{u}} F(\mathbf{p}) = 0$ if and only if $\mathbf{u}$ is perpendicular to $\vec{\nabla} F(\mathbf{p})$. This tells us that, at $\mathbf{p}$, the rate of change of F is zero if and only if we move in directions perpendicular to the gradient vector. But the rate of change of F being zero means that F is constant, i.e., that we're staying inside the level set. Therefore, $\vec{\nabla} F(\mathbf{p})$ is perpendicular to all of the directions in which $\mathbf{p}$ can move in order to stay inside the level set where $F = F(\mathbf{p})$.

Combining this with the Implicit Function Theorem, Theorem 2.7.16, we obtain:

Theorem 2.7.7. *Suppose that $F = F(\mathbf{x})$ is a continuously differentiable, real-valued function on an open neighborhood of $\mathbf{p}$ in $\mathbb{R}^{n+1}$, and that $\vec{\nabla}F(\mathbf{p}) \neq \mathbf{0}$.*

Then, near $\mathbf{p}$, the level set M where $F = F(\mathbf{p})$ looks smooth and is n-dimensional, i.e., has dimension 1 less than the dimension of the surrounding Euclidean space.

*In addition, a vector $\mathbf{v}$, based at $\mathbf{p}$, is tangent to M at $\mathbf{p}$, i.e., is a **tangent vector to M at $\mathbf{p}$**, if and only if*

$$\vec{\nabla}F(\mathbf{p}) \cdot \mathbf{v} = 0.$$

Therefore, at $\mathbf{p}$, there is a well-defined tangent set to M, and that tangent set consists of those $\mathbf{x}$ in $\mathbb{R}^{n+1}$ such that

$$\vec{\nabla}F(\mathbf{p}) \cdot (\mathbf{x} - \mathbf{p}) = 0,$$

i.e., the tangent set at $\mathbf{p}$ contains the point $\mathbf{p}$ and is perpendicular to the gradient vector $\vec{\nabla}F(\mathbf{p})$.

Furthermore, if M is the graph of a continuously differentiable function, then this tangent set agrees with the one defined in Definition 2.3.8.

The cases of level curves and level surfaces will be of particular interest to us.

Corollary 2.7.8. *Suppose that $F = F(x, y)$ is a continuously differentiable, real-valued function on an open neighborhood of $\mathbf{p} = (a, b)$ in $\mathbb{R}^2$, and that $\vec{\nabla}F(\mathbf{p}) \neq \mathbf{0}$.*

Then, near $\mathbf{p}$, the level set where $F = F(\mathbf{p})$ looks smooth and is 1-dimensional, i.e., is a level curve, and its tangent line is given by

$$F_x(a, b)(x - a) + F_y(a, b)(y - b) = 0.$$

Suppose that $F = F(x, y, z)$ is a continuously differentiable, real-valued function on an open neighborhood of $\mathbf{p} = (a, b, c)$ in $\mathbb{R}^3$, and that $\vec{\nabla}F(\mathbf{p}) \neq \mathbf{0}$.

Then, near $\mathbf{p}$, the level set where $F = F(\mathbf{p})$ looks smooth and is 2-dimensional, i.e., is a level surface, and its tangent plane is given by

$$F_x(a, b, c)(x - a) + F_y(a, b, c)(y - b) + F_z(a, b, c)(z - c) = 0.$$

Here, we have written "looks smooth", rather than "is smooth", because "smooth" has a technical definition, which we gave in Definition 2.6.9. We write "looks smooth" as a more intuitive way of describing a C^1 submanifold. If, in fact, F is smooth, i.e., C^∞, then the level set really is smooth.

The technical meaning of "n-dimensional" is also given in Definition 2.6.9.

In Definition 1.6.24, we gave a rigorous definition of a tangent vector to M as a vector which is tangent to a parameterized path inside of M.

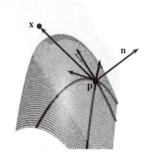

Figure 2.7.10: Paths in M, tangent vectors, the tangent plane, and $\mathbf{n} = \vec{\nabla}F(\mathbf{p})$.

The tangent set is an affine linear subspace of dimension n; recall Definition 1.4.22.

In the More Depth portion of Section 2.12, we shall look at the equivalent definitions of the tangent set more carefully.

Example 2.7.9. Let's look again at the level curves of the smooth function $F(x,y) = x^2 - y^2$, as we did in Example 2.7.2.

We quickly calculate that $\vec{\nabla}F = (2x, -2y)$.

Thus, $\vec{\nabla}F = \mathbf{0}$ if and only if $(x,y) = (0,0)$, which is on the level curve where $F(x,y) = F(0,0) = 0$. It follows from Theorem 2.7.16 that all of the level curves of F, other than the level curve where $F = 0$, are smooth everywhere, but the level curve where $F = 0$ is possibly not smooth, i.e., possibly has a singularity, at one point: $(x,y) = (0,0)$.

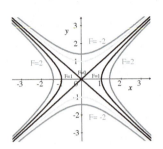

Figure 2.7.11: Some level curves of $F(x,y) = x^2 - y^2$.

Understand: Theorem 2.7.16 does not tell us that the level curve where $F = 0$ **must** have a singularity at $(0,0)$; it tells us that the only possible singularity on a level curve of F is at $(0,0)$, on the level curve where $F = F(0,0)$, i.e., where $x^2 - y^2 = 0$. From the graph, we know that, in fact, there is a singularity at $(0,0)$.

Let's look a little more carefully at the level curve where $F(x,y) = x^2 - y^2 = 1$. This describes a hyperbola; see Figure 2.7.12. Also in Figure 2.7.12, we sketched in the gradient vectors at the points $(1,0)$ and $(\sqrt{5}, 2)$, and included dotted lines for the corresponding tangent lines to the level curve; the gradient vectors are $\vec{\nabla}F(1,0) = (2,0)$ and $\vec{\nabla}F(\sqrt{5}, 2) = (2\sqrt{5}, -4)$. Finally, we included a displacement vector from $(a,b) = (\sqrt{5}, 2)$ to a point (x,y) on the corresponding tangent line, so that you can see that this vector is perpendicular to the gradient vector.

Using Corollary 2.7.8, it's trivial to find equations for the tangent lines of the level curve where $x^2 - y^2 = 1$ at the points $(1,0)$ and $(\sqrt{5}, 2)$; they are

$$2(x-1) + 0(y-0) = 0 \quad \text{and} \quad 2\sqrt{5}(x - \sqrt{5}) - 4(y-2) = 0.$$

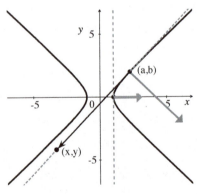

Figure 2.7.12: The level curve $x^2 - y^2 = 1$, with two gradient vectors and tangent lines.

Note that the gradient vectors do appear to be perpendicular to the curves at the given points. In fact, our results on directional derivatives, Theorem 2.5.3, tell us that, in addition to being perpendicular to the level set, the gradients vectors point in the directions in which F increases as rapidly as possible; in particular, they point in directions in which F increases. So, you should see the values of the level sets of F get bigger as you move in the direction of the gradient vectors.

To emphasize this point, we can look at a figure which includes more levels curves of F, and many more gradient vectors of F.

In Figure 2.7.13, we have once again included a few level curves of F, together with many (scaled) gradient vectors of F, based at the points at which they're calculated.

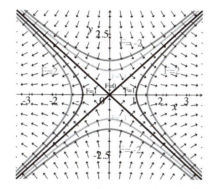

Figure 2.7.13: Level curves and scaled gradient vectors of $F(x, y) = x^2 - y^2$.

The gradient vectors have all been scaled to the same length for readability; so you should pay attention to their direction only. Note that the gradient vectors are perpendicular to the level curves (provided that the gradient vectors start on one of the graphed level curves), and point in directions in which F increases. In fact, recall from Theorem 2.5.3 that the gradient vectors point in the direction in which F increases most rapidly.

> The collection of gradient vectors, calculated at each point, is the *gradient vector field*. We shall look at vector fields in depth in Chapter 4.

As we saw in the previous example, another corollary of Theorem 2.7.7 is:

Corollary 2.7.10. *Suppose that $F = F(\mathbf{x})$ is a continuously differentiable, real-valued function on an open subset $\mathcal{U}$ of $\mathbb{R}^{n+1}$.*

If $\mathbf{p}$ is a singularity on the level set where $F = F(\mathbf{p})$, then $\vec{\nabla}F(\mathbf{p}) = \mathbf{0}$, i.e., all of the partial derivatives of F at $\mathbf{p}$ are zero.

> To be precise, by "singularity", we mean a point on the level set at which the level set is not a C^1 n-dimensional submanifold of $\mathbb{R}^{n+1}$. See Definition 2.6.9.

Consequently, we give a name to points at which the gradient vector is zero.

We should really refer to this as an $(n+1)$-dimensional **critical point**, because, later, we will want to restrict functions to boundary **submanifolds** of regions and look at smaller-dimensional critical points. However, as the context will always make this clear, we shall not use this more cumbersome terminology.

Definition 2.7.11. *Suppose that $F = F(\mathbf{x})$ is a real-valued function, whose domain is a subset E of $\mathbb{R}^{n+1}$. A point $\mathbf{p}$ in E is a **critical point of F** if $\mathbf{p}$ is an interior point of E at which $\vec{\nabla} F(\mathbf{p}) = \mathbf{0}$ or is a point where F is not differentiable; note that this includes all boundary points in E.*

Thus, a critical point of a continuously differentiable function F on an open subset of $\mathbb{R}^{n+1}$ is a point $\mathbf{p}$ at which $\vec{\nabla} F(\mathbf{p}) = \mathbf{0}$.

Example 2.7.12. Let's look back at Example 2.7.5, in which we had $S(x, y, z) = x^2 + y^2 + z^2$. As we saw, the type of level set where $S(x, y, z) = c$ depends on the value of c. If $c < 0$, there are no points in the level set. If $c = 0$, the level set consists solely of the origin. If $c > 0$, then the level set is, in fact, a surface: a sphere of radius $\sqrt{c}$, centered at the origin.

The gradient vector of S is easy to calculate:

$$\vec{\nabla} S = (2x, 2y, 2z) = 2(x, y, z).$$

Now it's simple to find the critical points: $\vec{\nabla} S = \mathbf{0}$ if and only if $(x, y, z) = (0, 0, 0)$, which is a point on the level set where $S = S(0, 0, 0) = 0$. Thus, there is a single critical point of F ; this explains why $S = 0$ is "allowed" to not be a smooth surface (it's a single point – not a surface) at the origin. Note that Corollary 2.7.10, by itself, does not imply that the level set where $S = 0$ is not a smooth surface; it takes other work to know that $S = 0$ actually defines a singularity at the origin.

You may wonder why the gradient vector doesn't look strange at points where $S(x, y, z) = c < 0$. After all, the level set where $c = 0$ is not a smooth surface either; it's an empty set. Keep in mind, though, that there are no points on the level set where $S = c < 0$; so, of course, there are no points on the level set at which the gradient vector is the zero vector.

Suppose we want an equation for the tangent plane to the level surface of S at $(1, 2, 3)$. Since we're handed a point on the plane, what we need to produce is a vector that's normal to the plane, which is what the gradient vector will give you.

The level surface in question is where $S(x, y, z) = x^2 + y^2 + z^2 = S(1, 2, 3) = 14$, i.e., a sphere of radius $\sqrt{14}$ centered at the origin.

We don't really need our work on gradient vectors to find a non-zero vector that's perpendicular to this sphere; at the point $(1, 2, 3)$, the radial line segment from $(0, 0, 0)$ out to $(1, 2, 3)$ is perpendicular to the sphere $x^2 + y^2 + z^2 = 14$. This means that $\mathbf{v} = (1, 2, 3)$ is a normal vector, and an equation for the tangent plane is

$$1(x - 1) + 2(y - 2) + 3(z - 3) = 0.$$

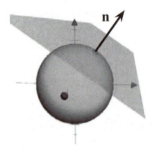

Figure 2.7.14: The level surface where $x^2 + y^2 + z^2 = 14$, the tangent plane at $(1, 2, 3)$, and a non-zero normal vector $\mathbf{n} = (1, 2, 3) = 0.5\,\vec{\nabla} S(1, 2, 3)$.

This agrees with what we'd get by using the gradient vector, since

$$\vec{\nabla}S(1,2,3) \;=\; 2(1,\,2,\,3),$$

and multiplying by any non-zero scalar, like $1/2$, to obtain the vector $(1,2,3)$, would still yield a non-zero vector that's normal to the tangent plane.

Example 2.7.13. Consider the function

$$G \;=\; z^2 - 2z - x^2 - y^2 \;=\; (z-1)^2 - x^2 - y^2 - 1$$

from Example 2.7.6.

a. Find all of the points where $\vec{\nabla}G = \mathbf{0}$ (i.e., find the critical points of G) and, hence, find those points at which the level set of G may have a singularity, i.e., at which the level set of G may not be smooth.

b. Find the points on the level set where $G = 0$ at which the tangent plane is horizontal.

c. Find an equation for the tangent plane to the level surface of G at $(3,4,6)$.

Figure 2.7.15: The level surface $z^2 - 2z - x^2 - y^2 = -1$.

Solution:

a. $\vec{\nabla}G = (-2x, -2y, 2(z-1))$. This is the zero vector if and only if $(x,y,z) = (0,0,1)$, which is on the level surface where $G = G(0,0,1) = -1$. As we saw earlier, the level surface where $G = -1$ does, in fact, have a singularity at $(0,0,1)$. See Figure 2.7.15

b. The tangent plane is horizontal if and only if $\vec{\nabla}G$ is of the form $(0,0,c)$, for any value of c (other than 0, which we just saw can't happen on the level surface where $G = 0$). From our gradient vector calculation above, the tangent plane to $G = 0$ is horizontal if $(x,y) = (0,0)$, which would mean that $(z-1)^2 - 0^2 - 0^2 - 1 = 0$, i.e., $(z-1)^2 = 1$. Therefore, $z = 0$ or $z = 2$, and the two points on the $G = 0$ level surface at which the tangent planes are horizontal are $(0,0,0)$ and $(0,0,2)$. See Figure 2.7.16

Figure 2.7.16: The level surface $z^2 - 2z - x^2 - y^2 = 0$, with two horizontal tangent planes.

c. The gradient vector of G at $(3,4,6)$ is $\vec{\nabla}G(3,4,6) = (-6,-8,10)$, and $G(3,4,6) = -1$. Therefore, by Corollary 2.7.8, the tangent plane to the level surface where $G = -1$ at $(3,4,6)$ is given by the equation

$$-6(x-3) - 8(y-4) + 10(z-6) \;=\; 0.$$

See Figure 2.7.17. Note that the tangent plane contains an entire line on the cone.

Figure 2.7.17: The level surface $G = -1$, with the tangent plane at $(3,4,6)$.

More Depth:

So far, we have looked at the tangent set of a level set, at a point $\mathbf{p}$, as the affine linear subspace that contains $\mathbf{p}$ and which is orthogonal to the gradient vector; then, using this, we defined a tangent vector to the level set at $\mathbf{p}$ to be a vector which is orthogonal to the gradient vector.

Here, however, we will take a different approach. We will define tangent vectors in terms of parameterized curves, and then move these tangent vectors out to $\mathbf{p}$ in order to define the tangent set.

We also want to discuss how the Inverse Function Theorem, Theorem 2.6.6, leads to the version of the Implicit Function Theorem that we have used in this section.

Recall our definition of a tangent vector from Definition 1.6.24.

Definition 2.7.14. *Let E be a subset of $\mathbb{R}^n$, and let $\mathbf{p}$ be a point in E. A **tangent vector to E at $\mathbf{p}$** is the velocity vector at $\mathbf{p}$ of a continuously differentiable path in E.*

*More precisely, a vector $\mathbf{v}$, based at $\mathbf{p}$, in $\mathbb{R}^n$ is a **tangent vector to E at $\mathbf{p}$** if and only if there exist $\epsilon > 0$, t_0, and a C^1 function*

$$\mathbf{r} : (t_0 - \epsilon, t_0 + \epsilon) \to \mathbb{R}^n$$

such that, for all t in the interval $(t_0 - \epsilon, t_0 + \epsilon)$, $\mathbf{r}(t)$ is in E, $\mathbf{p} = \mathbf{r}(t_0)$, and $\mathbf{v} = \mathbf{r}'(t_0)$.

Example 2.7.15. Consider the circular paraboloid M given by $z = 9 - x^2 - y^2$. What are some of the tangent vectors to M at the point $\mathbf{p} = (1, 2, 4)$?

Well, certainly, there's the zero vector, $\mathbf{0}$. **The zero vector is always a tangent vector.** Why? Because you can always take the constant position function $\mathbf{r}(t) = \mathbf{p}$. Then, $\mathbf{r}(t)$ is a C^1 curve that's always in M, $\mathbf{r}(0) = \mathbf{p}$, and $\mathbf{r}'(0) = \mathbf{0}$.

How do we get other tangent vectors to M at $\mathbf{p}$? We just produce curves $\mathbf{r}(t)$ in M, such that $\mathbf{r}(0) = (1, 2, 4)$, and then we calculate $\mathbf{r}'(0)$.

Alright. What's another C^1 curve $\mathbf{r}(t)$ in M such that $\mathbf{r}(0) = (1, 2, 4)$? Actually, they're pretty easy to produce; take any line in the xy-plane that passes through $(1, 2)$, parametrize it so that you're at the point $(1, 2)$ at $t = 0$, and then calculate the corresponding z-parameterization by using $z = 9 - x^2 - y^2$.

So, we could let $x = 1$ (constantly), $y = 2+t$, and find $z = 9-1^2-(2+t)^2 = 4-4t-t^2$. Thus, a C^1 curve in M such that $\mathbf{r}(0) = (1,2,4)$ is given by

$$\mathbf{r}(t) \;=\; (1,\; 2+t,\; 4-4t-t^2).$$

We find $\mathbf{r}'(t) \;=\; (0,\; 1,\; -4-2t)$, and so $\mathbf{v} \;=\; \mathbf{r}'(0) \;=\; (0,\; 1,\; -4)$ is a tangent vector to M at $\mathbf{p}$.

In fact, we can parametrize every line through $(1,2)$, and so produce an infinite number of $\mathbf{r}(t)$ and a corresponding infinite number of tangent vectors. We pick any constants a and b, and let $x = 1+at$, $y = 2+bt$, and so

$$z \;=\; 9 - (1+at)^2 - (2+bt)^2 \;=\; 4 - (2a+4b)t - (a^2+b^2)t^2.$$

We calculate the derivative of the corresponding C^1 (in fact, smooth) function

$$\mathbf{r}(t) \;=\; (1+at,\; 2+bt,\; 4-(2a+4b)t-(a^2+b^2)t^2),$$

and find that

$$\mathbf{v} \;=\; \mathbf{r}'(0) \;=\; (a,\; b,\; -2a-4b)$$

is a tangent vector to M at $\mathbf{p}$, for all a and b.

Writing $\mathbf{v} = (v_1,\, v_2,\, v_3)$, we find that these tangent vectors are precisely the "points" on the plane given by

$$v_3 \;=\; -2v_1 \;-\; 4v_2 \qquad \text{or, equivalently,} \qquad 2v_1 \;+\; 4v_2 \;+\; v_3 \;=\; 0. \qquad (2.9)$$

If we think of $(v_1,\, v_2,\, v_3)$ as a point, instead of a vector, then the above equation describes a plane through the origin, but, considering that it was formed from velocity vectors at $\mathbf{p}$, we typically want to consider this plane translated out to the point $\mathbf{p}$. Thus, to obtain the points in what we picture as the plane of tangent vectors, we look at all **points** $(x,y,z) \;=\; \mathbf{p} + \mathbf{v}$, where $\mathbf{v}$ is a tangent vector to M at $\mathbf{p}$.

Rewriting this last equality, we find

$$\mathbf{v} \;=\; (v_1, v_2, v_3) \;=\; (x,y,z) - \mathbf{p} \;=\; (x-1, y-2, z-4).$$

Inserting this into our equation for the plane through the origin in Formula 2.9, we find that (x,y,z) is on our tangent plane if and only if

$$2(x-1) \;+\; 4(y-2) \;+\; (z-4) \;=\; 0.$$

Does this agree with Corollary 2.7.8? The set M is the set where $x^2 + y^2 + z = 9$. Thus, if we let $F(x,y,z) = x^2 + y^2 + z$, then M is the level set where $F = 9$. We calculate

$$\vec{\nabla}F(1,2,4) \;=\; (2x,2y,1)\big|_{(1,2,4)} \;=\; (2,4,1),$$

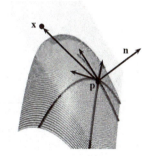

Figure 2.7.18: Paths in M, tangent vectors, the tangent plane, and $\mathbf{n} = \vec{\nabla}F(\mathbf{p})$.

and apply Corollary 2.7.8 to conclude that the tangent plane is, indeed, defined by our previous equation:

$$2(x - 1) \; + \; 4(y - 2) \; + \; (z - 4) \; = \; 0.$$

In Theorem 2.7.7, we claimed that the tangent set that we obtain for level sets is the same as the one that we defined in Definition 2.3.8, for graphs of functions. Let's check that here.

Looking back at Definition 2.3.8, if we have a continuously differentiable function $f(x, y)$, then the tangent set to the graph of $z = f(x, y)$ at $(a, b, f(a, b)))$ is the set of points (x, y, z) such that

$$z \; = \; f(a, b) \; + \; f_x(a, b)(x - a) \; + \; f_y(a, b)(y - b).$$

In our present example, where $f(x, y) = 9 - x^2 - y^2$ and $(a, b) = (1, 2)$, we find that an equation for the tangent set is

$$z \; = \; 4 \; + \; (-2 \cdot 1)(x - 1) \; + (-2 \cdot 2)(y - 2),$$

which, after rearranging, is the same equation that we found for the tangent plane that we found in terms of tangent vectors and by using Corollary 2.7.8.

Before we leave this example, there is one more important thing that we need to mention. We have shown that we get the entire tangent plane from tangent vectors to M at $\mathbf{p}$, but how do we know that we obtained **all** of the tangent vectors? We used a very specific collection of smooth curves on M to produce our tangent vectors; maybe some other weird paths would yield some more tangent vectors to M at $\mathbf{p}$.

In fact, this doesn't happen, but it takes the Implicit Function Theorem, Theorem 2.7.16, to know it.

As we stated earlier, if F is continuously differentiable and $\vec{\nabla} F(\mathbf{p}) \neq \mathbf{0}$, then the *Implicit Function Theorem* implies that, near $\mathbf{p}$, the level set where $F = F(\mathbf{p})$ is a C^1 hypersurface.

We wish to state this result more carefully, and show how to prove it using the Inverse Function Theorem, Theorem 2.6.6.

Recall the definition of a C^m submanifold from Definition 2.6.9; if $m \geq 1$, then, intuitively, a C^m submanifold is a set which, near each point, looks like a smoothly bent, or deformed, open subset of $\mathbb{R}^m$, but, technically, is only "smooth" up to a C^m change of coordinates.

> **Theorem 2.7.16. (Implicit Function Theorem, geometric form)** *Suppose that $1 \leq r \leq \infty$. Suppose that $F = F(\mathbf{x})$ is a real-valued function of class C^r on an open neighborhood of $\mathbf{p}$ in $\mathbb{R}^{n+1}$, and that $\vec{\nabla}F(\mathbf{p}) \neq \mathbf{0}$. Note, in particular, that F is at least of class C^1.*

> *Then, in an open neighborhood of $\mathbf{p}$, the level set given by $F(\mathbf{x}) = F(\mathbf{p})$ is a C^r submanifold at $\mathbf{p}$, of dimension n (i.e., a C^r hypersurface)*

Proof. Suppose $r \geq 1$ and that we have a C^r real-valued function $F = F(x_1, \ldots, x_n)$, and a point $\mathbf{p} = (p_1, \ldots, p_n)$. Let M denote the level set of F at $\mathbf{p}$. Suppose further that $\dfrac{\partial F}{\partial x_n}\Big|_{\mathbf{p}} \neq 0$.

Define a new multi-component function from a subset of $\mathbb{R}^n$ to a subset of $\mathbb{R}^n$ by

$$\mathbf{\Phi}(x_1, \ldots, x_n) = \big(x_1 - p_1, \, x_2 - p_2, \, \ldots \, x_{n-1} - p_{n-1}, \, F(x_1, \ldots, x_n) - F(\mathbf{p})\big).$$

Note that $\mathbf{\Phi}(\mathbf{p}) = \mathbf{0}$ and that a point $\mathbf{x}$ is in M, the level set where $F = F(\mathbf{p})$, if and only if the last coordinate of $\mathbf{\Phi}(\mathbf{x})$ is 0. Now, it is easy to show that $d_{\mathbf{p}}\mathbf{\Phi}$ is a bijection and, hence, the Inverse Function Theorem, Theorem 2.6.6, tells us that $\mathbf{\Phi}$ is a local C^r change of coordinates, which takes M, in an open neighborhood of $\mathbf{p}$ to an open neighborhood of the origin, intersected with $\mathbb{R}^{n-1} \times \{0\}$.

Therefore, by Definition 2.6.9, near $\mathbf{p}$, M is a C^r hypersurface in $\mathbb{R}^n$, i.e., a C^r submanifold, which looks smooth, and is of one dimension less than the surrounding space. $\qquad \square$

We have picked x_n for notational convenience; the argument that we give is the same, regardless of which non-zero partial derivative you choose, and one of them must be non-zero, since the theorem assumes that $\vec{\nabla}F(\mathbf{p}) \neq \mathbf{0}$.

Remark 2.7.17. You may have noticed that nothing in the statement of the Implicit Function Theorem in this section says that something is actually implicitly defined as a function of something else. That's true, and is why we called our statement the **geometric form** of the Implicit Function Theorem. We decided to place the true "implicit function" part, and the associated method of implicit differentiation in a separate section, Section 2.12.

Recall now the definition of the tangent space to C^1 submanifold from Definition 2.6.16.

Corollary 2.7.18. *Suppose that $F = F(\mathbf{x})$ is a real-valued function of class C^1 (or higher) on an open neighborhood of $\mathbf{p}$ in $\mathbb{R}^{n+1}$, and that $\vec{\nabla} F(\mathbf{p}) \neq \mathbf{0}$.*

Then the tangent space at $\mathbf{p}$ to the level set given by $F(\mathbf{x}) = F(\mathbf{p})$ exists, and is the n-dimensional linear subspace of $\mathbb{R}^{n+1}$ consisting of those vectors which are perpendicular to $\vec{\nabla} \mathbf{F}(\mathbf{p})$. In particular, any vector which is perpendicular to all of the tangent vectors to the level set at $\mathbf{p}$ is a scalar multiple of $\vec{\nabla} \mathbf{F}(\mathbf{p})$.

Proof. That the tangent space exists and is n-dimensional follows from the Implicit Function Theorem and Theorem 2.6.15. As the space of vectors perpendicular to $\vec{\nabla} F(\mathbf{p}) \neq \mathbf{0}$ is also n-dimensional, it suffices to show that each tangent vector is perpendicular to $\vec{\nabla} F(\mathbf{p})$

Suppose that $\mathbf{v}$ is a tangent vector, at $\mathbf{p}$, to the level set of F through $\mathbf{p}$. Thus, there exists a continuously differentiable function $\mathbf{r}(t)$, for t in some open interval around some t_0, such that $\mathbf{r}(t)$ is always in the level set, $\mathbf{p} = \mathbf{r}(t_0)$, and $\mathbf{v} = \mathbf{r}'(t_0)$. The fact that $\mathbf{r}(t)$ is always in the level set means that $F(\mathbf{r}(t)) = F(\mathbf{p})$ for all t in our interval. We can differentiate both sides of this equation, using the Chain Rule, Theorem 2.4.9, on the left, to obtain

$$\vec{\nabla} F(\mathbf{r}(t)) \cdot \mathbf{r}'(t) \;=\; 0.$$

At $t = t_0$, this gives us what we want: $\vec{\nabla} F(\mathbf{p}) \cdot \mathbf{v} \;=\; 0.$ $\qquad\square$

As we discussed in Example 2.7.15, the collection of tangent vectors, considered as points, is a set which passes through the origin, not (necessarily) through the point $\mathbf{p}$. You have to add the point $\mathbf{p}$ to each tangent vector to get the *tangent set*.

Definition 2.7.19. *Suppose that $F = F(\mathbf{x})$ is a C^1 function on an open neighborhood of $\mathbf{p}$ in $\mathbb{R}^n$, and that $\vec{\nabla} F(\mathbf{p}) \neq \mathbf{0}$. The set of all points of the form $\mathbf{p} + \mathbf{v}$, where $\mathbf{v}$ is in the tangent space to the level set of F at $\mathbf{p}$, is the **tangent set** of the level set of F at $\mathbf{p}$; this agrees with our earlier definition, Definition 2.3.8, of the tangent set in the case of the graph of a C^1 function.*

> A point $\mathbf{x} = (x_1, \ldots, x_n)$ is in the tangent set to the level set of F at $\mathbf{p} = (p_1, \ldots, p_n)$ if and only if
>
> $$\vec{\nabla} F(\mathbf{p}) \cdot (\mathbf{x} - \mathbf{p}) = 0,$$
>
> that is, if and only if
>
> $$\frac{\partial F}{\partial x_1}\bigg|_{\mathbf{p}} (x_1 - p_1) + \cdots + \frac{\partial F}{\partial x_n}\bigg|_{\mathbf{p}} (x_n - p_n) = 0.$$

> If $F = F(x, y)$, so that the level set of F near $\mathbf{p}$ is a curve, then the tangent set at $\mathbf{p}$ is referred to as the **tangent line** to the level set of F at $\mathbf{p}$. Naturally, if $F = F(x, y, z)$, so that the level set of F near $\mathbf{p}$ is a surface, then the tangent set at $\mathbf{p}$ is referred to as the **tangent surface** to the level set of F at $\mathbf{p}$.
>
> As the gradient vector $\vec{\nabla} F(\mathbf{p})$ is perpendicular to all of the tangent vectors to the level set of F at $\mathbf{p}$, we frequently say that $\vec{\nabla} F(\mathbf{p})$ is **perpendicular** (or **orthogonal**, or **normal**) to the level set of F at $\mathbf{p}$.

Remark 2.7.20. As we stated earlier, given a continuously differentiable function $f = f(x_1, \ldots, x_n)$, you now have two ways to find an equation for the tangent set to the graph of $z = f(\mathbf{x})$ at a point $(\mathbf{p}, f(\mathbf{p}))$.

You can use the graph of the linearization, as we did in Definition 2.3.8. The tangent set is the set of points $(\mathbf{x}, z)$ such that

$$z = f(\mathbf{p}) + \vec{\nabla} f(\mathbf{p}) \cdot (\mathbf{x} - \mathbf{p}).$$

However, we could also define

$$F(\mathbf{x}, z) = z - f(\mathbf{x}),$$

and note that the level set where $F = 0$ is the same as the graph of $z = f(\mathbf{x})$. Furthermore,

$$\vec{\nabla} F(\mathbf{p}, f(\mathbf{p})) = (-\vec{\nabla} f(\mathbf{p}), 1).$$

We then use Definition 2.6.16, and obtain that the tangent set at $(\mathbf{p}, f(\mathbf{p}))$ is the set of $(\mathbf{x}, z)$ such that

$$\vec{\nabla} F(\mathbf{p}, f(\mathbf{p})) \cdot \big((\mathbf{x}, z) - (\mathbf{p}, f(\mathbf{p}))\big) = 0.$$

This is the same as

$$(-\vec{\nabla} f(\mathbf{p}), 1) \cdot \big(\mathbf{x} - \mathbf{p}, z - f(\mathbf{p})\big) = -\vec{\nabla} f(\mathbf{p}) \cdot (\mathbf{x} - \mathbf{p}) + z - f(\mathbf{p}) = 0,$$

i.e.,

$$z = f(\mathbf{p}) + \vec{\nabla} f(\mathbf{p}) \cdot (\mathbf{x} - \mathbf{p}).$$

Therefore, for the graph of a C^1 function, you'll obtain the same tangent set, regardless of whether you use the definition in Definition 2.3.8 or the definition in Definition 2.6.16; that's good, otherwise we should use different terminology for the two sets.

Note that $\vec{\nabla} F$ is **never** zero in this setting, since there's a 1 in the final component. Looking back to Theorem 2.7.7 or ahead to Theorem 2.7.16, this tells us that the graph of a C^1 function *looks* smooth, and the graph of a smooth function *is* smooth.

Example 2.7.21. Let's look at a higher-dimensional example, involving a more-complicated function.

Consider the smooth function

$$G(w, x, y, z) = 7w - 3x + xze^{x \sin y} + wy \tan^{-1}(w + 5z).$$

Find an equation for the tangent set to the level set of G at the point $\mathbf{p} = (w, x, y, z) = (\pi/4, 1, 0, 0)$, and show that this level set is smooth at $\mathbf{p}$.

Solution: In theory, this is easy; an equation for the tangent set is

$$\vec{\nabla} G(\pi/4, 1, 0, 0) \cdot \big((w, x, y, z) - (\pi/4, 1, 0, 0)\big) = 0,$$

and, by Theorem 2.7.16, if $\vec{\nabla} G(\pi/4, 1, 0, 0) \neq \mathbf{0}$, then the level set given by $G = G(\pi/4, 1, 0, 0)$ is smooth at $(\pi/4, 1, 0, 0)$. Of course, actually calculating $\vec{\nabla} G(\pi/4, 1, 0, 0)$ is somewhat painful.

We find:

$$\frac{\partial G}{\partial w} = 7 + y \left(w \cdot \frac{1}{1 + (w + 5z)^2} + \tan^{-1}(w + 5z) \right),$$

$$\frac{\partial G}{\partial x} = -3 + z \left(xe^{x \sin y} \sin y + e^{x \sin y} \right),$$

$$\frac{\partial G}{\partial y} = xze^{x \sin y} x \cos y + w \tan^{-1}(w + 5z),$$

and

$$\frac{\partial G}{\partial z} = xe^{x \sin y} + wy \cdot \frac{5}{1 + (w + 5z)^2}.$$

Therefore,

$$\vec{\nabla} G(\pi/4, 1, 0, 0) = (7, -3, \pi/4, 1) \neq \mathbf{0},$$

so that the level set is smooth at $(\pi/4, 1, 0, 0)$, and our equation for the tangent set to the level set of G at $(\pi/4, 1, 0, 0)$ is

$$7\left(w - \frac{\pi}{4}\right) - 3(x-1) + \frac{\pi y}{4} + z = 0.$$

2.7.1 Exercises

Online answers to select exercises are here.

In each of the following exercises, you are given a continuously differentiable function $F(x, y)$. **(a)** Sketch by hand, or have technology sketch, the level sets where F equals -1, -0.5, 0, 0.5, and 1. **(b)** On each connected piece of your level curves (assuming you actually get a curve), sketch the direction of one gradient vector of f.

Basics:

1. $F(x, y) = xy$.

2. $F(x, y) = y^2 + \sin x$.

3. $F(x, y) = y^2 - x^3$.

4. $F(x, y) = y^2 - x^3 - x^2$.

5. $F(x, y) = xe^{y^2}$.

6. $F(x, y) = \dfrac{x^2}{4} + \dfrac{y^2}{9}$.

In each of the following exercises, you are given level curves of a continuously differentiable function. On each connected piece of each level curve, sketch the direction of one gradient vector of F. Note that, in ascending order of the value of F, the level curves are colored red, magenta, dark blue, light blue, and green.

7.

8.

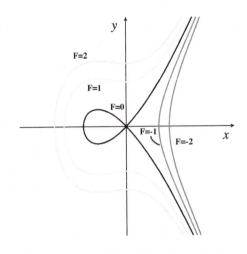

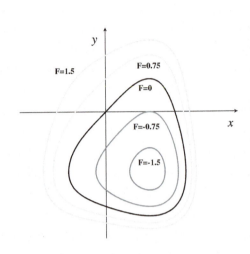

9.

10.

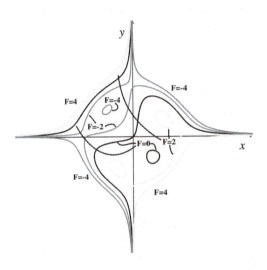

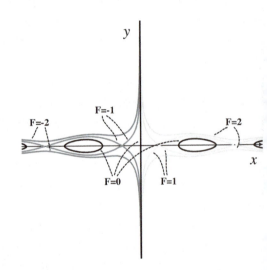

In each of the following exercises, you are given a continuously differentiable function $F(x, y)$ and a point $\mathbf{p} = (a, b)$. (a) Calculate $F(\mathbf{p})$, sketch, or have technology sketch, the level curve where $F = F(\mathbf{p})$, (b) sketch the gradient vector $\vec{\nabla} F(\mathbf{p})$, based at $\mathbf{p}$, (c) sketch in the tangent line to the level curve at $\mathbf{p}$, and (d) find an equation for the tangent line to the level curve at the point $\mathbf{p}$.

11. $F(x, y) = xy$, $\mathbf{p} = (1, 2)$.

12. $F(x, y) = y^2 + \sin x$, $\mathbf{p} = (0, 1)$.

13. $F(x, y) = y^2 - x^3$, $\mathbf{p} = (1, 1)$.

14. $F(x, y) = y^2 - x^3 - x^2$, $\mathbf{p} = (-1, 1)$.

15. $F(x, y) = xe^{y^2}$, $\mathbf{p} = (2, 0)$.

16. $F(x, y) = \dfrac{x^2}{4} + \dfrac{y^2}{9}$, $\mathbf{p} = (0, 3)$.

In each of the following exercises, you are given a continuously differentiable function $F(x, y, z)$ and a point $\mathbf{p} = (a, b, c)$. (a) Calculate $F(\mathbf{p})$, sketch, or have technology sketch, the level surface where $F = F(\mathbf{p})$, (b) sketch the gradient vector $\vec{\nabla}F(\mathbf{p})$, based at $\mathbf{p}$, (c) sketch the tangent plane to the level surface at $\mathbf{p}$, and (d) find an equation for the tangent plane to the level surface at the point $\mathbf{p}$.

17. $F(x, y, z) = x^2 + y^2 + z$, $\mathbf{p} = (1, 2, 3)$.

18. $F(x, y, z) = x^2 + y^2 + z^2$, $\mathbf{p} = (1, 2, 3)$.

19. $F(x, y, z) = x^2 + y^2 - z^2$, $\mathbf{p} = (3, 4, 5)$.

20. $F(x, y, z) = x^2 - y^2 + z$, $\mathbf{p} = (1, 2, 3)$.

21. $F(x, y, z) = xy - \ln z$, $\mathbf{p} = (0, 2, 1)$.

22. $F(x, y, z) = x^2 + z^2 - \tan^{-1} y$, $\mathbf{p} = (0, 1, 0)$.

23. Let $F(x, y, z) = (x + y + z - 1)^2$. Where does F have critical points? Are there singularities on any of the level sets of F? Explain.

24. Suppose that F is a continuously differentiable, real-valued function. Let M denote the level set where $F = c$. Suppose that $\mathbf{p}$ is a point in M.

    (a) What is the value of $F(\mathbf{p})$?

    (b) If $\vec{\nabla}F(\mathbf{p}) = \mathbf{0}$, is it true that M must have a singularity at $\mathbf{p}$?

    (c) If M has a singularity at $\mathbf{p}$, is it true that $\vec{\nabla}F(\mathbf{p})$ must equal $\mathbf{0}$?

    (d) If $\mathbf{p} = (2, -3, 5)$ and $\vec{\nabla}F(\mathbf{p}) = (7, 4, -1)$, give an equation for the tangent plane to M at $\mathbf{p}$.

In each of the following exercises, you are given a continuously differentiable function F and a point $\mathbf{p}$. Show that $\vec{\nabla}F(\mathbf{p}) \neq \mathbf{0}$, and find an equation for the tangent set to the level set where $F = F(\mathbf{p})$ at the point $\mathbf{p}$.

25. $F(x_1, x_2, x_3, x_4) = x_1 x_4 - x_2 x_3$, and $\mathbf{p} = (1, 2, -1, -2)$.

26. $F(x_1, x_2, x_3, x_4) = x_1 \cos(x_1 + x_4) + x_2 \sin(x_2 + x_3)$, and $\mathbf{p} = (0, 3, -3, 0)$.

27. $F(x_1, x_2, x_3, x_4, x_5) = x_1 + x_2^2 + x_3^3 + x_4^4 + x_5^5$, and $\mathbf{p} = (1, 1, 1, 1, 1)$.

28. $F(x_1, x_2, x_3, x_4, x_5) = x_1^2 + x_2 e^{5x_3} + x_4^2 \ln x_5$, and $\mathbf{p} = (1, 1, 0, 1, 1)$.

29. In Example 2.7.4, we saw that the level curve where $F(x, y) = x^3 + y^2 - x + y = 0.135$ appears to have a singularity near $(-0.5, -0.5)$. Find the critical points of F, and decide whether or not the level curve where $F = 0.135$ does, in fact, have a singularity. If not, explain why it appears to.

30. In the above exercise, you should have found a second critical point, $\mathbf{p}$, of $F(x, y) = x^3 + y^2 - x + y$, other than the one near $(0.5, -0.5)$. Using technology, graph the corresponding level curve $F = F(\mathbf{p})$. Does this level curve appear to have a singularity at $\mathbf{p}$? If so, in what sense is the curve singular at $\mathbf{p}$? It may help to graph some level curves where $F = c$ for values of c very close to $F(\mathbf{p})$.

More Depth: In each of the following exercises, you given a real-valued, continuously differentiable function F and a point p. **(a) Parameterize three different continuously differentiable paths** $\mathbf{r} = \mathbf{r}(t)$, **inside the level set where** $F = F(\mathbf{p})$, **such that** $\mathbf{r}(0) = \mathbf{p}$ **and** $\mathbf{r}'(0) \neq \mathbf{0}$. **(b) Describe the tangent space and the tangent set, at p, to the level set where** $F = F(\mathbf{p})$.

31. $F(x, y, z) = x^2 + y^2 + z$, $\mathbf{p} = (1, 2, 3)$.

32. $F(x, y, z) = x^2 + y^2 - z^2$, $\mathbf{p} = (3, 4, 5)$.

33. $F(x, y, z) = x^2 - y^2 + z$, $\mathbf{p} = (1, 2, 3)$.

34. $F(x, y, z) = xy - \ln z$, $\mathbf{p} = (0, 2, 1)$.

35. Suppose that $F = F(\mathbf{x})$ is a real-valued, continuously differentiable function, and let $\mathbf{p}$ be a point on the level set E where $F = c$. Explain what a tangent vector to E at $\mathbf{p}$ means, and explain why the Chain Rule implies that any such tangent vector is perpendicular to $\vec{\nabla} F(\mathbf{p})$.

36. Suppose that E is a subset of $\mathbb{R}^n$ and $\mathbf{p}$ is a point in E. Explain the difference between the tangent space and the tangent set to E at $\mathbf{p}$. Give an explicit example.

37. In your own words, what does the geometric form of the Implicit Function Theorem tell you?

38. Suppose that $F = F(\mathbf{x})$ is a real-valued, continuously differentiable function of n variables, and that F has no critical points on the level set M where $F = 0$. Thus, M has no singularities. Define $G(\mathbf{x}) = \big(F(\mathbf{x})\big)^2$. Does G have any critical points on the level set N where $G = 0$? Does N have any singularities?

2.8 Parameterizing Surfaces

So far, we have looked at surfaces in $\mathbb{R}^3$ in two ways: as graphs of functions of two variables, and as level surfaces of functions of three variables. Corresponding to these two manners of describing surfaces, we had two ways of obtaining equations for tangent planes.

In this section, we will discuss a third way of describing surfaces in space; we will *parameterize* surfaces. This is an analog of parametrizing curves in the plane and in space. Moreover, given a parameterization of a surface, we will have a third method of describing the tangent plane.

This section will be particularly important to us later, in Section 3.11, after we have discussed multivariable integration, and want to calculate surface area.

Basics:

Let's leap right into an example.

Example 2.8.1. Recall from Section 1.6 that, when we parameterize a curve in $\mathbb{R}^2$ or $\mathbb{R}^3$, the most basic thing that we mean is that we specify the x-, y- and, possibly z-coordinates of points on the curve in terms of a *parameter*, that is, another variable, such as t, where t is in some interval. Something like

$$x = \cos t, \quad y = \sin t, \quad \text{and} \quad z = t/5,$$

where t is any real number. This would describe an infinite spiral in space.

In fact, since $x^2 + y^2 = \cos^2 t + \sin^2 t = 1$, all of the points on this parameterized curve lie on a right circular cylinder of radius 1; see Figure 2.8.1.

Okay. So, we know how to parameterize a curve. Can we parameterize the cylinder that the above curve lies on? What does that even mean?

At the most basic level, a parameterization of a surface in $\mathbb{R}^3$ means specifying the points on the surface in terms of **two** variables. So, for instance, one possible parameterization of the cylinder where $x^2 + y^2 = 1$ (and z can be anything) would be

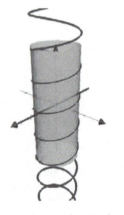

Figure 2.8.1: A spiral on a cylinder.

$$x = x(u,v) = \cos u, \quad y = y(u,v) = \sin u, \quad \text{and} \quad z = z(u,v) = v,$$

where u and v can be any real numbers. This means that v is specifying your z-coordinate, and u is specifying an angle in the copy of the xy-plane in the given z cross section.

Note that giving x, y, and z in terms of u and v is the same as giving the function

$$\mathbf{r}(u, v) \;=\; (x(u,v),\, y(u,v),\, z(u,v)) \;=\; (\cos u,\, \sin u,\, v),$$

from $\mathbb{R}^2$ to $\mathbb{R}^3$.

Hence, our most general notion of parameterizing a surface in $\mathbb{R}^3$ is to give a function from a subset of $\mathbb{R}^2$ into $\mathbb{R}^3$, and to refer to the image as the *parameterized surface*.

However, we need to impose more conditions before we can actually do any Calculus with such parameterizations. In particular, it requires more conditions on $\mathbf{r}$ to even know that the image of such a function is 2-dimensional, and so deserves to be called a "surface".

Example 2.8.2. There's no reason that we have to use u and v for our surface parameters (though they are the favorite surface parameter names).

Let's look at the surface parameterized by

$$x = x(t, \theta) = t\cos\theta + \sin\theta, \quad y = y(t,\theta) = t\sin\theta - \cos\theta, \quad \text{and} \quad z = z(t,\theta) = t,$$

i.e.,

$$\mathbf{r}(t,\theta) \;=\; (t\cos\theta + \sin\theta,\; t\sin\theta - \cos\theta,\; t),$$

where t and θ can be any real numbers. Note that, if you fix the value of θ, then the parameterization gives you a line, parameterized by t.

If you graph the image of $\mathbf{r}$, you obtain Figure 2.8.2, in which you can see the lines given by fixed values of θ. As you should notice, we labeled the graph as a "hyperboloid of one sheet, as a ruled surface". Why is the image of $\mathbf{r}$ a hyperboloid of one sheet? What is a *ruled surface*?

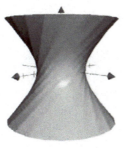

Figure 2.8.2: Hyperboloid of one sheet, as a ruled surface.

We leave it as an exercise for you to verify that

$$[x(t,\theta)]^2 + [y(t,\theta)]^2 - [z(t,\theta)]^2 \;=\; 1,$$

and so, the image of $\mathbf{r}$ is definitely contained in the hyperboloid of one sheet given by $x^2 + y^2 - z^2 = 1$. It takes more work to verify that you actually get the **entire** hyperboloid in the image, but it's true.

That the image of $\mathbf{r}$ is a ruled surface means what you may have guessed, and what we said was true in this example: when you fix one variable, the parameterization $\mathbf{r}$ gives you a parameterized line in the other variable. Continuing to use the parameters t and θ, a ruled surface is the image of a parameterization of the form

$$\mathbf{r}(t, \theta) = \mathbf{a}(\theta) + t\,\mathbf{b}(\theta).$$

Written in this form, our parameterization above is

$$\mathbf{r}(t, \theta) = (\sin\theta,\ -\cos\theta,\ 0) + t\,(\cos\theta,\ \sin\theta,\ 1).$$

Example 2.8.3. Suppose that $\mathcal{W}$ is an open subset of $\mathbb{R}^2$ and that $f = f(x, y)$ is a continuously differentiable function from $\mathcal{W}$ into $\mathbb{R}^3$. Then, we know from Remark 2.7.20 that the graph of f, call it M, is a surface which looks smooth.

Can we parameterize this surface?

Actually, graphs of functions are extremely easy to parameterize. You can let simply let x and y be the parameters. However, in order to make this look like a parameterization, we usually let $x = u$, and let $y = v$, i.e., we use as our parametrization:

$$x = u, \quad y = v, \quad \text{and} \quad z = f(u, v),$$

for all (u, v) in $\mathcal{W}$. Equivalently, our parameterization is the function

$$\mathbf{r}(u, v) = (u,\ v,\ f(u, v)).$$

In order to produce surfaces that are locally smooth or, at least, "look smooth" (i.e., are C^1 submanifolds; see Definition 2.6.9), and to be able to calculate equations for the tangent plane, we have to put some technical conditions on our surface parameterizations, similar to what we did for parameterizations of curves in Section 1.6.

We're going to define a *local regular parameterization of a surface* in $\mathbb{R}^3$ to be a continuously differentiable function from an open subset of $\mathbb{R}^2$ into $\mathbb{R}^3$, which satisfies something analogous to $\mathbf{r}'(t) \neq \mathbf{0}$ for local regular parameterizations of curves. Note that the condition that $\mathbf{r}'(t_0) \neq \mathbf{0}$ is equivalent to: if $b\mathbf{r}'(t) = \mathbf{0}$, then $b = 0$. This strange way of rewriting $\mathbf{r}'(t) \neq \mathbf{0}$ is what generalizes most nicely.

Now we're ready to define:

Definition 2.8.4. *A* **local regular parameterization of a surface** *in* $\mathbb{R}^3$ *consists of:*

1. *a non-empty, connected, open subset* $\mathcal{W}$ *in* $\mathbb{R}^2$,

2. *a continuously differentiable function* $\mathbf{r} = \mathbf{r}(u,v) = (x(u,v), y(u,v), z(u,v))$ *from* $\mathcal{W}$ *into* $\mathbb{R}^3$, *such that*

3. *at all points* (u,v) *in* $\mathcal{W}$, *the partial derivatives* $\mathbf{r}_u$ *and* $\mathbf{r}_v$ *are such that, if there are constants* a *and* b *such that* $a\mathbf{r}_u + b\mathbf{r}_v = \mathbf{0}$, *then* $a = b = 0$.

As in the case of curves, for a mathematician, a *regular curve* is **not** the image of the parameterization, but is the parameterization itself. We shall stick with the more intuitive notion of a surface as a set of points.

The **surface defined by the local regular parameterization** $\mathbf{r}$ *is simply the image of* $\mathbf{r}$.

Remark 2.8.5. In the language of linear algebra, Condition (3), above, is that, at all points in $\mathcal{W}$, $\mathbf{r}_u$ and $\mathbf{r}_v$ are linearly independent.

For two vectors, this linear independence is equivalent to: $\mathbf{r}_u \neq \mathbf{0}$ and $\mathbf{r}_v$ is not a scalar multiple of $\mathbf{r}_u$. This is the same as saying that neither $\mathbf{r}_u$ nor $\mathbf{r}_v$ is a scalar multiple of the other.

Finally, since we are dealing with two vectors in $\mathbb{R}^3$, we know from Theorem 1.5.5 that:

$\mathbf{r}_u$ and $\mathbf{r}_v$ being linearly independent is equivalent to: $\mathbf{r}_u \times \mathbf{r}_v \neq \mathbf{0}$.

This last manner of looking at linear independence is very important to us, as we'll see below. One nice feature of this characterization is that, since we're assuming that $\mathbf{r}$ is continuously differentiable, the vector $\mathbf{r}_u \times \mathbf{r}_v$ varies continuously, and so, being non-zero at one point implies that it's non-zero at all nearby points.

In other words, if $\mathbf{r}$ is a continuously differentiable function, and $\mathbf{r}_u$ and $\mathbf{r}_v$ are linearly independent at (u_0, v_0), then there exists an open neighborhood $\mathcal{W}$ of (u_0, v_0) such that the restriction of $\mathbf{r}$ to $\mathcal{W}$ is a local regular parameterization of a surface in $\mathbb{R}^3$.

Thus, if $\mathbf{r}$ is a continuously differentiable function, and $\mathbf{r}_u$ and $\mathbf{r}_v$ are linearly independent at (u_0, v_0), then we say that $\mathbf{r}$ is **regular at** (u_0, v_0) or **regular near** (u_0, v_0).

Example 2.8.6. Let's try to parameterize the cone, C, which is the graph of $z^2 = x^2 + y^2$, and see what the problem is with producing a local regular parameterization. As you might suspect, the cone point, the point where the surface isn't smooth, is the source of the difficulty.

Figure 2.8.3: The cone given by $z^2 = x^2 + y^2$.

You could try to parameterize by something like

$$\mathbf{r}(u,v) \; = \; (u,\, v,\, \sqrt{u^2 + v^2}),$$

where u and v can be all real numbers.

Certainly all of the (x, y, z) triples that you get from this satisfy $z^2 = x^2 + y^2$; so the image of this parameterization lies inside the cone C. However, the z-coordinate is always ≥ 0; you get only the top half of the cone. In many types of problems, parameterizing a surface in pieces, would be fine, and we could use one parametrization for the top half of the cone, and another for the bottom half.

However, we have another problem: as we saw in Example 2.2.15, $\sqrt{u^2 + v^2}$ is not differentiable at $(0,0)$, and so $\mathbf{r}$ is not continuously differentiable.

On the other hand, if we restrict $\mathbf{r}$ to the uv-plane with the origin removed, we obtain a local regular parameterization of the top half of the cone, minus the cone point. Furthermore, we could negate the z-coordinate in $\mathbf{r}$ to obtain a local regular parameterization of the bottom half of the cone, minus the cone point.

How about the parameterization

$$\mathbf{p}(u,v) \; = \; (u\cos v,\, u\sin v,\, u),$$

where u and v can be all real numbers?

Again, we see that all the points in the image of this parameterization satisfy $z^2 = x^2 + y^2$, so that the image definitely lies inside the cone C. In fact, this time, we get all of C in the image, since u determines the z-coordinate, and v determines the angle in the z cross section $z = u$. Furthermore, all of the component functions of $\mathbf{p}$ are smooth, and so $\mathbf{p}$ is continuously differentiable.

So, is this new parameterization $\mathbf{p}$ a local regular parameterization? No. We calculate

$$\mathbf{p}_u \; = \; (\cos v,\, \sin v,\, 1) \quad \text{and} \quad \mathbf{p}_v \; = \; (-u\sin v,\, u\cos v,\, 0).$$

Now, $\mathbf{p}_u$ is never the zero vector, but, if $u = 0$, then $\mathbf{p}_v = \mathbf{0}$ and, hence, $\mathbf{p}_v = 0\,\mathbf{p}_u$. Thus, where $u = 0$, $\mathbf{p}_u$ and $\mathbf{p}_v$ are not linearly independent, and so $\mathbf{p}$ is not regular at any point of the form $(0, b)$. Hence, $\mathbf{p}$ is not a local **regular** parameterization.

Notice that, if $u = 0$, then, regardless of the value of v, $\mathbf{p}(u, v) = \mathbf{0}$. Once again, the problem point on the surface is the non-smooth point at the origin.

And again, if we delete the cone point, our problems go away, though we must use two parameterizations if we want to have connected domains. For $u > 0$ (and v is any real number), the restriction of $\mathbf{p}$ yields a local regular parameterization of the top half of the cone, minus the cone point. For $u < 0$, the restriction of $\mathbf{p}$ yields a local regular parameterization of the bottom half of the cone, minus the cone point.

Example 2.8.7. In the previous example, we had a surface, and we wanted to produce a parameterization of it. However, frequently, you don't start with a surface given to you in some other way; you start with a parameterization, and use it to **define** a surface.

Let's consider the parameterization

$$\mathbf{r}(u, v) \; = \; (u^2 - v, v^3 + u, uv).$$

This is a smooth (C^∞) function. We calculate

$$\mathbf{r}_u \; = \; (2u, 1, v) \quad \text{and} \quad \mathbf{r}_v \; = \; (-1, 3v^2, u).$$

At the origin in $\mathbb{R}^2$, we find

$$\mathbf{r}_u(0, 0) \; = \; (0, 1, 0) \quad \text{and} \quad \mathbf{r}_v(0, 0) \; = \; (-1, 0, 0).$$

Clearly, $\mathbf{r}_u(0, 0)$ and $\mathbf{r}_v(0, 0)$ are linearly independent, and so this must also be true for all (u, v) in some (small) open neighborhood of the origin in $\mathbb{R}^2$.

Therefore, there exists an open neighborhood $\mathcal{W}$ of $\mathbf{0}$ in $\mathbb{R}^2$ such that the restriction $\mathbf{r}_{|_\mathcal{W}}$ of $\mathbf{r}$ to $\mathcal{W}$ is a local regular parameterization. In Figure 2.8.4, we have given the graph of the image of $\mathbf{r}$ restricted to where $-0.5 < u < 0.5$ and $-0.5 < v < 0.5$.

Figure 2.8.4: The image of $\mathbf{r}$ restricted to $|u| < 0.5$ and $|v| < 0.5$.

We want a theorem which tells us that, in some sense, the surfaces given by local regular parameterizations look smooth, and we want to be able to describe the tangent plane to such surfaces at "smooth" points.

Suppose that you have a local regular parameterization $\mathbf{r} = \mathbf{r}(u, v)$ of a surface M in $\mathbb{R}^3$, and $\mathbf{p} = \mathbf{r}(u_0, v_0)$. Then, we obtain two special curves in M, given by regular parameterizations of curves (recall Definition 1.6.10 and Theorem 1.6.11); namely, $\boldsymbol{\alpha}(t) = \mathbf{r}(u_0 + t, v_0)$ and $\boldsymbol{\beta}(t) = \mathbf{r}(u_0, v_0 + t)$ are both parameterizations of curves in M, both are regular at $t = 0$, and $\boldsymbol{\alpha}(0) = \boldsymbol{\beta}(0) = \mathbf{p}$. Thus, the vectors $\boldsymbol{\alpha}'(0)$ and $\boldsymbol{\beta}'(0)$ are tangent vectors to M at $\mathbf{p}$.

Recall the definition of a tangent vector to a surface or a more general subset of $\mathbb{R}^n$ from Theorem 2.7.7 and Definition 1.6.24.

Now, from the definition of the partial derivatives, $\boldsymbol{\alpha}'(0) = \mathbf{r}_u(u_0, v_0)$ and $\boldsymbol{\beta}'(0) = \mathbf{r}_v(u_0, v_0)$ and, since we're assuming that $\mathbf{r}_u(u_0, v_0)$ and $\mathbf{r}_v(u_0, v_0)$ are linearly independent, a vector $\mathbf{v}$ is perpendicular to both $\mathbf{r}_u(u_0, v_0)$ and $\mathbf{r}_v(u_0, v_0)$ if and only if $\mathbf{v}$ is a scalar multiple of the cross product $\mathbf{r}_u(u_0, v_0) \times \mathbf{r}_v(u_0, v_0)$.

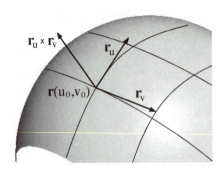

Figure 2.8.5: A parameterized surface, with curves where u and v are constant, and the tangent vectors $\mathbf{r}_u$ and $\mathbf{r}_v$.

In fact, we can say much more. The following theorem follows easily from the Inverse Function Theorem, Theorem 2.6.6, just as it does for regular curves.

Theorem 2.8.8. *Suppose that $1 \leq k \leq \infty$, E is a subset of $\mathbb{R}^2$, and that (u_0, v_0) is a point in the interior of E. Suppose that $\hat{\mathbf{r}}$ is a function from E into $\mathbb{R}^3$, which is of class C^k on an open neighborhood of (u_0, v_0) and is such that $\hat{\mathbf{r}}_u(u_0, v_0)$ and $\hat{\mathbf{r}}_v(u_0, v_0)$ are linearly independent, i.e., is such that $\hat{\mathbf{r}}_u(u_0, v_0) \times \hat{\mathbf{r}}_v(u_0, v_0) \neq \mathbf{0}$.*

Then, there exists an open neighborhood $\mathcal{W}$ of (u_0, v_0) in E such that the restriction $\mathbf{r}$ of $\hat{\mathbf{r}}$ to $\mathcal{W}$ is one-to-one, and the surface M, parameterized by the local regular parameterization $\mathbf{r}$, is a 2-dimensional C^k submanifold of $\mathbb{R}^3$, i.e., a C^k surface in $\mathbb{R}^3$.

> The theorem refers to a function of class C^k, where $k \geq 1$, even though we have defined a local regular parameterization as needing to be only continuously differentiable, i.e., of class C^1. The point is that, if the function is even more differentiable than C^1, then you know the analogous strong result for the class of the submanifold defined by the parameterization.

Finally, if we let $\mathbf{p} = \mathbf{r}(u_0, v_0)$, then the set of tangent vectors to M at $\mathbf{p}$ is the set of all vectors of the form

$$a\mathbf{r}_u(u_0, v_0) + b\mathbf{r}_v(u_0, v_0),$$

where a and b are constants. This is precisely the plane of vectors in $\mathbb{R}^3$ which contains the origin, and which is normal to $\mathbf{r}_u(u_0, v_0) \times \mathbf{r}_v(u_0, v_0)$.

However, we want to look at the tangent **set** – here, the tangent plane – which contains the point $\mathbf{p}$. As we discussed in the previous section, to transport the tangent vectors to the point $\mathbf{p}$ at which they're tangent to M, that is, to define the tangent set, we must add the point $\mathbf{p}$ to each of the tangent vectors.

Definition 2.8.9. *Suppose we have a local regular parameterization $\mathbf{r}$, from $\mathcal{W}$ into $\mathbb{R}^3$, and that (u_0, v_0) is a point in $\mathcal{W}$. Let $\mathbf{p} = \mathbf{r}(u_0, v_0)$.*

*The **tangent space** of $\mathbf{r}$ at (u_0, v_0) is the set of all vectors of the form*

$$a\mathbf{r}_u(u_0, v_0) + b\mathbf{r}_v(u_0, v_0),$$

where a and b are constants.

*The **tangent plane** of $\mathbf{r}$ at (u_0, v_0) consists of the points (x, y, z) of the form*

$$(x, y, z) = \mathbf{p} + a\mathbf{r}_u(u_0, v_0) + b\mathbf{r}_v(u_0, v_0),$$

where a and b are constants. This is the plane containing the point $\mathbf{p}$, which has a non-zero normal vector given by the cross product $\mathbf{r}_u(u_0, v_0) \times \mathbf{r}_v(u_0, v_0)$.

If you restrict $\mathbf{r}$ to a small enough open set around (u_0, v_0) so that you have a one-to-one function whose image is a C^1 submanifold M, then, naturally, we refer to the *tangent space of M at* $\mathbf{p}$ and the **tangent plane to** M **at** $\mathbf{p}$.

Theorem 2.8.8 implies that these notions of the tangent space and tangent plane agree with those in Definition 2.3.8 and Definition 2.6.16, when you look at the surface M as a graph of a function or as a level surface.

Remark 2.8.10. Note that

$$(x, y, z) = \mathbf{p} + a\mathbf{r}_u(u_0, v_0) + b\mathbf{r}_v(u_0, v_0)$$

is, itself, a parameterization of the tangent plane, with parameters a and b.

On the other hand, if we let $\mathbf{n} = \mathbf{r}_u(u_0, v_0) \times \mathbf{r}_v(u_0, v_0)$, then

$$\mathbf{n} \cdot \big((x, y, z) - \mathbf{p}\big) = 0$$

describes the tangent plane as a level surface.

Example 2.8.11. Let's look again at Example 2.8.7. We had the parameterization

$$\mathbf{r}(u, v) = (u^2 - v, v^3 + u, uv).$$

and calculated

$$\mathbf{r}_u = (2u, 1, v) \quad \text{and} \quad \mathbf{r}_v = (-1, 3v^2, u).$$

At the origin in $\mathbb{R}^2$, we found

$$\mathbf{r}_u(0, 0) = (0, 1, 0) \quad \text{and} \quad \mathbf{r}_v(0, 0) = (-1, 0, 0).$$

Figure 2.8.6: The image of $\mathbf{r}$ restricted to $|u| < 0.5$ and $|v| < 0.5$.

Since $\mathbf{r}_u(0,0)$ and $\mathbf{r}_v(0,0)$ are linearly independent, we concluded that there exists an open neighborhood $\mathcal{W}$ of $\mathbf{0}$ in $\mathbb{R}^2$ such that the restriction $\mathbf{r}_{|\mathcal{W}}$ of $\mathbf{r}$ to $\mathcal{W}$ is a local regular parameterization.

Theorem 2.8.8 allows us to conclude more. In fact, the neighborhood $\mathcal{W}$ can also be chosen so that the image of $\mathbf{r}_{|\mathcal{W}}$, is a smooth surface, M, as you can see in Figure 2.8.6. Furthermore, it tells us that we can describe the tangent plane to M (i.e., the tangent plane of $\mathbf{r}$) at $\mathbf{r}(0,0) = (0,0,0)$ in two different ways: parametrically, by

$$(x, y, z) \; = \; (0, 0, 0) + a\mathbf{r}_u(0,0) + b\mathbf{r}_v(0,0) \; = \; a(0, 1, 0) + b(-1,0,0) \; = \; (-b, a, 0),$$

or, after calculating $\mathbf{n} = (0, 1, 0) \times (-1, 0, 0) = (0, 0, 1) = \mathbf{k}$, as the level surface defined by

$$(0, 0, 1) \cdot \big((x, y, z) - (0, 0, 0) \big) \; = \; 0, \quad \text{i.e.,} \quad z = 0.$$

The tangent plane of $\mathbf{r}$ at $(0,0)$ is kind of boring. Let's look at the tangent plane of $\mathbf{r}$ at $(1,1)$. We find

$$\mathbf{r}(1,1) = (0,2,1), \quad \mathbf{r}_u(1,1) \; = \; (2,1,1), \quad \text{and} \quad \mathbf{r}_v(1,1) \; = \; (-1,3,1).$$

We calculate

$$\mathbf{r}_u(1,1) \times \mathbf{r}_v(1,1) \; = \; (2,1,1) \times (-1,3,1) \; = \; (-2,-3,7).$$

Since this cross product is not $\mathbf{0}$, the tangent plane of $\mathbf{r}$ at $(1,1)$ exists, and is given parametrically by

$$(x, y, z) \; = \; \mathbf{r}(1,1) + a\mathbf{r}_u(1,1) + b\mathbf{r}_v(1,1),$$

that is,

$$(x, y, z) \; = \; (0,2,1) + a(2,1,1) + b(-1,3,1).$$

The tangent plane as a level set is given by

$$(-2, -3, 7) \cdot \big((x, y, z) - (0, 2, 1) \big) \; = \; 0,$$

i.e.,

$$-2x - 3(y-2) + 7(z-1) \; = \; 0.$$

We should point out that, on a larger domain, the image of $\mathbf{r}$ is definitely **not** even a C^1 submanifold of $\mathbb{R}^3$; the graph crosses itself, and there is a single point (u_0, v_0) at which $\mathbf{r}$ is not regular, i.e., at which $\mathbf{r}_u$ and $\mathbf{r}_v$ are **not** linearly independent. We leave it as an exercise for you to find this unique point. You can see the corresponding point in the image in Figure 2.8.7; it's above the fourth quadrant, where the graph looks "pinched".

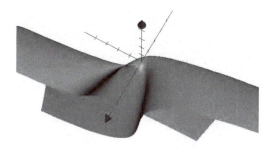

Figure 2.8.7: The image of $\mathbf{r}$ restricted to $|u| < 4$ and $|v| < 4$.

Example 2.8.12. Let's return to the case of the graph of a continuously differentiable function $f = f(x, y)$, which is defined on a connected, open subset of $\mathbb{R}^2$.

As we discussed in Example 2.8.3, we can easily parameterize the graph of f, which is where $z = f(x, y)$, by letting

$$x = u, \quad y = v, \quad \text{and} \quad z = f(u, v),$$

i.e., by letting

$$\mathbf{r}(u, v) = (u, v, f(u, v)).$$

It's easy to see that $\mathbf{r}$ is one-to-one. In addition, the image of $\mathbf{r}$ is the graph of $z = f(x, y)$, and Remark 2.7.20 tells us that this set, M, is a C^1 submanifold of $\mathbb{R}^3$, and so the graph looks smooth.

Furthermore, as we discussed in Remark 2.7.20 and defined in Definition 2.3.8, the tangent plane to M at $\mathbf{p}$ is the graph of

$$z = f(\mathbf{p}) + \vec{\nabla} f(\mathbf{p}) \cdot \big((x, y) - \mathbf{p}\big).$$

Do we gain any new information from Theorem 2.8.8? Well...not really. As we stated before, all of the notions of tangent plane give you the same thing for submanifolds. Nonetheless, the precise descriptions of the tangent plane may look a bit different from each other.

Considering M as a parameterized surface, we proceed as follows:

The vectors

$$\frac{\partial \mathbf{r}}{\partial u} = \mathbf{r}_u = (1, 0, f_u) \quad \text{and} \quad \frac{\partial \mathbf{r}}{\partial v} = \mathbf{r}_v = (0, 1, f_v)$$

are linearly independent at all points. Thus, at any point $\mathbf{p} = (u, v) = (x_0, y_0)$, we have

$$\mathbf{r}(\mathbf{p}) \;=\; (\mathbf{p}, f(\mathbf{p})), \quad \mathbf{r}_u(\mathbf{p}) \;=\; (1, 0, f_u(\mathbf{p})), \quad \text{and} \quad \mathbf{r}_v(\mathbf{p}) \;=\; (1, 0, f_v(\mathbf{p})),$$

and the tangent plane of $\mathbf{r}$ at $\mathbf{p}$, which is the same as the tangent plane to M at $\mathbf{p}$, is parameterized by

$$(x, y, z) \;=\; (\mathbf{p}, f(\mathbf{p})) + a(1, 0, f_u(\mathbf{p})) + b(0, 1, f_v(\mathbf{p})),$$

that is,

$$x = x_0 + a, \quad y = y_0 + b, \quad \text{and} \quad z = f(\mathbf{p}) + a f_u(\mathbf{p}) + b f_v(\mathbf{p}).$$

If we rewrite the first two equations above as $a = x - x_0$ and $b = y - y_0$, then the last equation above can be written as

$$z = f(\mathbf{p}) + \vec{\nabla} f(\mathbf{p}) \cdot (a, b) = f(\mathbf{p}) + \vec{\nabla} f(\mathbf{p}) \cdot (x - x_0, y - y_0) = f(\mathbf{p}) + \vec{\nabla} f(\mathbf{p}) \cdot \big((x, y) - \mathbf{p}\big),$$

which is what we had before.

You can also verify that you get the same equation by considering

$$\mathbf{n} \;=\; \mathbf{r}_u(\mathbf{p}) \times \mathbf{r}_v(\mathbf{p}) \;=\; (-f_u(\mathbf{p}), -f_v(\mathbf{p}), 1),$$

and looking at the tangent plane as the level set where

$$(-f_u(\mathbf{p}), -f_v(\mathbf{p}), 1) \cdot \big((x, y, z) - (\mathbf{p}, f(\mathbf{p}))\big) \;=\; 0.$$

More Depth:

Remark 2.8.13. It is important to note that the surface which is the image of a local regular parameterization $\mathbf{r}$ need **not** be a C^1 submanifold. The surface can easily cross itself, since $\mathbf{r}$ need not be one-to-one. Only after you restrict $\mathbf{r}$ to small open neighborhoods in the domain do you know that image (of the restricted) function is a submanifold of $\mathbb{R}^3$.

Even if $\mathbf{r}$ is a one-to-one, local regular parameterization, the image need not be a submanifold of $\mathbb{R}^3$; the surface can come infinitesimally close to itself at various points. We saw the analogous problem for curves in Exercise 50 in Section 1.6, and we can modify that example to produce a surface example:

The surface in Figure 2.8.8 is the image of the one-to-one local regular parameterization

$$\mathbf{r}(u, v) \;=\; \big(u, \, (1 - e^{-v})^2 - 1, \, -(1 - e^{-v})\big((1 - e^{-v})^2 - 1\big)\big).$$

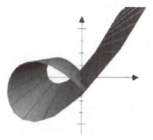

Figure 2.8.8: The image of a one-to-one local regular parameterization need not be a submanifold.

Example 2.8.14. Let's look again at the swallowtail function

$$\mathbf{p}(u, v) \;=\; \left(u, \; -2v^3 + uv, \; 3v^4 - uv^2\right),$$

from Example 2.1.28 and Exercise 49 in Section 2.1.

In Example 2.1.28, we showed that

$$\mathbf{p}_u \times \mathbf{p}_v \;=\; (-6v^2 + u)(-v^2\,\mathbf{i} + 2v\,\mathbf{j} + \mathbf{k})$$

and, thus, concluded that the cross product is $\mathbf{0}$ if and only if $-6v^2 + u = 0$, i.e., $u = 6v^2$.

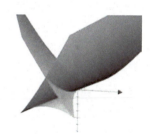

Figure 2.8.9: The surface parameterized by $\mathbf{p}$ is the swallowtail.

Now we know what this calculation is telling us. It tells us that $\mathbf{p}$ is a local regular parametrization, except along the points where $u = 6v^2$.

What about the obvious curve of points where the swallowtail crosses itself? These are not the result of $\mathbf{p}$ not being regular; these result from $\mathbf{p}$ not being one-to-one. Recall that, in Exercise 49 in Section 2.1, you were asked to show that the curve along which the swallowtail crosses itself, the image of the points where $\mathbf{p}$ is not one-to-one, is the curve given by $z = x^2/4$ and $y = 0$, i.e., a parabola in the xz-plane.

We can certainly parameterize surfaces inside $\mathbb{R}^n$, where $n > 3$. We can't picture the image of the parameterized surface in those high-dimensional cases, and the tangent plane is no longer a level hypersurface, but, aside from that, the results are the "same". All that we've changed below is: we changed $\mathbb{R}^3$ to $\mathbb{R}^n$, and we dropped all references to the cross product.

Theorem 2.8.15. *Suppose that $1 \leq k \leq \infty$, E is a subset of $\mathbb{R}^2$, and that (u_0, v_0) is a point in the interior of E. Suppose that $\hat{\mathbf{r}}$ is a function from E into $\mathbb{R}^n$, which is of class C^k on an open neighborhood of (u_0, v_0) and is such that $\hat{\mathbf{r}}_u(u_0, v_0)$ and $\hat{\mathbf{r}}_v(u_0, v_0)$ are linearly independent.*

Then, there exists an open neighborhood $\mathcal{W}$ of (u_0, v_0) in E such that the restriction $\mathbf{r}$ of $\hat{\mathbf{r}}$ to $\mathcal{W}$ is one-to-one, and the surface M, parameterized by the local regular parameterization $\mathbf{r}$, is a 2-dimensional C^k submanifold of $\mathbb{R}^n$, i.e., a C^k surface in $\mathbb{R}^n$.

Finally, if we let $\mathbf{p} = \mathbf{r}(u_0, v_0)$, then the set of tangent vectors to M at $\mathbf{p}$ is the set of all vectors of the form

$$a\mathbf{r}_u(u_0, v_0) + b\mathbf{r}_v(u_0, v_0),$$

where a and b are constants.

Definition 2.8.16. *Suppose we have a local regular parameterization $\mathbf{r}$, from $\mathcal{W}$ into $\mathbb{R}^n$, and that (u_0, v_0) is a point in $\mathcal{W}$. Let $\mathbf{p} = \mathbf{r}(u_0, v_0)$.*

*The **tangent space** of $\mathbf{r}$ at (u_0, v_0) is the set of all vectors of the form*

$$a\mathbf{r}_u(u_0, v_0) + b\mathbf{r}_v(u_0, v_0),$$

where a and b are constants.

*The **tangent plane** of $\mathbf{r}$ at (u_0, v_0) is the set of points $\mathbf{x} = (x_1, \ldots, x_n)$ of the form*

$$\mathbf{x} = \mathbf{p} + a\mathbf{r}_u(u_0, v_0) + b\mathbf{r}_v(u_0, v_0),$$

where a and b are constants.

Example 2.8.17. Find a parameterization, at $(1, -1)$, of the tangent plane of

$$\mathbf{r}(u, v) = (ue^v, \, ve^{-u}, \, u^2 - v^2, \, uv).$$

Solution:

We find

$$\mathbf{r}_u = (e^v, \, -ve^{-u}, \, 2u, \, v) \quad \text{and} \quad \mathbf{r}_v = (ue^v, \, e^{-u}, \, -2v, \, u),$$

and so,

$$\mathbf{r}_u(1,-1) \;=\; (e^{-1}, e^{-1}, 2, -1) \quad \text{and} \quad \mathbf{r}_v(1,-1) \;=\; (e^{-1}, e^{-1}, 2, 1).$$

If you look at the first three components, $\mathbf{r}_u(1,-1)$ and $\mathbf{r}_v(1,-1)$ are the same; hence, if one were a scalar multiple of the other, that scalar would have to be 1. But the last components of each are different; so the vectors are not scalar multiples of each other, i.e., they are linearly independent.

Therefore, in an open neighborhood of $(1,-1)$, the image of $\mathbf{r}$ is a smooth surface in $\mathbb{R}^4$, and the tangent plane to this smooth surface at $\mathbf{r}(1,-1) = (e^{-1}, -e^{-1}, 0, -1)$ is parameterized by

$$(w, x, y, z) \;=\; (e^{-1}, -e^{-1}, 0, -1) \;+\; a(e^{-1}, e^{-1}, 2, -1) \;+\; b(e^{-1}, e^{-1}, 2, 1).$$

+ Linear Algebra:

Theorem 2.8.15 is a theorem about functions from subsets of $\mathbb{R}^2$ into $\mathbb{R}^n$, and so, is a theorem about surfaces in any dimension. In fact, we don't need to start with subsets of $\mathbb{R}^2$. The only real added difficulty is that linear independence of more than two vectors is more difficult to define and to check.

A condition equivalent to a collection of j vectors in $\mathbb{R}^n$ being linearly independent is that the matrix which has the vectors as its columns (or rows) has rank j. This enables us to state things nicely in terms of the Jacobian matrix (recall Definition 2.2.22).

Definition 2.8.18. *A* **local, regular parameterization** *into $\mathbb{R}^n$ consists of:*

1. *a non-empty, connected, open subset $\mathcal{W}$ in $\mathbb{R}^j$,*

2. *a continuously differentiable function $\mathbf{r} = \mathbf{r}(\mathbf{w})$ from $\mathcal{W}$ into $\mathbb{R}^n$, such that*

3. *at all points $\mathbf{w}$ in $\mathcal{W}$, the $n \times j$ Jacobian matrix of $\mathbf{r}$ at $\mathbf{w}$ has rank j.*

Note that this rank condition implies, in particular, that $j \le n$.

The proof of the following theorem, once again, follows quickly from the Inverse Function Theorem, Theorem 2.6.6.

Theorem 2.8.19. *Suppose that $1 \le k \le \infty$, E is a subset of $\mathbb{R}^j$, and that $\mathbf{w}$ is a point in the interior of E. Suppose that $\hat{\mathbf{r}} = \hat{\mathbf{r}}(u_1, \ldots, u_j)$ is a function from E into $\mathbb{R}^n$, which is of class C^k on an open neighborhood of $\mathbf{w}$ and is such that the $n \times j$ Jacobian matrix of $\hat{\mathbf{r}}$ at $\mathbf{w}$ has rank j.*

Then, there exists an open neighborhood $\mathcal{W}$ of $\mathbf{w}$ in E such that the restriction $\mathbf{r}$ of $\hat{\mathbf{r}}$ to $\mathcal{W}$ is a one-to-one local, regular parameterization, and the set, M, parameterized by $\mathbf{r}$, is a j-dimensional C^k submanifold of $\mathbb{R}^n$.

Finally, if we let $\mathbf{p} = \mathbf{r}(\mathbf{w})$, then the set of tangent vectors to M at $\mathbf{p}$ is the set of all vectors in the column space *of the Jacobian matrix $[d_{\mathbf{w}}\mathbf{r}]$, which is the image of the linear map $d_{\mathbf{w}}\mathbf{r}$.*

Definition 2.8.20. *Suppose we have a local, regular parameterization $\mathbf{r}$, from $\mathcal{W}$ into $\mathbb{R}^n$, and that $\mathbf{w}$ is a point in $\mathcal{W}$. Let $\mathbf{p} = \mathbf{r}(\mathbf{w})$.*

*The **tangent space** of $\mathbf{r}$ at $\mathbf{w}$ is the column space of the Jacobian matrix $[d_{\mathbf{w}}\mathbf{r}]$, which is the image of the linear map $d_{\mathbf{w}}\mathbf{r}$*

*The **tangent set** of $\mathbf{r}$ at $\mathbf{w}$ is the set of points $\mathbf{x} = (x_1, \ldots, x_n)$ such that $\mathbf{x} - \mathbf{p}$ is in the tangent space of $\mathbf{r}$ at $\mathbf{w}$, i.e., $\mathbf{x}$ is in the tangent set if and only if there exists a k-dimensional column vector $[\mathbf{a}]$ such that*

$$[\mathbf{x}] = [\mathbf{p}] + [d_{\mathbf{w}}\mathbf{r}][\mathbf{a}].$$

Example 2.8.21. Consider the parameterization

$$\mathbf{r}(u, v, w) = (u^2 - v,\ w\ln v,\ e^{w-u},\ \sin(uv),\ uvw),$$

from $\mathbb{R}^3$ into $\mathbb{R}^5$. Show that $\mathbf{r}$ is a local, regular parameterization in an open neighborhood of $(0, 1, 2)$, and find a parameterization of the tangent set of $\mathbf{r}$ at $(0, 1, 2)$.

Solution:

First, we find $\mathbf{r}(0, 1, 2) = (1, 0, e^2, 0, 0)$.

The Jacobian matrix of $\mathbf{r}$ is

$$[d\mathbf{r}] = \begin{bmatrix} 2u & -1 & 0 \\ 0 & w/v & \ln v \\ -e^{w-u} & 0 & e^{w-u} \\ v\cos(uv) & u\cos(uv) & 0 \\ vw & uw & uv \end{bmatrix}.$$

Note that $\mathbf{r}_u$, $\mathbf{r}_v$, and $\mathbf{r}_w$ form the columns of $[d\mathbf{r}]$.

The Jacobian matrix of $\mathbf{r}$, evaluated at $(0,1,2)$, is

$$
\left[d_{(0,1,2)}\mathbf{r}\right] = \left.\begin{bmatrix} 2u & -1 & 0 \\ 0 & w/v & \ln v \\ -e^{w-u} & 0 & e^{w-u} \\ v\cos(uv) & u\cos(uv) & 0 \\ vw & uw & uv \end{bmatrix}\right|_{(0,1,2)} = \begin{bmatrix} 0 & -1 & 0 \\ 0 & 2 & 0 \\ -e^2 & 0 & e^2 \\ 1 & 0 & 0 \\ 2 & 0 & 0 \end{bmatrix}.
$$

It is easy to see that this matrix has rank 3, i.e., that the columns are linearly independent, for the third column is not the zero vector, the second column is not a scalar multiple of the third, because of the top two non-zero entries in the second column; the first column is not a linear combination of the second and third, because of the bottom two non-zero entries in the first column.

Hence, Theorem 2.8.19 tells us that there exists an open neighborhood $\mathcal{W}$ of $(0,1,2)$ in $\mathbb{R}^3$ such that the restriction $\mathbf{r}$ of $\hat{\mathbf{r}}$ to $\mathcal{W}$ is one-to-one, and the set, M, parameterized by the local regular parameterization $\mathbf{r}$, is a 3-dimensional smooth submanifold of $\mathbb{R}^5$, and the tangent set to this manifold, at $\mathbf{r}(0,1,2)$, which is the tangent set of $\mathbf{r}$ at $(0,1,2)$, is parameterized by

$$
\begin{bmatrix} x_1 \\ x_2 \\ x_3 \\ x_4 \\ x_5 \end{bmatrix} = \begin{bmatrix} 1 \\ 0 \\ e^2 \\ 0 \\ 0 \end{bmatrix} + a_1 \begin{bmatrix} 0 \\ 0 \\ -e^2 \\ 1 \\ 2 \end{bmatrix} + a_2 \begin{bmatrix} -1 \\ 2 \\ 0 \\ 0 \\ 0 \end{bmatrix} + a_3 \begin{bmatrix} 0 \\ 0 \\ e^2 \\ 0 \\ 0 \end{bmatrix}.
$$

2.8.1 Exercises

Online answers to select exercises are here.

Basics:

In each of the following exercises, you are given a continuously differentiable parameterization $\mathbf{r} = \mathbf{r}(u,v)$, and a point (u_0, v_0). **(a)** Determine whether or not $\mathbf{r}$ is regular at (u_0, v_0) and, if it is, **(b)** describe the tangent plane of $\mathbf{r}$ at (u_0, v_0) parametrically, and **(c)** describe the tangent plane as a level set.

1. $\mathbf{r}(u,v) = (u^2 + v^2, u, v)$ and $(u_0, v_0) = (1,2)$.

2. $\mathbf{r}(u,v) = (u, u\cos v, u\sin v)$ and $(u_0, v_0) = (1,0)$.

3. $\mathbf{r}(u,v) = (5u + 3v + 1, u - v + 2, 2u + v - 4)$ and $(u_0, v_0) = (-1, 2)$.

4. $\mathbf{r}(u,v) = (5u + 3v + 1, u - v + 2, 2u + v - 4)$ and $(u_0, v_0) = (6, -7)$.

5. $\mathbf{r}(u,v) = (uv, \ln u, uv \ln u)$ and $(u_0, v_0) = (e, 1)$.

6. $\mathbf{r}(u,v) = (u, v^2, \tan^{-1}(u))$ and $(u_0, v_0) = (\pi/4, 1)$.

7. $\mathbf{r}(u,v) = (e^u \cos v, e^u \sin v, u)$ and $(u_0, v_0) = (0, 0)$.

8. $\mathbf{r}(u,v) = (ue^{5v}, 2u^3, u^3 + ue^{5v} + 1)$ and $(u_0, v_0) = (1, 0)$.

In the following exercises, you are given the same parameterizations $\mathbf{r} = \mathbf{r}(u,v)$, and points (u_0, v_0) as in the previous exercises. In each exercise, sketch by hand, or using technology, the image of $\mathbf{r}$, near $\mathbf{r}(u_0, v_0)$, and sketch in the tangent vectors $\mathbf{r}_u(u_0, v_0)$ and $\mathbf{r}_v(u_0, v_0)$, based at $\mathbf{r}(u_0, v_0)$.

9. $\mathbf{r}(u,v) = (u^2 + v^2, u, v)$ and $(u_0, v_0) = (1, 2)$.

10. $\mathbf{r}(u,v) = (u, u \cos v, u \sin v)$ and $(u_0, v_0) = (1, 0)$.

11. $\mathbf{r}(u,v) = (5u + 3v + 1, u - v + 2, 2u + v - 4)$ and $(u_0, v_0) = (-1, 2)$.

12. $\mathbf{r}(u,v) = (5u + 3v + 1, u - v + 2, 2u + v - 4)$ and $(u_0, v_0) = (6, -7)$.

13. $\mathbf{r}(u,v) = (uv, \ln u, uv \ln u)$ and $(u_0, v_0) = (e, 1)$.

14. $\mathbf{r}(u,v) = (u, v^2, \tan^{-1}(u))$ and $(u_0, v_0) = (\pi/4, 1)$.

15. $\mathbf{r}(u,v) = (e^u \cos v, e^u \sin v, u)$ and $(u_0, v_0) = (0, 0)$.

16. $\mathbf{r}(u,v) = (ue^{5v}, 2u^3, u^3 + ue^{5v} + 1)$ and $(u_0, v_0) = (1, 0)$.

In each of the following exercises, you are given an equation which defines a surface in $\mathbb{R}^3$. (a) Find a continuously differentiable parameterization $(x, y, z) = \mathbf{r}(u, v)$ of the surface. (b) Determine whether or not your parameterization is a local regular parameterization.

17. $z = x^2 + y^2$.

18. $x^3 - y^2 - z^2 = 1$.

19. $y = \sin x$. (Remember: we are looking at surfaces in $\mathbb{R}^3$.)

20. $x^2 + z^2 = 9$.

21. $x^2 + y^2 = e^z$.

22. $5x^2 + 3y^2 = z^2$.

23. $2x^2 z^2 + y^3 - x^4 - z^4 = 3$.

24. $y^2 - x^3 - z^2 x^2 = 0$. (Hint: Look for a parameterization in which $y = ux$.)

   ▶

In each of the following exercises, you are given a parameterization $\mathbf{r}(\theta, t) = \mathbf{a}(\theta) + t\,\mathbf{b}(\theta)$ of a ruled surface and a point (θ_0, t_0). (a) Show that $\mathbf{r}$ is regular at (θ_0, t_0), and parameterize the tangent plane of $\mathbf{r}$ at (θ_0, t_0). (b) Sketch, or have technology sketch, the surface which is the image of $\mathbf{r}$ near $\mathbf{r}(\theta_0, t_0)$, and include several lines from the ruling.

25. $\mathbf{r}(\theta, t) = (2\theta, -3\theta + 1, 5\theta - 2) + t(1, 2, 3)$ and $(\theta_0, t_0) = (1, 0)$.   ▶

26. $\mathbf{r}(\theta, t) = (\cos\theta, \sin\theta, 0) + t(0, 0, \theta)$ and $(\theta_0, t_0) = (\pi, 0)$.

27. $\mathbf{r}(\theta, t) = (\theta, \theta^2, \theta^3) + t(\theta^3, \theta^4, \theta^5)$ and $(\theta_0, t_0) = (1, 2)$.

28. $\mathbf{r}(\theta, t) = (\cos\theta, \sin\theta, \theta) + t(-\sin\theta, \cos\theta, 1)$ and $(\theta_0, t_0) = (0, 1)$.

29. In Example 2.8.11, we had the parameterization

$$\mathbf{r}(u, v) = (u^2 - v, v^3 + u, uv),$$

   and said that there is a unique point (u_0, v_0) at which $\mathbf{r}$ is not regular. Find that point.

30. Consider a ruled surface, parameterized by $\mathbf{r}(\theta, t) = \mathbf{a}(\theta) + t\mathbf{b}(\theta)$, where $\mathbf{a}$ and $\mathbf{b}$ are continuously differentiable. Show that $\mathbf{r}$ is regular at $(0, 0)$ if and only if $\mathbf{b}(0) \neq \mathbf{0}$ and $\mathbf{a}'(0)$ is not a scalar multiple of $\mathbf{b}(0)$.   ▶

| **More Depth:** |

In the following exercises, you are given a parameterization $(x, y, z) = \mathbf{r}(u, v)$. Determine all of those triples (a, b, c) in the image of $\mathbf{r}$ which come from, at least, two different (u, v) pairs.

31. $\mathbf{r}(u, v) = (u^2 - v, u(u^2 - v), v)$.   ▶

32. $\mathbf{r}(u, v) = (u, v^2 - 1, v(v^2 - 1))$.

33. $\mathbf{r}(u, v) = (uv(u^2 - 1)(v^2 - 1), u^2, v^2)$.

34. $\mathbf{r}(u, v) = (3 - u^{-2}, (3 - u^{-2})u^{-1}, v)$.

In each of the following exercises, you are given a parameterization $\mathbf{r}(u, v)$ of a surface in $\mathbb{R}^n$, where $n \geq 4$, and a point (u_0, v_0) at which $\mathbf{r}$ is regular. Give a parameterization of the tangent plane of $\mathbf{r}$ at (u_0, v_0).

35. $\mathbf{r}(u,v) = (u^2 + v, v^3 + u, e^u \cos v, e^u \sin v)$ and $(u_0, v_0) = (0,0)$.

36. $\mathbf{r}(u,v) = (u^3, 3u^2v, 3uv^2, v^3)$ and $(u_0, v_0) = (1,1)$.

37. $\mathbf{r}(u,v) = (u\ln v, ve^u, v^3 + \tan^{-1} u, u + v, \sin(uv))$ and $(u_0, v_0) = (0,1)$.

38. $\mathbf{r}(u,v) = (u^7, u^6v, u^5v^2, u^4v^3, u^3v^4, u^2v^5)$ and $(u_0, v_0) = (1,1)$.

39. Discuss the various notions of the tangent plane to a surface in $\mathbb{R}^3$.

40. Suppose that $\mathbf{r} = \mathbf{r}(u,v)$ is a continuously differentiable function into $\mathbb{R}^3$, which is regular at (u_0, v_0). Suppose that Δu and Δv are positive real numbers that are close to zero. Explain why $\mathbf{r}_u(u_0, v_0)\Delta u$ and $\mathbf{r}_v(u_0, v_0)\Delta v$ are approximately equal to

$$\mathbf{r}(u_0 + \Delta u, v_0) - \mathbf{r}(u_0, v_0) \quad \text{and} \quad \mathbf{r}(u_0, v_0 + \Delta v) - \mathbf{r}(u_0, v_0),$$

respectively.

Using your work above, give an interpretation, in terms of area, to the quantity

$$\big|\mathbf{r}_u(u_0, v_0) \times \mathbf{r}_v(u_0, v_0)\big|\Delta u\Delta v.$$

It may be helpful to recall Theorem 1.5.5 and to look at Figure 2.8.5.

+ Linear Algebra:

In each of the following exercises, you are given a continuously differentiable parameterization r from $\mathbb{R}^j$ into $\mathbb{R}^n$, and a point w in $\mathbb{R}^j$. Verify that r is a local, regular parameterization in an open neighborhood of w, and parameterize the tangent set of the image at $\mathbf{p} = \mathbf{r}(\mathbf{w})$.

41. $\mathbf{r}(w_1, w_2, w_3) = (w_1, w_2e^{w_3}, w_3^2, w_1 + w_2 + w_3)$ at $(1, 0, 1)$.

42. $\mathbf{r}(w_1, w_2, w_3) = (w_1w_2, w_2w_3, 5w_3, w_1^2)$ at $(1, 1, 1)$.

43. $\mathbf{r}(w_1, w_2, w_3, w_4) = (w_1, w_2e^{w_3}, w_4 + \ln w_1, w_3^2, w_1 + w_2 + w_3)$ at $(1, 0, 1, 0)$.

44. $\mathbf{r}(w_1, w_2, w_3, w_4) = (w_1w_2, w_2w_3, w_3w_4, w_4^2, w_1^2)$ at $(1, 1, 1, 1)$.

45. $\mathbf{r}(w_1, w_2, w_3, w_4, w_5) = (w_1, w_2, w_3, w_4, w_5, w_1 + 2w_2 + 3w_3 + 4w_4 + 5w_5)$ at $(5, 4, 3, 2, 1)$.

46. $\mathbf{r}(w_1, w_2, w_3, w_4, w_5) = (w_5, w_1 \cos w_2, w_1 \sin w_2, w_3 \sin w_4, w_3 \cos w_4, w_2 + w_4)$ at $(2, 0, 3, 0 - 1)$.

2.9 Local Extrema

Determining where functions are as big or small as possible, i.e., where they attain maximum or minimum values, is important for understanding the geometry of the graph and in many physical applications. Frequently, we are interested in when the value of the function at a point is the biggest or smallest among all **nearby** points; this is the question of *local* maximum and minimum values. We usually shorten *maximum and minimum values* to *maxima* and *minima*, and lump both terms together by referring to *extrema*.

If you have a real-valued single-variable function $f = f(x)$, and f attains a local maximum or minimum value – a local *extreme* value – at $x = a$, then you should recall that either $f'(a)$ does not exist or that $f'(a) = 0$. Thus, we called a point where the derivative was undefined or zero, a *critical point* of f, and the relevant theorem states that, if f attains a local extreme value at a, then a has to be a critical point of f; the function doesn't **have** to attain a local extreme value at a critical point, but the critical points are the only places where you need to look for local extreme values.

The situation is essentially the same for a real-valued multivariable function $f = f(\mathbf{x})$: if the function f attains a local extreme value at $\mathbf{x} = \mathbf{a}$, then either the total derivative is undefined at $\mathbf{a}$ or is the zero function, i.e., the gradient vector satisfies $\vec{\nabla} f(\mathbf{a}) = \mathbf{0}$. Recalling that we defined a *critical point* of f to be a place where the total derivative doesn't exist or is the zero function, the relevant theorem once again states that, if f attains a local extreme value at $\mathbf{a}$, then $\mathbf{a}$ must be a critical point of f.

As in the single-variable case, there is a *First Derivative Test* and a *Second Derivative Test* for a function $f = f(x, y)$; these tests can tell us if a critical point yields a local maximum or a local minimum value at a point. However, in the multivariable setting, there is a third possible conclusion from the Second Derivative Test; a point can be a *saddle point*.

Basics:

Suppose that we have a real-valued function $f = f(\mathbf{x})$, whose domain is a subset E of $\mathbb{R}^n$. There are two types of *extreme value* questions that we are interested in.

We would like to know where in E the function f attains its biggest and smallest values, or determine that no such points exist; this is the *global* extreme value question, which we shall address in the next section.

We would also like to know those points $\mathbf{p}$ in E such that the value of $f(\mathbf{p})$ is the biggest or smallest that f attains among all points of E *near* the point $\mathbf{p}$. This is the question of *local extreme values*.

Example 2.9.1. Consider, for instance, the function given by

$$f(x,y) \;=\; e^{-x^2}\left(3 - \frac{y^4}{4} + \frac{5y^3}{3} - 3y^2\right).$$

The graph appears in Figure 2.9.1, where we have included some level curves.

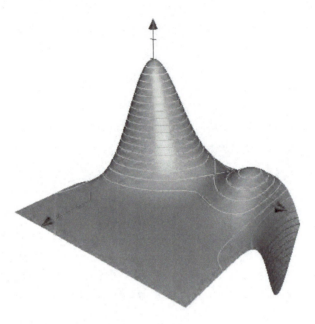

Figure 2.9.1: The graph of f, with level curves.

From the graph, it appears that f attains a global maximum value at $(0,0)$; that value is $f(0,0) = 3$. You can also see that the function f attains a local maximum value of $f(0,3) = 3/4$. The function attains no global minimum value, as it gets arbitrarily negative as the absolute value of y increases. Does f attain any local minimum values? It's difficult to say from the graph. It's hard to see exactly what's happening at the "valley" between the two "peaks". And maybe we're not looking in the right places or closely enough.

In fact, the function f has no local minima. But, the question is: How do we see that? For that matter, without having the graph, how can we find the local maxima?

First, we need to give the technical definitions of global and local extrema.

Definition 2.9.2. *Suppose that $f = f(\mathbf{x})$ is a real-valued function, whose domain is a subset E of $\mathbb{R}^n$. Let $\mathbf{p}$ be a point in E.*

Then, we say that f **attains a global maximum (respectively, minimum) value of $f(\mathbf{p})$ at $\mathbf{p}$** *if and only if, for all $\mathbf{x}$ in E, $f(\mathbf{p}) \geq f(\mathbf{x})$ (respectively, $f(\mathbf{p}) \leq f(\mathbf{x})$).*

If f attains a global maximum or minimum value at $\mathbf{p}$, then we say that f **attains a global extreme value of $f(\mathbf{p})$ at $\mathbf{p}$**. *We say that the global maximum/minimum/extreme value is* **strict** *provided that the strict inequalities $f(\mathbf{p}) > f(\mathbf{x})$ or $f(\mathbf{p}) < f(\mathbf{x})$ hold in the definition above for $\mathbf{x} \neq \mathbf{p}$.*

We say that f attains a **local** *maximum/minimum/extreme value at $\mathbf{p}$ provided that there is an open neighborhood $\mathcal{U}$ of $\mathbf{p}$ in $\mathbb{R}^n$ such that the restriction of f to the intersection $E \cap \mathcal{U}$ attains a maximum/minimum/extreme value at $\mathbf{p}$.*

Global/local extreme values are sometimes referred to as *absolute/relative* extreme values. We shall not use this terminology.

Remark 2.9.3. Note that the inequalities in Definition 2.9.2 include the possibility that nearby points give an **equal** value. This implies, for example, that the constant function $f(x, y) = 17$ attains a local maximum value and a local minimum value of 17 at every single point, but does not obtain **strict** local extreme values anywhere.

Another remark is that it is **extremely** important to know what domain is being used for a function f. This is important even for functions of a single variable. Suppose that you have the function $f(x) = x$. Unless explicitly told otherwise, you would probably assume that the domain of f is as big as possible – all of $\mathbb{R}$. Then f would have no local extreme values, for f gets smaller as x gets smaller and bigger as x gets bigger.

But what if you were asked for the local extreme values of $f(x) = x$ on the interval $[0, \infty)$? This could lead to confusion. Does this mean the local extreme values of f as a function on $\mathbb{R}$, but we only consider those values that correspond to points in $[0, \infty)$? If so, we would once again have no local extreme values. However, usually the phrase "f on the interval $[0, \infty)$" should be interpreted as "the new function obtained from f by restricting the domain to $[0, \infty)$". Interpreted this way, f (restricted to $[0, \infty)$) suddenly attains a local (in fact, global) minimum value of 0 at $x = 0$.

The point is that, even though we are sometimes sloppy about specifying the domain of a function, it really is part of the definition of the function, and sometimes we have to be careful. In general, throughout this textbook, we will be precise, and make it clear that we have restricted the domains of our functions for which we are trying to find local (or global) extreme values.

Example 2.9.4. If you have the graph of a function $f = f(x, y)$, it is frequently easy to spot the points at which the function attains local maximum and minimum values; they correspond to the highest points, the "peaks", and the lowest points, the "pits", on the graph.

Consider, for example, our old friend from Example 2.3.16:

$$z = f(x, y) = \frac{5x}{x^2 + y^2 + 1}.$$

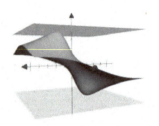

Figure 2.9.2: The graph of $z = f(x, y)$.

The graph of f appears in Figure 2.9.2, where we have included the two horizontal tangent planes that we found in Example 2.3.16. Note that those two horizontal tangent planes seem to be located at the peak and pit of the graph, the places where f attains a local maximum and local minimum value (in fact, it looks like f may attain **global** extreme values at the two points in question, but that's irrelevant to us here). Thus, using our results from Example 2.3.16, we suspect that the local extreme values of $-5/2$ and $5/2$ occur at the points $(-1, 0)$ and $(1, 0)$.

Is this right? Yes, but it takes some work to prove it. The next theorem tells us that, if $f = f(x, y)$ is differentiable, then the graph of f must have a horizontal tangent plane at any point where f attains a local extreme value; this doesn't guarantee that f attains local extreme values where the tangent plane is horizontal, but it does mean that those are the only possible points where extrema can be attained.

Before we get to the theorem, we want to remind you of our definition of a *critical point* from Definition 2.7.11.

Note that $\vec{\nabla} f(\mathbf{p}) = \mathbf{0}$ is equivalent to $d_{\mathbf{0}} f$ being the zero function (the function which is always 0).

Definition 2.9.5. *Suppose that $f = f(\mathbf{x})$ is a real-valued function, whose domain is a subset E of $\mathbb{R}^n$. A **critical point of** f is a point $\mathbf{p}$ in E such that either f is not differentiable at $\mathbf{p}$ or $\vec{\nabla} f(\mathbf{p}) = \mathbf{0}$.*

Now, our first theorem on local extreme values is:

It is important that, if $\mathbf{p}$ is a point in the domain of f, but is not an **interior** point of the domain, then f is **not** differentiable at $\mathbf{p}$, and so is a critical point. Thus, any boundary points of the domain of a function, which are contained in the domain, are always places that have to be checked when looking for local extreme values.

Theorem 2.9.6. *Suppose that $f = f(\mathbf{x})$ is a real-valued function, whose domain is a subset E of $\mathbb{R}^n$. If f attains a local extreme value at a point $\mathbf{p}$ in E, then $\mathbf{p}$ is a critical point of f (which includes all non-interior points of E).*

Proof. The proof of this is quite easy, given what we know from single-variable Calculus. Suppose that $f = f(x_1, \ldots, x_n)$ is differentiable at $\mathbf{p} = (p_1, \ldots, p_n)$ and that f attains a local extreme value at $\mathbf{p}$. Then, there exists an open ball $\mathcal{U}$, centered at $\mathbf{p}$, in $\mathbb{R}^n$ such that $f(\mathbf{p})$ is the largest or smallest value $f(\mathbf{x})$, for all $\mathbf{x}$ in $\mathcal{U}$.

Fix an index i, where $1 \leq i \leq n$. Fix all of the variables other than x_i; that is, look at the function of a single variable given by

$$g(x_i) = f(p_1, \ldots, p_{i-1}, x_i, p_{i+1}, \ldots, p_n),$$

where the domain is the open subset $\mathcal{W}$ of $\mathbb{R}$ consisting of those x_i such that the point $(p_1, \ldots, p_{i-1}, x_i, p_{i+1}, \ldots, p_n)$ is in $\mathcal{U}$. Then, as f obtained a local extreme value at $\mathbf{p}$, g obtains a local extreme value at p_i and is differentiable at p_i, since f is differentiable at $\mathbf{p}$ and

$$g'(p_i) = \left. \frac{\partial f}{\partial x_i} \right|_{\mathbf{p}}.$$

From single-variable Calculus, we conclude that

$$0 = g'(p_i) = \left. \frac{\partial f}{\partial x_i} \right|_{\mathbf{p}},$$

i.e., that each partial derivative of f at $\mathbf{p}$ must be 0, which is what we needed to show. $\square$

Remark 2.9.7. If $f = f(x, y)$ and $\vec{\nabla} f(\mathbf{p}) = \mathbf{0}$, then Definition 2.3.8 tells us that the tangent plane to the graph of f at $\mathbf{p}$ is horizontal. Thus, at places where f is differentiable and attains a local extreme value, there must be a horizontal tangent plane. We saw an example of this in Example 2.9.4.

It is possible for a function $f = f(\mathbf{x})$ to be differentiable at $\mathbf{p}$, with $\vec{\nabla} f(\mathbf{p}) = \mathbf{0}$, and for f to attain neither a local maximum nor a local minimum value at $\mathbf{p}$. Given what you know from single-variable Calculus, this shouldn't be surprising; you know, for instance that, if $g(x) = x^3$, then $g'(0) = 0$, even though g does not attain a local extreme value at 0. We give a name to such critical points.

> **Definition 2.9.8.** *Suppose that $f = f(\mathbf{x})$ is a real-valued function, whose domain is a subset E of $\mathbb{R}^n$. A **saddle point** of f is a critical point of f at which f is differentiable, so that $\vec{\nabla} f(\mathbf{p}) = \mathbf{0}$, but at which f does not attain a local extreme value. We also say that the **graph of f has a saddle point** at $(\mathbf{p}, f(\mathbf{p}))$.*

We have defined a saddle point in the most general way possible. Some references require more conditions on the critical point, such as requiring the graph to be concave up in some "directions" and concave down in others, where the definition of "direction" involves using smooth cross sections or smooth parameterized curves with non-zero velocity. We prefer to avoid this complication in the definition, even though it means that, for us, the function $f(x, y) = x^3$ has saddle points at all points of the form $(0, b)$ even though the graph doesn't look very "saddle-like".

Example 2.9.9. Let's return to Example 2.9.1, where we had the function given by

$$f(x, y) \;=\; e^{-x^2}\left(3 - \frac{y^4}{4} + \frac{5y^3}{3} - 3y^2\right).$$

This function is differentiable everywhere, so the critical points are where the gradient vector is zero. We find

$$f_x \;=\; -2xe^{-x^2}\left(3 - \frac{y^4}{4} + \frac{5y^3}{3} - 3y^2\right)$$

and

$$f_y \;=\; e^{-x^2}(-y^3 + 5y^2 - 6y) \;=\; -e^{-x^2}y(y^2 - 5y + 6) \;=\; -e^{-x^2}y(y-2)(y-3).$$

As e^{-x^2} is never 0, we quickly solve $f_x = 0$ and $f_y = 0$ simultaneously to find that $x = 0$ and $y = 0$, 2, or 3. Thus, there are three critical points: $(x, y) = (0, 0)$, $(0, 2)$, and $(0, 3)$. These yield the two peaks, corresponding to local maxima, and the valley, a saddle point, that we saw in Figure 2.9.1.

It is easy to find the Second Derivative Test in other textbooks; the First Derivative Test is not so well-known.

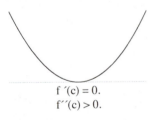

f ′(c) = 0.
f ″(c) > 0.

Figure 2.9.3: A typical local minimum of a single-variable function.

f ′(c) = 0.
f ″(c) < 0.

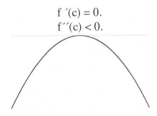

Figure 2.9.4: A typical local maximum of a single-variable function.

Of course, our question is: can we use Calculus to find the points at which functions attain local extrema (local extreme values), **without** graphing the function? In particular, for functions of more than two variables, graphing becomes problematic.

As in the case of single-variable functions, there are two results which can tell us when a critical point of a function actually yields a local extreme value: the *First* and *Second Derivative Tests*.

While the First Derivative Test yields a stronger conclusion and can be used at points where the function is not differentiable, it is typically much more complicated to use than the Second Derivative Test. So, we will present the Second Derivative Test first, and cover the First Derivative Test in the More Depth portion of this section. There is also a Second Derivative Test for functions of any number of variables, but it is difficult to state; we shall cover only the two-variable case $f = f(x, y)$ here, and present the general case in the + Linear Algebra portion of this section.

First, let's recall the Second Derivative Test for a single-variable function. Suppose that $f = f(x)$ and that $f'(c) = 0$. Then, if $f''(c) > 0$, the graph of f is concave up at $(c, f(c))$ and f attains a local minimum value at c. On the other hand, if $f''(c) < 0$, the graph of f is concave down at $(c, f(c))$ and f attains a local maximum value at c.

Now suppose that we have $f = f(x, y)$ and f is differentiable at $\mathbf{p} = (a, b)$. The single-variable condition that $f'(c) = 0$ is replaced by the condition that $\vec{\nabla} f(\mathbf{p}) = \mathbf{0}$.

But how do we replace the second derivative condition? We need something that implies that a surface looks concave up or concave down, like the graphs of $f(x,y) = x^2 + y^2$ and $f(x,y) = -x^2 - y^2$. Or has a saddle point, like the graph of $f(x,y) = x^2 - y^2$. See Figure 2.9.5, Figure 2.9.6, and Figure 2.9.7.

In fact, the Second Derivative Test will involve all of the second partial derivatives: f_{xx}, f_{yy}, and $f_{xy} = f_{yx}$ (we'll assume that f has continuous second partial derivatives so that this last equality holds). We'll consider a matrix containing these second partial derivatives, the *Hessian matrix*.

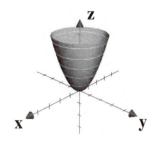

Figure 2.9.5: The graph of $z = x^2 + y^2$.

Definition 2.9.10. *Suppose that $f = f(x,y)$ has continuous second partial derivatives (i.e., is of class C^2) in an open neighborhood of a point* **p**. *Then, the* **Hessian matrix** *of f at* **p** *is the matrix*

$$\left[\begin{array}{cc} f_{xx} & f_{xy} \\ f_{xy} & f_{yy} \end{array}\right]_{\Big|_{\mathbf{p}}}.$$

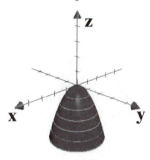

Figure 2.9.6: The graph of $z = -x^2 - y^2$.

Definition 2.9.11. *Suppose that $f = f(x,y)$ has continuous second partial derivatives (i.e., is of class C^2) in an open neighborhood of a point* **p**. *Then, we say that* **p** *is a* **non-degenerate critical point of** *f if and only if $\vec{\nabla}f(\mathbf{p}) = \mathbf{0}$ and the determinant D of the Hessian matrix of f at* **p** *is not zero, i.e.,*

$$D = \left(f_{xx}f_{yy} - f_{xy}^2\right)_{\Big|_{\mathbf{p}}} \neq 0.$$

How should you think of the determinant of the Hessian matrix at a non-degenerate critical point of f? If you think about the case where $f_{xy}(\mathbf{p}) = 0$, you'll find it easier to understand and remember the Second Derivative Test.

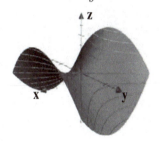

Figure 2.9.7: The graph of $z = x^2 - y^2$.

If $f_{xy}(\mathbf{p}) = 0$ and $D > 0$, then either f_{xx} and f_{yy} are both positive at **p** or both negative, so that the curves that you get for the x- and y-cross sections are either both concave up or both concave down. While it's not easy to show, the presence of the possibly non-zero f_{xy}^2 term in D implies that, if the x- and y-cross sections are both concave up or both concave down, then **all** of the cross sections in all directions are correspondingly concave up or down; this will correspond to the surface being "concave up" or "concave down". To determine which concavity you have, you need to check whether $f_{xx}(\mathbf{p})$ is positive or negative (you could, instead, check the sign of $f_{yy}(\mathbf{p})$).

On the other hand, if $D < 0$ and $f_{xy}(\mathbf{p}) = 0$, then $f_{xx}(\mathbf{p})$ and $f_{yy}(\mathbf{p})$ must have different signs, so that the x- and y-cross sections have different concavity, which implies that the value of f decreases in one cross section and increases in the other. Thus, there is a saddle point at $(\mathbf{p}, f(\mathbf{p}))$. When $D < 0$, the presence of the possibly non-zero f_{xy}^2 term in D implies that there are two directions, but maybe not the x- and y-directions, such that f decreases in one direction and increases in the other, so that $\mathbf{p}$ does, in fact, correspond to a saddle point.

If you keep in mind the case where $f_{xy}(\mathbf{p}) = 0$, you should find the following theorem fairly easy to remember.

Theorem 2.9.12. (Second Derivative Test) *Suppose that $f = f(x, y)$ has continuous second partial derivatives in an open neighborhood of a point $\mathbf{p}$, and that $\mathbf{p}$ is a non-degenerate critical point of f, i.e., $\vec{\nabla} f(\mathbf{p}) = \mathbf{0}$ and $D = \left(f_{xx} f_{yy} - f_{xy}^2 \right)\Big|_{\mathbf{p}} \neq 0$.*

Then, there are three possible cases:

1. *If $D > 0$ and $f_{xx}(\mathbf{p}) > 0$, then f attains a local minimum value at $\mathbf{p}$.*

2. *If $D > 0$ and $f_{xx}(\mathbf{p}) < 0$, then f attains a local maximum value at $\mathbf{p}$.*

3. *If $D < 0$, there are cross sections of the graph of f at $(\mathbf{p}, f(\mathbf{p}))$ which are concave up and cross sections which are concave down; hence, $\mathbf{p}$ corresponds to a saddle point, and f attains neither a local minimum nor a local maximum value at $\mathbf{p}$.*

> For the proof, see Theorem 3.4 of [3].

> For convenience, we frequently say that f, itself, possesses a saddle point at $\mathbf{p}$, rather than saying that the graph of f has a saddle point at $(\mathbf{p}, f(\mathbf{p}))$.

In cases (1) and (2) above, you could check the sign of f_{yy}, instead of f_{xx}; these two unmixed second partial derivatives must have the same sign if $D > 0$.

Example 2.9.13. Consider the function $f(x, y) = 3x - x^3 - 2y^2 + y^4$. Find the critical points of f, show that they are all non-degenerate, and classify the critical points of f, according to whether f attains a local maximum value, a local minimum value, or has a saddle point.

Solution:

First, we need to find the critical points, and then we'll look at the second partial derivatives and the value of D.

We find
$$f_x = 3 - 3x^2 \qquad \text{and} \qquad f_y = -4y + 4y^3.$$

Thus, to find the critical points, we need to simultaneously solve the equations:
$$3 - 3x^2 = 3(1 - x^2) = 0 \qquad \text{and} \qquad -4y + 4y^3 = 4y(-1 + y^2) = 0.$$

We quickly find that $x = \pm 1$, and $y = 0$ or $y = \pm 1$. Therefore, there are 6 critical points:
$$(-1, -1), \quad (-1, 0), \quad (-1, 1), \quad (1, -1), \quad (1, 0), \quad \text{and} \quad (1, 1).$$

To classify the critical points, we need the second derivatives. We calculate:

$$f_{xx} = -6x, \qquad f_{yy} = -4 + 12y^2, \qquad \text{and} \qquad f_{xy} = 0;$$

thus,

$$D = f_{xx}f_{yy} - f_{xy}^2 = -6x(-4 + 12y^2).$$

Now, we have to consider each critical point separately. We shall see that each critical point is non-degenerate along the way.

At $(x, y) = (-1, -1)$:

$D = -6(-1)(-4 + 12(-1)^2) > 0$ and $f_{xx} = -6(-1) > 0$. Since $D > 0$, f attains a local maximum or minimum value at $(-1, -1)$. As $f_{xx} > 0$, you should think "concave up", which means that f attains a **local minimum** value at $(-1, -1)$.

At $(x, y) = (-1, 0)$:

$D = -6(-1)(-4) < 0$. There is no more work to do; since $D < 0$, there is a **saddle point** at $(-1, 0)$.

At $(x, y) = (-1, 1)$:

$D = (-6)(-1)(-4 + 12) > 0$ and $f_{xx} = -6(-1) > 0$. So, f attains another **local minimum** value at $(-1, 1)$.

At $(x, y) = (1, -1)$:

$D = -6(1)(-4 + 12) < 0$. So, there is **saddle point** at $(1, -1)$.

At $(x, y) = (1, 0)$:

$D = -6(1)(-4) > 0$ and $f_{xx} = -6(1) < 0$. So, f attains a **local maximum** value at $(1, 0)$.

At $(x, y) = (1, 1)$:

$D = -6(1)(-4 + 12) < 0$. So, there is a **saddle point** at $(1, 1)$.

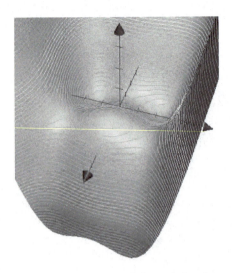

Figure 2.9.8: The graph of f, with level curves.

We didn't want to ruin the surprise by giving the graph first, but now you should look at our results from above and compare with what you see in Figure 2.9.8.

Without the graph of f in three dimensions, if you're given a collection of level curves, then you can still see where the local maxima, local minima, and saddle points occur; you see smoothly distorted circles/ovals near local maxima and minima, and crossing curves at saddle points. We will explain this in the More Depth portion of this section.

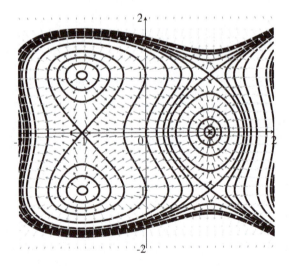

Figure 2.9.9: Level curves and scaled gradient vectors of $f(x, y) = 3x - x^3 - 2y^2 + y^4$.

If you're not given the values of f for the level curves, then you need some other data to distinguish between local maxima and minima; gradient vectors (even scaled ones) are enough, since the gradient vectors point in directions in which the value of f increases (in fact, increases most rapidly).

The arrows point towards points where f attains local maxima, and point away from points where f attains local minima. See Figure 2.9.9.

Example 2.9.14. Consider the function $f(x, y) = 4xy - x^4 - y^4$. Find the critical points of f, show that they are all non-degenerate, and classify the critical points where f attains a local maximum value, a local minimum value, or has a saddle point.

Solution:

As in the previous example, we first need to find the critical points, and then we'll look at the second partial derivatives and the value of D.

We find

$$f_x = 4y - 4x^3 \quad \text{and} \quad f_y = 4x - 4y^3.$$

Thus, to find the critical points, we need to simultaneously solve the equations:

$$4y - 4x^3 = 4(y - x^3) = 0 \quad \text{and} \quad 4x - 4y^3 = 4(x - y^3) = 0.$$

This means that we need $y = x^3$ and $x = y^3$; substituting $x = y^3$ into the first equation, we find that $y = (y^3)^3 = y^9$. Therefore, $y = 0$ or $1 = y^8$. Hence, $y = 0$ or ± 1. We also must have $x = y^3$. So, there are three critical points:

$$(-1, -1), \quad (0, 0), \quad \text{and} \quad (1, 1).$$

To classify the critical points, we need the second derivatives. We calculate:

$$f_{xx} = -12x^2, \quad f_{yy} = -12y^2, \quad \text{and} \quad f_{xy} = 4;$$

thus,

$$D = f_{xx}f_{yy} - f_{xy}^2 = 144x^2y^2 - 16.$$

Now, we have to consider each critical point separately. Again, we shall see that each critical point is non-degenerate along the way.

At $(x, y) = (-1, -1)$:

$D = 144 - 16 > 0$ and $f_{xx} = -12(-1)^2 < 0$. So, f attains a local maximum value at $(-1, -1)$.

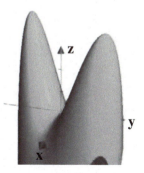

Figure 2.9.10: The graph of $z = 4xy - x^4 - y^4$.

At $(x, y) = (0, 0)$:

$D = -16 < 0$. So, there is a saddle point at $(0, 0)$.

At $(x, y) = (1, 1)$:

$D = 144 - 16 > 0$ and $f_{xx} = -12(1)^2 < 0$. So, f attains a local maximum value at $(1, 1)$.

Remark 2.9.15. You might think that, given the graph of $z = f(x, y)$, it would be relatively easy to see, approximately, where f attains local maxima, local minima, or has a saddle point. However, the problem is that parts of the graph may obscure other parts, and the whole graph frequently blocks your view of the coordinate axes.

Consider, for instance, the graph of $z = f(x, y) = x^3 + y^3 + 3x^2 - 3y^2$.

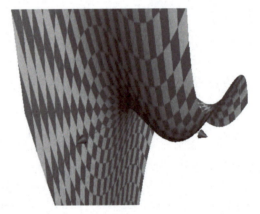

Figure 2.9.11: The graph of $z = x^3 + y^3 + 3x^2 - 3y^2$, from the usual perspective.

Can you see where the critical points of f are from the graph? How about if we look from a different perspective?

Figure 2.9.12 is better, but still not so good. Of course, if we actually had a 3D model in front of us, it would be easier. So what do we do, given that we have to draw something on a 2-dimensional page or blackboard?

Actually, level curves, as in Figure 2.9.13, marked with the corresponding values of f are much more informative.

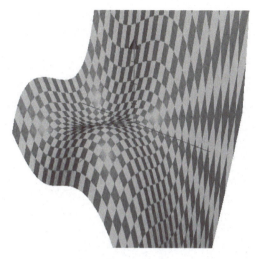

Figure 2.9.12: The graph of $z = x^3 + y^3 + 3x^2 - 3y^2$, from the "negative" perspective.

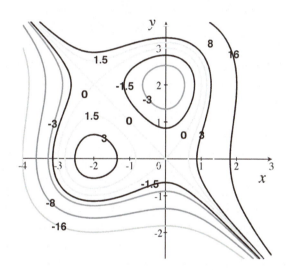

Figure 2.9.13: Level curves of $f(x, y) = x^3 + y^3 + 3x^2 - 3y^2$.

Remember that you should see distorted circles/ellipses near points where f attains local maxima or minima, and crossing lines/curves at saddle points. Thus, from Figure 2.9.13, you should be able to see that:

- f attains a local maximum value near the point $(-2, 0)$ and that value is somewhere between 3 and 8;

- f attains a local minimum value near the point $(0, 2)$ and that value is somewhere between -8 and -3; and

- f has saddle points near the points $(-2, 2)$ and $(0, 0)$, and the value of f at these saddle points is approximately 0.

You can use the Second Derivative Test to verify our estimates, but the point is that, if you're going to use graphical data to analyze critical points of functions $f = f(x, y)$, you'd rather have level curves of f, with the values of f, instead of the graph of $z = f(x, y)$. Alternatively, as in Figure 2.9.9, level curves, without values, but with scaled gradient vectors, are useful (though you can't then estimate the values of f at the critical points).

More Depth:

Example 2.9.16. Let's apply the Second Derivative Test to Example 2.9.1, where the function was given by

$$f(x, y) = e^{-x^2}\left(3 - \frac{y^4}{4} + \frac{5y^3}{3} - 3y^2\right).$$

In Example 2.9.9, we found

$$f_x = -2xe^{-x^2}\left(3 - \frac{y^4}{4} + \frac{5y^3}{3} - 3y^2\right)$$

and

$$f_y = e^{-x^2}(-y^3 + 5y^2 - 6y) = -e^{-x^2}y(y^2 - 5y + 6) = -e^{-x^2}y(y - 2)(y - 3),$$

and three critical points: $(x, y) = (0, 0)$, $(0, 2)$, and $(0, 3)$.

We now need the second derivatives:

$$f_{xx} = -2(-2x^2e^{-x^2} + e^{-x^2})\left(3 - \frac{y^4}{4} + \frac{5y^3}{3} - 3y^2\right) =$$

$$-2e^{-x^2}(-2x^2 + 1)\left(3 - \frac{y^4}{4} + \frac{5y^3}{3} - 3y^2\right),$$

$$f_{yy} = e^{-x^2}(-3y^2 + 10y - 6),$$

and

$$f_{xy} = -2xe^{-x^2}(-y^3 + 5y^2 - 6y).$$

Therefore,

$$D = f_{xx}f_{yy} - f_{xy}^2 =$$

$$-2e^{-2x^2}(-2x^2+1)\left(3-\frac{y^4}{4}+\frac{5y^3}{3}-3y^2\right)(-3y^2+10y-6)-4x^2e^{-2x^2}(-y^3+5y^2-6y)^2.$$

Now, we will apply the Second Derivative Test and, for ease, check the sign of f_{yy}, instead of f_{xx}.

At $(x, y) = (0, 0)$:

$D = -2 \cdot 1 \cdot 1 \cdot 3 \cdot (-6) - 0 > 0$ and $f_{yy} = 1 \cdot (-6) < 0$. So, f attains a local maximum value at $(0, 0)$.

At $(x, y) = (0, 2)$:

$D = -2 \cdot 1 \cdot 1 \cdot (3 - 4 + 40/3 - 12)(-12 + 20 - 6) - 0 < 0$. So, there is a saddle point at $(0, 2)$.

At $(x, y) = (0, 3)$:

$D = -2 \cdot 1 \cdot 1 \cdot (3 - 81/4 + 45 - 27)(-27 + 30 - 6) - 0 > 0$ and $f_{yy} = 1 \cdot (-27 + 30 - 6) < 0$. So, f attains a local maximum value at $(0, 3)$.

Thus, we recover that there are two peaks and the saddle point (which we called a "valley"), as we saw in Figure 2.9.1.

For a function $f = f(x, y)$, the standard, basic examples of the three types of non-degenerate critical points that can occur are those that we graphed in Figure 2.9.5, Figure 2.9.6, and Figure 2.9.7: $f(x, y) = x^2 + y^2$, for a local minimum, $f(x, y) = -x^2 - y^2$ for a local maximum, and $f(x, y) = x^2 - y^2$, for a saddle point. We tend to picture the graph of any function of two variables, near a non-degenerate critical point, as looking like one of these three canonical examples. We also picture the level curves near non-degenerate critical points as looking like the level curves of these three functions. Finally, when we have a saddle point at a non-degenerate critical point, we picture $x-$ and y-cross sections which look like parabolas, which are concave up in some directions and concave down in others.

The question is: **should** we picture/think of graphs at non-degenerate critical points in this way – as being just like the graphs of the three basic examples, except, possibly, smoothly distorted somewhat? The answer is: yes.

For functions which have (at least) continuous third-order partial derivatives (i.e., are of class C^3), the *Morse Lemma* tells us precisely that, after a local change of coordinates, a function of two variables with a non-degenerate critical point becomes one of the three basic examples, possibly shifted up or down.

We give a reference for the proof alongside Theorem 2.9.24, where we give the result for any number of variables.

λ_1 and λ_2 are the *eigenvalues* of the Hessian matrix. Eigenvalues are of fundamental importance in linear algebra.

Theorem 2.9.17. **(Morse Lemma)** *Suppose that $f = f(x, y)$ is real-valued function of class C^r, where $r \geq 3$, on an open subset $\mathcal{U}$ of $\mathbb{R}^2$. Suppose further that $\mathbf{p}$, in $\mathcal{U}$, is a non-degenerate critical point of f.*

Define numbers λ_1 and λ_2 by requiring:

$$\lambda_1 \lambda_2 = f_{xx}(\mathbf{p}) f_{yy}(\mathbf{p}) - f_{xy}^2(\mathbf{p}) \quad \text{and} \quad \lambda_1 + \lambda_2 = f_{xx}(\mathbf{p}) + f_{yy}(\mathbf{p}).$$

Note that, since $\mathbf{p}$ is a non-degenerate critical point, the first equation implies that neither λ_1 nor λ_2 is zero.

It is irrelevant which of the two numbers is called λ_1 and which is called λ_2. The two equations combine to yield a quadratic equation, which has two roots, or one repeated root, which would be used for both values.

In the smooth case, where $r = \infty$, we mean that $r - 2$ is also ∞, i.e., if f is smooth, so is the change of coordinates.

Then, there exists a C^{r-2} local change of coordinates at $\mathbf{0}$, $(x, y) = \Phi(u, v)$, such that $\Phi(\mathbf{0}) = \mathbf{p}$ and such that:

a. if $\lambda_1 > 0$ and $\lambda_2 > 0$, then $(f \circ \Phi)(u, v) = f(\mathbf{p}) + u^2 + v^2$;

b. if $\lambda_1 < 0$ and $\lambda_2 < 0$, then $(f \circ \Phi)(u, v) = f(\mathbf{p}) - u^2 - v^2$;

c. if λ_1 and λ_2 have different signs, then $(f \circ \Phi)(u, v) = f(\mathbf{p}) + u^2 - v^2$.

Remark 2.9.18. The Morse Lemma provides the technical reason why graphs and level curves of functions of two variables look the way they do near non-degenerate critical points; they should look like the graphs and level curves of $u^2 + v^2$, $-u^2 - v^2$, and $u^2 - v^2$, all shifted upward by $f(\mathbf{p})$.

However, we still need to provide a explanation as to why cases (1), (2), and (3) of the Second Derivative Test correspond to cases (a), (b), and (c) of the Morse Lemma.

We have

$$\lambda_1 \lambda_2 = f_{xx}(\mathbf{p}) f_{yy}(\mathbf{p}) - f_{xy}^2(\mathbf{p}) = D.$$

Therefore, $D < 0$ if and only if λ_1 and λ_2 have different signs; hence, case (3) of the Second Derivative Test is case (c) of the Morse Lemma.

We also see that $D > 0$ if and only if λ_1 and λ_2 have the same sign, i.e., are both positive or both negative. Now note that $f_{xy}^2(\mathbf{p}) \geq 0$, so that $D > 0$ implies

$f_{xx}(\mathbf{p})f_{yy}(\mathbf{p}) = D + f_{xy}^2(\mathbf{p}) > 0$, and so $f_{xx}(\mathbf{p})$ and $f_{yy}(\mathbf{p})$ also have the same sign. Since $\lambda_1 + \lambda_2 = f_{xx}(\mathbf{p}) + f_{yy}(\mathbf{p})$, the common sign of the λ_i's is the same as the common sign of the f_{xx} and f_{yy}. It follows that cases (1) and (2) of the Second Derivative Test correspond to cases (a) and (b) of the Morse Lemma.

We wish to develop a multivariable First Derivative Test, but, first, let's recall what the single-variable First Derivative Test says.

Suppose that we have $f = f(x)$ which is continuous on the open interval (a, b), and that p is a point in (a, b) such that $f'(x) > 0$ for all x in the interval (a, p), and $f'(x) < 0$ for all x in the interval (p, b). Then, even if f is not differentiable at p, f increases on the interval $(a, p]$ and decreases on the interval $[p, b)$; thus, f attains a local (actually, global, on (a, b)) maximum value at p. If f' is first negative, then positive, there is the analogous result for a local minimum value.

How do we generalize this to a statement for multivariable functions? At first, it may seem difficult to produce a multivariable statement that corresponds to the fact that the sign of the derivative "switches as x passes through p".

Recall that, for a function $f = f(x)$ of a single variable, $d_x f(v) = f'(x) \cdot v$. Thus, we can write the condition that "$x < p$ implies $f'(x) > 0$" as "$x < p$ implies $d_x f(p-x) > 0$", and the condition that "$p < x$ implies $f'(x) < 0$" as "$p < x$ implies $d_x f(p-x) > 0$". Therefore, we can state the single-variable First Derivative Test for a local maximum as: **Suppose that f is continuous on the open interval (a, b), that p is in (a, b), and that, for all $x \neq p$ in (a, b), $d_x f(p-x) > 0$. Then, f attains a local maximum value at p.**

In words, what this says is that f attains a local maximum value at p provided that, at all points x near (but unequal to) p, the instantaneous rate of change of f, as you head from x, straight towards p, is positive. This seems like it should also be true for multivariable functions and, in fact, it is.

We give the proof of this theorem, since it is not too difficult, and it is instructive to see how single-variable results lead to multivariable results.

> **Theorem 2.9.19.** **(First Derivative Test)** *Let $f = f(\mathbf{x})$ be a continuous function on an open ball $\mathcal{U}$ in $\mathbb{R}^n$, where $\mathcal{U}$ is centered at a point $\mathbf{p}$.*
>
> *Suppose that, for all $\mathbf{x} \neq \mathbf{p}$ in $\mathcal{U}$, $d_{\mathbf{x}} f$ exists and $d_{\mathbf{x}} f(\mathbf{p} - \mathbf{x}) > 0$ (respectively, $d_{\mathbf{x}} f(\mathbf{p} - \mathbf{x}) < 0$). Then, f attains a strict global maximum (respectively, minimum) value at $\mathbf{p}$.*

Note that, in particular, since f attains an extreme value at $\mathbf{p}$, $\mathbf{p}$ must be a critical point of f. However, it may be that f is not differentiable at $\mathbf{p}$.

Proof. We shall prove the maximum case; the minimum case is completely analogous.

Suppose that, for all $\mathbf{x} \neq \mathbf{p}$ in $\mathcal{U}$, $d_{\mathbf{x}}f(\mathbf{p} - \mathbf{x}) > 0$. Let $\mathbf{a}$ be a point in $\mathcal{U}$, such that $\mathbf{a} \neq \mathbf{p}$. We wish to show that $f(\mathbf{p}) > f(\mathbf{a})$.

Consider the function $g : [0,1] \to \mathbb{R}$ given by $g(t) = f\big(\mathbf{a} + t(\mathbf{p} - \mathbf{a})\big)$. As f is defined on an open ball centered at $\mathbf{p}$, and $\mathbf{a} + t(\mathbf{p} - \mathbf{a})$, for $0 \leq t \leq 1$, parameterizes the line segment from $\mathbf{a}$ to $\mathbf{p}$, g is well-defined. In addition, g is continuous on $[0,1]$ and, by the Chain Rule, for $0 < t < 1$, $g'(t)$ exists and

$$g'(t) \;=\; \vec{\nabla}f\big(\mathbf{a} + t(\mathbf{p} - \mathbf{a})\big) \cdot (\mathbf{p} - \mathbf{a}).$$

If we let $\mathbf{x} = \mathbf{a} + t(\mathbf{p} - \mathbf{a})$, then $\mathbf{x}$ is in $\mathcal{U}$ and

$$\mathbf{p} - \mathbf{x} \;=\; \mathbf{p} - \big(\mathbf{a} + t(\mathbf{p} - \mathbf{a})\big) \;=\; (\mathbf{p} - \mathbf{a}) - t(\mathbf{p} - \mathbf{a}) \;=\; (1-t)(\mathbf{p} - \mathbf{a}).$$

Therefore, for $0 < t < 1$, $\mathbf{x} \neq \mathbf{p}$ and

$$\mathbf{p} - \mathbf{a} \;=\; \frac{1}{1-t}(\mathbf{p} - \mathbf{x}),$$

where $1/(1-t) > 0$; hence, for $0 < t < 1$,

$$g'(t) \;=\; \vec{\nabla}f(\mathbf{x}) \cdot \frac{1}{1-t}(\mathbf{p} - \mathbf{x}) \;=\; \frac{1}{1-t}\,\vec{\nabla}f(\mathbf{x}) \cdot (\mathbf{p} - \mathbf{x}),$$

which, by assumption, is positive.

Therefore, g is continuous on $[0,1]$, differentiable on $(0,1)$, with $g'(t) > 0$ on $(0,1)$. By the Mean Value Theorem, there exists c in $(0,1)$ such that

$$f(\mathbf{p}) - f(\mathbf{a}) \;=\; g(1) - g(0) \;=\; g'(c)(1-0) \;=\; g'(c) > 0,$$

i.e., $f(\mathbf{p}) > f(\mathbf{a})$, as we wished to show. $\square$

Remark 2.9.20. We want to make three remarks.

First, the open ball in the First Derivative Test can have infinite radius, i.e., be all of $\mathbb{R}^n$; you may check that the proof is still valid in that case.

Second, while we have stated the First Derivative Test as a test for **global** extreme values, it is typically used to verify that you have a **local extreme value**; you usually start with a function with a large domain, restrict the domain to a small enough open ball around the point $\mathbf{p}$, and verify that the restricted function attains a global extreme value at $\mathbf{p}$, which implies that the original unrestricted function attains a **local** extreme value at $\mathbf{p}$.

Finally, while the First Derivative Test applies in many situations where the Second Derivative Test does not apply, the First Derivative Test is usually very complicated to use. If the Second Derivative Test works, you almost always want to use it, rather than use the First Derivative Test.

Having written above that the First Derivative Test is usually very complicated to use, we'll now give an example where it's easy to use, while the Second Derivative Test cannot be applied.

Example 2.9.21. Let $f(x,y) = \sqrt{x^2 + y^2}$. Clearly, the global minimum value of this function is 0, which is attained at $(x,y) = (0,0)$. However, we would like to see that the First Derivative Test would also tell us that there's a global minimum value at $(0,0)$.

We quickly find

$$\vec{\nabla} f(x,y) = \left(\frac{x}{\sqrt{x^2 + y^2}}, \frac{y}{\sqrt{x^2 + y^2}} \right).$$

Note that $\vec{\nabla} f$ does not exist at $(0,0)$. Thus, $(0,0)$ is a critical point of f. However, f is continuous, and even though $\vec{\nabla} f$ does not exist at $(0,0)$, we can, nonetheless, apply the First Derivative Test.

We find, for all $(x,y) \neq (0,0)$,

$$d_{(x,y)}f\big((0,0) - (x,y)\big) = \vec{\nabla} f(x,y) \cdot -(x,y) = \left(\frac{x}{\sqrt{x^2 + y^2}}, \frac{y}{\sqrt{x^2 + y^2}} \right) \cdot -(x,y) =$$

$$-\frac{x^2 + y^2}{\sqrt{x^2 + y^2}} = -\sqrt{x^2 + y^2} < 0.$$

Hence, the First Derivative Test tells us (again) that f attains a global minimum value at $(0,0)$.

We want to generalize the Second Derivative Test so that we can investigate local extrema of functions of any number of variables. The nicest way to do this is to give a Morse Lemma, like that in Theorem 2.9.17, but one that works for any number of variables. The reason this is relegated to the + Linear Algebra portion of this section is because it uses, in a crucial manner, the notion of the eigenvalues of a matrix. If f is a function of n variables, then the matrix in question will be the $n \times n$ Hessian matrix of f at a critical point.

So far, we have defined the Hessian matrix only for functions of two variables. Hopefully, the following definition seems obvious.

+ **Linear Algebra:**

Definition 2.9.22. *Suppose that $f = f(x_1, \ldots, x_n)$ has continuous second partial derivatives (i.e., is of class C^2) on an open subset of R^n. Then, the* **Hessian matrix**, H_f, *of f is the matrix of second partial derivatives of f, i.e., H_f is the $n \times n$ matrix whose (i,j)-th entry is $\partial^2 f / \partial x_i \partial x_j$.*

Note that, since f has continuous second partial derivatives, H_f is a symmetric matrix, i.e., the (i,j)-th entry and the (j,i)-the entry are the same. We denote the Hessian matrix, evaluated at a point $\mathbf{p}$ by $H_f(\mathbf{p})$.

Naturally, we also make the definition:

Definition 2.9.23. *Suppose that $f = f(x_1, \ldots, x_n)$ has continuous second partial derivatives (i.e., is of class C^2) on an open subset $\mathcal{U}$ of R^n.*

Then, a point $\mathbf{p}$, in $\mathcal{U}$, is a **non-degenerate critical point of** f *if and only if $\vec{\nabla} f(\mathbf{p}) = \mathbf{0}$ and the determinant of $H_f(\mathbf{p})$ is not zero, i.e., $\det H_f(\mathbf{p}) \neq 0$. The condition that $\det H_f(\mathbf{p}) \neq 0$ is equivalent to requiring that 0 is not an eigenvalue of $H_f(\mathbf{p})$, and is also equivalent to $H_f(\mathbf{p})$ being invertible.*

The generalization of our two-variable Morse Lemma in Theorem 2.9.17, which also serves as a generalization of the Second Derivative Test, is:

Theorem 2.9.24.   (Morse Lemma) *Suppose that $f = f(x_1, \ldots, x_n)$ is real-valued function of class C^r, where $r \geq 3$, on an open subset $\mathcal{U}$ of $\mathbb{R}^n$. Suppose further that $\mathbf{p}$, in $\mathcal{U}$, is a non-degenerate critical point of f.*

Then, there exists a C^{r-2} local change of coordinates at $\mathbf{0}$, $(x_1, \ldots, x_n) = \Phi(u_1, \ldots, u_n)$, such that $\Phi(\mathbf{0}) = \mathbf{p}$ and such that

$$(f \circ \Phi)(\mathbf{u}) \;=\; f(\mathbf{p}) - u_1^2 - u_2^2 - \cdots - u_i^2 + u_{i+1}^2 + \cdots + u_{n-1}^2 + u_n^2,$$

where i equals the number of negative eigenvalues (counted with multiplicities) of the matrix $H_f(\mathbf{p})$.

For the proof in the smooth case, see Lemma 2.2 of [6]. The C^r case follows the same proof; you just have to keep track of the differentiability class throughout.

The Morse Lemma holds when $r = \infty$; in this case, $r - 2$ also equals ∞. The Morse lemma is also true if f is real analytic at $\mathbf{p}$; in this case, the change of coordinates is also real analytic.

Remark 2.9.25. Note that f must be at least C^3 to apply the Morse Lemma, while the Second Derivative Test required only a C^2 function. Thus, For a C^2 function which is not C^3, we cannot derive the Second Derivative Test from the Morse Lemma. In fact, in the C^2 case, the Second Derivative Test is an easy consequence of the *Taylor-Lagrange Theorem*, Theorem 2.13.11.

2.9.1 Exercises

Online answers to select exercises are here.

Basics:

In each of the following exercises, you are given a function $f = f(x, y)$. (a) Determine the critical points of f. (b) Use technology to graph $z = f(x, y)$, and identify the points on the graph that correspond to your critical points, if possible.

1. $f(x, y) = (x - 1)^2 + (y + 2)^2 + x - 3y$.

2. $f(x, y) = x^3 + 3xy - y^3 + 2$.

3. $f(x, y) = x^3 + y^3 - 3x - 12y$.

4. $f(x, y) = x^2 + y^2 + x^2 y$.

5. $f(x, y) = (x^2 + 2x)(y - 1) + y^3 - 11y$.

6. $f(x, y) = y^2 + x \ln y - x + ey$.

7. $f(x, y) = x^2 + 1 - 2x \cos y$, where $-\pi < y < \pi$.

8. $f(x, y) = xy + \dfrac{25}{x} + \dfrac{5}{y}$.

In each of the following exercises, you are given a function $f = f(x, y)$ from the previous exercises. Verify that the critical points of f are non-degenerate, and classify each one as a point where f has a local maximum value, a local minimum value, or a saddle point.

9. $f(x, y) = (x - 1)^2 + (y + 2)^2 + x - 3y$.

10. $f(x, y) = x^3 + 3xy - y^3 + 2$.

11. $f(x, y) = x^3 + y^3 - 3x - 12y$.

12. $f(x, y) = x^2 + y^2 + x^2 y$.

13. $f(x, y) = (x^2 + 2x)(y - 1) + y^3 - 11y$.

14. $f(x, y) = y^2 + x \ln y - x + ey$.

15. $f(x, y) = x^2 + 1 - 2x \cos y$, where $-\pi < y < \pi$.

16. $f(x, y) = xy + \dfrac{25}{x} + \dfrac{5}{y}$.

In each of the following exercises, you are given level curves of $z = f(x, y)$ and scaled gradients vectors of f. Curves which are the same color are part of the same level curve. (a) Approximate the critical points of f, and classify each one as a point where f has a local maximum value, a local minimum value, or a saddle point. (b) Suppose that the values of f for the given level curves are $-2, -1, 0, 1$, and 2; label the level curves with the corresponding values of f.

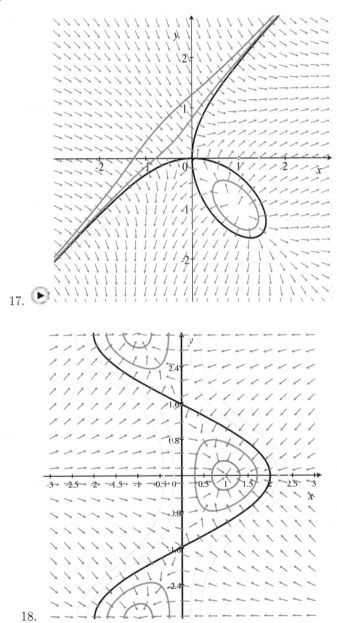

17.

18.

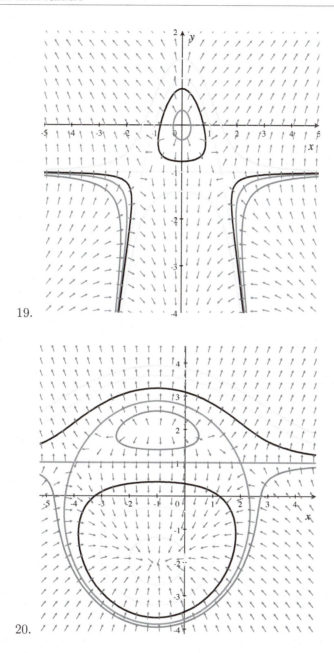

19.

20.

More Depth:

In each of the following exercises, you are given a function $f = f(x,y)$ from the previous exercises. For each critical point p, use the Morse Lemma, Theorem 2.9.17, to write f, after a coordinate change, as $f(\mathbf{p}) + u^2 + v^2$, $f(\mathbf{p}) - u^2 - v^2$, or $f(\mathbf{p}) + u^2 - v^2$, and then describe what this implies about the level curves of f near p.

21. $f(x,y) = (x-1)^2 + (y+2)^2 + x - 3y$.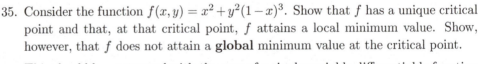

22. $f(x,y) = x^3 + 3xy - y^3 + 2$.

23. $f(x,y) = x^3 + y^3 - 3x - 12y$.

24. $f(x,y) = x^2 + y^2 + x^2 y$.

25. $f(x,y) = (x^2 + 2x)(y-1) + y^3 - 11y$.

26. $f(x,y) = y^2 + x \ln y - x + ey$.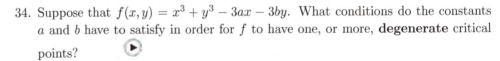

27. $f(x,y) = x^2 + 1 - 2x \cos y$, where $-\pi < y < \pi$.

28. $f(x,y) = xy + \dfrac{25}{x} + \dfrac{5}{y}$.

In each of the following exercises, you are given a function $f = f(x,y)$ and a point p in $\mathbb{R}^2$. Use the First Derivative Test, Theorem 2.9.19, to show that f attains a local extreme value at p.

29. $f(x,y) = \sqrt{x^4 + y^4}$ and $\mathbf{p} = (0,0)$.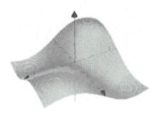

30. $f(x,y) = x^4 + x^2 + y^2 - 2xy + 2x - 2y + 1$ and $\mathbf{p} = (0,1)$.

31. $f(x,y) = xye^{-x-y}$ and $\mathbf{p} = (1,1)$.

32. $f(x,y) = xe^{-x}(4 - y^2)$, and $\mathbf{p} = (1,0)$.

33. Determine the 5 critical points of the function

$$f(x,y) \;=\; xye^{-x^2 - y^2},$$

and apply the Second Derivative Test to classify them. Verify that your results agree with what you see in Figure 2.9.14.

Figure 2.9.14: The graph of $z = xye^{-x^2-y^2}$.

34. Suppose that $f(x,y) = x^3 + y^3 - 3ax - 3by$. What conditions do the constants a and b have to satisfy in order for f to have one, or more, **degenerate** critical points?

35. Consider the function $f(x,y) = x^2 + y^2(1-x)^3$. Show that f has a unique critical point and that, at that critical point, f attains a local minimum value. Show, however, that f does not attain a **global** minimum value at the critical point.

This should be contrasted with the case of a single-variable differentiable function $h : \mathbb{R} \to \mathbb{R}$: If h has a unique critical point and, at that critical point, h attains a local minimum value, then h must attain a global minimum value at the critical point.

36. Suppose that a function $f = f(x, y)$ has a unique critical point and that, at that critical point, f has a saddle point. Can f attain a local maximum value anywhere? A local minimum value? Explain, or give examples.

37. Suppose that $f : \mathbb{R}^2 \to \mathbb{R}$ and $g : \mathbb{R}^2 \to \mathbb{R}$ are differentiable, and that f and g each attain local maximum values at a point **p**. Must $f + g$ attain a local maximum value at **p**? Explain, or give a counterexample.

38. Suppose that $f : \mathbb{R}^2 \to \mathbb{R}$ and $g : \mathbb{R}^2 \to \mathbb{R}$ are differentiable, and that f and g each attain local maximum values at a point **p**. Must fg (f times g) attain a local maximum value at **p**? Explain, or give a counterexample.

+ **Linear Algebra:**

In each of the following exercises, determine the critical points p of the given f, show that they are all non-degenerate, determine the number (with multiplicities) of negative eigenvalues of the Hessian matrix of f at each critical point p, and use the Morse Lemma, Theorem 2.9.24, to rewrite f, after a coordinate change, in the form $f(\mathbf{p}) - u_1^2 - u_2^2 - \cdots - u_i^2 + u_{i+1}^2 + \cdots + u_{n-1}^2 + u_n^2$.

39. $f(x, y, z) = x^3 + y^3 + z^3 - 3x - 12y - 27z$.

40. $f(x, y, z) = x^3 + 3xy - y^3 + z^2$.

41. $f(w, x, y, z) = w^2 - x^2 + 4yz - y^4 - z^4$.

42. $f(w, x, y, z) = w^2 + x^2 + w^2x + y^2 + z \ln y - z + ey$.

2.10   Optimization

Optimization refers to problems which involve maximizing or minimizing the value of a function, especially a function which come from modeling a physical or real-world problem. Unsurprisingly, this is related to our work on local extrema from the last section. However, here, we are interested solely in global extrema, and it requires some assumptions to even know that global extreme values exist. We either need for our function to be continuous and for its domain to be compact, or we need to appeal to physical/intuitive reasoning to decide that a global maximum or minimum exists.

Basics:

You might think that the Second Derivative Test for local maxima and minima would be useful here. However, knowing that a point yields a **local** extreme value doesn't answer the global question.

It's true that **some** useful information might be obtained from the Second Derivative Test, provided that the critical points are non-degenerate. If a function attains a local minimum (respectively, maximum) value at a point **p**, then certainly the function doesn't attain a global maximum (respectively, minimum) value at **p**, and if f has a saddle point at **p**, then f does not attain a global extreme value at **p**.

Generally, the information obtained from the Second Derivative Test is of such limited value that we do not use this test when looking at global extrema questions.

In many physical or real-world problems, you want to maximize or minimize the value of a function. You might want to minimize the amount of material required to make a container of a specific shape, or charge the right price for some manufactured items to maximize profit, or minimize the time it takes to travel from one place to another. There are many examples that we shall look at in this section and in its exercises.

Of course, a big part of such problems is to take the words that you're given and translate them into a mathematics problem. This is a question of *modeling*, taking a real-world problem and writing a mathematics problem which seems to describe, or model, the situation well.

In this section, all of our problems will be modeled by finding the global maximum or minimum value of a continuous real-valued function f, defined on a subset E of $\mathbb{R}^n$. We will deal with two types of problems:

1. ones in which E is a compact set (Definition 1.1.14), so that we can apply the Extreme Value Theorem (Theorem 1.7.13) to conclude that f attains both a global maximum and a global minimum value; or

2. ones in which physical or intuitive reasoning implies the existence of either a global maximum or a global minimum value, and there's only one critical point, so that must be where the global extreme value occurs.

Let's look at some examples.

Example 2.10.1. Suppose that a circular metal plate occupies the points in the xy-plane which satisfy $x^2 + y^2 \leq 1$, where x and y are measured in meters. Assume that the temperature of the plate, in °C, is given by

$$T = 2x^2 + 2y^2 - y.$$

Determine the hottest and coldest points on the plate.

Solution:

The function T is continuous, and the region under consideration, which we'll call D (for disk), is compact, i.e., closed and bounded (D is closed, since it contains its boundary, and bounded, since it doesn't extend out infinitely far). Thus, the Extreme Value Theorem guarantees that T attains both a global maximum and a global minimum value on D.

How do we actually locate the points at which T attains its global maximum and minimum values? Well...global extrema are certainly local extrema; so, by Theorem 2.9.6, the global extreme values must occur at critical points of the function T restricted to the domain D. This means that we need to check all of the interior critical points, plus all of the points on the boundary.

The function T is differentiable at each point in the interior of D. Thus, interior critical points are interior points where $\vec{\nabla}T = \mathbf{0}$. We need to solve:

$$\frac{\partial T}{\partial x} = 4x = 0 \qquad \text{and} \qquad \frac{\partial T}{\partial y} = 4y - 1 = 0.$$

These are certainly easy to solve simultaneously; we find one critical point

$$(x, y) = (0, 1/4).$$

We should remark here that, **had a solution to $\vec{\nabla}T = \mathbf{0}$ occurred outside the region D, we would have ignored it.**

However, $(0, 1/4)$ **is** in the region D, and so it's a possible location for a global extreme value of T on D. In fact, this is the **only** possible location for a global extreme value of T on the interior of D. But now we need to deal with the boundary.

We need to look for extreme values of T restricted to the boundary of D, i.e., restricted to the circle where $x^2 + y^2 = 1$. Notice that, if $x^2 + y^2 = 1$, then

$$T = 2x^2 + 2y^2 - y = 2(x^2 + y^2) - y = 2 \cdot 1 - y = 2 - y,$$

where y can be any y-coordinate of a point on the circle $x^2 + y^2 = 1$, i.e., $-1 \leq y \leq 1$.

Thus, on the boundary of D, we are reduced to a single-variable optimization problem; we need to find the global extreme values of $T = 2 - y$ on the closed, bounded interval $[-1, 1]$. We need to check for interior critical points, points in $(-1, 1)$, where $T' = 0$; but there aren't any. But we must also check the boundary of $[-1, 1]$, that is, we have to check where $y = -1$ and where $y = 1$. As we are looking at (x, y) pairs which satisfy $x^2 + y^2 = 1$, if $y = \pm 1$, then $x = 0$. Hence, the (x, y) pairs on the boundary of D, where global extrema of the original T can occur, are

$$(0, -1) \qquad \text{and} \qquad (0, 1).$$

When we restrict the function T from having domain D to a function with domain $x^2 + y^2 = 1$, we are technically producing a new function, and could/should give it a new name. However, for notational ease, we'll continue to denote this new restricted function by T, but keep in mind that it's the original rule for T, but with a new domain.

Okay. Great. Now we have three points in the region D at which $T = 2x^2 + 2y^2 - y$ could possibly attain its global extreme values. But how do we decide which of these points yield the global maximum value and which yield the global minimum value?

It's easy at this point; you simple evaluate the function T at these three points, and see where you get the biggest number and the smallest number.

(x, y)	$T = 2x^2 + 2y^2 - y$
$\left(0, \dfrac{1}{4}\right)$	$0 + 2\left(\dfrac{1}{4}\right)^2 - \dfrac{1}{4} = -\dfrac{1}{8}$
$(0, -1)$	$0 + 2(-1)^2 - (-1) = 3$
$(0, 1)$	$0 + 2(1)^2 - 1 = 1$

Figure 2.10.1: The hottest and coldest points on the circular plate.

Thus, the minimum temperature on the plate is $-1/8$ °C, which occurs at only one point $(0, 1/4)$, and the maximum temperature on the plate is 3 °C, which also occurs at only one point $(0, -1)$.

Example 2.10.2. What happens if we keep the "same" function from the previous example, but now look at a triangular plate, which occupies the region where $x \geq 0$, $y \geq 0$, and $x + y \leq 1$?

The temperature of the plate, in °C, is once again given by

$$T = 2x^2 + 2y^2 - y,$$

but the boundary now consists of 3 line segments. We still have a continuous function on a compact set, so global extreme values definitely exist.

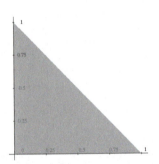

Figure 2.10.2: Our new triangular metal plate.

We proceed to find the global extrema in the same general way that we did in the last example; we first look for interior points where $\vec{\nabla} T = \mathbf{0}$, and then we have to analyze the boundary separately, by reducing ourselves to single-variable optimization problems.

In the interior:

Our gradient vector calculation hasn't changed; we find the same partial derivatives that we did before, and the same point where they're simultaneously 0, namely

$$(x, y) = (0, 1/4).$$

But now, the point $(0, 1/4)$ is **not** an interior point of our region; it's in the region, but it's on the boundary. Therefore, there are no interior critical points, and both the global maximum value and the global minimum value must be attained at points on the boundary.

You may be wondering "do I need to include $(0, 1/4)$ in my list of points to check"? The answer is: "at this point, no", but, on the other hand, if you did include it, it's no big deal. We'll find it again when we check the boundary.

On the boundary:

There are three distinct pieces to the boundary, and we must consider each one separately. This time, to be completely clear, we will denote the restricted versions of T by T_B, T_L, and T_S, when restricting to the bottom line segment, B, left-hand line segment, L, and slanted line segment, S, respectively.

On the boundary portion B, the bottom piece of the boundary, we have $y = 0$ and $0 \leq x \leq 1$. The function T, restricted to where $y = 0$, gives us

$$T_B \;=\; 2x^2 + 2y^2 - y \;=\; 2x^2.$$

Thus, we have the single-variable optimization problem: find the points where $T_B = 2x^2$ attains global maximum and global minimum values on the compact interval $0 \leq x \leq 1$. We look for critical points of T_B on the open interval $(0, 1)$, and we have to check the endpoints of the interval separately. But $T_B' = 4x = 0$ implies that $x = 0$, which is not in $(0, 1)$.

Thus, global extreme values of T_B, i.e., global extreme values of T restricted to B, can occur only at the endpoints of $[0, 1]$, that is, where $x = 0$ or $x = 1$. Of course, B is where $y = 0$, and so the (x, y) pairs that we need to put in our list to check are:

$$(x, y) = (0, 0) \qquad \text{and} \qquad (x, y) = (1, 0).$$

On the boundary portion L, the left piece of the boundary, we have $x = 0$ and $0 \leq y \leq 1$. The function T, restricted to where $x = 0$, gives us

$$T_L \;=\; 2x^2 + 2y^2 - y \;=\; 2y^2 - y.$$

Thus, we again have a single-variable optimization problem: find the points where $T_L = 2y^2 - y$ attains global maximum and global minimum values on the compact interval $0 \leq y \leq 1$. We look for critical points of T_L on the open interval $(0, 1)$, and we have to check the endpoints of the interval separately. Solving $T_L' = 4y - 1 = 0$ gives us $y = 1/4$. We also need to include the endpoints of the interval as points to check: $y = 0$ and $y = 1$.

The line segment B is where $x = 0$, and so the (x, y) pairs that we need to put in our list to check are:

$$(x, y) = (0, 1/4), \qquad (x, y) = (0, 0), \qquad \text{and} \qquad (x, y) = (0, 1).$$

As promised, we found the point $(0, 1/4)$ again. Note that $(0, 0)$ was already on our list of points to check. Not surprisingly, we've found that we need to check all three corner points of the boundary. It's always true that you have to check all sharp points on the boundary.

––––––––––

On the boundary portion S, the slanted piece of the boundary, we have $x + y = 1$, i.e., $x = 1 - y$, and $0 \le y \le 1$. The function T, restricted to where $x = 1 - y$, gives us

$$T_S \;=\; 2x^2 + 2y^2 - y \;=\; 2(1 - y)^2 + 2y^2 - y \;=$$

$$2(1 - 2y + y^2) + 2y^2 - y \;=\; 4y^2 - 5y + 2.$$

Thus, we have a third single-variable optimization problem: find the points where $T_S = 4y^2 - 5y + 2$ attains global maximum and global minimum values on the compact interval $0 \le y \le 1$. We look for critical points of T_S on the open interval $(0, 1)$, and we have to check the endpoints of the interval separately. Solving $T'_S = 8y - 5 = 0$ gives us $y = 5/8$. We also, again, need to include the endpoints of the interval as points to check: $y = 0$ and $y = 1$.

The line segment S is where $x = 1 - y$, and so the (x, y) pairs that we need to put in our list to check are:

$$(x, y) = (3/8, 5/8), \qquad (x, y) = (1, 0), \qquad \text{and} \qquad (x, y) = (0, 1).$$

Of course, we already had these two corner points; the only new point for our list is $(3/8, 5/8)$.

––––––––––

Finally, we take all of the possible points where T could attain global extrema, and plug them into T:

(x, y)	$T = 2x^2 + 2y^2 - y$
$(0, 0)$	0
$(1, 0)$	$2(1)^2 + 0 = 2$
$\left(0, \dfrac{1}{4}\right)$	$0 + 2\left(\dfrac{1}{4}\right)^2 - \dfrac{1}{4} = -\dfrac{1}{8}$
$(0, 1)$	$0 + 2(1)^2 - 1 = 1$
$(3/8, 5/8)$	$2(3/8)^2 + 2(5/8)^2 - 5/8 = 7/16$

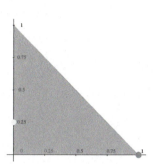

Figure 2.10.3: The hottest and coldest points on the triangular plate.

Therefore, the minimum temperature on the plate is (again) $-1/8\,°\mathrm{C}$, which occurs at only one point $(0, 1/4)$, and the maximum temperature on the plate is now $2\,°\mathrm{C}$, which also occurs at only one point $(1, 0)$.

Example 2.10.3. Let E be the solid region in $\mathbb{R}^3$ which is above the half-cone where $z = \sqrt{x^2 + y^2}$ and below the plane where $z = 5$; this is a solid half-cone. Let's find the maximum and minimum values of

$$f(x, y, z) = x^2 + y^2 - z^2 - x - y$$

on E.

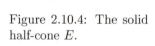

Figure 2.10.4: The solid half-cone E.

Before we go any further, note that the boundary of E breaks up into two pieces: the top piece T, where $z = 5$ and $\sqrt{x^2 + y^2} \leq 5$, and the bottom piece B, where $z = \sqrt{x^2 + y^2}$ and $0 \leq z \leq 5$. Furthermore, each of T and B have a boundary, which is the intersection of T and B, i.e., the circle C where T and B intersect. The circle C is where $z = 5$ and $\sqrt{x^2 + y^2} = 5$, i.e., $z = 5$ and $x^2 + y^2 = 25$.

Now, we look for critical points of f on the interior of E, and look at f restricted to pieces of the boundary of E.

On the interior of E, we look at points where

$$f_x = 2x - 1 = 0, \qquad f_y = 2y - 1 = 0, \qquad \text{and} \qquad f_z = -2z = 0.$$

This last equation tells us that $z = 0$, and the only point in E, where $z = 0$, is the origin. But the origin is on the boundary of E. Thus, f has no interior critical points, and so the maximum and minimum values of f are obtained on the boundary of E.

We restrict our function f to the top piece T, and obtain

$$f_T(x, y) = x^2 + y^2 - 25 - x - y,$$

where (x, y) satisfies $\sqrt{x^2 + y^2} \leq 5$, i.e., is in the closed disk D of radius 5, centered at the origin.

Critical points of f_T in the interior of D occur where

$$\frac{\partial f_T}{\partial x} = 2x - 1 = 0 \quad \text{and} \quad \frac{\partial f_T}{\partial y} = 2y - 1 = 0;$$

that is, at $(x, y) = (1/2, 1/2)$. We're restricted to where $z = 5$, so the triple that we need to check is

$$(x, y, z) = (1/2, 1/2, 5),$$

which is in the interior of D.

But we must also look at f_T restricted to the boundary of D. The boundary of D is the circle C that we defined at the beginning, where $z = 5$ and $x^2 + y^2 = 25$. So, we write f_C for f_T (or f), restricted to C, and we have $f_C(x, y) = -x - y$.

If we parameterize C by $x = 5 \cos t$, $y = 5 \sin t$, $z = 5$, and $0 \leq t \leq 2\pi$, then we find

$$f_C(t) = -5 \cos t - 5 \sin t.$$

Now, we calculate that $f_C'(t) = 5 \sin t - 5 \cos t$, and this equals 0 when $\sin t = \cos t$, i.e., at $t = \pi/4$ and $t = 5\pi/4$. Thus, two more points that we need to check for extreme values are

$$(x, y, z) = (5 \cos(\pi/4), 5 \sin(\pi/4), 5) = (5/\sqrt{2}, 5/\sqrt{2}, 5)$$

and

$$(x, y, z) = (5 \cos(5\pi/4), 5 \sin(5\pi/4), 5) = (-5/\sqrt{2}, -5/\sqrt{2}, 5).$$

However, we also need to check at the endpoints (the boundary) of the interval $0 \leq t \leq 2\pi$; as $t = 0$ and $t = 2\pi$ give us the same point, we have one new point to check:

$$(x, y, z) = (5 \cos(0), 5 \sin(0), 5) = (5, 0, 5).$$

It remains for us to look at f_B, the restriction of f to B, where $z = \sqrt{x^2 + y^2}$ and $0 \leq z \leq 5$. This gives us

$$f_B(x, y) = -x - y,$$

defined on the disk D where $x^2 + y^2 \leq 25$. Note that we already looked at f_C, which is the restriction of f_B to the boundary circle C of the disk. Thus, we need worry about only critical points of f_B on the interior of D. But both partial derivatives of f_B are -1, and so are never 0.

Therefore, we have no additional points to add to our list of points that we need to check.

Thus, we plug the points $(1/2, 1/2, 5)$, $(5/\sqrt{2}, 5/\sqrt{2}, 5)$, $(-5/\sqrt{2}, -5/\sqrt{2}, 5)$, and $(5, 0, 5)$ into $f(x, y, z) = x^2 + y^2 - z^2 - x - y$, and find that the largest value is

$$f(-5/\sqrt{2}, -5/\sqrt{2}, 5) \ = 5\sqrt{2},$$

and the smallest value is

$$f(1/2, 1/2, 5) \ = \ -51/2.$$

Example 2.10.4. Here, we will consider a problem like some of those from Section 1.3, which we could handle without Calculus. But, now, we'd like to approach the problem as an optimization problem.

The problem is to find the point on the plane $2x + y - z = 1$ which is closest to the point $(0, 1, 3)$; that is, we want to find the point which minimizes the function "distance to $(0, 1, 3)$, restricted to points where $2x + y - z = 1$".

Thus, we consider the function

$$D(x, y, z) \ = \ \sqrt{(x - 0)^2 + (y - 1)^2 + (z - 3)^2},$$

with its domain restricted to those (x, y, z) such that $2x + y - z = 1$. The condition that we must have $2x + y - z = 1$ is referred to as a *constraint*, and we say that "we want to minimize D, subject to the constraint that $2x + y - z = 1$".

However, we can get slightly tricky, and simplify our problem by noting that a point minimizes D if and only if it minimizes D^2, and squaring D gets rid of the messy square root.

So, our problem is to find the point which minimizes

$$D^2 \ = \ x^2 + (y - 1)^2 + (z - 3)^2,$$

subject to the constraint that $2x + y - z = 1$.

It may be tempting to start taking partial derivatives with respect to x, y, and z, and finding where $\vec{\nabla} D^2 = \mathbf{0}$, **but you can't do that yet.** Why not? Because a partial derivative means a derivative with respect to one variable while all other independent variables are held constant and, **because of the constraint**, x, y, and z cannot vary independently; you can't keep y and z constant, and let x change, if you always have to have $2x + y - z = 1$.

How do we get around this? We solve the constraint for one of the variables, in terms of the other two, and rewrite D^2 in terms of those two independent variables. Here, for instance, we write that the constraint is equivalent to

$$z \; = \; 2x + y - 1,$$

and so, in terms of x and y, we have

$$f(x, y) \; = \; D^2 \; = \; x^2 + (y-1)^2 + (2x + y - 4)^2,$$

where x and y now vary independently, and can be any real numbers.

Notice that this function $f(x, y) = D^2$ definitely does **not** have a domain which is compact; the domain is all of $\mathbb{R}^2$. However, we appeal to geometric intuition that there is, in fact, a point on the plane $2x + y - z = 1$ which is closest to $(0, 1, 3)$ and, hence, if we find that f has a single critical point, then it must be where D^2 (and, thus D) attains a global minimum value.

We find

$$f_x \; = \; 2x + 4(2x + y - 4) \; = \; 10x + 4y - 16$$

and

$$f_y \; = \; 2(y - 1) + 2(2x + y - 4) \; = \; 4x + 4y - 10.$$

Thus, critical points occur precisely where

$$10x + 4y \; = \; 16 \qquad \text{and} \qquad 4x + 4y = 10.$$

Subtracting the two equations yields

$$6x = 6 \qquad \text{and so } x = 1,$$

from which we conclude that $y = 6/4 = 3/2$. These are the x- and y-coordinates of the point on the plane $2x + y - z = 1$ which is closest to $(0, 1, 3)$. How do we find the z-coordinate? Use that $z = 2x + y - 1$ to obtain that

$$z \; = \; 2(1) + (3/2) - 1 \; = \; 5/2.$$

Therefore, we find that the point on the plane $2x + y - z = 1$ which is closest to $(0, 1, 3)$ is the point $(1, 3/2, 5/2)$.

Example 2.10.5. Suppose that a cardboard box is to be constructed with no top and a volume of 4000 cubic inches. Find the dimensions of the box which minimize the amount of cardboard required, i.e., which minimize the surface area.

Solution: We'll call the length, width, and height of the box L, W, and H, respectively. Note that the box pictured in Figure 2.10.5 is **not** drawn with dimensions that correspond to a box with minimum surface area.

Because of the volume requirement, we need

$$LWH \; = \; 4000,$$

where L, W and H are measured in inches. As in Example 2.10.4, this type of equation, which prevents the variables from being able to change independently, is called a constraint.

We will denote the surface area, in square inches, by A. We have two sides of area WH, two sides of area LH, and the bottom of area LW; thus

$$A \; = \; 2WH \; + \; 2LH \; + \; LW, \tag{2.10}$$

and we want to minimize A, subject to the constraint that $LWH = 4000$. We assume that it is physically obvious that such a minimum area exists.

We cannot simply take the partial derivatives of A with respect to L, W, and H and set them equal to 0 to find critical points; our constraint prevents the variables from varying independently. As in Example 2.10.4, we solve the constraint equation for one of the variables, and then substitute into the area function to obtain an expression for the area in terms of two variables which are independent.

We solve the constraint for H to obtain

$$H \; = \; \frac{4000}{LW},$$

and substitute into the area equation, Formula 2.10, to find

$$A \; = \; 2W \cdot \frac{4000}{LW} \; + \; 2L \cdot \frac{4000}{LW} \; + \; LW \; = \; 8000L^{-1} \; + \; 8000W^{-1} \; + \; LW,$$

in which L and W may now vary independently.

We find the critical point(s) of A by solving

$$\frac{\partial A}{\partial L} \; = \; -8000L^{-2} + W = 0 \quad \text{and} \quad \frac{\partial A}{\partial W} \; = \; -8000W^{-2} + L = 0.$$

We find

$$W \; = \; \frac{8000}{L^2} \quad \text{and} \quad L \; = \; \frac{8000}{W^2}$$

and so

$$W \; = \; \frac{8000}{(8000)^2/W^4} \; = \; \frac{W^4}{8000}.$$

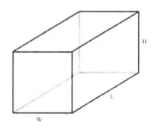

Here, by "box", we mean what we typically think of as a box shape, i.e., the base and sides are rectangles, and the sides are perpendicular to the base.

Figure 2.10.5: A box with no top.

Therefore, either $W = 0$ or $W^3 = 8000$. We can't have $W = 0$ and satisfy the constraint (or have a box). Hence, we need $W^3 = 8000$, i.e., $W = 20$. Thus, we find that the unique critical point of A, which satisfies the constraint, occurs when $W = 20$.

Plugging this into $L = 8000/W^2$, we find that $L = 20$. Plugging $W = 20$ and $L = 20$ into the constraint equation, we find that

$$H = \frac{4000}{20 \cdot 20} = 10.$$

Thus, the dimensions of the box, with a volume of 4000 cubic inches, with no top, that has minimum surface area are $L = W = 20$ inches and $H = 10$ inches.

More Depth:

In the first two problems of this section, our functions were continuous and the domains were compact, i.e., closed and bounded. This enabled us to use the Extreme Value Theorem, Theorem 1.7.13, to conclude that the functions in question attained global extreme values on E. However, in each of the last two problems, we did not have compact domains, and we appealed to "physical obviousness" to know that a global minimum existed; then we found only one critical point, so that had to be where the global minimum occurred. What we would like to do now is show how you can eliminate the appeal to "physical obviousness" and, instead, prove carefully that a global minimum exists in each of our last two examples above.

If we were trying to maximize the value of f, we would want to find a compact region E, outside of which the value of f is small.

Our attack on each of the problems is to produce a nice, compact region, which we'll call E, such that, outside of E, the function f in question is huge. We then do the optimization procedure for f restricted to E, and then show that the minimum value of f on E is smaller than the large values of f outside of E.

Example 2.10.6. In Example 2.10.4, we reduced our problem to minimizing the function

$$f(x, y) = D^2 = x^2 + (y - 1)^2 + (2x + y - 4)^2,$$

where x and y can be any real numbers. We found a single critical point of $(x, y) = (1, 3/2)$ and, since we assumed that a global minimum value of f existed, it had to occur at this single critical point. Here, we will **prove** that a minimum exists.

Because $(2x + y - 4)^2 \geq 0$, we know that, if $x^2 + (y - 1)^2 > 100$, then

$$f = x^2 + (y - 1)^2 + (2x + y - 4)^2 > 100.$$

Therefore, we define the set E to consist of the ordered pairs (x, y) such that

$$x^2 + (y - 1)^2 \leq 100,$$

and then we know that, for points (x, y) outside of E, $f(x, y) > 100$.

The set E is compact; it is the disk whose boundary is the circle where $x^2 + (y-1)^2 = 100$. We let $\hat{f}$ denote the function f restricted to the set E. Then, the Extreme Value Theorem tells us that $\hat{f}$ attains a global minimum value on E (and a global maximum value, which we don't care about). As we know that f is greater than 100 outside of E, if the global minimum value of $\hat{f}$ on E is ≤ 100, then we know that the global minimum of the original function f, on all of $\mathbb{R}^2$, is the same as the global minimum value of $\hat{f}$ on E.

The global minimum value of $\hat{f}$ can occur only at critical points in the interior of E or on the boundary of E, where $x^2 + (y-1)^2 = 100$. But, again, because $(2x + y - 4)^2 \geq 0$, $\hat{f} \geq 100$ on the boundary of E (where $x^2 + (y-1)^2 = 100$). The only critical point of $\hat{f}$ on the interior of E is the same point that we found in Example 2.10.4: $(x, y) = (1, 3/2)$, where the value of $\hat{f}$ is

$$\hat{f}(1, 3/2) = f(1, 3/2) = 1 + (1/2)^2 + (-1/2)^2 = 3/2.$$

Since $3/2 < 100$, the global minimum value of $\hat{f}$ on E and, hence, the global minimum value of f occurs at the single point $(1, 3/2)$ and the value is $3/2$.

At this point, if you want the closest point on the plane $2x + y - z = 1$, as was really the problem in Example 2.10.4, you proceed as we did in that example and produce the z-coordinate by using $(x, y) = (1, 3/2)$ and $z = 2x + y - 1 = 5/2$.

Example 2.10.7. In Example 2.10.5, we boiled our problem down to needing to find the global minimum value of

$$A = 8000L^{-1} + 8000W^{-1} + LW,$$

where $L > 0$ and $W > 0$. We shall prove rigorously that that global minimum value is attained at the unique point $(L, W) = (20, 20)$.

Consider the compact region E of all of those points (L, W) such that $1 \leq L \leq 10,000$ and $1 \leq W \leq 10,000$. We wish to see that A is big outside of E.

If $0 < L < 1$ and $W > 0$, then certainly $A > 8000$. Similarly, if $0 < W < 1$ and $L > 0$, then $A > 8000$. If both L and W are ≥ 1 and either one is $> 10,000$, then certainly $A > 10,000$. Therefore, outside of E (still assuming that $L > 0$ and $W > 0$), A is definitely greater than the smaller of 8000 and 10,000, i.e., $A > 8000$. On the boundary of E, the same reasoning tells us that $A \geq 8000$. We could get better lower bounds for A outside of E and on the boundary, but we don't need the best lower bounds; we just need **some** large lower bound.

In the interior of E, the only critical point is the one we found in Example 2.10.5: $(L, W) = (20, 20)$. The value of A at $(20, 20)$ is $A(20, 20) = 400 + 400 + 400 = 1200$.

How did we come up with this region E? We just picked **some** region, outside of which we could easily tell that A was big. In more-complicated problems, coming up with a suitable region E can be much more difficult.

The global minimum of A on E must occur at an interior critical point or a point on the boundary, but, at points on the boundary $A \geq 8000$. Hence, the global minimum value of A on E is 1200, which occurs at the single point $(L, W) = (20, 20)$, and this is the global minimum value of A for all $L > 0$, $W > 0$, since, outside of E, $A > 8000$.

Online answers to select exercises are here.

2.10.1   Exercises

Basics:

In each of the following problems, you are given a function $f(x, y)$ on a compact region E in $\mathbb{R}^2$. Find the maximum and minimum values of f on E, and the points at which these extreme values are attained.

1. $f(x, y) = xy$, and E is the filled rectangle where $-2 \leq x \leq 2$ and $-1 \leq y \leq 1$.
   ▶

2. $f(x, y) = xy$, and E is the filled circle (so, a disk) of radius 2, centered at the origin.

3. $f(x, y) = xy$, and E is the filled triangle with vertices $(0, 1)$, $(1, 1)$, and $(1, 2)$.

4. $f(x, y) = x^2 \sin y + x$, and E is the filled rectangle where $-1 \leq x \leq 1$ and $0 \leq y \leq \pi$.

5. $f(x, y) = y e^{xy} + e^{xy}$, and E is the filled triangle with vertices $(0, 0)$, $(-2, -2)$, and $(-2, 2)$.

6. $f(x, y) = x^2 \ln y - 4y$, and E is the filled rectangle where $-3 \leq x \leq 3$ and $0.5 \leq y \leq 1.5$.

7. $f(x, y) = x^2 \ln y + 4y$, and E is the filled "curved triangle" where $0 \leq x \leq 4$ and $1 \leq y \leq e^x$. (Hint: Sketch the region E.)

8. $f(x, y) = 3x^2 - x^3 - 4y^2 + y^4$, and E is the (closed) disk of radius 2, centered at the origin.   ▶

In each of the following problems, you are given a function $f(x, y, z)$ on a compact region E in $\mathbb{R}^3$. Find the maximum and minimum values of f on E, and the points at which these extreme values are attained.

9. $f(x, y, z) = x^2 + y^2 - z^2 - x - y$, and E is the closed ball of radius 3, centered at the origin, i.e., E is where $x^2 + y^2 + z^2 \leq 3^2$.

10. $f(x, y, z) = x^2 + y^2 - z^2 - x - y$, and E is the closed right, circular cylinder where $x^2 + y^2 \leq 3^2$ and $-4 \leq z \leq 4$.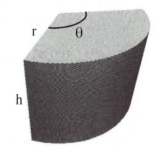

11. $f(x, y, z) = x^2 + yz - 2x - y - 2z$, and E is the solid cube where $0 \leq x \leq 3$, $0 \leq y \leq 3$, and $0 \leq z \leq 3$.

12. $f(x, y, z) = x \ln y + ze^{-z} - y$, and E is the rectangular solid where $0 \leq x \leq 2$, $0.5 \leq y \leq 1.5$, and $0 \leq z \leq 3$.

13. Suppose that a cardboard box is to be constructed with no top and a volume of 4000 cubic inches. Also, suppose that the cardboard for the bottom costs 5 cents per square inch, while the cardboard for the sides costs 1 cent per square inch. Find the dimensions of the box which minimize the cost of the cardboard required.

14. Suppose that a cardboard box is to be constructed with no top and a volume of 4000 cubic inches. Also, suppose that the cardboard for the bottom costs 5 cents per square inch, while the cardboard for the sides costs 1 cent per square inch. Find the dimensions of the box which minimize, and those which maximize, the cost of the cardboard required, if the length L, width W, and height H, all in inches, must satisfy $1 \leq L \leq 20$, $1 \leq W \leq 20$, and $1 \leq H \leq 20$.

15. Suppose that a cardboard box (with a top) is to be constructed with a volume of 4000 cubic inches. Also, suppose that the cardboard for the bottom costs 4 cents per square inch, the cardboard for the sides costs 2 cents per square inch, and the cardboard for the top costs 1 cent per square inch. Find the dimensions of the box which minimize the cost of the cardboard required.

16. Suppose that a cardboard box (with a top) is to be constructed, at a cost of \$5.40. Also, suppose that the cardboard for the bottom costs 4 cents per square inch, the cardboard for the sides costs 2 cents per square inch, and the cardboard for the top costs 1 cent per square inch. What is the maximum volume of such a box?

17. A curved prism has a base which is a sector of a disk of radius r inches and central angle θ radians. The prism extends vertically to a height of h inches.

    (a) Derive formulas for the volume V, in cubic inches, and the total surface area (including the top and bottom) A, in square inches.

    (b) If V is required to be 27, what values of r, θ, and h minimize the surface area A of the prism?

18. Now, suppose that the prism from the previous exercise is required to have $V = 27$, $1 \leq r \leq 4$, and $1 \leq \theta \leq 3$. Find the dimensions of the prism that maximize the total surface area of the five sides.

19. Near a small airport, there is a ridge. A cross section of the surface of the ridge is modeled by the equation $y = 64 - x^2$, where x and y are measured in yards. The standard flight path over the ridge is given by $y = 100 - (x/2)$. The airport wishes to place a warning beacon on the ridge at the point closest to the flight path. At what x-coordinate on the ridge should they place the beacon?

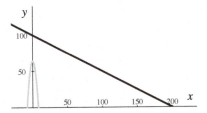

Figure 2.10.6: Ridge: $y = 64 - x^2$. Flight path: $y = 100 - (x/2)$.

More Depth:

20. A non-conductive plate occupies the square region where $1 \leq x \leq 2$ and $1 \leq y \leq 2$, with x and y measured in meters. Suppose that the electric potential at each point on the plate produced by point-charges at $(0,0)$ and $(3,3)$ is

$$P(x,y) = -\frac{1}{\sqrt{x^2 + y^2}} - \frac{2.25}{\sqrt{(x-3)^2 + (y-3)^2}} \quad \text{joules.}$$

Show that the global maximum and minimum values of the potential must occur on the boundary. (Hint: This is one of those unusual global optimization problems where the Second Derivative Test, Theorem 2.9.12, is useful.)

21. In Exercise 15, above, you were (implicitly) told that the cost function C attained a global minimum value. You wrote C as a function of two variables, $C = C(x,y)$, where x and y were two of the dimensions (of course, you may have used other variable names). Then, you found only one critical point for $C(x,y)$; hence, your unique critical point had to yield the global minimum value for C. Call this minimum value m.

    Now, prove that your answer to Exercise 15 was correct by producing a compact region E in $\mathbb{R}^2$ such that:

    • the minimum value of $C = C(x,y)$, restricted to E, is m;

    • for all $x > 0$ and $y > 0$ such that (x,y) is not in E, $C(x,y) > m$.

22. In Exercise 16, above, you were (implicitly) told that the volume function V attained a global maximum value. You wrote V as a function of two variables, $V = V(x,y)$, where x and y were two of the dimensions (of course, you may have used other variable names). Then, you found only one critical point for $V(x,y)$;

hence, your unique critical point had to yield the global maximum value for V. Call this maximum value M.

Now, prove that your answer to Exercise 16 was correct by producing a compact region E in $\mathbb{R}^2$ such that:

- the maximum value of $V = V(x, y)$, restricted to E, is M;

- for all $x > 0$ and $y > 0$ such that (x, y) is not in E, $V(x, y) < M$.

23. In Exercise 19, above, you were (implicitly) told that the distance, or distance-squared, function D attained a global minimum value. You wrote D as a function of two variables, $D = D(a, b)$, where a and b were x-coordinates of points on the two different curves (of course, you may have used other variable names). Then, you found only one critical point for $D(a, b)$; hence, your unique critical point had to yield the global minimum value for D. Call this minimum value m.

Now, prove that your answer to Exercise 19 was correct by producing a compact region E in $\mathbb{R}^2$ such that:

- the minimum value of $D = D(a, b)$, restricted to E, is m;

- for all a and b such that (a, b) is not in E, $D(a, b) > m$.

24. Consider the function $f(x, y) = x^2 + y^2(1 - x)^3$.

   (a) What is $\lim_{x \to \infty} f(x, 0)$?

   (b) What is $\lim_{x \to \infty} f(x, 1)$?

   (c) Show that f has a unique critical point.

   (d) How do you know that the unique critical point of f does not yield a global extreme value of f?

   (e) Produce a compact subset E of $\mathbb{R}^2$ so that the unique critical point $\mathbf{p}$ of f is in the interior of E, and so that the restriction $\hat{f}$ of f to E attains a global minimum value at $\mathbf{p}$.

2.11   Lagrange Multipliers

As we have seen in earlier sections, we sometimes want to maximize or minimize a function $f = f(\mathbf{x})$, where we wish to consider only those $\mathbf{x}$'s which satisfy a *constraint* equation $g(\mathbf{x}) = c$. In those earlier sections, we solved the constraint equation for one of the x_i's, and substituted that into f, in order to produce a function in which the variables could vary independently.

In this section, we will present a different approach: the method of *Lagrange multipliers*. One advantage to this method is that it does not require us to be able to solve algebraically for any of the variables in the constraint equation.

In the More Depth portion of the section, we shall deal with the case in which you have more than one constraint equation.

Basics:

In the previous section, we looked at two problems of the form: minimize the value of a function $f = f(\mathbf{x})$, subject to the constraint $g(\mathbf{x}) = c$, where c was a constant. The difficulty is that requiring that the constraint be satisfied means that the individual variables in $\mathbf{x} = (x_1, \ldots, x_n)$ are not allowed to vary independently.

To have a specific example to work with, let's look at one of our earlier problems: minimize the value of

$$f = f(x, y, z) = x^2 + (y - 1)^2 + (z - 3)^2,$$

subject to the constraint that $2x + y - z = 1$.

This means that we want to find the minimum value of $f(x, y, z)$ among all of those values that we get from points (x, y, z) which satisfy the constraint equation. In the previous section, we solved the constraint equation for z to find $z = 2x + y - 1$, and then substituted this into the function f to reduce our problem to minimizing the value of the function $x^2 + (y - 1)^2 + (2x + y - 4)^2$, where x and y do, in fact, vary independently, while x, y and z did not.

But what would you do if you couldn't explicitly solve the constraint equation for one of the variables? Is there another way to deal with constrained optimization problems? The answer is "yes", and the new method that we will discuss here is called the method of *Lagrange multipliers*.

Joseph-Louis Lagrange (1736 - 1813) was born in Turin, Italy as Giuseppe Lodovico Lagrangia. Lagrange had French ancestry and, even as a youth, used the French form of his family name, which is how he is now commonly known. Lagrange made significant contributions to many, many areas of mathematics and physics. In particular, he invented the prime notation for derivatives, proved that the Taylor remainder could be written in the Lagrange Form, and developed the method of Lagrange multipliers.

We will first consider a special case, to make the general case more comprehensible.

Suppose that we have $f = f(x, y, z)$, and let $g(x, y, z) = z$. We want to find local (or, possibly, global) maxima and minima of f, subject to the constraint that $g(x, y, z) =$

$z = c$, where c is a constant. This is easy. We look at the function of two variables $\hat{f} = f(x, y, c)$ and need to find where it attains local extrema. We know that these can occur only at critical points. Therefore, we look at points (x, y) such that

$$\frac{\partial \hat{f}}{\partial x} = 0 \quad \text{and} \quad \frac{\partial \hat{f}}{\partial y} = 0.$$

In terms of the original f, this means that we look for points (x, y, z) such that

$$z = c \quad \text{and} \quad \vec{\nabla} f(x, y, z) = (0, 0, \lambda) = \lambda(0, 0, 1),$$

where λ could be any real number.

Writing the conditions in terms of g, and noting that $\vec{\nabla} z = (0, 0, 1)$, we find that the points we're interested in are those points (x, y, z) such that

$$g(x, y, z) = c \quad \text{and} \quad \vec{\nabla} f(x, y, z) = \lambda \vec{\nabla} g(x, y, z),$$

where λ could be any real number.

Do these formulas work with more general constraints? Yes. In fact, the Implicit Function Theorem, Theorem 2.12.2, tells us that, if $\vec{\nabla} g \neq \mathbf{0}$, then, after a change of coordinates, the level set where $g(x, y, z) = c$ "becomes" the level set where one coordinate is constant.

Intuitively, what's going on here? Of course we need the constraint equation $g(x, y, z) = c$ to be satisfied – it's a requirement in the problem. But why do we need that the gradient vector of f is a scalar multiple of the gradient vector of g? And do we really need that $\vec{\nabla} g \neq \mathbf{0}$, as we wrote in the previous paragraph?

Suppose that f and g are smooth functions. Then, the Implicit Function Theorem, in its geometric form (Theorem 2.7.16), tells us that $\vec{\nabla} g$ being non-zero at a point $\mathbf{p}$ implies that the level set of g through $\mathbf{p}$ is smooth at $\mathbf{p}$. Suppose now that $\vec{\nabla} g(\mathbf{p}) \neq \mathbf{0}$. If $\vec{\nabla} f(\mathbf{p}) = \lambda \vec{\nabla} g(\mathbf{p})$ and $\lambda = 0$, then $\vec{\nabla} f(\mathbf{p}) = \mathbf{0}$ and $\mathbf{p}$ is a critical point of f, even if you ignore the constraint; it's not surprising that such a point would be a candidate for where f could attain a local extreme value. If $\vec{\nabla} f(\mathbf{p}) = \lambda \vec{\nabla} g(\mathbf{p})$ and $\lambda \neq 0$, then $\vec{\nabla} f(\mathbf{p}) \neq \mathbf{0}$ and, since the gradient vectors are perpendicular to the tangent sets to $f = f(\mathbf{p})$ and $g = g(\mathbf{p})$ at $\mathbf{p}$, we see that the level sets $f = f(\mathbf{p})$ and $g = g(\mathbf{p})$ have the same tangent set at $\mathbf{p}$; in other words, the level sets $f = f(\mathbf{p})$ and $g = g(\mathbf{p})$ just glance off of each other.

To picture this, consider Figure 2.11.1, in which we have graphed, in gray, the circular paraboloid P given by $g(x, y, z) = -x^2 + y - (z - 1)^2 = 1$ and three level sets of $f(x, y, z) = x^2 + y^2 + z^2$, which is the distance squared from the origin. Notice how the orange level set of f just glances off of P, and does so at the point on P which is closest to the origin, i.e., at which f attains local (in fact, global) minimum.

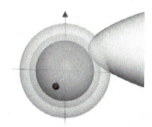

Figure 2.11.1: A fixed level set of g, and three level sets of f.

Why should points on P at which f attains local extrema also be points where the level set of f just glances off of P? If you're at a point $\mathbf{p}$ on P at which f attains a local minimum, then, if you decrease f slightly, the new level set of f can't hit P anymore, near $\mathbf{p}$ anyway. If you're at a point $\mathbf{p}$ on P at which f attains a local maximum, then, if you increase f slightly, the new level set of f can't hit P anymore, near $\mathbf{p}$. Thus, if f attains a local extreme value at a point $\mathbf{p}$, then changing the value of f slightly makes the level set of f not intersect P near $\mathbf{p}$; in this sense, the level set of f glances off of P at $\mathbf{p}$. While it may seem geometrically "clear" that this means that the level set of f and P should have the same tangent set at $\mathbf{p}$, this really requires a proof; see the More Depth portion of this section.

We should point out that we don't really need *smooth* functions for our optimization purposes; though it's handy in geometric discussions to discuss smooth functions and smooth graphs. Requiring our functions to be continuously differentiable is enough.

In light of this discussion, we will make a definition of a critical point of a function, subject to a constraint, and state the "obvious" theorem that relates critical points and local extrema.

Definition 2.11.1. *Suppose that $f = f(\mathbf{x})$ and $g = g(\mathbf{x})$ are continuously differentiable functions on an open subset of $\mathbb{R}^n$, where $n \geq 2$. Let M denote the level set given by $g(\mathbf{x}) = c$. Suppose that, for all $\mathbf{x}$ in M, $\vec{\nabla}g(\mathbf{x}) \neq \mathbf{0}$.*

---

*Then, a **critical point of f, subject to the constraint** $g(\mathbf{x}) = c$ is a point $\mathbf{x}$ such that $g(\mathbf{x}) = c$ and such that there exists a scalar λ such that*

$$\vec{\nabla}f(\mathbf{x}) \;=\; \lambda\vec{\nabla}g(\mathbf{x}).$$

*The scalar λ is called a **Lagrange multiplier**, and the set M is called the **constraint set** or **constraint submanifold**.*

---

*A critical point of f, subject to the constraint $g(\mathbf{x}) = c$, is also called a **critical point of $f_{|_M}$**, the restriction of f to the constraint submanifold.*

Given our earlier definitions, results, and the discussion above, it should come as no surprise that we have the following theorem.

Theorem 2.11.2. *Suppose that $f = f(\mathbf{x})$ and $g = g(\mathbf{x})$ are continuously differentiable functions on an open subset of $\mathbb{R}^n$, where $n \geq 2$. Let M denote the level set given by $g(\mathbf{x}) = c$. Suppose that, for all $\mathbf{x}$ in M, $\vec{\nabla}g(\mathbf{x}) \neq \mathbf{0}$.*

Then, if $\mathbf{p}$ is a point at which f, subject to the constraint $g(\mathbf{x}) = c$, attains a local extreme value, then $\mathbf{p}$ must be a critical point of f, subject to the constraint $g(\mathbf{x}) = c$.

In other words, if the restriction $f_{|M}$ attains a local extreme value at a point $\mathbf{p}$ in M, then $\mathbf{p}$ must be a critical point of $f_{|M}$.

Example 2.11.3. Use Lagrange multipliers to find the global maximum and minimum values of $f(x, y) = 4x^2 + y^2$ on the circle given by $x^2 + y^2 = 1$, i.e., subject to the constraint that $g(x, y) = x^2 + y^2 = 1$.

Solution: The function f is continuous, and the circle is compact; so, the Extreme Value Theorem, Theorem 1.7.13, tells us that f does, in fact, attain global extreme values on the circle. These global extreme values are certainly local extreme values, and so, by Theorem 2.11.2, they must occur at critical points of f, subject to the constraint $g(x, y, z) = x^2 + y^2 = 1$.

Thus, we want to find the points (x, y) for which there exists a number λ such that

$$x^2 + y^2 = 1, \quad \text{and} \quad \vec{\nabla}f(x, y) = \lambda\vec{\nabla}g(x, y),$$

that is, points which satisfy

$$x^2 + y^2 = 1, \quad \frac{\partial f}{\partial x} = \lambda\frac{\partial g}{\partial x}, \quad \text{and} \quad \frac{\partial f}{\partial y} = \lambda\frac{\partial g}{\partial y}.$$

Thus, we need to solve:

$$x^2 + y^2 = 1, \quad 8x = \lambda \cdot 2x, \quad \text{and} \quad 2y = \lambda \cdot 2y.$$

The last equation tells us that $y = 0$ or $\lambda = 1$.

If $y = 0$, the first equation tells us that $x = \pm 1$ (the second equation then tells us that $\lambda = 4$, but we don't really care).

If $\lambda = 1$, then the second equation tells us that $x = 0$, and so the first equation tells us that $y = \pm 1$.

Therefore, there are four critical points of f, subject to the constraint: $(x, y) = (-1, 0), (1, 0), (0, -1), \text{ and } (0, 1)$.

How do we decide what the global maximum and minimum values are at this point? We do what did in Section 2.10: plug the critical points into f and look for the biggest and smallest values.

(x, y)	$f = 4x^2 + y^2$
$(-1, 0)$	4
$(1, 0)$	4
$(0, -1)$	1
$(0, 1)$	1

Therefore, the global maximum value of f on the circle is 4, which occurs at two points $(-1, 0)$ and $(1, 0)$. The global minimum value of f on the circle is 1, which occurs at two points $(0, -1)$ and $(0, 1)$.

Example 2.11.4. Use Lagrange multipliers to find the global maximum and minimum values of $f(x, y) = 4x^2 + y^2$ on the disk given by $x^2 + y^2 \leq 1$.

Solution: As in the previous example, the function f is continuous, and the disk is compact; so, the Extreme Value Theorem, Theorem 1.7.13, tells us that f does, in fact, attain global extreme values on the disk. These global extreme values either occur in the interior of the disk, or on its boundary, which is the circle given by $x^2 + y^2 = 1$.

Thus, the problem naturally breaks up into two distinct pieces:

• find the critical points of f in the interior of the disk, and

• find the critical points of f restricted to the circle where $x^2 + y^2 = 1$, i.e., subject to the constraint that $x^2 + y^2 = 1$.

We then take both sets of critical points, make a list, and evaluate f at each critical point, looking for the largest and smallest values of f.

Of course, in the previous example, we already used Lagrange multipliers to find all of critical points of f, restricted to the circle where $x^2 + y^2 = 1$. We found four points; $(\pm 1, 0)$ and $(0, \pm 1)$.

What about critical points of f on the interior of the disk? These are simply the points (x, y) where $x^2 + y^2 < 1$ and $\vec{\nabla} f(x, y) = \mathbf{0}$. The only solution of

$$\vec{\nabla} f(x, y) = (8x, 2y) = (0, 0)$$

is $(x, y) = (0, 0)$ which does, in fact, lie in the open disk where $x^2 + y^2 < 1$.

Therefore, we add $(0, 0)$ to the list of points that we need to check. We find:

(x, y)	$f = 4x^2 + y^2$
$(-1, 0)$	4
$(1, 0)$	4
$(0, -1)$	1
$(0, 1)$	1
$(0, 0)$	0

Therefore, the global maximum value of f on the disk is 4, which occurs at two points $(-1, 0)$ and $(1, 0)$. The global minimum value of f on the disk is 0, which occurs at $(0, 0)$.

Example 2.11.5. Let's look back at Example 2.10.4, and see how you handle the problem using Lagrange multipliers.

Our problem was to find the point which minimizes

$$f(x, y, z) = D^2 = x^2 + (y - 1)^2 + (z - 3)^2,$$

subject to the constraint that $g(x, y, z) = 2x + y - z = 1$.

We assumed that f, subject to the constraint, in fact, attained a global minimum. By Theorem 2.11.2, this global minimum, which is certainly also a local minimum, must occur at a critical point of f, subject to the constraint $g(x, y, z) = 2x + y - z = 1$.

Thus, the minimum occurs at a point (x, y, z) for which there exists a number λ such that

$$2x + y - z = 1 \quad \text{and} \quad \vec{\nabla} f(x, y, z) = \lambda \vec{\nabla} g(x, y, z),$$

that is,

$$2x + y - z = 1, \quad \frac{\partial f}{\partial x} = \lambda \frac{\partial g}{\partial x}, \quad \frac{\partial f}{\partial y} = \lambda \frac{\partial g}{\partial y}, \quad \text{and} \quad \frac{\partial f}{\partial z} = \lambda \frac{\partial g}{\partial z}.$$

Thus, to find the critical point(s), subject to the constraint, we need to solve the following four equations and four unknowns or, at least, find the values of x, y, and z:

$$2x + y - z = 1, \quad 2x = \lambda \cdot 2, \quad 2(y - 1) = \lambda \cdot 1, \quad \text{and} \quad 2(z - 3) = \lambda \cdot -1.$$

There is no one best way to solve these equations, and different people may solve for x, y, and z in different ways. One obvious approach is to solve for x, y, and z in terms of λ in the last three equations, insert those into the first equation to solve for λ, and then determine x, y, and z.

We find

$$x = \lambda, \quad y = 1 + \frac{\lambda}{2}, \quad \text{and} \quad z = 3 - \frac{\lambda}{2},$$

and, substituting into the first equation

$$2\lambda + 1 + \frac{\lambda}{2} - \left(3 - \frac{\lambda}{2}\right) = 1.$$

This reduces to $3\lambda = 3$, and so $\lambda = 1$. Substituting back into the equations for x, y, and z in terms of λ, we find what we found in Example 2.10.4:

$$x = 1, \quad y = 3/2, \quad \text{and} \quad z = 5/2.$$

Example 2.11.6. Now we'll use Lagrange multipliers on our earlier problem from Example 2.10.5: minimize

$$A = A(L, W, H) = 2WH + 2LH + LW,$$

subject to the constraint that $g(L, W, H) = LWH = 4000$, assuming that such a minimum exists.

We want to solve

$$LWH = 4000 \quad \text{and} \quad \vec{\nabla}A = \lambda \vec{\nabla}g,$$

i.e.,

$$LWH = 4000, \quad 2H + W = \lambda WH, \quad 2H + L = \lambda LH, \quad \text{and} \quad 2W + 2L = \lambda LW.$$

Since L, W, and H were the length, width, and height of a box, respectively, they are all > 0. It follows from any of the last three equations, above, that $\lambda > 0$. Hence, we may divide the second equation by the third equation without worrying about dividing by zero; we find

$$\frac{2H + W}{2H + L} \;=\; \frac{\lambda W H}{\lambda L H} \;=\; \frac{W}{L},$$

and so

$$2HL + WL \;=\; 2HW + WL \quad \text{and, hence,} \quad W = L.$$

Plugging this into the last of our four equations, we find

$$2L + 2L = \lambda L^2 \quad \text{and so} \quad 4 = \lambda L.$$

Plugging this into the third of our four equations yields

$$2H + L \;=\; 4H, \quad \text{and so} \quad H = L/2.$$

Substituting $W = L$ and $H = L/2$ into our constraint gives us

$$\frac{L^3}{2} \;=\; 4000 \quad \text{and so} \quad L = 20.$$

Finally, we find $W = L = 20$ and $H = L/2 = 10$, which, of course, is what we found in Example 2.10.5.

| More Depth: |

While we motivated our definition of a critical point of a function, subject to a constraint, Definition 2.11.1, we wish to explain things more rigorously here.

Suppose that $\mathcal{U}$ is an open subset of $\mathbb{R}^n$ and let $f : \mathcal{U} \to \mathbb{R}$ be a continuously differentiable function (see Definition 2.1.23). Let M be a C^1 k-dimensional submanifold of $\mathcal{U}$ (see Definition 2.6.9). Recall the definition of the tangent space $T_{\mathbf{p}}M$ from Definition 2.6.16. Also recall that the restriction of f to M is denoted by $f_{|M}$.

Definition 2.11.7. *A **critical point** of $f_{|M}$ is a point $\mathbf{p}$ in M such that, for all $\mathbf{v}$ in $T_{\mathbf{p}}M$,*

$$d_{\mathbf{p}}f(\mathbf{v}) \;=\; \vec{\nabla}f(\mathbf{p}) \cdot \mathbf{v} \;=\; 0.$$

Example 2.11.8. Suppose that M is, in fact, equal to the whole open subset $\mathcal{U}$. Then we want to see that a critical point of f restricted to M is just an ordinary critical point of f, i.e., a point where the gradient vector is zero.

This is easy, for if $M = \mathcal{U}$, then, for all $\mathbf{p}$ in M, $T_{\mathbf{p}}M = \mathbb{R}^n$. In particular, the standard basis elements $\mathbf{e}_i$ (recall Definition 1.2.21) are all in $T_{\mathbf{p}}M$. Thus, Definition 2.11.7 requires that a critical point $\mathbf{p}$ of f restricted to M has $\overrightarrow{\nabla} f(\mathbf{p}) \cdot \mathbf{e}_i = 0$, for all i, i.e., that the gradient vector of f is zero.

Now we can explain Definition 2.11.1 with the following theorem:

Theorem 2.11.9. *Suppose that $f = f(\mathbf{x})$ and $g = g(\mathbf{x})$ are continuously differentiable functions on an open subset of $\mathbb{R}^n$, where $n \geq 2$. Let M denote the level set given by $g(\mathbf{x}) = c$. Suppose that, for all $\mathbf{x}$ in M, $\overrightarrow{\nabla} g(\mathbf{x}) \neq \mathbf{0}$.*

Then, $f_{|M}$ has a critical point at $\mathbf{p}$ in M if and only if there exists a scalar λ such that $\overrightarrow{\nabla} f(\mathbf{p}) = \lambda \overrightarrow{\nabla} g(\mathbf{p})$.

Proof. The Implicit Function Theorem, Theorem 2.7.16, implies that M is a C^1 hypersurface. Now, Corollary 2.7.18 tells us that, for all $\mathbf{p}$ in M, any vector which to perpendicular to all of the tangent vectors in $T_{\mathbf{p}}M$ is a scalar multiple of $\overrightarrow{\nabla} g(\mathbf{p})$. Thus, $\mathbf{p}$ is a critical point of f restricted to M if and only $\overrightarrow{\nabla} f(\mathbf{p})$ is a scalar multiple of $\overrightarrow{\nabla} g(\mathbf{p})$. $\square$

The question now is why we care about the concept of a critical point of f restricted to M. The answer is that we want Theorem 2.11.2; we want to know where to look for local extreme value of a constrained function.

We want to give a proof of this.

Theorem 2.11.10. *Suppose that $f = f(\mathbf{x})$ and $g = g(\mathbf{x})$ are continuously differentiable functions on an open subset of $\mathbb{R}^n$, where $n \geq 2$. Let M denote the level set given by $g(\mathbf{x}) = c$. Suppose that, for all $\mathbf{x}$ in M, $\overrightarrow{\nabla} g(\mathbf{x}) \neq \mathbf{0}$.*

Then, if the restriction $f_{|M}$ attains a local extreme value at a point $\mathbf{p}$ in M, then $\mathbf{p}$ must be a critical point of $f_{|M}$.

Proof. This is quite easy. If $f_{|M}$ attains a local extreme value at a point $\mathbf{p}$ in M, then for any parameterized continuously differentiable curve $\mathbf{r}(t)$ in M such that $\mathbf{r}(t_0) = \mathbf{p}$, the composed function $f \circ \mathbf{r}$ attains a local extreme value at t_0. This means that

$$0 = (f \circ \mathbf{r})'(t_0) = d_{\mathbf{p}} f(\mathbf{r}'(t_0)).$$

In other words, $d_{\mathbf{p}}f(\mathbf{v}) = 0$ for all tangent vectors $\mathbf{v}$ to M at $\mathbf{p}$. Thus, $\mathbf{p}$ is a critical point of $f_{|_M}$. $\qquad\qquad\qquad\qquad\qquad\qquad\qquad\qquad\qquad\qquad\qquad\qquad\square$

We want to to discuss what happens with more constraints. The proof is analogous to the one-constraint case.

Theorem 2.11.11. *Suppose that f and g_1, g_2, ..., g_r are continuously differentiable functions on an open subset of $\mathbb{R}^n$, where $n \geq r + 1$, and that $c_1, \ldots, c_r$ are constants. Let M denote the set of $\mathbf{x}$ such that*

$$g_1(\mathbf{x}) = c_1, \ g_2(\mathbf{x}) = c_2, \ \ldots, \ g_{r-1}(\mathbf{x}) = c_{r-1}, \ \text{and} \ g_r(\mathbf{x}) = c_r.$$

Finally, suppose that the restricted/constrained function $f_{|_M}$ attains a local extreme value at a point $\mathbf{p}$ in M.

Then, either

1. *there exist constants $a_1, \ldots, a_r$, not all 0 (but, some of which may be 0) such that*

$$a_1 \vec{\nabla} g_1(\mathbf{p}) \ + \ a_2 \vec{\nabla} g_2(\mathbf{p}) \ + \ \cdots \ + \ a_r \vec{\nabla} g_r(\mathbf{p}) \ = \ \mathbf{0},$$

*in which case, we say that $\mathbf{p}$ is a **critical point of the constraint function** $\mathbf{g} = (g_1, \ldots, g_r)$, or*

2. *$\mathbf{p}$ is not a critical point of the constraint function, but there exist constants $\lambda_1, \ldots, \lambda_r$ (possibly all 0) such that*

$$\vec{\nabla} f(\mathbf{p}) \ = \ \lambda_1 \vec{\nabla} g_1(\mathbf{p}) \ + \ \lambda_2 \vec{\nabla} g_2(\mathbf{p}) \ + \ \cdots \ + \ \lambda_r \vec{\nabla} g_r(\mathbf{p}),$$

*in which case, we say that $\mathbf{p}$ is a **critical point of f restricted to the constraint submanifold where $g_1(\mathbf{x}) = c_1$, ..., $g_r(\mathbf{x}) = c_r$**.*

For the proof, see section 4.8 of [3].

That the constraint set is a C^1 submanifold of $\mathbb{R}^n$ near a non-critical point of $\mathbf{g}$ follows from the Implicit Function Theorem.

The λ_i, above, are, of course, called *Lagrange multipliers*.

Example 2.11.12. Find the global maximum and minimum values of

$$f(x, y, z) = 2x - y + 3z,$$

constrained by the conditions that $x^2 + y^2 + z^2 = 25$ and $x + y - z = 0$.

Solution:

Let $g_1(x, y, z) = x^2 + y^2 + z^2$ and $g_2(x, y, z) = x + y - z$. Our constraint set is the closed, bounded, i.e., compact, subset M of $\mathbb{R}^3$ defined by $g_1(x, y, z) = 25$ and $g_2(x, y, z) = 0$. As M is compact and f is continuous, the restriction of f to M, $f_{|_M}$, attains both a global maximum and global minimum value.

These global extreme values are, of course, local extreme values, and so Theorem 2.11.11 tells us that a point $\mathbf{p}$ at which f attains one of its global extreme values must be either a critical point of the multi-component constraint function $(g_1(x, y, z), g_2(x, y, z))$ or a critical point of f restricted to the constraint submanifold. We will locate all critical points of either kind, and then plug them all into f, to see where f is the biggest and the smallest.

We first look for critical points (x, y, z) of the function (g_1, g_2) which are in the constraint set M where $g_1 = 25$ and $g_2 = 0$. These points occur where we can solve

$$x^2 + y^2 + z^2 = 25, \quad x + y - z = 0, \quad \text{and} \quad a_1 \vec{\nabla} g_1(x, y, z) + a_2 \vec{\nabla} g_2(x, y, z) = \mathbf{0},$$

where a_1 and a_2 are not both 0.

Thus, we need to solve

$$x^2 + y^2 + z^2 = 25, \quad x + y - z = 0, \quad \text{and} \quad a_1(2x, 2y, 2z) + a_2(1, 1, -1) = \mathbf{0}.$$

If $a_1 = 0$, then we see immediately that we must have $a_2 = 0$, which is not allowed (we can't have both a_1 and a_2 being 0). Thus, we would have to have $a_1 \neq 0$ and so, dividing the entire righthand equation above by $2a_1$, and letting $c = -a_2/(2a_1)$, we are reduced to solving

$$x^2 + y^2 + z^2 = 25, \quad x + y - z = 0, \quad \text{and} \quad (x, y, z) = c(1, 1, -1),$$

where c is allowed to be 0.

The last equation tells us that we must have $x = c$, $y = c$, and $z = -c$; substituting these values into the middle equation yields that $c = 0$, which, in turn, implies that $x = y = z = 0$. However, $x = y = z = 0$ doesn't satisfy $x^2 + y^2 + z^2 = 25$. Therefore, the constraint function (g_1, g_2) has no critical points in the constraint set M.

Now we look for critical points of f restricted to the constraint submanifold M. We need to solve:

$$x^2 + y^2 + z^2 = 25, \quad x + y - z = 0, \quad \text{and} \quad \vec{\nabla} f(x, y, z) = \lambda_1 \vec{\nabla} g_1(x, y, z) + \lambda_2 \vec{\nabla} g_2(x, y, z),$$

that is,

$$x^2 + y^2 + z^2 = 25, \quad x + y - z = 0, \quad \text{and} \quad (2, -1, 3) = \lambda_1(2x, 2y, 2z) + \lambda_2(1, 1, -1).$$

Now, λ_1 cannot be equal to 0, for then $(2, -1, 3)$ would be a scalar multiple of $(1, 1, -1)$, which it's not. Thus, we can divide the last equation by $2\lambda_1$, rearrange, and let $b = 1/(2\lambda_1)$ and $c = -\lambda_2/(2\lambda_1)$ to obtain

$$(x, y, z) \;=\; b\,(2, -1, 3) \;+\; c\,(1, 1, -1).$$

Substituting $x = 2b + c$, $y = -b + c$, and $z = 3b - c$ into $x + y - z = 0$, we conclude that $-2b + 3c = 0$. Hence, $b = 3c/2$, and

$$(x, y, z) \;=\; \frac{3c}{2}\,(2, -1, 3) \;+\; c\,(1, 1, -1) \;=\; c\left(4, -\frac{1}{2}, \frac{7}{2}\right). \tag{2.11}$$

Plugging this into $x^2 + y^2 + z^2 = 25$, we find:

$$16c^2 + \frac{1}{4}c^2 + \frac{49}{4}c^2 \;=\; 25.$$

Therefore,

$$c^2 \;=\; \frac{100}{114} \qquad \text{and so} \qquad c \;=\; \pm\frac{10}{\sqrt{114}}.$$

Returning to Formula 2.11, we find that there are precisely two critical points of f restricted to the constraint submanifold M:

$$(x, y, z) \;=\; \pm\frac{10}{\sqrt{114}}\left(4, -\frac{1}{2}, \frac{7}{2}\right).$$

One of these must yield the global maximum value of $f_{|M}$ and one must yield the global minimum value. To decide which is which, and to determine the values, we simply insert each into $f(x, y, z) = 2x - y + 3z$ and find

$$f\left(\frac{10}{\sqrt{114}}\left(4, -\frac{1}{2}, \frac{7}{2}\right)\right) \;=\; \frac{10}{\sqrt{114}}\left(8 + \frac{1}{2} + \frac{21}{2}\right) \;=\; \frac{190}{\sqrt{114}},$$

and

$$f\left(-\frac{10}{\sqrt{114}}\left(4, -\frac{1}{2}, \frac{7}{2}\right)\right) \;=\; -\frac{190}{\sqrt{114}}.$$

We wish to revisit Theorem 2.11.11 and, now, using the language and techniques of linear algebra, restate the theorem and look at two examples.

$\boxed{\textbf{+ Linear Algebra:}}$

Recall the setup:

f and g_1, g_2, ..., g_r are continuously differentiable functions on an open subset of $\mathbb{R}^n$, and $c_1, \ldots, c_r$ are constants. Let M denote the set of $\mathbf{x}$ such that

$$g_1(\mathbf{x}) = c_1, \; g_2(\mathbf{x}) = c_2, \; \ldots, \; g_{r-1}(\mathbf{x}) = c_{r-1}, \text{ and } g_r(\mathbf{x}) = c_r.$$

Finally, suppose that the restricted/constrained function $f_{|M}$ attains a local extreme value at a point $\mathbf{p}$ in M.

Cases (1) and (2) from Theorem 2.11.11 can be combined to yield:

there exist constants $a_0, a_1, \ldots, a_r$, not all 0 (but, some of which may be 0) such that

$$a_0 \vec{\nabla} f(\mathbf{p}) \;+\; a_1 \vec{\nabla} g_1(\mathbf{p}) \;+\; a_2 \vec{\nabla} g_2(\mathbf{p}) \;+\; \cdots \;+\; a_r \vec{\nabla} g_r(\mathbf{p}) \;=\; \mathbf{0};$$

if $a_0 = 0$, you're in Case (1) of Theorem 2.11.11, and if $a_0 \neq 0$, you may divide the whole equation by a_0, put the $\vec{\nabla} g_i$ terms on the other side, and rename the constants to put yourself in Case (2) of Theorem 2.11.11.

This, by definition, is what it means for the collection of vectors $\vec{\nabla} f(\mathbf{p}), \vec{\nabla} g_1(\mathbf{p}), \ldots, \vec{\nabla} g_r(\mathbf{p})$ to be linearly dependent, and is equivalent to saying that the Jacobian matrix of $(f, g_1, \ldots, g_r)$,

$$[d(f, g_1, \ldots, g_r)] \;=\; \begin{bmatrix} \frac{\partial f}{\partial x_1} & \frac{\partial f}{\partial x_2} & \cdots & \frac{\partial f}{\partial x_n} \\ \frac{\partial g_1}{\partial x_1} & \frac{\partial g_1}{\partial x_2} & \cdots & \frac{\partial g_1}{\partial x_n} \\ \vdots & \vdots & \cdots & \vdots \\ \frac{\partial g_r}{\partial x_1} & \frac{\partial g_r}{\partial x_2} & \cdots & \frac{\partial g_r}{\partial x_n} \end{bmatrix},$$

has rank $\leq r$ at $\mathbf{p}$. If $n \geq r+1$, this is what it means for the function $(f, g_1, \ldots, g_r)$ to have a *critical point* at $\mathbf{p}$.

We can now restate Theorem 2.11.11 in matrix terms:

Theorem 2.11.13. *Suppose that f and g_1, g_2, $\ldots$, g_r are continuously differentiable functions on an open subset of $\mathbb{R}^n$, where $n \geq r+1$, and that $c_1, \ldots, c_r$ are constants. Let M denote the set of $\mathbf{x}$ such that*

$$g_1(\mathbf{x}) = c_1, \; g_2(\mathbf{x}) = c_2, \; \ldots, \; g_{r-1}(\mathbf{x}) = c_{r-1}, \; \text{and } g_r(\mathbf{x}) = c_r.$$

Finally, suppose that the restricted/constrained function $f_{|M}$ attains a local extreme value at a point $\mathbf{p}$ in M.

> Then, $\mathbf{p}$ is a critical point of the function $(f, g_1, \ldots, g_r)$, i.e., at $\mathbf{p}$, the Jacobian matrix of $(f, g_1, \ldots, g_r)$,
>
> $$[d(f, g_1, \ldots, g_r)] = \begin{bmatrix} \frac{\partial f}{\partial x_1} & \frac{\partial f}{\partial x_2} & \cdots & \frac{\partial f}{\partial x_n} \\ \frac{\partial g_1}{\partial x_1} & \frac{\partial g_1}{\partial x_2} & \cdots & \frac{\partial g_1}{\partial x_n} \\ \vdots & \vdots & \cdots & \vdots \\ \frac{\partial g_r}{\partial x_1} & \frac{\partial g_r}{\partial x_2} & \cdots & \frac{\partial g_r}{\partial x_n} \end{bmatrix},$$
>
> has rank $\leq r$.

Does reformulating the method of Lagrange multipliers in linear algebra terms actually help us calculate in specific examples? Yes, at least in two cases: when there are two variables (so, $n = 2$) and one constraint (so, $r = 1$), or when there are three variables (so, $n = 3$) and there are two constraints (so, $r = 2$). What's special about these cases?

There is a characterization of the rank of a matrix in terms of minors – determinants of square submatrices. A matrix has rank $\leq r$ if and only if every $(r + 1) \times (r + 1)$ minor of the matrix is 0. Typically, calculating the rank of a matrix by using minors is horribly inefficient, compared to using Gaussian elimination.

However, when $n = r + 1$, so that $r = n - 1$, the Jacobian matrix of $(f, g_1, \ldots, g_{n-1})$ **has only one** $(r + 1) \times (r + 1) = n \times n$ **minor, namely the whole Jacobian matrix** and, if this matrix is relatively small, like when $n = 2$ or 3, then calculating the determinant of the Jacobian matrix and setting it equal to 0 is a nice way to proceed to find the critical points.

Example 2.11.14. Let's return to Example 2.11.3. The problem was to find the global maximum and minimum values of $f(x, y) = 4x^2 + y^2$ subject to the constraint that $g(x, y) = x^2 + y^2 = 1$.

From our discussion above, we know that the points at which the constrained f can attain local extreme values are points at which the constraint is satisfied and at which the determinant of the Jacobian matrix $[d(f, g)]$ is zero. Therefore, we need to simultaneously solve

$$x^2 + y^2 = 1$$

and

$$\det \begin{bmatrix} f_x & f_y \\ g_x & g_y \end{bmatrix} = \det \begin{bmatrix} 8x & 2y \\ 2x & 2y \end{bmatrix} = 16xy - 4xy = 12xy = 0.$$

The second equation tells us that either $x = 0$ or $y = 0$; substituting these into the first equation tells us that the four critical points which need to be checked are

$$(0,1), \quad (0,-1), \quad (1,0), \quad \text{and} \quad (-1,0).$$

These, of course, are precisely the points we found in Example 2.11.3, and the rest of the solution proceeds as it did there.

Example 2.11.15. Now let's look again at Example 2.11.12. We wanted to find the global maximum and minimum values of

$$f(x,y,z) = 2x - y + 3z,$$

constrained by the conditions that $g_1(x,y,z) = x^2 + y^2 + z^2 = 25$ and $g_2(x,y,z) = x + y - z = 0$. This boiled down to finding the critical points of the function (f, g_1, g_2), such that $x^2 + y^2 + z^2 = 25$ and $x + y - z = 0$, and then evaluating f at these critical points, looking for the largest and smallest values.

We now know that we can find the needed critical points by simultaneously solving

$$x^2 + y^2 + z^2 = 25, \quad x + y - z = 0,$$

and

$$\det \begin{bmatrix} 2 & -1 & 3 \\ 2x & 2y & 2z \\ 1 & 1 & -1 \end{bmatrix} = -4y + 6x - 2z - 6y - 2x - 4z = 4x - 10y - 6z = 0.$$

Dividing this last equation by 2, we find that the points at which the constrained f can possibly attain global extrema are the simultaneous solutions of

$$x^2 + y^2 + z^2 = 25, \quad x + y - z = 0, \quad \text{and} \quad 2x - 5y - 3z = 0.$$

We leave it to you to show that the last two equations imply that $y = -x/8$ and $z = 7x/8$ and, after substituting these into the first equation, you find the same points that we found in Example 2.11.12:

$$(x,y,z) = \pm \frac{10}{\sqrt{114}} \left(4, -\frac{1}{2}, \frac{7}{2} \right).$$

2.11.1 Exercises

In each of the following exercises, you are given a function f, and a constraint equation $g = c$. **(a)** Show that there are no points p, satisfying the constraint, and such that $\vec{\nabla}g(\mathbf{p}) = \mathbf{0}$. **(b)** Use Lagrange multipliers to find the critical points of f, subject to the constraint.

Online answers to select exercises are here.

Basics:

1. $f(x, y) = 3x^2 + 2y^2$, and $x - y = 30$. ▶

2. $f(x, y) = x - y$, and $3x^2 + 2y^2 = 30$.

3. $f(x, y) = xy$, and $x^4 + y^2 = 3/4$.

4. $f(x, y) = x^4 + y^2$, and $xy = 3/4$.

5. $f(x, y) = \frac{2}{3}e^{3x} + \frac{4}{3}e^{3y}$, and $e^{2x} + e^{2y} = 20$.

6. $f(x, y) = xy + \ln y$, and $2x + y = 3$.

7. $f(x, y) = x - y$, and $x^3 + xy - y^3 = 1$. (Hint: $2x^3 - x^2 - 1$ has only one real root.)

8. $f(x, y, z) = 2x - 3y + z$, and $x^2 + 2y^2 + 3z^2 = 318$. ▶

9. $f(x, y, z) = 2x - 3y + z$, and $xyz = 1$.

10. $f(x, y, z) = xy - z^2$, and $x^2 + y^2 + z^2 = 1$.

11. $f(x, y, z) = x^3 + y^4 + z^4$, and $x + y + z = 0$.

12. $f(x, y, z) = x^3 + y^4 + z^4$, and $x^2 + y^2 + z^2 = 1$.

In each of the following exercises, you are given a function f, and a constraint inequality. Use Lagrange multipliers to find the critical points of f, subject to the constraint.

13. $f(x, y) = 3x^2 + 2y^2$, and $x - y \le 30$. ▶

14. $f(x, y) = 3x^2 + 2y^2$, and $x - y \ge 30$.

15. $f(x, y) = x - y$, and $3x^2 + 2y^2 \le 30$.

16. $f(x, y) = xy$, and $x^4 + y^2 \le 3/4$.

17. $f(x, y) = \frac{2}{3}e^{3x} + \frac{4}{3}e^{3y}$, and $e^{2x} + e^{2y} \le 20$.

18. $f(x, y, z) = xy - z^2$, and $x^2 + y^2 + z^2 \le 1$.

In each of the following exercises, you are given the graph of a constraint equation $g(x, y) = c$, in red. You are also given level curves of a function $f(x, y)$, in blue. Estimate the critical points of f, subject to the constraint.

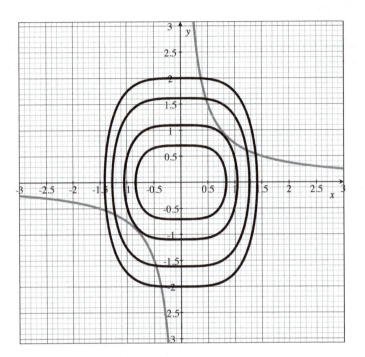

19.

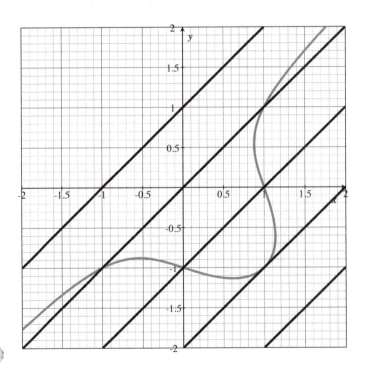

20.

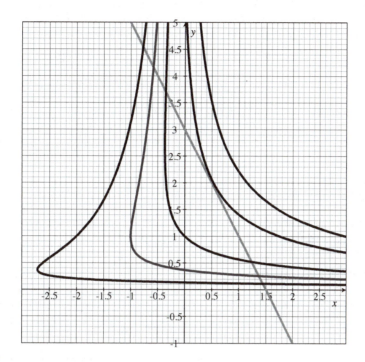

21.

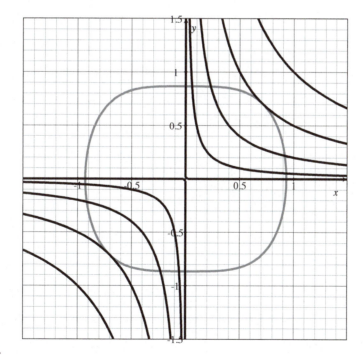

22.

In each of the following exercises, you are given a function f, and a constraint equality, or inequality, which defines a compact region E in $\mathbb{R}^2$ or $\mathbb{R}^3$. Use Lagrange multipliers to find the global maximum and minimum values of the restriction of f to E, i.e., the global extreme values of f, subject to the constraint.

23. $f(x,y) = x - y$, and $3x^2 + 2y^2 \le 30$.

24. $f(x,y) = xy$, and $x^4 + y^2 \le 3/4$.

25. $f(x,y,z) = xy - z^2$, and $x^2 + y^2 + z^2 \le 1$.

26. $f(x,y,z) = 2x - 3y + z$, and $x^2 + 2y^2 + 3z^2 \le 318$.

The following exercises were taken from the previous section. Here, however, you are required to use Lagrange multipliers to solve the problems.

27. Near a small airport, there is a ridge. A cross section of the surface of the ridge is modeled by the equation $y = 64 - x^2$, where x and y are measured in yards. The standard flight path over the ridge is given by $y = 100 - (x/2)$. The airport wishes to place a warning beacon on the ridge at the point closest to the flight path. At what x-coordinate on the ridge should they place the beacon?

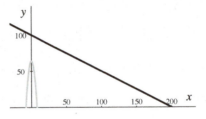

Figure 2.11.2: Ridge: $y = 64 - x^2$. Flight path: $y = 100 - (x/2)$.

28. Suppose that a cardboard box (with a top) is to be constructed with a volume of 4000 cubic inches. Also, suppose that the cardboard for the bottom costs 4 cents per square inch, the cardboard for the sides costs 2 cents per square inch, and the cardboard for the top costs 1 cent per square inch. Find the dimensions of the box which minimize the cost of the cardboard required.

29. Suppose that a cardboard box (with a top) is to be constructed, at a cost of $5.40. Also, suppose that the cardboard for the bottom costs 4 cents per square inch, the cardboard for the sides costs 2 cents per square inch, and the cardboard for the top costs 1 cent per square inch. What is the maximum volume of such a box?

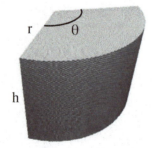

30. A curved prism has a base which is a sector of a disk of radius r inches and central angle θ radians. The prism extends vertically to a height of h inches.

(a) Derive formulas for the volume V, in cubic inches, and the total surface area (including the top and bottom) A, in square inches.

(b) If V is required to be 27, what values of r, θ, and h minimize the surface area A of the prism?

In each of the following exercises, you are given two constraint equations $g_1(x, y, z) = c_1$ and $g_2(x, y, z) = c_2$, which define a compact subset E of $\mathbb{R}^3$, and a function $f(x, y, z)$. Use Lagrange multipliers to find the global maximum and minimum values of the restriction of f to E, i.e., the global extreme values of f, subject to the constraints.

More Depth:

31. $f(x, y, z) = xy - z^2$; $\dfrac{x^2}{2^2} + \dfrac{y^2}{3^2} + \dfrac{z^2}{5^2} = 1$ and $x + y = 1$.

32. $f(x, y, z) = y$; $x^2 + y^2 + z^2 = 2$ and $x^2 - yz = 1$.

33. $f(x, y, z) = xy$; $x^4 + z = 0$ and $z - z^2 - y^2 = 0$.

34. $f(x, y, z) = xe^{-x}$; $x + y + z = 3$ and $x^2 + x + y^2 + y + z^2 + z = 8$.

35. Suppose that $f(x, y)$ and $g(x, y)$ are continuously differentiable, and are each symmetric in x and y, i.e., for all (x, y), $f(x, y) = f(y, x)$ and $g(x, y) = g(y, x)$. Is it true that every critical point (a, b) of f, subject to the constraint $g(x, y) = c$, will satisfy $a = b$, i.e., have equal x- and y-coordinates? Give an argument or produce a counterexample.

36. Suppose that $f(x, y)$ and $g(x, y)$ are continuously differentiable, and that $\vec{\nabla} f(a, b) \neq \mathbf{0}$ and $\vec{\nabla} g(a, b) \neq \mathbf{0}$. Suppose further that (a, b) is a critical point of f, subject to the constraint $g = c$. Let $m = f(a, b)$. Show that (a, b) is also a critical point of g, subject to the constraint $f = m$.

In each of the following problems, you are given either $f(x, y)$ and a single constraint equation $g = c$, or $f(x, y, z)$ and two constraint equations $g_1 = c_1$ and $g_2 = c_2$. Use determinants to find the critical points of (f, g) or (f, g_1, g_2) which satisfy the constraint(s).

+ Linear Algebra:

37. $f(x, y) = x + 3y$; $x^2 + xy + y^2 = 1$.

38. $f(x, y) = x + y$; $x^3 + 6x^2 y + 4y^3 = 36$.

39. $f(x, y, z) = xy - z^2$; $\dfrac{x^2}{2^2} + \dfrac{y^2}{3^2} + \dfrac{z^2}{5^2} = 1$ and $x + y = 1$.

40. $f(x, y, z) = y$; $x^2 + y^2 + z^2 = 2$ and $x^2 - yz = 1$.

41. $f(x, y, z) = xy$; $x^4 + z = 0$ and $z - z^2 - y^2 = 0$.

42. $f(x, y, z) = xe^{-x}$; $x + y + z = 3$ and $x^2 + x + y^2 + y + z^2 + z = 8$.

2.12   Implicit Differentiation

In Section 2.7, we looked at level sets, given by $F(x_1, \ldots, x_n) = c$, and discussed a geometric form of the *Implicit Function Theorem*; we commented that that form of the theorem made it unclear where the name comes from, since the statement of the theorem didn't seem to say anything about one variable being an implicit function of other variables.

In this section, we give a more traditional form of the Implicit Function Theorem, one that says that $F(x_1, \ldots, x_n) = c$ implicitly defines one of the variables as a function of the others at any point $\mathbf{p}$ at which $\vec{\nabla} F(\mathbf{p}) \neq \mathbf{0}$. The theorem also tells us how to differentiate this dependent variable – this implicitly defined function – with respect to the other variables.

Basics:

There are two fundamental questions that we will consider in this section.

1. Given $F = F(x_1, \ldots, x_n)$ and a constant c, when can we, at least in theory, solve the equation $F(x_1, \ldots, x_n) = c$ for one of the x_i's in terms of the others, near a given point $\mathbf{p}$ in the level set $F = c$?

2. If we **can** solve $F(x_1, \ldots, x_n) = c$ for one of the x_i's in terms of the others, can we give nice formulas for the partial derivatives of the dependent x_i, with respect to the other variables?

The answer to both of these is provided by the *Implicit Function Theorem*. However, before we state the theorem, let's look at a simple, but illuminating, example.

Example 2.12.1. Consider the function $F(x, y) = x^2 + y^2$. For $c > 0$, the level set given by $F(x, y) = c$ is a circle of radius $\sqrt{c}$, centered at the origin. Let's fix $c = 1$, so that we are looking at the level curve $x^2 + y^2 = 1$. Can we consider y as a function of x and, if so, how do we find $\partial y / \partial x = dy/dx$?

If you solve for y in terms of x, you obtain

$$y = \pm\sqrt{1 - x^2}.$$

Figure 2.12.1: The graph of $x^2 + y^2 = 1$.

This is not a **function**. Functions are required to give back a single value at every point in the domain; $y = \pm\sqrt{1 - x^2}$ gives two different y values for each x in the interval $(-1, 1)$. On the graph, this is reflected by fact that the *vertical line test* fails; that is, the graph of $x^2 + y^2 = 1$ is not the graph of a function because there are vertical lines which intersect the graph more than once.

However, in this example, we can at least "split" y into two separate functions. If we have $x^2 + y^2 = 1$ **and require that** $y \geq 0$, then we obtain that $y = \sqrt{1 - x^2}$. The graph of this function is the blue upper-half of the graph in Figure 2.12.1.

If we have $x^2 + y^2 = 1$ **and require that** $y \leq 0$, then we obtain that $y = -\sqrt{1 - x^2}$. The graph of this function is the red lower-half of the graph in Figure 2.12.1.

In order to distinguish which y we are discussing, we will let $u(x) = \sqrt{1 - x^2}$ and $l(x) = -\sqrt{1 - x^2}$. As we now have **functions**, we can ask about the differentiability of $u(x)$ and $l(x)$, and ask for formulas for the derivatives.

We find

$$u'(x) \;=\; \left[(1 - x^2)^{1/2}\right]' \;=\; \frac{1}{2}\,(1 - x^2)^{-1/2}(-2x) \;=\; \frac{-x}{\sqrt{1 - x^2}}.$$

Note that the denominator on the right above is precisely $u(x)$ itself. Therefore, if we have $y = u(x) = \sqrt{1 - x^2}$, we find that $y = u(x)$ is differentiable wherever $y = \sqrt{1 - x^2} \neq 0$ and, at all such points

$$dy/dx = -x/y.$$

Now let's look at $y = l(x) = -\sqrt{1 - x^2}$. Then, $l(x)$ is simply $-u(x)$, and our work above tells us that $y = l(x)$ is differentiable wherever $\sqrt{1 - x^2} \neq 0$ and, at all such points $l'(x) = -u'(x) = x/\sqrt{1 - x^2}$. We can rewrite this; if $y = -\sqrt{1 - x^2}$, then, what we just found was that

$$dy/dx = x/(-y) = -x/y.$$

This is exactly the formula we found for $y = u(x) = \sqrt{1 - x^2}$!

What we have just seen is that the equation $x^2 + y^2 = 1$ can be used to define y as a function of x, provided that we specify that $y \geq 0$ or $y \leq 0$, and, in either case, if $y \neq 0$, you find

$$\frac{dy}{dx} \;=\; -\frac{x}{y}.$$

Could we have found this one common formula without splitting things up into the cases where $y \geq 0$ and $y \leq 0$?

Yes, and it's why we want the Implicit Function Theorem.

We would like to look at the same sort of situation for functions $F = F(x_1, \ldots, x_n)$ of any number of variables. When can we say that the equation $F = c$, for a given constant c, defines one of the variables as a function of the other variables, at least **near** a point $\mathbf{p}$ that's in the level set? And, if we can say that one of the variables **is** implicitly a function of the others, can we find formulas for the partial derivatives of the implicitly defined function?

To address both of these questions, we have:

We have assumed that F is continuously differentiable, i.e., of class C^1. We could have assumed that F was of class C^r, for any $r \geq 1$, including the C^∞ or real analytic cases. Then, the Implicit Function Theorem looks precisely the same, except that the implicitly defined function g would also be of the same class.

This proof is instructive, and we shall sketch it in the More Depth portion.

Theorem 2.12.2. (Implicit Function Theorem) *Suppose that $F = F(\mathbf{x})$ is continuously differentiable on an open neighborhood of a point $\mathbf{p}$ in $\mathbb{R}^n$, and that*
$$\left. \frac{\partial F}{\partial x_i} \right|_{\mathbf{p}} \neq 0.$$

Then, there exists an open neighborhood $\mathcal{U}$ of $\mathbf{p}$, and a continuously differentiable function g on an open neighborhood of $(p_1, \ldots, p_{i-1}, p_{i+1}, \ldots, p_n)$ in $\mathbb{R}^{n-1}$ such that, for all $(x_1, \ldots, x_n)$ in $\mathcal{U}$,

$$F(x_1, \ldots, x_n) = F(\mathbf{p}) \qquad \text{if and only if} \qquad x_i = g(x_1, \ldots, x_{i-1}, x_{i+1}, \ldots, x_n).$$

Any two such functions g must be the same on an open neighborhood of $(p_1, \ldots, p_{i-1}, p_{i+1}, \ldots, p_n)$. Furthermore, if x_i denotes the value of the function g from above, then, for all j unequal to i,

$$\frac{\partial x_i}{\partial x_j} = \frac{\partial g}{\partial x_j} = -\frac{\partial F/\partial x_j}{\partial F/\partial x_i},$$

where the left-hand side is evaluated at points $(x_1, \ldots, x_{i-1}, x_{i+1}, \ldots, x_n)$ close to $(p_1, \ldots, p_{i-1}, p_{i+1}, \ldots, p_n)$ and the right-hand side is evaluated at

$$(x_1, \ldots, x_{i-1}, g(x_1, \ldots, x_{i-1}, x_{i+1}, \ldots, x_n), x_{i+1}, \ldots, x_n).$$

What the above theorem says is that you can, in theory, solve the equation

$$F(x_1, \ldots, x_n) = c$$

for x_i in terms of the other variables, near a point $\mathbf{p}$ that's in the level set, **provided** that $\partial F/\partial x_i$ is non-zero at $\mathbf{p}$; in addition, the solution is a continuously differentiable function, and its partial derivatives can be calculated from the partial derivatives of F.

Remark 2.12.3. If you read the statement of the Implicit Function Theorem, and look at the formula

$$\frac{\partial x_i}{\partial x_j} = -\frac{\partial F/\partial x_j}{\partial F/\partial x_i},$$

you may worry that the formula works only at points that are close to the point $\mathbf{p}$ where $\partial F/\partial x_i \neq 0$, and you may not know when you're "close enough" to $\mathbf{p}$. So, how can you know when you're calculating correctly?

Assuming that F is continuously differentiable, the good news is that the formulas for the partial derivatives are valid **anywhere that the quantities in the formulas exist**, because the denominators are all $\partial F/\partial x_i$, which must be non-zero for the quotients to be defined. But $\partial F/\partial x_i$ being non-zero is precisely what you need in order to apply the Implicit Function Theorem.

In light of the Implicit Function Theorem, we make the following definition:

> **Definition 2.12.4.** *Suppose that $F = F(\mathbf{x})$ is continuously differentiable on an open neighborhood of a point $\mathbf{p}$ in $\mathbb{R}^n$, and that $\partial F/\partial x_i \neq 0$ at $\mathbf{p}$.*
>
> *Then, we say that the equation $F(\mathbf{x}) = F(\mathbf{p})$ **implicitly defines** x_i **as a function of** $(x_1, \ldots, x_{i-1}, x_{i+1}, \ldots, x_n)$ **near** $\mathbf{p}$, and the **implicitly defined function** is precisely the function g that we described in Theorem 2.12.2.*

Technically, the **function** g which is described in the Implicit Function Theorem is what's known as the *germ of a function* at a point. The notion of a germ is a formal way of saying that two functions, defined on open neighborhoods of a fixed point, are equivalent, in a strong sense "the same", if they agree on an open set which contains the point in question.

Example 2.12.5. Let's return to our previous example, in which we had $F(x, y) = x^2 + y^2$, and we were looking at the level curve where $F = 1$, i.e., the circle where $x^2 + y^2 = 1$.

To use the Implicit Function Theorem and conclude that y can be written as a function of x near the point $\mathbf{p} = (a, b)$ on the circle, we need to know that the partial derivative $\partial F/\partial y \neq 0$ at (a, b). Of course, we easily find that

$$\frac{\partial F}{\partial y}\bigg|_{(a,b)} = 2y\big|_{(a,b)} = 2b,$$

which is non-zero if and only if $b \neq 0$.

Therefore, the Implicit Function Theorem tells us that, near each point (a, b) on the unit circle, other than the points where $y = 0$, i.e., other than at $(-1, 0)$ and $(1, 0)$, $x^2 + y^2 = 1$ implicitly defines y as a function of x, i.e., we can solve uniquely for y as a function of x (near the point (a, b)); furthermore, when y is considered as a function of x,

$$\frac{dy}{dx} = \frac{\partial y}{\partial x} = -\frac{\partial F/\partial x}{\partial F/\partial y} = -\frac{2x}{2y} = -\frac{x}{y},$$

which is precisely what we concluded in the previous example.

Example 2.12.6. Show that, near the point $(x, y, z) = (1, 0, e)$, the equation

$$2z + xye^y - z\ln\left(\frac{z}{x}\right) = e :$$

a. Implicitly defines x as a function of y and z, and calculate $\partial x/\partial y$ and $\partial x/\partial z$, in terms of x, y and z.

b. Implicitly defines y as a function of x and z, and calculate $\partial y/\partial x$ and $\partial y/\partial z$, in terms of x, y and z.

c. Does **not** implicitly define z as a function of x and y (at least, not in a way that can be concluded from the Implicit Function Theorem).

Solution:

Let

$$G(x, y, z) = 2z + xye^y - z\ln\left(\frac{z}{x}\right).$$

It would be good to first check that the point $\mathbf{p} = (1, 0, e)$ is, in fact, in the level set where $G = e$; it's true that, otherwise, this would be a trick question, or an error, but it's still a good thing to check. We find

$$G(1, 0, e) = 2e + 1 \cdot 0 \cdot e^0 - e\ln(e/1) = e.$$

Also note that we are near a point where $x = 1 > 0$ and $z = e > 0$, so that both x and z can be assumed to be positive throughout the problem; hence, $\ln(z/x)$ is defined and G is continuously differentiable.

For parts (a), (b), and (c), we need the partial derivatives of G, and we need to know if they're zero at $\mathbf{p}$. We find:

$$\left.\frac{\partial G}{\partial x}\right|_{\mathbf{p}} = \left.\left(ye^y + \frac{z}{x}\right)\right|_{(1,0,e)} = e \neq 0;$$

$$\left.\frac{\partial G}{\partial y}\right|_{\mathbf{p}} = \left.(xye^y + xe^y)\right|_{(1,0,e)} = 1 \neq 0;$$

and

$$\left.\frac{\partial G}{\partial z}\right|_{\mathbf{p}} = \left.(2 - 1 - \ln z + \ln x)\right|_{(1,0,e)} = 0.$$

Since $\partial G/\partial z$ is 0 at $(1, 0, e)$, we **cannot** use the Implicit Function Theorem to conclude that $G = e$ implicitly defines z as a function of x and y near the point $(1, 0, e)$; this answers part (c).

However, as the partial derivatives $\partial G/\partial x$ and $\partial G/\partial y$ are non-zero at $\mathbf{p}$, we know, by the Implicit Function Theorem, that the equation $G(x, y, z) = e$ implicitly defines x as a function of y and z, and implicitly defines y as a function of x and z, near the point $\mathbf{p}$.

Near any point (not just near $\mathbf{p}$) where $\partial G/\partial x \neq 0$, the Implicit Function Theorem applies, and

$$\frac{\partial x}{\partial y} = -\frac{\partial G/\partial y}{\partial G/\partial x} = -\frac{xye^y + xe^y}{ye^y + \frac{z}{x}},$$

and

$$\frac{\partial x}{\partial z} = -\frac{\partial G/\partial z}{\partial G/\partial x} = -\frac{1 - \ln z + \ln x}{ye^y + \frac{z}{x}},$$

Also, near any point (not just near $\mathbf{p}$) where $\partial G/\partial y \neq 0$, the Implicit Function Theorem applies, and

$$\frac{\partial y}{\partial x} = -\frac{\partial G/\partial y}{\partial G/\partial y} = -\frac{ye^y + \frac{z}{x}}{xye^y + xe^y},$$

and

$$\frac{\partial y}{\partial z} = -\frac{\partial G/\partial z}{\partial G/\partial y} = -\frac{1 - \ln z + \ln x}{xye^y + xe^y},$$

More Depth:

Suppose that, near the point $\mathbf{p}$, $F(\mathbf{x}) = F(\mathbf{p})$ implicitly defines x_i as a continuously differentiable function g of the other variables. This means that we are assuming that F is continuously differentiable at $\mathbf{p}$, and that $\partial F/\partial x_i \neq 0$ at $\mathbf{p}$.

Then, near $\mathbf{p}$, the level set where $F(\mathbf{x}) = F(\mathbf{p})$ is the same as the graph of the function $x_i = g(x_1, \ldots, x_{i-1}, x_{i+1}, \ldots, x_n)$. Thus, we have two "competing" notions of the tangent set, at $\mathbf{p}$, to the level set where $F(\mathbf{x}) = F(\mathbf{p})$; we could write an equation for the tangent set to a level set, as in Definition 2.6.16, or using the tangent set to the graph of a function, as in Definition 2.3.8.

Of course, to have no confusion or contradiction in our terminology, the two "different" equations that we obtain for the tangent set had better be equivalent, and we stated that this was the case in Theorem 2.7.7. We can prove this now; our formulas for the implicit partial derivatives yield this equivalence quickly.

Let $\hat{\mathbf{p}} = (p_1, \ldots, p_{i-1}, p_{i+1}, \ldots, p_n)$. Definition 2.3.8 tells us that the tangent set to the graph of $x_i = g(x_1, \ldots, x_{i-1}, x_{i+1}, \ldots, x_n)$ at the point $(x_1, \ldots, x_n) = (p_1, \ldots, p_n)$ is given by the equation

$$x_i = p_i + \left.\frac{\partial x_i}{\partial x_1}\right|_{\hat{\mathbf{p}}} (x_1 - p_1) + \left.\frac{\partial x_i}{\partial x_2}\right|_{\hat{\mathbf{p}}} (x_2 - p_2) + \cdots + \left.\frac{\partial x_i}{\partial x_{i-1}}\right|_{\hat{\mathbf{p}}} (x_{i-1} - p_{i-1}) +$$

$$\frac{\partial x_i}{\partial x_{i+1}}\bigg|_{\hat{\mathbf{p}}} (x_{i+1} - p_{i+1}) + \cdots + \frac{\partial x_i}{\partial x_{n-1}}\bigg|_{\hat{\mathbf{p}}} (x_{n-1} - p_{n-1}) + \frac{\partial x_i}{\partial x_n}\bigg|_{\hat{\mathbf{p}}} (x_n - p_n),$$

and the Implicit Function Theorem tells us that this is equivalent to

$$x_i = p_i - \frac{\partial F/\partial x_1}{\partial F/\partial x_i}\bigg|_{\mathbf{p}} (x_1 - p_1) - \frac{\partial F/\partial x_2}{\partial F/\partial x_i}\bigg|_{\mathbf{p}} (x_2 - p_2) - \cdots - \frac{\partial F/\partial x_{i-1}}{\partial F/\partial x_i}\bigg|_{\mathbf{p}} (x_{i-1} - p_{i-1}) -$$

$$\frac{\partial F/\partial x_{i+1}}{\partial F/\partial x_i}\bigg|_{\mathbf{p}} (x_{i+1} - p_{i+1}) - \cdots - \frac{\partial F/\partial x_{n-1}}{\partial F/\partial x_i}\bigg|_{\mathbf{p}} (x_{n-1} - p_{n-1}) - \frac{\partial F/\partial x_n}{\partial F/\partial x_i}\bigg|_{\mathbf{p}} (x_n - p_n).$$

Multiplying throughout by $(\partial F/\partial x_i)|_{\mathbf{p}}$ and putting all the terms on one side, we obtain the same equation for the tangent set to a level set that we gave in Definition 2.6.16.

We now wish to prove, or sketch the proof of, the Implicit Function Theorem, Theorem 2.12.2. Not surprisingly, this proof is essentially the same as the proof of the geometric form of the Implicit Function Theorem that we gave in Section 2.7.

Suppose that we have a C^1 real-valued function $F = F(x_1, \ldots, x_n)$, and a point $\mathbf{p} = (p_1, \ldots, p_n)$. Let M denote the level set of F at $\mathbf{p}$. Suppose further that

$$\frac{\partial F}{\partial x_n}\bigg|_{\mathbf{p}} \neq 0.$$

We will prove Theorem 2.12.2 in the case where $i = n$; this clearly proves the statement regardless of the value of the index i.

Define a new multi-component function from a subset of $\mathbb{R}^n$ to a subset of $\mathbb{R}^n$ by

$$\boldsymbol{\Phi}(x_1, \ldots, x_n) = \big(x_1, x_2, \ldots x_{n-1}, F(x_1, \ldots, x_n)\big) = (y_1, \ldots, y_n).$$

Note that a point $\mathbf{x}$ is in M, the level set where $F = F(\mathbf{p})$, if and only if $y_n = F(\mathbf{p})$. Now, it is easy to show that $d_{\mathbf{p}}\boldsymbol{\Phi}$ is a bijection and, hence, the Inverse Function Theorem, Theorem 2.6.6, tells us that $\boldsymbol{\Phi}$ is a local C^1 change of coordinates. If we let $\hat{\mathbf{p}} = (p_1, \ldots, p_{n-1})$, then Φ^{-1} is a C^1 function from an open neighborhood of $(\hat{\mathbf{p}}, F(\mathbf{p}))$ to an open neighborhood of $\mathbf{p}$.

From the definition of Φ, we have that Φ^{-1} must be of the form

$$\Phi^{-1}(y_1, \ldots, y_n) = \big(y_1, \ldots, y_{n-1}, h(y_1, \ldots, y_n)\big) = (x_1, \ldots, x_n),$$

where h is a C^1 function and $(x_1, \ldots, x_n)$ is in M if and only if $y_n = F(\mathbf{p})$. Note that, for $1 \le i \le n-1$, $x_i = y_i$. Define g on an open neighborhood of $\hat{\mathbf{p}}$ by

$$g = g(y_1, \ldots, y_{n-1}) = h(y_1, \ldots, y_{n-1}, F(\mathbf{p})) = x_n.$$

This g is the function given in Theorem 2.12.2: for all $(x_1, \ldots, x_n)$ in an open neighborhood of $\mathbf{p}$, $F(x_1, \ldots, x_n) = F(\mathbf{p})$ if and only if $x_n = g(x_1, \ldots, x_{n-1})$.

The rest of the Implicit Function Theorem is easy.

If we had two such g, then $x_n = g(x_1, \ldots, x_{n-1})$ implies that they have the same values, i.e., are the same function, at least when restricted to an open neighborhood of $\hat{\mathbf{p}}$ on which they're both defined.

Since, for all $(x_1, \ldots, x_{n-1})$ in an open neighborhood of $\hat{\mathbf{p}}$,

$$F\big(x_1, \ldots, x_{n-1}, g(x_1, \ldots, x_{n-1})\big) = F(\mathbf{p}),$$

we can differentiate both sides with respect to x_i, for $1 \leq j \leq n-1$, and use the Chain Rule to obtain

$$\vec{\nabla} F\big(x_1, \ldots, x_{n-1}, g(x_1, \ldots, x_{n-1})\big) \cdot \frac{\partial}{\partial x_j}\big(x_1, \ldots, x_{n-1}, g(x_1, \ldots, x_{n-1})\big) = 0.$$

Dropping the explicit reference to the points where each quantity is evaluated, this is equivalent to:

$$\left(\frac{\partial F}{\partial x_1}, \ldots, \frac{\partial F}{\partial x_{n-1}}, \frac{\partial F}{\partial x_n}\right) \cdot \left(0, \ldots, 0, 1, 0, \ldots, 0, \frac{\partial g}{\partial x_j}\right) = 0,$$

where the 1 is in the j-th position. Hence, near $\mathbf{p}$, we have

$$\frac{\partial F}{\partial x_j} + \frac{\partial F}{\partial x_n} \cdot \frac{\partial g}{\partial x_j} = 0,$$

i.e.,

$$\frac{\partial g}{\partial x_j} = -\frac{\partial F/\partial x_j}{\partial F/\partial x_n},$$

which is what we wanted to show.

2.12.1 Exercises

Online answers to select exercises are here.

Basics:

In each of the following exercises, you are given an equation, in two or three variables, and a point p which satisfies the equation. Verify that the equation implicitly defines x as a function of the remaining variable(s), near p, and calculate the partial derivative(s) of x, at p, with respect to the other variable(s).

1. $x^2 + xy + y^2 = 3$, and $\mathbf{p} = (1, 1)$.

2. $\sin(xy) + xy^2 = 0$, and $\mathbf{p} = (0, 1)$.

3. $xe^y - \tan^{-1}(x + y) + \frac{\pi}{4} = 1$, and $\mathbf{p} = (1, 0)$.

4. $x^5 - x - y = 20$, and $\mathbf{p} = (2, 10)$.

5. $x^3 + ye^z - xyz = 9$, and $\mathbf{p} = (2, 1, 0)$.

6. $\cos(xy) + xz = 7$, $\mathbf{p} = (2, \pi, 3)$.

7. $x^3 + y^3 = x + z$, and $\mathbf{p} = (1, 1, 1)$.   ▶

8. $z = y^2 + x \ln x$, and $\mathbf{p} = (1, 2, 4)$.

In each of the following exercises, you are given an equation, involving z and other variables. (a) Rewrite (if necessary) the equation in the form $F = c$, where F is a function of the given variables, and c is a constant. Clearly identify your function F. (b) Assume that your equation $F = c$ implicitly defines z as a function of the other variables, and calculate the partial derivative(s) of z, with respect to the other variable(s). (c) Then, say what condition needs to hold for you to know, from the Implicit Function Theorem, that z is, in fact, implicitly defined in terms of the other variables.

9. $x + z + y^2 = 2 + z^3$.   ▶

10. $ye^{-z} + xz = 5 + 3 \sin y$.

11. $\dfrac{z \ln y - x \ln w}{xz + wy} = 1$.

12. $\tan^{-1} z = x^2 + y^2 + z^2 - 1$.

13. $x^5 + y^5 + z^5 = x + y + z + xyz$.

14. $\sin(w - x) + \cos(y - z) = w - x + y - z$.

More Depth: In each of the following exercises, you are given an equation $F(x, y, z) = c$, defining a level surface E. You are also given a point $\mathbf{p} = (a, b, c)$, which satisfies the equation. (a) Verify that the equation implicitly defines z as a function of x and y, near $\mathbf{p}$. (b) Find an equation for the tangent plane to E at $\mathbf{p}$ in the form $z = c + A(x - a) + B(y - b)$.

15. $x - y^2 - z^2 = 0$, and $\mathbf{p} = (5, 1, 2)$.

16. $y - xe^z + ze^x = 0$, and $\mathbf{p} = (-1, 0, -1)$.   ▶

17. $x + z + y^2 - z^3 = 2$, and $\mathbf{p} = (1, 1, 1)$.

18. $ye^{-z} + xz - 3 \sin y + 3 = \frac{\pi}{2}$, and $\mathbf{p} = (1, \pi/2, 0)$.

19. $x^2 + y^2 + z^2 - \tan^{-1} z = 1$, and $\mathbf{p} = (1/\sqrt{2}, 1/\sqrt{2}, 0)$.

20. $x^5 + y^5 + z^5 - x - y - z - xyz = -1$, and $\mathbf{p} = (1, 1, 1)$.

21. In Figure 2.12.2, you are given the graph of a level curve $F(x, y) = c$.

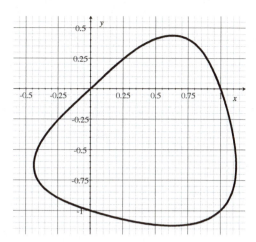

Figure 2.12.2: The level curve where $F(x, y) = c$.

(a) Approximate those points near which the equation cannot possibly implicitly define y as a function of x.

(b) Approximate those points near which the equation cannot possibly implicitly define x as a function of y.

22. In Figure 2.12.3, you are given the graph of a level curve $G(x, y) = c$.

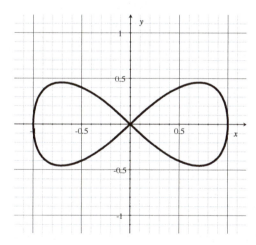

Figure 2.12.3: The level curve where $G(x, y) = c$.

(a) Approximate those points near which the equation cannot possibly implicitly define y as a function of x.

(b) Approximate those points near which the equation cannot possibly implicitly define x as a function of y.

23. Let $F(x, y) = y^2 - 2x^2 y + x^4$.

   (a) Verify that $(2, 4)$ is a point on the level curve where $F = 0$.

   (b) Show that $\partial F / \partial y = 0$ at $(2, 4)$, so that we cannot use the Implicit Function Theorem to conclude that $F = 0$ implicitly defines y as a function of x near $(2, 4)$.

   (c) Show that, nonetheless, $F = 0$ does define y as a function of x, near **every** point where $F = 0$. (Hint: Use algebra. It may be helpful to have technology graph the level curve.)

24. Explain why part (c) of the previous problem does not contradict the Implicit Function Theorem.

2.13 Multivariable Taylor Polynomials and Series

In single-variable Calculus, given a function $f = f(x)$ and a point a, you should have looked at $T_f^n(x; a)$, the n-th order *Taylor polynomial* of f at a (provided that the n-th derivative of f exists at a). The importance of the Taylor polynomials is that, while the function f itself may be extremely complicated, we can frequently approximate it very well by something simple, a polynomial; in addition, we expect the approximation to be "good" when x is "close" to a, and we expect the approximation to get even better for larger values of n. If the error in the approximation goes to zero as n goes to infinity, then we say the f equals its *Taylor series* near a, and so can represent the function f by what is essential a polynomial of infinite degree.

The question in this section is:

How do we do all of this for a function $f = f(x, y)$ of two variables or, more generally, for a function of any number of variables?

As we shall see, describing what a Taylor polynomial should be for a multivariable function is relatively straightforward. However, writing down general formulas is pretty unattractive because the multivariable versions of things get very unwieldy. This is why we will spend most of our time on the two-variable case; the situation is already complicated there, but much easier to present than the situation for more variables.

Basics:

We'll begin by quickly summarizing the basics of Taylor polynomials and series for single-variable functions, and use that as a guide for how to deal with the multivariable case.

Suppose you have a function $f = f(x)$, that is differentiable n times at a point a. Then, $T_f^n(x; a)$, the *n-th order Taylor polynomial of f centered at a* is the unique polynomial

$$T_f^n(x; a) = c_0 + c_1(x - a) + c_2(x - a)^2 + \cdots + c_{n-1}(x - a)^{n-1} + c_n(x - a)^n = \sum_{k=0}^{n} c_k(x - a)^k$$

such that the value of $T_f^n(x; a)$ and its derivatives, up to the n-th one, match the values of f and its derivatives, up to the n-th one, at the point a.

Quick calculations then tell you that, for $0 \leq k \leq n$, we must have

$$c_k = \frac{f^{(k)}(a)}{k!},$$

and so the n-th order Taylor polynomial is

$$T_f^n(x; a) = \sum_{k=0}^{n} \frac{f^{(k)}(a)}{k!}(x - a)^k$$

The expectation is that, since f and all of its derivatives, up to the n-th one, match $T_f^n(x; a)$ and it derivatives, at a, then, in fact, for all x **close** to a, not just **at** a, the values of $f(x)$ and $T_f^n(x; a)$ should be approximately the same.

In order to state precise results on how well $T_f^n(x; a)$ approximates $f(x)$, we define the *n-th order remainder*, $R_f^n(x; a)$ to be the difference between f and $T_f^n(x; a)$, i.e.,

$$R_f^n(x; a) = f(x) - T_f^n(x; a).$$

What we want to know is that, when n is big and x is close to a, the value of $R_f^n(x; a)$ is close to zero. To prove this in different situations, there are various theorems related to showing that the remainder is close to zero.

We don't actually need that f is of class C^n. We need that f is of class C^{n-1}, that all of the $(n-1)$-order partial derivatives are differentiable, and that the order in which we take the partial derivatives of order n is irrelevant. However, the most common way of obtaining all of these conditions is to assume that f is of class C^n, even though, in the single-variable case, this means we are assuming not just that f is n times differentiable, but also that the n-th derivative is continuous.

If $\lim_{n \to \infty} R_f^n(x; a) = 0$, then $f(x)$ equals its *Taylor series*, a power series, which is essentially a Taylor polynomial of infinite order.

We want to develop Taylor polynomials and series for multivariable functions. We will do this for a function $f = f(x, y)$ of two variables, and save the case of more variables for the More Depth portion of this section.

We will suppose that f is of class C^n, that is, that all of the partial derivatives of f of all orders $\leq n$ exist and are continuous. This implies that the order in which we take mixed partial derivatives is irrelevant.

What should the n-th order Taylor polynomial $T_f^n(x, y; a, b)$ of f centered at (a, b) mean? It should be the polynomial, in powers of $(x - a)$ and $(y - b)$, that goes up to the n-th degree, which has the same values as f for all of its partial derivatives and higher-order partial derivatives at (a, b).

Notice how the Taylor polynomial quickly becomes complicated, with three terms that have degree 2. Similarly, in the higher-order Taylor polynomials, with two variables, there are 4 terms of degree 3, 5 terms of degree 4, and so on.

Before we introduce some necessarily complicated notation for Taylor polynomials of arbitrary order n, let's first look at the second Taylor polynomial:

$$T = T_f^2(x, y; a, b) = c + p(x - a) + q(y - b) + r(x - a)^2 + s(x - a)(y - b) + t(y - b)^2,$$

where c, p, q, r, s, and t are constants, determined by the requirement that all of the partial derivatives of f and T, up to the second order, must be the same at (a, b).

We find that we must have

$$f(a, b) = T(a, b) = c,$$

$$\frac{\partial f}{\partial x}\Big|_{(a,b)} = \frac{\partial T}{\partial x}\Big|_{(a,b)} = \left(p + 2r(x - a) + s(y - b)\right)\Big|_{(a,b)} = p;$$

$$\frac{\partial f}{\partial y}\Big|_{(a,b)} = \frac{\partial T}{\partial y}\Big|_{(a,b)} = \left(q + s(x - a) + 2t(y - b)\right)\Big|_{(a,b)} = q;$$

$$\frac{\partial^2 f}{\partial x^2}\Big|_{(a,b)} = \frac{\partial^2 T}{\partial x^2}\Big|_{(a,b)} = 2r\Big|_{(a,b)} = 2r;$$

$$\frac{\partial^2 f}{\partial x \partial y}\bigg|_{(a,b)} = \frac{\partial^2 T}{\partial x \partial y}\bigg|_{(a,b)} = s\bigg|_{(a,b)} = s;$$

and

$$\frac{\partial^2 f}{\partial y^2}\bigg|_{(a,b)} = \frac{\partial^2 T}{\partial y^2}\bigg|_{(a,b)} = 2t\bigg|_{(a,b)} = 2t;$$

Therefore, inserting the values for p, q, r, s, and t, we find

$$T_f^2(x, y; a, b) = \tag{2.12}$$
$$f(a,b) + \frac{\partial f}{\partial x}\bigg|_{(a,b)} (x - a) + \frac{\partial f}{\partial y}\bigg|_{(a,b)} (y - b) +$$
$$\frac{1}{2}\left(\frac{\partial^2 f}{\partial x^2}\bigg|_{(a,b)} (x-a)^2 + 2\frac{\partial^2 f}{\partial x \partial y}\bigg|_{(a,b)} (x-a)(y-b) + \frac{\partial^2 f}{\partial y^2}\bigg|_{(a,b)} (y-b)^2 \right).$$

Looking at this formula for the 2nd-order Taylor polynomial for f centered at (a, b), notice that we can write the expression

$$\frac{\partial f}{\partial x}\bigg|_{(a,b)} (x-a) + \frac{\partial f}{\partial y}\bigg|_{(a,b)} (y-b)$$

more compactly as $d_{(a,b)}f(x-a, y-b)$.

This makes us want to adopt a similar compact notation for the second partial derivatives piece of our 2nd-order Taylor polynomial. Thus, we define the *2nd total derivative of f at (a, b)* to be

$$d_{(a,b)}^{(2)}f(v, w) = \frac{\partial^2 f}{\partial x^2}\bigg|_{(a,b)} v^2 + 2\frac{\partial^2 f}{\partial x \partial y}\bigg|_{(a,b)} vw + \frac{\partial^2 f}{\partial y^2}\bigg|_{(a,b)} w^2.$$

With this new notation, our formula for the 2nd-order Taylor polynomial of f at (a, b) becomes very short:

$$T_f^2(x, y; a, b) = f(a, b) + d_{(a,b)}f(x-a, y-b) + \frac{1}{2}d_{(a,b)}^{(2)}f(x-a, y-b). \tag{2.13}$$

If you compare this with the single-variable formula for the 2nd-order Taylor polynomial, you may initially be concerned by the fact that you don't explicitly see the increasing powers of $(x - a)$ and $(y - b)$; notice, however, that the second powers are present in the definition of $d_{(a,b)}^{(2)}f$, and so are somewhat hidden in Formula 2.13.

Example 2.13.1. Calculate the 2nd-order Taylor polynomial for

$$f(x,y) = \sin(x+y) + xe^{-y}$$

at $(0,0)$.

Solution:

We calculate

$$f(0,0) \;=\; 0,$$

$$\frac{\partial f}{\partial x}\Big|_{(0,0)} \;=\; \big(\cos(x+y)+e^{-y}\big)\Big|_{(0,0)} \;=\; 2, \qquad \frac{\partial f}{\partial y}\Big|_{(0,0)} \;=\; \big(\cos(x+y)-xe^{-y}\big)\Big|_{(0,0)} \;=\; 1,$$

$$\frac{\partial^2 f}{\partial x^2}\Big|_{(0,0)} \;=\; \big(-\sin(x+y)\big)\Big|_{(0,0)} \;=\; 0,$$

$$\frac{\partial^2 f}{\partial x \partial y}\Big|_{(0,0)} \;=\; \big(-\sin(x+y)-e^{-y}\big)\Big|_{(0,0)} \;=\; -1,$$

and

$$\frac{\partial^2 f}{\partial y^2}\Big|_{(0,0)} \;=\; \big(-\sin(x+y)+xe^{-y}\big)\Big|_{(0,0)} \;=\; 0.$$

Inserting these numbers into Formula 2.12 or Formula 2.13, we find

$$T_f^2(x,y;0,0) \;=\; 0+2\cdot x+1\cdot y \;+\; \frac{1}{2}\big(0+2(-1)xy+0\big) \;=\; 2x+y-xy.$$

Thus, we expect that, for (x,y) close to $(0,0)$, we have a good approximation

$$\sin(x+y) + xe^{-y} \;\approx\; 2x+y-xy.$$

In order to write relatively nice formulas for the higher-order Taylor polynomials, it will be helpful to introduce the following compact notation for the other pieces of the formula:

Definition 2.13.2. *Suppose that f is of class C^r in an open neighborhood of (a, b). Then, the r-th **total derivative of f at** (a, b) is the function from $\mathbb{R}^2$ to $\mathbb{R}$ given by*

$$d_{(a,b)}^{(r)} f(v, w) = \sum_{k=0}^{r} \binom{r}{k} \frac{\partial^r f}{\partial x^k \partial y^{r-k}} \Bigg|_{(a,b)} v^k w^{r-k}.$$

When $r = 0$, we mean that $d_{(a,b)}^{(r)} f(v, w) = f(a, b)$.

Note that when $r = 1$, we have

$$d_{(a,b)}^{(1)} f(v, w) = \frac{\partial f}{\partial x}\Bigg|_{(a,b)} \cdot v + \frac{\partial f}{\partial y}\Bigg|_{(a,b)} \cdot w = d_{(a,b)} f(v, w),$$

so that the first total derivative is just the same as the total derivative. If you proceed with the calculation of the coefficients of the n-th order Taylor polynomial of f centered at (a, b), as we did above in the case when $n = 2$, what you find is easy to write in terms of higher-order total derivatives. You obtain:

Definition 2.13.3. *Suppose that $f = f(x, y)$ is of class C^n in an open neighborhood of (a, b). Then, the n-th **order Taylor polynomial of f centered at** (a, b) is the polynomial*

$$T_f^n(x, y; a, b) = \sum_{r=0}^{n} \frac{1}{r!} d_{(a,b)}^{(r)} f(x - a, y - b).$$

It is the unique polynomial of total degree $\leq n$, in powers of $(x-a)$ and $(y-b)$, such that, at (a, b), the functions f and $T_f^n(x, y; a, b)$ have the same values and partial derivatives up to order n.

The *total degree* of a polynomial in powers of $(x - a)$ and $(y - b)$ is the largest sum of exponents, $k + j$, for any summand $c(x - a)^k(y - b)^j$ in the polynomial that appears with a non-zero coefficient after simplifying by combining terms with the same powers of $(x - a)$ and $(y - b)$.

Remark 2.13.4. Notice that, with our new notation, the formula for the Taylor polynomials of $f = f(x, y)$ centered at (a, b), is an "obvious" analog of the formula for the Taylor polynomials of $g = g(x)$ centered at a:

$$T_g^n(x; a) = \sum_{r=0}^{n} \frac{g^{(r)}(a)}{r!}(x - a)^r.$$

Example 2.13.5. Calculate the 3rd-order Taylor polynomial for

$$f(x,y) = \sin(x+y) + xe^{-y}$$

at $(0,0)$, and use technology to graph f, its 1st-order Taylor polynomial, and its 3rd-order Taylor polynomial at $(0,0)$ on the same graph, in order to see how the functions compare.

Solution:

We calculated $T_f^2(x,y;0,0) = 2x + y - xy$ in Example 2.13.1, and there's no need for us to redo that work; we can use that

$$T_f^3(x,y;0,0) = T_f^2(x,y;0,0) + \frac{1}{3!} d_{(0,0)}^{(3)} f(x,y),$$

where

$$d_{(0,0)}^{(3)} f(x,y) = \sum_{k=0}^{3} \binom{3}{k} \frac{\partial^3 f}{\partial x^k \partial y^{3-k}}\bigg|_{(0,0)} x^k y^{3-k}.$$

We need all of the 3rd-order partial derivatives at $(0,0)$. Recalling that

$$\frac{\partial^2 f}{\partial x^2} = -\sin(x+y) \quad \text{and} \quad \frac{\partial^2 f}{\partial y^2} = -\sin(x+y) + xe^{-y},$$

we find

$$\frac{\partial^3 f}{\partial x^3}\bigg|_{(0,0)} = \big(-\cos(x+y)\big)\big|_{(0,0)} = -1,$$

$$\frac{\partial^3 f}{\partial x^2 \partial y}\bigg|_{(0,0)} = \big(-\cos(x+y)\big)\big|_{(0,0)} = -1,$$

$$\frac{\partial^3 f}{\partial x \partial y^2}\bigg|_{(0,0)} = \big(-\cos(x+y) + e^{-y}\big)\big|_{(0,0)} = 0,$$

and

$$\frac{\partial^3 f}{\partial y^3}\bigg|_{(0,0)} = \big(-\cos(x+y) - xe^{-y}\big)\big|_{(0,0)} = -1.$$

Hence,

$$d_{(0,0)}^{(3)} f(x,y) = -1x^3 + 3(-1)x^2 y + 3(0)xy^2 + -1y^3,$$

and so

$$T_f^3(x,y;0,0) = T_f^2(x,y;0,0) + \frac{1}{3!} d_{(0,0)}^{(3)} f(x,y) = 2x + y - xy + \frac{1}{6}\big(-x^3 - 3x^2 y - y^3\big).$$

In Figure 2.13.1 and Figure 2.13.2, we give two different views of the graph of f (in blue), the graph of T_f^1 (in green/gray mesh), which is the tangent plane to the graph of

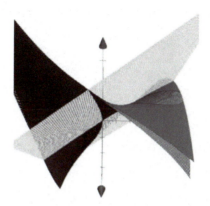

Figure 2.13.1: The graphs of f, T_f^1, and T_f^3.

Figure 2.13.2: Another view of the graphs of f, T_f^1, and T_f^3.

f at $(0,0)$, and the graph of T_f^3 (in red mesh). Notice how closely the graph of T_f^3 hugs the graph of f. The approximation of f by T_f^3 is certainly much better than the linear approximation.

$\boxed{\textbf{More Depth:}}$

The case of Taylor polynomials for more than 2 variables becomes even more cumbersome to discuss. The discussion of Taylor *series* and the remainder, even for the 2-variable case, is also quite messy. We shall state the relevant definitions and theorems, and forgo providing what would be the necessarily lengthy, complicated examples. The number of definitions that we make may seem a bit overwhelming, but are necessary to discuss Taylor polynomials and Taylor series for functions of any number of variables.

Suppose that we have a function $f = f(\mathbf{x})$ which is of class C^n on an open subset of $\mathbb{R}^m$. The n-th order Taylor polynomial of f, $T_f^n(\mathbf{x}; \mathbf{a})$, centered at a point $\mathbf{a}$, is, of course, the unique polynomial of total degree $\leq n$, in powers of all of the $(x_i - a_i)$, such that, at $\mathbf{a}$, the functions f and $T_f^n(\mathbf{x}; \mathbf{a})$ have the same values and partial derivatives up to order n.

A formula for the n-th order Taylor polynomial of f, centered at a point $\mathbf{a}$, is given essentially as in Definition 2.13.3; the hard part is to correctly define $d_{\mathbf{a}}^{(r)} f(\mathbf{v})$.

Definition 2.13.6. *Suppose that $f = f(\mathbf{x})$ is of class C^r in an open neighborhood of $\mathbf{a}$ in $\mathbb{R}^m$. Then, the r-th* **total derivative** *of f at $\mathbf{a}$ is the function from $\mathbb{R}^m$ to $\mathbb{R}$ given by*

$$d_{\mathbf{a}}^{(r)} f(\mathbf{v}) \;=\; \sum_{k_1, k_2, \ldots, k_m} \left[\frac{r!}{k_1! k_2! \ldots k_m!} \cdot \frac{\partial^r f}{\partial x_1^{k_1} x_2^{k_2} \ldots x_m^{k_m}} \bigg|_{\mathbf{a}} \cdot v_1^{k_1} v_2^{k_2} \ldots v_m^{k_m} \right],$$

where the summation is over all m-tuples $(k_1, \ldots, k_m)$ of non-negative integers such that $k_1 + \cdots + k_m = r$.
When $r = 0$, we mean that $d_{\mathbf{a}}^{(r)} f(\mathbf{v}) = f(\mathbf{a})$.

Definition 2.13.7. *Suppose that $f = f(\mathbf{x})$ is of class C^n in an open neighborhood of $\mathbf{a}$ in $\mathbb{R}^m$. Then, the n-th order* **Taylor polynomial** *of f centered at $\mathbf{a}$ is the polynomial*

$$T_f^n(\mathbf{x}; \mathbf{a}) \;=\; \sum_{r=0}^{n} \frac{1}{r!} d_{\mathbf{a}}^{(r)} f(\mathbf{x} - \mathbf{a}).$$

It is the unique polynomial of total degree $\leq n$, in powers of all of the $(x_i - a_i)$ such that, at $\mathbf{a}$, the functions f and $T_f^n(\mathbf{x}; \mathbf{a})$ have the same values and partial derivatives up to order n.

Definition 2.13.8. *Suppose that $f = f(\mathbf{x})$ is of class C^n in an open neighborhood of $\mathbf{a}$ in $\mathbb{R}^m$. Then, the n-th order* **Taylor remainder** *of f centered at $\mathbf{a}$ is*

$$R_f^n(\mathbf{x}; \mathbf{a}) \;=\; f(\mathbf{x}) \;-\; T_f^n(\mathbf{x}; \mathbf{a}).$$

Rather than give a separate definition of what it means to have a power series in any number of variables, and what it means for it to converge, and then having a theorem relating the Taylor remainder and convergence, we combine these things in the following definition.

Definition 2.13.9. *Suppose that $f = f(\mathbf{x})$ is of class C^∞ in an open neighborhood $\mathcal{W}$ of $\mathbf{a}$ in $\mathbb{R}^m$. Let $\mathbf{x}$ be a point in $\mathcal{W}$. If $\lim_{n \to \infty} R_f^n(\mathbf{x}; \mathbf{a}) = 0$, then the* **Taylor series** *of f centered at $\mathbf{a}$,*

$$T_f^\infty(\mathbf{x}; \mathbf{a}) \;=\; \sum_{r=0}^{\infty} \frac{1}{r!} d_{\mathbf{a}}^{(r)} f(\mathbf{x} - \mathbf{a}),$$

exists and converges to $f(\mathbf{x})$.

Definition 2.13.10. *Suppose that $f = f(\mathbf{x})$ is of class C^∞ in an open neighborhood of $\mathbf{a}$ in $\mathbb{R}^m$, and that, for all $\mathbf{x}$ in a, possibly smaller, open neighborhood of $\mathbf{a}$, $T_f^\infty(\mathbf{x}; \mathbf{a})$ converges to $f(\mathbf{x})$. Then, we say that f is **real analytic at a**.*

*If f is of class C^∞ on an open subset $\mathcal{W}$ of $\mathbb{R}^m$, and is real analytic at each point $\mathbf{a}$ in $\mathcal{W}$, then we simply say that f is a **real analytic function**.*

The usual way of getting a handle on the Taylor remainder is:

Theorem 2.13.11. (**Taylor-Lagrange Theorem**) *Suppose that f is of class C^{n+1} in an open neighborhood $\mathcal{W}$ of $\mathbf{a}$ in $\mathbb{R}^m$, let $\mathbf{x}$ be a point in $\mathcal{W}$, unequal to $\mathbf{a}$, such that the line segment, L, between $\mathbf{x}$ and $\mathbf{a}$ is contained in $\mathcal{W}$.*

Then, there exists a point $\mathbf{c}$ on L, unequal to either of $\mathbf{x}$ and $\mathbf{a}$, such that

$$R_f^n(\mathbf{x}; \mathbf{a}) = \frac{1}{(n+1)!} \, d_{\mathbf{c}}^{(n+1)} f(\mathbf{x} - \mathbf{a}).$$

For the proof, see Trench, [S], Theorem 5.4.8.

We will now show how you use the Taylor-Lagrange Theorem to prove the Second Derivative Test, Theorem 2.9.12, for functions $f = f(x, y)$ of class C^2.

Suppose that $f = f(x, y)$ is a real-valued function of class C^2 in an open neighborhood $\mathcal{U}$ of (a, b) in $\mathbb{R}^2$.

For all (x, y) in $\mathcal{U}$, let $D(x, y) = f_{xx} f_{yy} - f_{xy}^2$. Assume (a, b) is a non-degenerate critical point of f, i.e., that $d_{(a,b)} f = 0$ and that $D(a, b) \neq 0$.

If $D(a, b) < 0$, then, since f is C^2, there exists an open disk $\mathcal{W}$, centered at (a, b), on which D is always < 0. If $D(a, b) > 0$ and $f_{xx}(a, b) > 0$, then, since f is C^2, there exists an open disk $\mathcal{W}$, centered at (a, b), on which D and f_{xx} are always > 0. If $D(a, b) > 0$ and $f_{xx}(a, b) < 0$, then, since f is C^2, there exists an open disk $\mathcal{W}$, centered at (a, b), on which D is always > 0 and f_{xx} is always < 0. The point is: whichever of the three cases we're in, we let $\mathcal{W}$ be an open disk, centered at (a, b), such that the corresponding inequalities hold at all points of $\mathcal{W}$.

Let $(x, y) \neq (a, b)$ be a point in $\mathcal{W}$. Note that, since $\mathcal{W}$ is a disk, for all $(x, y) \neq (a, b)$ in $\mathcal{W}$, the line segment connecting (x, y) and (a, b) in contained in W. Then, applying the Taylor-Lagrange Theorem in the case when $n = 1$, we conclude that there exists a point $\mathbf{c}$ on the line segment between (x, y) and (a, b) such that

$$f(x, y) - f(a, b) = \frac{1}{2} \left[f_{xx}(\mathbf{c})(x - a)^2 + 2f_{xy}(\mathbf{c})(x - a)(y - b) + f_{yy}(\mathbf{c})(y - b)^2 \right].$$

Let $A = f_{xx}(\mathbf{c})$, $B = f_{xy}(\mathbf{c})$, $C = f_{yy}(\mathbf{c})$, $u = x - a$, and $v = y - b$. Note that, as

$(x, y) \neq (a, b)$, it follows that $(u, v) \neq (0, 0)$. Finally, let

$$E = Au^2 + 2Buv + Cv^2,$$

so that

$$f(x, y) - f(a, b) = \frac{1}{2}E.$$

Then f attaining a local maximum or minimum value at (a, b) is all of question of E being always negative or always positive, respectively.

This is now a pure algebra problem. If A and C are both 0, then $D = AC - B^2 = -B^2$, and it is trivial to see that there exist pairs (u, v) arbitrarily close to $(0, 0)$ such that $E = 2Buv$ is negative and pairs such that E is positive, and hence $f(a, b)$ is neither a local maximum nor local minimum value.

If one of A or C is nonzero, then, by symmetry, without loss of generality, we may assume that $A \neq 0$. In this case, completing the square yields

$$E = A\left[\left(u + \frac{B}{A}v\right)^2 + \frac{AC - B^2}{A^2}v^2\right].$$

All of the cases of the Second Derivative Test follow easily now; we leave them as an exercise.

Online answers to select exercises are here.

2.13.1   Exercises

Basics:

In each of the following exercises, you are given a function $f = f(x, y)$ and a point $\mathbf{p} = (a, b)$. (a) Calculate $T_f^2(x, y; a, b)$. (b) Use technology to graph $f(x, y)$ and $T_f^2(x, y; a, b)$, together, near $(a, b, f(a, b))$.

1. $f(x, y) = xe^{xy} + e^y$, and $\mathbf{p} = (0, 0)$.

2. $f(x, y) = \sin(xy) + \cos x$, and $\mathbf{p} = (0, 0)$.

3. $f(x, y) = x^3 + y \ln x$, and $\mathbf{p} = (1, 2)$.

4. $f(x, y) = e^{x^2 + y^2}$, and $\mathbf{p} = (1, 0)$.

5. $f(x, y) = e^{\sqrt{x^2 + y^2}}$, and $\mathbf{p} = (0, 1)$.

6. $f(x, y) = \dfrac{xy^2 + y^3}{x^3}$, and $\mathbf{p} = (1, -1)$.

7. $f(x, y) = xy + \sqrt{x} + \sqrt{y}$, and $\mathbf{p} = (1, 1)$.

8. $f(x, y) = \tan^{-1}(y/x)$, and $\mathbf{p} = (2, -2)$.

In each of the following exercises, you are given a function $f = f(x, y)$ and a point $\mathbf{p} = (a, b)$, from the previous exercises. Calculate $T_f^3(x, y; a, b)$.

9. $f(x, y) = xe^{xy} + e^y$, and $\mathbf{p} = (0, 0)$.

10. $f(x, y) = \sin(xy) + \cos x$, and $\mathbf{p} = (0, 0)$.

11. $f(x, y) = x^3 + y \ln x$, and $\mathbf{p} = (1, 2)$.

12. $f(x, y) = e^{x^2 + y^2}$, and $\mathbf{p} = (1, 0)$.

13. $f(x, y) = e^{\sqrt{x^2 + y^2}}$, and $\mathbf{p} = (0, 1)$.

14. $f(x, y) = \dfrac{xy^2 + y^3}{x^3}$, and $\mathbf{p} = (1, -1)$.

15. $f(x, y) = xy + \sqrt{x} + \sqrt{y}$, and $\mathbf{p} = (1, 1)$.

16. $f(x, y) = \tan^{-1}(y/x)$, and $\mathbf{p} = (2, -2)$.

In each of the following exercises, you are given a function $f = f(x, y, z)$ and a point a, from the previous exercises. (a) Calculate $T_f^2(x, y, z; \mathbf{a})$. (b) Calculate $T_f^3(x, y, z; \mathbf{a})$.

More Depth:

17. $f(x, y, z) = x^3 y + e^{z^2}$, and $\mathbf{a} = (1, 0, 0)$.

18. $f(x, y, z) = \sqrt{2x + 3y + 5z}$, and $\mathbf{a} = (1, -1, 2)$.

19. $f(x, y, z) = \dfrac{x + y}{z}$, and $\mathbf{a} = (0, 0, 1)$.

20. $f(x, y, z) = x \ln y + y \ln z$, and $\mathbf{a} = (-2, 1, 1)$.

21. $f(x, y, z) = \dfrac{\sin(x + y)}{\cos(y + z)}$, and $\mathbf{a} = (2, -2, 2)$.

22. $f(x, y, z) = e^{xy} \sin z$, and $\mathbf{a} = (0, 0, 0)$.

Chapter 3

Multivariable Integrals

If you recall how the Riemann (definite) integral of a function $f = f(x)$ on a closed interval $[a, b]$ is defined, you could give the following informal description:

Chop up the interval $[a, b]$ into a bunch of little subintervals. Take a sample point in each subinterval, evaluate f at each sample point, and multiply by the length of the corresponding subinterval. Then, add up all of those products. This is a *Riemann sum*, and the definite integral is the limit of the Riemann sums as the lengths of the subintervals approaches zero. If f is continuous on $[a, b]$, such a limit exists, and is independent of all of the choices made.

So, how do we define the integral of a function $f = f(x, y)$ over a region R in the xy-plane? We do the analogous thing. We chop up the region R into lots of little rectangles (with some small pieces leftover), we pick a sample point in each rectangle, we evaluate f at each sample point, multiply by the area of the corresponding rectangle, and add up all of these products. This is a Riemann sum, and the definite integral is the limit of the Riemann sums as the lengths of the sides of the rectangles approach zero. It turns out that, if R is a manageable region, and f is continuous on R, such a limit exists, and is independent of all of the choices made.

We also define, in a similar manner, the integral of a function of three variables over a solid region and, more generally, the integral of a function of n variables over manageable regions in $\mathbb{R}^n$.

There are many applications to calculating volume, mass, centers of mass and moments, and surface area. We will also consider integrating using non-Cartesian coordinate systems because some problems are much nicer to set up in polar, cylindrical, or spherical coordinates.

There is, of course, the question of how you actually **calculate** these higher-dimensional integrals. In single-variable Calculus, the Fundamental Theorem of Calculus tells us how we can use anti-derivatives, or *indefinite integrals*, to calculate the values of definite integrals. In multivariable Calculus, there is an analogous result; we use iterated anti-derivatives, usually called *iterated integrals*, to calculate the value of definite integrals. Iterated integrals consist simply of anti-differentiating with respect to each variable, while holding the others constant; thus, you perform a series of partial anti-derivatives. The main difficulty lies in producing the correct limits of integration.

3.1 Partial Anti-Derivatives and Iterated Integrals

In later sections of this chapter, we shall discuss definite integrals of multivariable functions and their applications. *Fubini's Theorem*, which we shall first encounter in Theorem 3.2.2, tells us that the practical calculation of multivariable integrals boils down to calculating an iterated collection of what are, essentially, *partial anti-derivatives*, anti-derivatives in which all variables, other than one, are held constant.

Such iterated partial anti-derivatives are referred to simply as *iterated integrals*, and we discuss them in this section.

Partial anti-derivatives are best described using an example.

Basics:

Example 3.1.1. Suppose you are told that there is a function $f = f(x, y)$ such that

$$\frac{\partial f}{\partial x} = 3x^2 - 5y^2.$$

To what extent does this determine the function f?

We need to undo partial differentiation. We naturally refer to this process as *partial anti-differentiation*; you anti-differentiate, with respect to x, treating y as a constant. That is, since $\partial f/\partial x = 3x^2 - 5y^2$, we look at the anti-derivative

It is customary to write dx in this "partial anti-derivative", not ∂x. If this seems inconsistent to you, we sympathize.

$$f = \int (3x^2 - 5y^2)\, dx;$$

this anti-differentiation is with respect to x, holding y constant.

Assuming that y is a constant, we find

$$f = \int (3x^2 - 5y^2)\, dx = x^3 - 5xy^2 + A(y), \tag{3.1}$$

where $A = A(y)$ is a "constant", as far as x is concerned, i.e., a function which does not depend on x, but rather depends only on y (including the possibility of being an actual constant).

If it helps, think about it this way: if $A = A(y)$ is any function of just y (or, possibly, an actual constant), then

$$\frac{\partial}{\partial x}\left(x^3 - 5xy^2 + A(y)\right) = 3x^2 - 5y^2,$$

and so, if we want to allow for every possible anti-derivative, with respect to x, in Formula 3.1, we must allow for $A(y)$ to be an arbitrary function of y.

Here, we discuss definite integrals via the Fundamental Theorem. In Theorem 3.1.9 of the More Depth portion, we will discuss the precise way in which this is a **definite** integral.

Example 3.1.2. What about the definite integral

$$\int_1^2 (3x^2 - 5y^2)\, dx?$$

What does this mean?

Again, it means that you treat y as a constant. Using the Fundamental Theorem of Calculus, and our earlier partial anti-derivative calculation, we find

$$\int_1^2 (3x^2 - 5y^2)\, dx \;=\; x^3 - 5xy^2 + A(y) \Big|_{x=1}^{x=2} \;=$$

$$\left(8 - 10y^2 + A(y)\right) - \left(1 - 5y^2 + A(y)\right) \;=\; 7 - 5y^2.$$

Note that our superscript and subscript on the evaluation bar are not just 2 and 1, but also include $x =$. We want to be very clear about which of the variables is taking on the values.

Note that, just as in a single-variable definite integral, the "constant" $A(y)$ disappears in the definite integral, and so you needn't have added it in the first place.

Example 3.1.3. Of course, if y is a constant in the integral, then the limits of integration can depend on y. Consider

$$\int_{\sin y}^{y^7} (3x^2 - 5y^2)\, dx.$$

We calculate as before, but without including the $A(y)$, since we know it cancels out:

$$\int_{\sin y}^{y^7} (3x^2 - 5y^2)\, dx \;=\; x^3 - 5xy^2 \Big|_{x=\sin y}^{x=y^7} \;=$$

$$(y^7)^3 - 5y^7 \cdot y^2 - \left((\sin y)^3 - 5(\sin y)y^2\right) \;=\; y^{21} - 5y^9 - \sin^3 y + 5y^2 \sin y.$$

Example 3.1.4. We can iterate this partial integration process. Consider

$$\int_0^2 \left[\int_y^{y^2} (5x + 2y)\, dx \right] dy.$$

The brackets make it clear that you evaluate the interior integral first, and find

$$\int_0^2 \left[\int_y^{y^2} (5x + 2y)\, dx \right] dy \;=\; \int_0^2 \left(\frac{5x^2}{2} + 2xy \;\Big|_{x=y}^{x=y^2} \right) dy \;=\;$$

$$\int_0^2 \left[\frac{5y^4}{2} + 2y^3 - \left(\frac{5y^2}{2} + 2y^2 \right) \right] dy \;=\; \int_0^2 \left(\frac{5y^4}{2} + 2y^3 - \frac{9y^2}{2} \right) dy \;=\;$$

$$\frac{y^5}{2} + \frac{y^4}{2} - \frac{3y^3}{2} \;\Big|_0^2 \;=\; 16 + 8 - 12 - 0 \;=\; 12.$$

This is referred to as an *iterated integral*.

We included the big brackets in the iterated integral above. However, it is more common to omit the brackets, and simply pair the integral signs and differentials from the inside and work your way out. Thus, the iterated integral above would normally be written simply as

$$\int_0^2 \int_y^{y^2} (5x + 2y)\, dx\, dy,$$

and you're just supposed to know that you perform the inside integral first.

Example 3.1.5. Evaluate the iterated integral

$$\int_1^3 \int_0^{\sin x} \left(\frac{1 + 2y}{\sin x} \right) dy\, dx.$$

The limits of integration on the inside integral are allowed to depend on the variable in the outside differential, but the outside limits of integration should not depend on the variable in the inside differential.

Solution:

We calculate

$$\int_1^3 \int_0^{\sin x} \left(\frac{1 + 2y}{\sin x} \right) dy\, dx \;=\; \int_1^3 \left(\frac{y + y^2}{\sin x} \;\Big|_{y=0}^{y=\sin x} \right) dx \;=\; \int_1^3 \left(\frac{\sin x + \sin^2 x}{\sin x} - 0 \right) dx \;=\;$$

$$\int_1^3 (1 + \sin x)\, dx \;=\; x - \cos x \;\Big|_1^3 \;=\; 3 - \cos 3 - (1 - \cos 1) \;=\;$$

$$2 - \cos 3 + \cos 1.$$

More Depth: Iterated integrals can involve functions of any number of variables, and inside limits of integration may depend on any variables that are farther out, i.e., that appear in differentials that are farther outside.

Example 3.1.6. Consider the iterated integral

$$\int_0^2 \int_y^1 \int_z^{yz} 8xyz \; dx \, dz \, dy.$$

Inserting parentheses and brackets, this integral means

$$\int_0^2 \left(\int_y^1 \left[\int_z^{yz} 8xyz \; dx \right] dz \right) dy,$$

where you perform the integrals from the inside out. Notice that the limits of integration for x depend on the variables y and z, which are farther out in the integral, and the limits of integration on z depend on y, which is farther out. However, if we wish to obtain a number for an answer, the limits of integration on the variable farthest outside – y, in this example –must be constants.

We find

$$\int_0^2 \int_y^1 \int_z^{yz} 8xyz \; dx \, dz \, dy \;=\; \int_0^2 \int_y^1 \left(4x^2yz \; \Big|_{x=z}^{x=yz} \right) dz \, dy \;=\;$$

$$\int_0^2 \int_y^1 \left(4y^3z^3 - 4yz^3 \right) dz \, dy \;=\; \int_0^2 \left(y^3z^4 - yz^4 \; \Big|_{z=y}^{z=1} \right) dy \;=\;$$

$$\int_0^2 \left(y^3 - y - \left(y^7 - y^5 \right) \right) dy \;=\; \frac{y^4}{4} - \frac{y^2}{2} - \frac{y^8}{8} + \frac{y^6}{6} \; \Big|_0^2 \;=\; -\frac{58}{3}.$$

Recall Theorem 2.1.29, which tells us that, on a connected open set, the partial derivatives of a function determine the original function up to the addition of a constant. But how do you actually **find** the original function (up to adding an arbitrary constant)? You use partial anti-derivatives.

Example 3.1.7. Suppose that $f : \mathbb{R}^2 \to \mathbb{R}$ is such that, for all x and y,

$$\frac{\partial f}{\partial x} = 3x^2 - 5y^2 \quad \text{and} \quad \frac{\partial f}{\partial y} = -10xy + 8y^3.$$

Can we determine f? The answer is: yes, up to the addition of an arbitrary constant.

Let's see why.

First, we see that f is a partial anti-derivative of $3x^2 - 5y^2$, with respect to x. As we saw in Example 3.1.1, we obtain

$$f = \int (3x^2 - 5y^2) \, dx = x^3 - 5xy^2 + A(y).$$

But how do we determine $A(y)$?

This isn't bad; we take the partial derivative of this last equation, with respect to y, and we require it to equal what we were initially given for $\partial f/\partial y$, namely, $-10xy + 8y^3$. Note that, since A is a function of only y, $\partial A/\partial y$ can be written as simply $A'(y)$.

We find that we need

$$\frac{\partial}{\partial y} \left(x^3 - 5xy^2 + A(y) \right) = -10xy + 8y^3,$$

and so,

$$-10xy + A'(y) = -10xy + 8y^3.$$

Subtracting $-10xy$ from each side, we find that we need to solve

$$A'(y) = 8y^3.$$

This is an easy single-variable Calculus problem; we obtain

$$A(y) = \int 8y^3 \, dy = 2y^4 + C,$$

where C is really a constant constant this time.

Combining Formula 3.1 with $A(y) = 2y^4 + C$, we conclude that

$$f(x, y) = x^3 - 5xy^2 + 2y^4 + C,$$

for some constant C.

> How would you know if there were **no** function f which satisfies the given two equations? You would see it at this point. The reference to x would not cancel out, and you would end up needing $A'(y)$ to equal a function that depends on x; a contradiction, since $A(y)$, itself, does not depend on x.

Example 3.1.8. Suppose $f = f(x, y, z)$ is such that

$$\frac{\partial f}{\partial y} = 2x^2 y - 8y^3 \sin z + 7z e^{xy}.$$

Determine f, as far as it's possible.

Solution:

The function f is the partial anti-derivative

$$f = \int (2x^2 y - 8y^3 \sin z + 7ze^{xy})\, dy,$$

where both x and z are constant.

Thus, we find

$$f = \int (2x^2 y - 8y^3 \sin z + 7ze^{xy})\, dy = x^2 y^2 - 2y^4 \sin z + \frac{7ze^{xy}}{x} + A(x, z),$$

where $A(x, z)$ is an arbitrary function of x and z, but does not depend on y.

Finally, we need to discuss a theoretical point.

If you look at the outside integral in the iterated integral

$$\int_0^2 \left[\int_y^{y^2} (5x + 2y)\, dx \right] dy.$$

you see that $0 \le y \le 2$. This means that the inside integral $\int_y^{y^2} (5x + 2y)\, dx$ is defining a function, call it F, from the interval $[0, 2]$ into $\mathbb{R}$; the function F is given by

$$F(y) = \int_y^{y^2} (5x + 2y)\, dx$$

and the value of the iterated integral means $\int_0^2 F(y)\, dy$, i.e.,

$$\int_0^2 \left[\int_y^{y^2} (5x + 2y)\, dx \right] dy = \int_0^2 F(y)\, dy.$$

Of course, we like/need for F to be continuous on $[0, 2]$ so that we know that the definite integral exists and can be calculated via the Fundamental Theorem of Calculus. In all of our examples, we have used continuous, easily anti-differentiable functions, so it was obvious that the integrals existed.

However, we would like to have a general theorem which tells us that, if

1. A is a compact subset of $\mathbb{R}^{n-1}$,

2. p and q are continuous real-valued functions on A, and

3. $f = f(x_1, \ldots, x_n)$ is a continuous function on the set of points in $\mathbb{R}^n$ such that $(x_1, \ldots, x_{n-1})$ is in A and x_n is in the closed interval between $p(x_1, \ldots, x_{n-1})$ and $q(x_1, \ldots, x_{n-1})$,

then the real-valued function $F = F(\mathbf{a})$ on A given by

$$F(\mathbf{a}) = \int_{p(\mathbf{a})}^{q(\mathbf{a})} f(\mathbf{a}, x_n)\, dx_n$$

is continuous. This is what would allow us to partially integrate $F(\mathbf{a})$, i.e., allow us to iterate the integral.

There is, indeed, such a theorem, which we state below. In the statement, we use the *projection* function, $\pi : \mathbb{R}^n \to R^{n-1}$ given by

$$\pi(x_1, \ldots, x_{n-1}, x_n) = (x_1, \ldots, x_{n-1});$$

this is referred to as *projection onto the first* $(n - 1)$ *coordinates*. By rearranging the coordinates, it is trivial to see that this theorem actually applies if we project onto **any** $(n - 1)$ coordinates and so let any x_i coordinate be the "special" coordinate which is omitted. The notation is easier if we let the special coordinate be x_n, and project onto the first $(n - 1)$ coordinates.

Theorem 3.1.9. *Let π denote projection onto the first $(n-1)$ coordinates. Suppose that A is a compact subset of $\mathbb{R}^{n-1}$, and that p and q are continuous functions from A to $\mathbb{R}$. Let C be the subset of $\mathbb{R}^n$ defined by $\mathbf{x} \in \mathbb{R}^n$ is in C if and only if*

$$\pi(\mathbf{x}) \in A, \text{ and } x_n \text{ is in the closed interval between } p(\pi(\mathbf{x})) \text{ and } q(\pi(\mathbf{x})).$$

Then, C is compact.

Suppose, further, that f is a continuous function on C. Then, the function $F : A \to \mathbb{R}$ given by

$$F(\mathbf{a}) = \int_{p(\mathbf{a})}^{q(\mathbf{a})} f(\mathbf{a}, x_n)\, dx_n$$

is defined and continuous.

It is easy to show that C is closed and bounded and, hence, compact. Continuous functions on compact sets are *uniformly continuous*; recall Definition 1.7.17 and Theorem 1.7.18. It is the uniform continuity of f on C that guarantees that the integral function F is continuous. See the discussion of Theorem 5.6 in Fleming, [3].

Online answers to select exercises are here.

Basics:

3.1.1   Exercises

In each of the following exercises, you are given the partial derivative f_x or f_y of a function $f(x, y)$. Determine f as far as possible, using $A(y)$ or $A(x)$ to denote an arbitrary function of y or x. In your solution, use the integral notation.

1. $f_x = x^2 y + y + 2$.

2. $f_y = x^2 y + y + 2$.

3. $f_y = e^{x^2} \cos y - x^3 + y^3$.

4. $f_x = \dfrac{\ln y}{1 + x^2}$.

5. $f_x = \cos(xy) + x e^y$.

6. $f_y = \cos(xy) + x e^y$.

7. $f_y = \dfrac{x}{x^2 + y^2}$.

8. $f_x = \dfrac{x}{x^2 + y^2}$.

In each of the following exercises, calculate the given (partial) integral. You may need to leave arbitrary functions of one of the variables in your answer.

9. $\displaystyle \int (xy + x + y)\, dx$

10. $\displaystyle \int (xy + x + y)\, dy$

11. $\displaystyle \int_1^3 (xy + x + y)\, dx$

12. $\displaystyle \int_{\ln y}^{\sin y} (xy + x + y)\, dx$

13. $\displaystyle \int (x \sin(xy) + y e^x + x^2)\, dy$

14. $\displaystyle \int_0^1 (x \sin(xy) + y e^x + x^2)\, dy$

15. $\displaystyle \int_0^x (x \sin(xy) + y e^x + x^2)\, dy$

16. $\displaystyle\int \frac{2y}{\sqrt{1-x^2}}\,dx$

17. $\displaystyle\int \frac{2y}{\sqrt{1-x^2}}\,dy$

18. $\displaystyle\int_{\ln x}^{x^2} \frac{2y}{\sqrt{1-x^2}}\,dy$

19. $\displaystyle\int_{1}^{x^2} \frac{2x}{y}\,dy$

20. $\displaystyle\int_{\sqrt{y}}^{y} \frac{2x}{y}\,dx$

In each of the following exercises, calculate the given iterated integral.

21. $\displaystyle\int_{0}^{2}\int_{y}^{2} xy\,dx\,dy$

22. $\displaystyle\int_{0}^{2}\int_{0}^{x} xy\,dy\,dx$

23. $\displaystyle\int_{0}^{1}\int_{x^2}^{x} 1\,dy\,dx$

24. $\displaystyle\int_{0}^{1}\int_{y}^{\sqrt{y}} 1\,dx\,dy$

25. $\displaystyle\int_{0}^{1}\int_{x^2}^{x} (x+y)\,dy\,dx$

26. $\displaystyle\int_{0}^{1}\int_{y}^{\sqrt{y}} (x+y)\,dx\,dy$

27. $\displaystyle\int_{1}^{2}\int_{0}^{2x} e^{x^2}\,dy\,dx$

28. $\displaystyle\int_{1}^{2}\int_{0}^{y} 2xye^{x^2}\,dx\,dy$

29. $\displaystyle\int_{1}^{e}\int_{0}^{\ln y} \frac{1}{y}\,dx\,dy$

30. $\displaystyle\int_{0}^{1}\int_{e^x}^{e} \frac{1}{y}\,dy\,dx$

In each of the following exercises, calculate the given iterated integral of the function. Assume that the function depends on only the variables that explicitly appear in the problem.

More Depth:

31. $\displaystyle\int_0^1 \int_1^2 \int_{-1}^1 (xy^2 + z^3)\, dz\, dy\, dx$

32. $\displaystyle\int_0^1 \int_1^x \int_x^y (xy^2 + z^3)\, dz\, dy\, dx$

33. $\displaystyle\int_0^1 \int_1^x \int_0^\pi (z\cos y + ye^x)\, dy\, dz\, dx$

34. $\displaystyle\int_0^1 \int_0^\pi \int_1^x (z\cos y + ye^x)\, dz\, dy\, dx$

35. $\displaystyle\int_4^9 \int_0^{\sqrt{y}} \int_0^z (e^{z^2} + yz + 1)\, dx\, dz\, dy$

36. $\displaystyle\int_0^3 \int_z^1 \int_{\sqrt{xz}}^{\sqrt{z}} \left(\frac{12xyz}{1 - 3z^2 + 2z^3}\right) dy\, dx\, dz$

37. $\displaystyle\int_4^8 \int_0^{z/2} \int_x^z \left(\frac{2y}{z^2 - x^2}\right) dy\, dx\, dz$

38. $\displaystyle\int_{-1}^1 \int_0^{\ln x} \int_0^{e^y} x\, dz\, dy\, dx$ $\;\blacktriangleright$

In each of the following exercises, you are given all of the partial derivatives of a function f. Determine f, up to adding an arbitrary constant.

39. $f_x = e^y + 2x$ and $f_y = xe^y + 1$.

40. $f_x = y\cos(xy) + e^x$ and $f_y = x\cos(xy) - \cos y$. $\;\blacktriangleright$

41. $f_x = \dfrac{y}{1 + x^2} - 2$ and $f_y = \tan^{-1} x + 3$.

42. $f_x = -\dfrac{x}{(x^2 + y^2)^{3/2}} + \dfrac{1}{x^2}$ and $f_y = -\dfrac{y}{(x^2 + y^2)^{3/2}} + \dfrac{1}{y^2}$

43. $f_x = 1$, $f_y = 2$, and $f_z = 3$.

44. $f_x = y + 2x$, $f_y = x + z + 3y^2$, and $f_z = y + 5z^4$.

45. $f_x = e^y$, $f_y = xe^y + z^2$, and $f_z = 2yz + \sec^2 z$.

46. $f_x = y^{-1} + e^x$, $f_y = -xy^{-2} + z^{-1} + e^y$, and $f_z = -yz^{-2} + e^z$.

3.2 Integration in $\mathbb{R}^2$

In the previous section, we discussed iterated integrals. Our primary interest in iterated integrals is that they allow us to calculate *definite integrals* over planar regions in $\mathbb{R}^2$ and solid regions in $\mathbb{R}^3$. As in single-variable Calculus, the definite integral is an "infinite sum of infinitesimal contributions" and, as in single-variable Calculus, the rigorous definition of the definite integral is a limit of *Riemann sums*.

To emphasize the 2-dimensional aspect of integrating over planar regions, we refer to such integrals as *double integrals*, and write a pair of integral signs.

Double integrals have many applications, such as calculating area, volume, mass, centers of mass, etc., which we shall consider in later sections.

<hr>

Let's begin with a motivating example, which should make it clear why we want to define integrals over 2-dimensional regions, i.e., why we want to define *double integrals*.

Basics:

Example 3.2.1. Suppose that we have a thin metal plate, a *lamina*, which we imagine as being so thin that it's reasonable to think of it as an idealized 2-dimensional object. Then, it is standard to discuss δ_{ar}, the *area-density* of such a plate; this is the mass per unit area. However, the area-density may vary at different points in the plate, and so, if we set up a Cartesian coordinate xy-plane on the plate, δ_{ar} may depend on x and y. We'll write $\delta_{\mathrm{ar}} = \delta_{\mathrm{ar}}(x, y)$ and, at least for now, we'll assume that δ_{ar} is continuous.

The question is: given the region R in the xy-plane which is occupied by the plate, and the area-density function, can we determine the (total) mass of the plate?

First, you should think about how you could **estimate** the mass of the plate. If you think about similar problems in the single-variable context, and how we estimated using *Riemann sums*, you should have a good idea of a reasonable approach.

We chop up R into lots of little rectangles, estimate the mass of each little rectangle, and add all of these little masses together, to approximate the mass of the whole plate. But how do we estimate the mass of each little rectangle? We take any point $\mathbf{p}$ (a *sample point*) in the little rectangle, evaluate $\delta_{\mathrm{ar}}(\mathbf{p})$, and assume – since the rectangle is small and since δ_{ar} is continuous – that $\delta_{\mathrm{ar}}(\mathbf{p})$ is a reasonable approximation of the area-density at every point in the small rectangle; hence, $\delta_{\mathrm{ar}}(\mathbf{p})$ times the area of the little rectangle should be a good estimate of the mass of the little rectangle.

Adding together all of these little mass contributions gives an **estimate** of the mass of the plate. Then, we do what we did for functions of one variable; we take the limit as the size (here, length and width) of the little pieces approaches 0. If this limit exists, and does not depend on precisely how the sizes of the little rectangles approach zero or on how the sample points are chosen, then we say that this limit is the mass of the plate and that the function δ_{ar} is *Riemann integrable* over the rectangle R.

In the discussion above, we ignored one problem, which you may have noticed yourself: if R has any curved edges, we won't be able to chop R into a bunch of little rectangles, without having some little rectangles sticking outside of R. See, for instance, Figure 3.2.1.

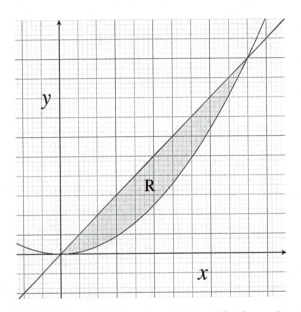

Figure 3.2.1: Little rectangles stick outside the region R.

What do you do about the little rectangles that cross the boundary? The answer is that, provided the boundary of the region is made up of curves, which have no area themselves, then, in the limit, as you use smaller and smaller rectangles, it doesn't matter whether you count the rectangles that cross the boundary or not. You get the same limit regardless.

The chopping up and estimating procedure that we discussed above should seem familiar. It should remind you of Riemann sums from single-variable Calculus.

To define the double integral, the Riemann integral over a region in $\mathbb{R}^2$, we must define partitions of plane regions, sample points, Riemann sums, take limits, and state a number of theorems. We cover all of this in the More Depth portion of this section. For now, we wish to give the quick intuition, state the theorems that we will usually use, and look at some examples.

If R is a subset of $\mathbb{R}^2$ (which will need to be reasonable in a sense that we'll discuss) and $f = f(x, y)$ is a real-valued function on R, we would like to define the *double integral* of f over R,

$$\iint_R f(x, y) \, dA,$$

which we think of as being **a continuous sum of the values of f multiplied by infinitesimal pieces of area, for points in the set R.**

What does this mean? It means what we already discussed in the context of the mass of a thin plate. You should think of dividing R into lots of little rectangles, where, near the boundary of R, you might have just pieces of rectangles, which we'll call partial rectangles. Any little rectangle has area, which we'll denote by ΔA; of course, $\Delta A = \Delta x \Delta y$, where Δx is the width of the rectangle and Δy is its height. Pick a point $\mathbf{s}$, a sample point, inside the little rectangle, and look at the product $f(\mathbf{s})\Delta A$.

You do this for each little rectangle in R, either including or excluding the partial rectangles near the boundary. Then, you add up all of these products $f(\mathbf{s})\Delta A$; this is a *Riemann sum*. Finally, you take the limit of this process, as the lengths and widths of all of the little rectangles approach zero, provided that the limit exists and doesn't depend on any of the choices that we made.

If, in fact, the limit exists and doesn't depend on any of the choices that we made, then we say that f is *Riemann integrable on* R, and the limit is the Riemann integral of f on (or, over) R; in $\mathbb{R}^2$, this Riemann integral is normally referred to as the double integral of f on R, and is denoted by

$$\iint_R f(x, y) \, dA.$$

You may wonder how a function can fail to be Riemann integrable. One way is that, if f is unbounded on a region R, then f is **not** Riemann integrable on R.

Of course, the discussion above doesn't tell us which functions on which regions are Riemann integrable, and it doesn't tell us how to calculate the double integral even when it exists.

The easiest case, which we deal with first, is where the region R is, itself, a rectangle. Suppose that $a < b$ and $c < d$. The closed rectangle of points (x, y) such that $a \leq x \leq b$ and $c \leq y \leq d$ is denoted by $[a, b] \times [c, d]$.

For the proof, see, for instance, Theorem 7.2.1 of Trench, [8].

Theorem 3.2.2. (Fubini's Theorem) *Suppose that $f = f(x, y)$ is continuous on the rectangle $R = [a, b] \times [c, d]$.*

Then, f is Riemann integrable on R, and the integral can be calculated via iterated integrals:

$$\iint_R f(x, y)\, dA \; = \; \int_a^b \int_c^d f(x, y)\, dy\, dx \; = \; \int_c^d \int_a^b f(x, y)\, dx\, dy.$$

Example 3.2.3. Let's look at a specific example. Suppose that x and y are measured in meters, and that we have a thin metal plate which occupies the rectangle $R = [0, 3] \times [0, 2]$. Suppose that the area-density function for the plate is $\delta_{ar} = \delta_{ar}(x, y) = 1 + xy$ kg/m^2. What is the mass of the plate?

Solution:

We can discuss this in terms of Riemann sums, but it is more intuitive to describe this using "infinitesimal" language.

At any point (x, y) in the plate, you should think that the infinitesimal amount of mass, dm, at (x, y) is the area-density $\delta_{ar}(x, y)$ times the infinitesimal area dA of an infinitesimal rectangle containing (x, y). Thus,

$$dm \; = \; \delta_{ar}(x, y)\, dA \; = \; (1 + xy)\, dy\, dx \; = \; (1 + xy)\, dx\, dy,$$

and the total mass is the continuous sum of all of the infinitesimal pieces of mass, i.e.,

$$\text{mass} \; = \; \iint_R \delta_{ar}(x, y)\, dA \; = \; \int_0^3 \int_0^2 (1 + xy)\, dy\, dx \; = \; \int_0^2 \int_0^3 (1 + xy)\, dx\, dy.$$

We may calculate either one of these iterated integrals. We'll actually do both, to verify that they yield the same result. We first choose the iterated integral with the dx on the outside, and find

$$\text{mass} \; = \; \int_0^3 \int_0^2 (1 + xy)\, dy\, dx \; = \; \int_0^3 \left(y + \frac{xy^2}{2} \, \Big|_{y=0}^{y=2} \right) dx \; =$$

$$\int_0^3 (2 + 2x)\, dx \; = \; 2x + x^2 \, \Big|_0^3 \; = \; 15 \;\; \text{kg}.$$

If we integrate in the reverse order, we find

$$\text{mass} = \int_0^2 \int_0^3 (1+xy)\,dx\,dy = \int_0^2 \left(x + \frac{x^2 y}{2}\Big|_{x=0}^{x=3} \right) dy =$$

$$\int_0^2 \left(3 + \frac{9y}{2} \right) dy = 3y + \frac{9y^2}{4}\Big|_0^2 = 15 \ \text{kg}.$$

Of course, we obtain the same answer either way, but notice that the steps in the $dy\,dx$ iterated integral are slightly easier than in the $dx\,dy$ iterated integral. It is an interesting aspect of iterated integrals that *reversing the order of the integration* can slightly, or even greatly, affect the level of difficulty.

Now let's look at integrating over a region that's not a rectangle; for instance, how do you integrate over the region in Figure 3.2.1, which is the region bounded by the graphs of $y = x$ and $y = x^2$?

For that, we have:

Theorem 3.2.4. *Suppose that p and q are continuous real-valued functions on the compact interval $[a,b]$ in $\mathbb{R}$ such that, for all x in $[a,b]$, $p(x) \leq q(x)$. Let R be the region of points (x,y) in $\mathbb{R}^2$ such that $a \leq x \leq b$ and $p(x) \leq y \leq q(x)$. Suppose that $f = f(x,y)$ is a continuous function on R.*

Then, f is Riemann integrable on R, and the double integral can be calculated in terms of iterated integrals:

$$\iint_R f(x,y)\,dA = \int_a^b \int_{p(x)}^{q(x)} f(x,y)\,dy\,dx.$$

Note that Theorem 3.1.9 tells us that this iterated integral is actually defined.

This result follows essentially by applying Theorem 3.2.2 to an extension of f by zero to an enclosing rectangle, except that you have to use a version of Theorem 3.2.2 which says that f may be discontinuous on a set of measure zero. See the More Depth portion of this section.

Of course, we can switch the roles of x and y in the preceding theorem.

Theorem 3.2.5. *Suppose that r and s are continuous real-valued functions on the compact interval $[c,d]$ in $\mathbb{R}$ such that, for all y in $[c,d]$, $r(y) \leq s(y)$. Let R be the region of points (x,y) in $\mathbb{R}^2$ such that $c \leq y \leq d$ and $r(y) \leq x \leq s(y)$. Suppose that $f = f(x,y)$ is a continuous function on R.*

Then, f is Riemann integrable on R, and the double integral can be calculated in terms of iterated integrals:

$$\iint_R f(x,y)\,dA = \int_c^d \int_{r(y)}^{s(y)} f(x,y)\,dx\,dy.$$

If the region R can be described in the both of the ways given in Theorem 3.2.4 and in Theorem 3.2.5, then the equalities in the theorems yield a generalized version of Fubini's Theorem from Theorem 3.2.2.

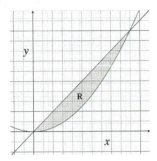

Figure 3.2.2: The region R, bounded by the graphs of $y = x^2$ and $y = x$.

Example 3.2.6. Let's integrate $f(x, y) = 2xy + 6y^2$ over the region R from Figure 3.2.1, which is also shown in Figure 3.2.2. The region R is bounded by the graphs of $y = x^2$ and $y = x$. Note that R can also be described as the region bounded by the graphs of $x = \sqrt{y}$ and $x = y$.

First, we look at R as being the points (x, y), where $0 \leq x \leq 1$ and where, for a given x, the y-coordinates are always between $y = x^2$ and $y = x$, i.e., $x^2 \leq y \leq x$. Thus, Theorem 3.2.4 tells us that

$$\iint_R (2xy + 6y^2)\, dA \;=\; \int_0^1 \int_{x^2}^x (2xy + 6y^2)\, dy\, dx \;=\;$$

$$\int_0^1 \left[(xy^2 + 2y^3)\, \Big|_{y=x^2}^{y=x} \right] dx \;=\; \int_0^1 (3x^3 - x^5 - 2x^6)\, dx \;=\;$$

$$\frac{3x^4}{4} - \frac{x^6}{6} - \frac{2x^7}{7}\, \Big|_0^1 \;=\; \frac{25}{84}.$$

Now, we wish to reverse the order of integration. This means that we look at R as being the points (x, y), where $0 \leq y \leq 1$ and where, for a given y, the x-coordinates are always between $x = y$ and $x = \sqrt{y}$, i.e., $y \leq x \leq \sqrt{y}$.

Thus, Theorem 3.2.5 tells us that

$$\iint_R (2xy + 6y^2)\, dA \;=\; \int_0^1 \int_y^{\sqrt{y}} (2xy + 6y^2)\, dx\, dy \;=\;$$

$$\int_0^1 \left[(x^2 y + 6xy^2)\, \Big|_{x=y}^{x=\sqrt{y}} \right] dy \;=\; \int_0^1 \left(y^2 + 6y^{5/2} - 7y^3 \right) dy \;=\;$$

$$\frac{y^3}{3} + \frac{12y^{7/2}}{7} - \frac{7y^4}{4}\, \Big|_0^1 \;=\; \frac{25}{84}.$$

Of course, the two different iterated integrals give the same result, because they both describe the Riemann integral of f on the region R.

Notice that reversing the order of integration over a region that is not a rectangle does **not** mean that you simply swap the limits of integration. That is, in our $dy\, dx$ iterated integral, we integrated y from x^2 to x; you certainly don't integrate y from x^2 to x in the $dx\, dy$ iterated integral. If you did, you'd end up with a function of x, not a number. Over a non-rectangular region, you really have to do some work to determine the new limits of integration when you reverse the order.

Example 3.2.7. Suppose that we just wanted to find the area of the region R in the previous example, that is, find the area of the region trapped between the graphs of $y = x$ and $y = x^2$. How do we do that?

The total area is simply the continuous sum of all of the infinitesimal chunks of area. Hence we simply want to add up all of the infinitesimal dA's, i.e., we use the constant 1 as our integrand. The limits of integration remain the same as in the previous problem, since we're integrating over the same region.

Writing dA in place of $1\,dA$, we find

$$\text{Area} = \iint_R dA = \int_0^1 \int_{x^2}^x dy\,dx =$$

$$\int_0^1 \left[y \Big|_{y=x^2}^{y=x} \right] dx = \int_0^1 (x - x^2)\,dx = \left. \frac{x^2}{2} - \frac{x^3}{3} \right|_0^1 = \frac{1}{6}.$$

Note that the integral $\int_0^1 (x - x^2)\,dx$ is precisely what you would have used in single-variable Calculus in order to calculate the area of the region R.

If you start with a region R, and break it up into two pieces R_1 and R_2, which intersect (at most) along a finite number of points or curves, i.e., along a set which has zero area, then you can split up the integral over R, provided that the integrals over R_1 and R_2 exist. That is, if the Riemann integrals of f over R_1 and R_2 exist, and R_1 and R_2 intersect along a set with zero area, then

> We give a more technical statement of this result in Theorem 3.2.21.

$$\iint_R f(x,y)\,dA = \iint_{R_1} f(x,y)\,dA + \iint_{R_2} f(x,y)\,dA.$$

Example 3.2.8. Integrate $f(x,y) = 2xy$ over the region R in the 1st quadrant that's bounded by the curves given by $y = 4 - x^2$ and $y = 3x$; see Figure 3.2.3 (and note that the scales are different on the two axes).

Solution: If you put the x integral on the outside, i.e., if you look at the region R as being where $0 \leq x \leq 1$ and $3x \leq y \leq 4 - x^2$, then you don't need to split this integral into pieces. You calculate

$$\iint_R 2xy\,dA = \int_0^1 \int_{3x}^{4-x^2} 2xy\,dy\,dx =$$

$$\int_0^1 \left[xy^2 \; \Big|_{y=3x}^{y=4-x^2} \right] dx \;\; = \;\; \int_0^1 \left[x(4-x^2)^2 - x(3x)^2 \right] dx \;\; =$$

$$\int_0^1 \left(16x - 17x^3 + x^5 \right) dx \;\; = \;\; 8x^2 - \frac{17x^4}{4} + \frac{x^6}{6} \; \Big|_0^1 \;\; = \;\; \frac{47}{12}.$$

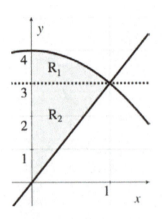

Figure 3.2.3: The region R is split into regions R_1 and R_2.

On the other hand, you should get the same result by adding the integrals of f over R_1 and R_2, and these integrals can be calculated with the y integrals on the outside, i.e., by reversing the order of integration. The subregion R_1 is where $3 \le y \le 4$ and x varies between $x = 0$ and $x = \sqrt{4-y}$, i.e., $0 \le x \le \sqrt{4-y}$.

The subregion R_2 is where $0 \le y \le 3$ and x varies between $x = 0$ and $x = y/3$, i.e., $0 \le x \le y/3$.

Therefore, we can calculate

$$\iint_R 2xy \, dA \;\; = \;\; \iint_{R_1} 2xy \, dA \; + \; \iint_{R_2} 2xy \, dA,$$

where

$$\iint_{R_1} 2xy \, dA \;\; = \;\; \int_3^4 \int_0^{\sqrt{4-y}} 2xy \, dx \, dy \;\; = \;\; \int_3^4 \left[x^2 y \; \Big|_{x=0}^{x=\sqrt{4-y}} \right] dy \;\; =$$

$$\int_3^4 (4y - y^2) \, dy \;\; = \;\; 2y^2 - \frac{y^3}{3} \; \Big|_3^4 \;\; = \;\; \frac{5}{3},$$

and

$$\iint_{R_2} 2xy \, dA \;\; = \;\; \int_0^3 \int_0^{y/3} 2xy \, dx \, dy \;\; = \;\; \int_0^3 \left[x^2 y \; \Big|_{x=0}^{x=y/3} \right] dy \;\; =$$

$$\int_0^3 \frac{y^3}{9} \, dy \;\; = \;\; \frac{y^4}{36} \; \Big|_0^3 \;\; = \;\; \frac{9}{4}.$$

And so, we recover that

$$\iint_R 2xy \, dA \;\; = \;\; = \;\; \frac{5}{3} + \frac{9}{4} \;\; = \;\; \frac{47}{12}.$$

In the previous example, we reversed the order of integration, but that seemed to make the problem more difficult or, at least, no easier. However, there are times when reversing the order of integration makes calculation much easier, or makes an otherwise impossible problem manageable.

Example 3.2.9. How do you evaluate the iterated integral

$$\int_0^1 \int_y^1 e^{-x^2} \, dx \, dy ?$$

As you should remember, e^{-x^2} has no elementary anti-derivative, so we can't perform the inside integral in any nice way. So, what do we do?

The cool answer is that we equate this iterated integral with the corresponding double integral over some region R, and then we reverse the order of integration to obtain an iterated integral that we can actually calculate.

What's the region R? The limits of integration in the iterated integral tell us that it's the region described by $0 \leq y \leq 1$ and $y \leq x \leq 1$, i.e., for each y value between 0 and 1, the x coordinates go from $x = y$ to $x = 1$. Thus, we draw the lines given by $x = y$ and $x = 1$, and look at the region in-between for $0 \leq y \leq 1$. This is the triangle shown in Figure 3.2.4.

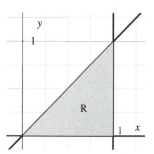

Figure 3.2.4: The region R.

But now we can easily describe R as the region where $0 \leq x \leq 1$ and y is between $y = 0$ and $y = x$, i.e., $0 \leq y \leq x$.

Therefore,

$$\int_0^1 \int_y^1 e^{-x^2} \, dx \, dy \;=\; \iint_R e^{-x^2} \, dA \;=\; \int_0^1 \int_0^x e^{-x^2} \, dy \, dx \;=$$

$$\int_0^1 \left[e^{-x^2} y \, \Big|_{y=0}^{y=x} \right] dx \;=\; \int_0^1 x e^{-x^2} \, dx.$$

And now, an easy substitution of $u = -x^2$ gives us

$$\int_0^1 x e^{-x^2} \, dx \;=\; -\frac{e^{-x^2}}{2} \Big|_0^1 \;=\; \frac{1}{2}\left(1 - \frac{1}{e}\right).$$

More Depth:

We want to give the rigorous definition of the double integral, the Riemann integral in $\mathbb{R}^2$.

We'll first give the definition of the Riemann integral for functions defined on rectangles and, later, we'll use the definition on rectangles to define the integral over more general regions.

Recall the following definition from single-variable Calculus.

Definition 3.2.10. *A **partition** of the interval $[a, b]$, into n subintervals, is an ordered set of numbers $x_0, x_1, \ldots, x_n$ such that $x_0 = a$, $x_n = b$, and $x_0 < x_1 < \cdots < x_n$.*

*The **mesh** of a partition of an interval is the maximum length of a subinterval of the partition.*

Now we need to define similar concepts for rectangles, instead of for intervals. Our goal is to define a Riemann sum of a function on a rectangle, and take the limit as the subrectangles become arbitrarily small. Everything that we define now should feel similar/analogous to what we did for single-variable functions.

Definition 3.2.11. *Suppose that $a < b$ and $c < d$. The closed rectangle of points (x, y) such that $a \leq x \leq b$ and $c \leq y \leq d$ is denoted by $[a, b] \times [c, d]$.*

*A **partition** $\mathcal{P}$ of the rectangle $[a, b] \times [c, d]$ consists of a partition $x_0, \ldots, x_i, \ldots, x_m$ for $[a, b]$ and a partition $y_0, \ldots, y_j, \ldots, y_k$ for $[c, d]$.*

*The (i, j)-**subrectangle** of $\mathcal{P}$ is the rectangle $[x_{i-1}, x_i] \times [y_{j-1}, y_j]$. Its area is $\Delta A_{(i,j)} = \Delta x_i \Delta y_j$.*

*The **mesh** of $\mathcal{P}$, denoted $\|\mathcal{P}\|$, is the maximum of the meshes of the partitions of the intervals $[a, b]$ and $[c, d]$.*

*A **sample set** $\mathcal{S}$ for the partition of the rectangle is a set of sample points $\mathbf{s}_{(i,j)}$, where $\mathbf{s}_{(i,j)}$ is in the (i, j)-subrectangle.*

Definition 3.2.12. *Suppose that f is a real-valued function, which is defined on the rectangle $[a, b] \times [c, d]$. A **Riemann sum** for f on $[a, b] \times [c, d]$ is a summation*

$$\sum_{i=1}^{m} \sum_{j=1}^{k} f(\mathbf{s}_{(i,j)}) \Delta A_{(i,j)} = \sum_{i=1}^{m} \sum_{j=1}^{k} f(\mathbf{s}_{(i,j)}) \Delta x_i \Delta y_j,$$

for some partition of $[a, b] \times [c, d]$ and sample points $\mathbf{s}_{(i,j)}$.

Definition 3.2.13. (The Double Integral) *Suppose that f is defined on the closed rectangle $R = [a, b] \times [c, d]$, and that there exists a real number L such that, for all $\epsilon > 0$, there exists $\delta > 0$ such that, for all partitions $\mathcal{P}$ of R, with mesh less than δ, and for all sample sets $\mathcal{S}$ for $\mathcal{P}$,*

$$\left| \sum_{i=1}^{m} \sum_{j=1}^{k} f(\mathbf{s}_{(i,j)}) \Delta A_{(i,j)} - L \right| < \epsilon.$$

Then, we say that f is **Riemann integrable** *on R, and that the* **double integral of f on R exists and equals** *L. We write*

$$\iint_R f(x, y) \, dA = L.$$

There is a more general, but more complicated, type of integral, known as the *Lebesgue integral*. See, for instance, Chapter 11 of Rudin, [7]. However, whenever the Riemann integral is defined, so is the Lebesgue integral and the values of the two integrals are equal.

We want to weaken the hypothesis that our function has to be continuous on the **whole** rectangle; we really need for the function to be continuous *almost everywhere*, where this seemingly imprecise term actually has a very serious rigorous definition. One of the reasons that we need to discuss this is so that we can use integration over rectangles to define integration over more general regions.

We need to briefly discuss sets of *measure zero*; see Chapter 11 of Rudin, [7], or Chapter 5 of Fleming, [3], for a complete treatment. The definition is difficult to work with. However, after the definition, we state a theorem which actually tells us how we'll always conclude that a set has measure zero. We should mention that the first sentence of the definition is essentially circular; nonetheless, we include it because it conveys the right intuitive idea.

We haves used closed rectangles in the definition. You can also use open rectangles, or open or closed disks. All of these yield the same notion of measure zero.

Definition 3.2.14. *A subset A of $\mathbb{R}^2$ has* **measure zero** *in $\mathbb{R}^2$ provided that the area of A is defined and the area of A is 0.*

Technically, a subset A of $\mathbb{R}^2$ has measure zero in $\mathbb{R}^2$ if and only if, for all $\epsilon > 0$, there exists a finite number, or infinite sequence, of closed rectangles whose union contains A and the sum of the areas of the rectangles is $< \epsilon$.

A property defined at points in $\mathbb{R}^2$ is said to hold **almost everywhere** *if and only if the set of points where the property does not hold has measure zero.*

If you allow only a finite number of rectangles in the definition, then you obtain the definition of *zero content*. A set of zero content has measure zero, but there exists sets of measure zero which do not have zero content.

The subsets of measure zero in $\mathbb{R}^2$ that we will consider in this book will all be finite collections of points and curves, which, of course, seem like they should obviously have no area. The precise theorem that we need is:

This follows from an easy generalization of Example 7.1.5 of Trench, [8].

Theorem 3.2.15. *Any subset of $\mathbb{R}^2$ which is contained in a union of a finite number, or infinite sequence, of points and graphs of continuous functions on compact subsets of $\mathbb{R}^1$ has measure zero.*

We want to give a name to the types of "nice" regions that we want to consider in $\mathbb{R}^2$.

Definition 3.2.16. *A **standard region** in $\mathbb{R}^2$ is a compact set whose boundary is contained in a union of a finite number of points and graphs of continuous functions on compact subsets of $\mathbb{R}^1$.*

Example 3.2.17. For instance, it will be important to us that a region such as the blue region in Figure 3.2.5 is a standard region. Why is this?

Because the boundary is the union of the graphs of:

- $x = p(y) = 1/2$, for y in the compact interval $[1/2, 113/64]$,

- $x = q(y) = 5/2$, for y in the compact interval $[1/2, 173/64]$,

- $y = r(x) = 1/2$, for x in the compact interval $[1/2, 5/2]$, and

- $y = s(x) = \dfrac{x^3}{8} - \dfrac{x}{2} + 2$, for x in the compact interval $[1/2, 5/2]$.

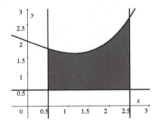

Figure 3.2.5: The boundary of the region has measure 0.

Now that we have an idea of what "measure zero" means, we can give an improved theorem about when a function on a rectangle is Riemann integrable.

Theorem 3.2.18. *Suppose that $f = f(x, y)$ is bounded on the rectangle $R = [a, b] \times [c, d]$, i.e., that there exists M such that, for all $\mathbf{x}$ in R, $|f(\mathbf{x})| \le M$.*

Then, f is Riemann integrable on R if and only if f is continuous almost everywhere in R, i.e., the set of points in R at which f is not continuous has measure zero.

Why do we care about Theorem 3.2.18? Do we really want to look at weird examples of functions which aren't continuous?

Well...in the end, not really. However, if we want to integrate $f = f(x, y)$ over a region R which is not a rectangle, we actually pick a rectangle $\hat{R}$ containing R, and define a new function, $\hat{f}$, which agrees with f at points in R, and is zero at points in the rectangle $\hat{R}$ that are not in R. We then integrate $\hat{f}$ on $\hat{R}$. Typically, the function $\hat{f}$ is **not** continuous on the boundary of R, but, if the boundary has measure zero, this isn't a problem because of Theorem 3.2.18.

So, let's do it. Let's define the Riemann integral over a non-rectangular region.

Suppose that R is a bounded region in $\mathbb{R}^2$, so that there is a rectangle $[a, b] \times [c, d]$ which contains R (in fact, there are an infinite number of rectangles that we could pick). It may help to once again look at Figure 3.2.1:

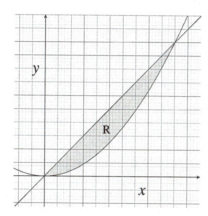

Figure 3.2.6: Little rectangles stick outside the region R.

Note that the entire background piece of the Cartesian plane **is** a rectangle $\hat{R}$ which encloses R. The grid inside of $\hat{R}$ yields a partition of $\hat{R}$, whether you look at just the bigger, darker rectangles/squares, or look at the smaller little rectangles/squares. As we mentioned earlier, given any partition of $\hat{R}$, some of the rectangles in the partition will cross the boundary, so that some of the subrectangles in the partition stick outside of the original region R.

The question is: when you calculate a Riemann sum, should you count or not count the contributions from subrectangles which lie partly outside of R? You would hope that, in the limit, as the sizes of the subrectangles approach zero, that it wouldn't matter, since the parts hanging outside would become negligible to total area and in the Riemann sum. In fact, if the boundary of R has measure zero, this line of reasoning is correct.

What's the easiest way to discuss Riemann sums for a function f, defined on R, which may include or exclude pieces that hang out over the boundary of a region R?

The answer is pretty sneaky: you define a new function, an extension $\hat{f}$ of the function f to a function on the entire enclosing rectangle, which is zero outside of R, i.e., we define $\hat{f}$ to be the function on $\hat{R}$ given by

$$\hat{f}(\mathbf{x}) = \begin{cases} f(\mathbf{x}), & \text{if } \mathbf{x} \text{ is in } R; \\ 0, & \text{if } \mathbf{x} \text{ is in } \hat{R}, \text{ but } \mathbf{x} \text{ is not in } R. \end{cases}$$

We call such an $\hat{f}$ an *extension of f by zero*.

So, what good does it do to define such an extension of f by zero?

If a subrectangle of area ΔA of our partition of $\hat{R}$ is completely outside of R, then, for any sample point $\mathbf{s}$ in that subrectangle, $\hat{f}(\mathbf{s}) = 0$, and so $\hat{f}(\mathbf{s})\Delta A = 0$ and there's no contribution to any Riemann sum.

If a subrectangle of area ΔA of our partition of $\hat{R}$ is completely inside of R, then, for any sample point $\mathbf{s}$ in that subrectangle, $\hat{f}(\mathbf{s}) = f(\mathbf{s})$, and so $\hat{f}(\mathbf{s})\Delta A = f(\mathbf{s})\Delta A$ and the contribution to a Riemann sum is what we want.

If a subrectangle of area ΔA of our partition of $\hat{R}$ is partially in R and partially outside of R, then, for sample points $\mathbf{s}$ outside of R, in that subrectangle, there's no contribution to a given Riemann sum, but for sample points inside of R, you do add a contribution of $f(\mathbf{s})\Delta A$ to a given Riemann sum.

Therefore, Riemann sums of $\hat{f}$ on $\hat{R}$ give us precisely what we want: Riemann sums of f on R, in which subrectangles which hang over the boundary are counted or are not counted, depending on the choice of sample points.

Definition 3.2.19. *Suppose that R is a bounded region in $\mathbb{R}^2$, and that f is a real-valued function, defined on R. Let $\hat{R}$ be any rectangle which encloses R, and let $\hat{f}$ be the function on $\hat{R}$ which is the extension of f by zero.*

It is important that this definition does not depend on the choice of the rectangle $\hat{R}$ which encloses the region R.

*Then, we say that f is **Riemann integrable on** R if and only if $\hat{f}$ is Riemann integrable on $\hat{R}$ and, in this case, we define the double integral of f on R by*

$$\iint_R f(x, y)\, dA \;=\; \iint_{\hat{R}} \hat{f}(x, y)\, dA.$$

The big theorem which tells us when a function is integrable is:

Theorem 3.2.20. *Suppose that R is a bounded subset of* $\mathbb{R}^2$ *and that the boundary of R has measure zero. Then, a function f on R is Riemann integrable if and only if f is bounded on R and f is continuous almost everywhere in the interior of R.*

In particular, a continuous function on a standard region is Riemann integrable.

See, for instance, page 185 of Fleming, [3].

Finally, the theorem that allows us to split integrals as we did in Example 3.2.8 is:

Theorem 3.2.21. *Suppose that* R_1 *and* R_2 *are bounded subsets of* $\mathbb{R}^2$ *that each have a boundary with measure zero, and that* R_1 *and* R_2 *intersect in a set of measure zero (which includes not intersecting at all). Suppose that f is a function which is Riemann integrable on* R_1 *and on* R_2.

Then, the union R of R_1 *and* R_2 *is a bounded subset of* $\mathbb{R}^2$ *whose boundary has measure 0, f is Riemann integrable on R, and*

$$\iint_R f(x,y)\, dA \;=\; \iint_{R_1} f(x,y)\, dA \;+\; \iint_{R_2} f(x,y)\, dA.$$

See part 7 of Theorem 5.4 of Fleming, [3].

Improper Integrals:

In the discussion and theorems above, we dealt with bounded regions and bounded functions. As you should remember from single-variable Calculus, *improper integrals* are an extension of the Riemann integral to unbounded regions or functions; we took limits of Riemann integrals of bounded functions on closed, bounded intervals $[a, b]$.

For double integrals, and later for triple integrals and higher-dimensional integrals, the situation is more complicated. There are different approaches that we could use, such as taking limits of Riemann integrals using sequences of compact regions that approach the given region, or by using *Lebesgue integration*. Both of these approaches lie beyond the scope of this textbook.

So, we shall simply give a couple of examples, without discussing the general theory behind them.

Example 3.2.22. Integrate $g(x,y) = y/\sqrt{x}$ over the rectangle $R = [0,1] \times [3,5]$, i.e., $0 \leq x \leq 1$ and $3 \leq y \leq 5$.

Solution:

This integral is improper, since the function g is unbounded as $x \to 0^+$. The integral of g over R means the limit of the integrals of g over $[a, 1] \times [3, 5]$, as $a \to 0^+$. We proceed as you would expect, and use a single-variable improper integral in our iterated integral.

We find

$$\iint_R \frac{y}{\sqrt{x}}\, dA \;=\; \lim_{a\to 0^+} \int_3^5 \int_a^1 yx^{-1/2}\, dx\, dy \;=\; \lim_{a\to 0^+} \int_3^5 \left[2y\sqrt{x}\; \Big|_{x=a}^{x=1} \right] dy \;=\;$$

$$\lim_{a\to 0^+} \int_3^5 (2y - 2y\sqrt{a})\, dy \;=\; \lim_{a\to 0^+} \left((y^2 - y^2\sqrt{a})\; \Big|_3^5 \right) \;=\; 25 - 9 \;=\; 16.$$

Example 3.2.23. Integrate the function $f(x,y) = xy$ over the region R which lies under the curve given by $y = e^{-x^2}$ and above the interval $[0, \infty)$ on the x-axis; see Figure 3.2.7.

Solution:

This integral is improper, since the region R is unbounded. We will proceed without explicitly writing the limit as x approaches ∞; but keep in mind that, when we write $\pm\infty$ as a limit of integration, taking a limit is what we mean.

We find

$$\iint_R xy\, dy\, dx \;=\; \int_0^\infty \int_0^{e^{-x^2}} xy\, dy\, dx \;=\; \int_0^\infty \left[\frac{xy^2}{2}\; \Big|_{y=0}^{y=e^{-x^2}} \right] dx \;=\;$$

$$\frac{1}{2} \int_0^\infty xe^{-2x^2}\, dx.$$

Making the substitution $u = -2x^2$, so that $du = -4x\, dx$, and noting that u goes from 0 to $-\infty$ as x goes from 0 to ∞, we find

$$\frac{1}{2} \int_0^{-\infty} e^u \cdot -\frac{1}{4}\, du \;=\; \frac{1}{8} \int_{-\infty}^0 e^u\, du \;=\; \frac{1}{8} e^u\; \Big|_{-\infty}^0 \;=\; \frac{1}{8}.$$

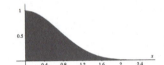

Figure 3.2.7: The region R tapers off very rapidly.

Online answers to select exercises are here.

3.2.1   Exercises

Basics:

In each of the following exercises, you are given a function $f(x,y)$, and a region R in the xy-plane is described in words or by inequalities. (a) Sketch the region R. (b) Give an iterated integral which calculates $\iint_R f(x,y)\, dA$. (c) Evaluate your integral from part (b).

1. $f(x,y) = x + y$, and R is the region where $-1 \le x \le 1$ and $1 \le y \le 2$.

2. $f(x, y) = xy + 1$, and R is the region where $0 \leq x \leq 1$ and $0 \leq y \leq 2x$.

3. $f(x, y) = xy + 1$, and R is the region where $0 \leq y \leq 2$ and $\frac{y}{2} \leq x \leq 1$.

4. $f(x, y) = xy + 1$, and R is the region in the first quadrant below the line where $y = 2x$ and to the left of the line where $x = 1$.

5. $f(x, y) = \dfrac{x \cos y}{y^2}$, and R is the region where $\frac{\pi}{2} \leq y \leq \pi$ and $0 \leq x \leq 2y$.

6. $f(x, y) = 2xy$, and R is the region where $0 \leq x \leq 2$ and $x^2 \leq y \leq 6 - x$. ▶

7. $f(x, y) = y^2 e^{xy}$, and R is the region where $0 \leq y \leq 3$ and $y \leq x \leq 5$.

8. $f(x, y) = 5$, and R is the region above the graph of $y = e^x$, below the graph of $x + y = 4$, for $0 \leq x \leq 1$.

9. $f(x, y) = 5$, and R is the region where $0 \leq x \leq 1$ and $e^x \leq y \leq 4 - x$.

10. $f(x, y) = 1$, and R is the region where $\frac{\pi}{6} \leq x \leq \frac{5\pi}{6}$ and $\frac{1}{2} \leq y \leq \sin x$.

11. $f(x, y) = x$, and R is the region where $\frac{\pi}{6} \leq x \leq \frac{5\pi}{6}$ and $\frac{1}{2} \leq y \leq \sin x$. ▶

12. $f(x, y) = \dfrac{y}{1 + x^2}$, and R is the region where $-1 \leq x \leq 1$ and $x \leq y \leq 2$.

In each of the following exercises, you are given an iterated integral, which is equal to a double integral $\iint_R f \, dA$. (a) Sketch the region of integration R in $\mathbb{R}^2$. If the variables are not x and y, you'll need to make a choice of which variable goes with which axis. (b) Evaluate the given iterated integral.

13. $\displaystyle\int_2^3 \int_1^5 (3x^2 - 2xy) \, dy \, dx$

14. $\displaystyle\int_1^5 \int_2^3 (3x^2 - 2xy) \, dx \, dy$

15. $\displaystyle\int_2^3 \int_1^5 xe^{xy} \, dy \, dx$ ▶

16. $\displaystyle\int_0^1 \int_w^2 (w + 2z) \, dz \, dw$

17. $\displaystyle\int_0^1 \int_{w^2}^2 (w + 2z) \, dz \, dw$

18. $\displaystyle\int_0^1 \int_{w^2}^w (w + 2z) \, dz \, dw$

19. $\displaystyle\int_0^1 \int_{e^q}^{4-q} 7 \, dp \, dq$

20. $\displaystyle\int_0^2 \int_{\alpha^2}^{6-\alpha} 2\alpha\beta \, d\beta \, d\alpha$

21. $\displaystyle\int_{\pi/2}^{\pi} \int_0^{2v} \frac{u \cos v}{v^2} \, du \, dv$

22. $\displaystyle\int_{-1}^1 \int_r^1 \frac{2s}{1+r^2} \, ds \, dr$

23. $\displaystyle\int_0^{\pi/4} \int_{\cos x}^{\sec x} \sec x \, dy \, dx$

24. $\displaystyle\int_0^2 \int_0^{w^{3/2}} 6te^{-w^4} \, dt \, dw$

In each of the following exercises, you are given an iterated integral, or a sum of iterated integrals, which is equal to a double integral $\iint_R f(x,y) \, dA$. **(a) Sketch the region of integration R. (b) Give an iterated integral for $\iint_R f(x,y) \, dA$, in which you reverse the order of integration from what you were originally given. (c) Evaluate your integral from part (b).**

25. $\displaystyle\int_{-2}^2 \int_{x^2}^4 1 \, dy \, dx$

26. $\displaystyle\int_{-2}^2 \int_{x^2}^4 \sqrt{y} \, dy \, dx$

27. $\displaystyle\int_1^{e^2} \int_{\ln x}^2 1 \, dy \, dx$

28. $\displaystyle\int_0^1 \int_0^2 1 \, dy \, dx + \int_1^{e^2} \int_{\ln x}^2 1 \, dy \, dx$

29. $\displaystyle\int_0^{2\pi} \int_{x/2}^{\pi} \frac{x \cos y}{y^2} \, dy \, dx$

30. $\displaystyle\int_0^{\pi} \int_{\pi/2}^{\pi} \frac{x \cos y}{y^2} \, dy \, dx + \int_{\pi}^{2\pi} \int_{x/2}^{\pi} \frac{x \cos y}{y^2} \, dy \, dx$

31. Write a brief essay, explaining, non-technically, the definition of the definite integral $\iint_R f(x,y) \, dA$.

32. Consider a definite integral $\iint_R f(x,y) \, dA$.

   (a) If $f(x,y) = 1$, what does this integral give you?

   (b) If $f(x,y)$ is the area density of a thin metal plate at the point (x,y), what does this integral give you?

(c) If $f(x, y)$ is the height of a solid object above the point (x, y) in the base R of the object, what does this integral give you?

More Depth:

33. In Figure 3.2.8, you are given a table of values of a continuous function $f(x, y)$, where, as usual, the x values are listed horizontally, and the y values are given vertically. The given x- and y-coordinates yield a partition $\mathcal{P}$ of $R = [-1, 0] \times [0, 2]$. Use this data to estimate the value of the definite integral $\iint_R f \, dA$:

   (a) Using upper-left corners of the subrectangles of the partition as sample points.

   (b) Using lower-right corners of the subrectangles of the partition as sample points.

y \ x	-1	-0.8	-0.6	-0.4	-0.2	0
2	4.000	3.280	2.720	2.320	2.080	2.000
1.6	3.200	2.624	2.176	1.856	1.664	1.600
1.2	2.400	1.968	1.632	1.392	1.248	1.200
0.8	1.600	1.312	1.088	0.928	0.832	0.800
0.4	0.800	0.656	0.544	0.464	0.416	0.400
0	0.000	0.000	0.000	0.000	0.000	0.000

Figure 3.2.8: The values of $f(x, y)$.

34. Consider the function $f(x, y) = xy + x^2$ on the rectangle $R = [0, 1] \times [0, 2]$. Let $\mathcal{P}$ be the partition of R which results from partitioning each of the intervals $[0, 1]$ and $[0, 2]$ into four subintervals of equal length.

   (a) Estimate $\iint_R f \, dA$ by using the Riemann sum with partition $\mathcal{P}$, in which the sample points are taken to be upper-left corners of the subrectangles of the partition.

   (b) Estimate $\iint_R f \, dA$ by using the Riemann sum with partition $\mathcal{P}$, in which the sample points are taken to be lower-right corners of the subrectangles of the partition.

   (c) Calculate the exact value of $\iint_R f \, dA$.

35. Consider the function $f(x, y) = xy - y^2$ on the rectangle $R = [-1, 1] \times [0, 2]$. Let $\mathcal{P}$ be the partition of R which results from partitioning each of the intervals $[-1, 1]$ and $[0, 2]$ into four subintervals of equal length.

(a) Estimate $\iint_R f \, dA$ by using the Riemann sum with partition $\mathcal{P}$, in which the sample points are taken to be lower-left corners of the subrectangles of the partition.

(b) Estimate $\iint_R f \, dA$ by using the Riemann sum with partition $\mathcal{P}$, in which the sample points are taken to be upper-right corners of the subrectangles of the partition.

(c) Calculate the exact value of $\iint_R f \, dA$.

36. In Figure 3.2.9, you are given a table of values of a continuous function $g(x, y)$, where, as usual, the x values are listed horizontally, and the y values are given vertically. The given x- and y-coordinates yield a partition $\mathcal{P}$ of $R = [1, 3] \times [-1.5, -1]$. Use this data to estimate the value of the definite integral $\iint_R g \, dA$:

(a) Using lower-left corners of the subrectangles of the partition as sample points.

(b) Using upper-right corners of the subrectangles of the partition as sample points.

x	1	1.4	1.8	2.2	2.6	3
y						
-1		-2.000	-2.960	-4.240	-5.840	-7.760	-10.000
-1.1		-2.200	-3.256	-4.664	-6.424	-8.536	-11.000
-1.2		-2.400	-3.552	-5.088	-7.008	-9.312	-12.000
-1.3		-2.600	-3.848	-5.512	-7.592	-10.088	-13.000
-1.4		-2.800	-4.144	-5.936	-8.176	-10.864	-14.000
-1.5		-3.000	-4.440	-6.360	-8.760	-11.640	-15.000

Figure 3.2.9: The values $g(x, y)$.

In each of the following exercises, you are given an improper iterated integral that corresponds to a double integral over a region R. (a) Sketch the region R of integration. (b) Evaluate the iterated integral.

37. $\displaystyle\int_0^1 \int_0^1 \frac{e^y}{\sqrt{x}} \, dy \, dx$

38. $\displaystyle\int_1^3 \int_0^{y-1} \frac{x}{(y-1)^{5/2}} \, dx \, dy$

39. $\displaystyle\int_0^1 \int_{\sqrt{1-x^2}}^1 y^{-2} \, dy \, dx$

40. $\displaystyle\int_1^\infty \int_1^x \frac{1}{1+x^2} \, dy \, dx$

41. $\displaystyle\int_{-\infty}^{-1}\int_{1}^{2} ye^{xy}\,dx\,dy$

42. $\displaystyle\int_{0}^{\infty}\int_{0}^{\infty} xye^{-x^2-y^2}\,dy\,dx$ ▶

43. Give an example of a region R in the xy-plane and a continuous real-valued function f on R such that the Riemann integral $\iint_R f\,dA$ does **not** exist.

44. Give an example of a compact region R in the xy-plane and a real-valued function f on R such that the Riemann integral $\iint_R f\,dA$ does **not** exist.

3.3   Integration with Polar Coordinates

In this section, we describe, and give examples of, computing double integrals in polar coordinates r and θ, instead of in Cartesian coordinates x and y.

In the More Depth portion of this section, we discuss the general case of integrating using non-Cartesian coordinates for $\mathbb{R}^2$.

Basics:

Some integrals simply don't look very nice in terms of the Cartesian coordinates x and y. For instance, consider integrating the function $f(x, y) = x^2 + y^2$ over the quarter disk in the first quadrant, inside the circle of radius 2, centered at the origin.

On the other hand, if you remember polar coordinates from single-variable Calculus (see, for instance, Section 3.4 of [5]), then, as we shall see, the problem becomes much easier.

First, though, we want to give a brief reminder of how polar coordinates work, and we need to see how to represent an infinitesimal chunk of area dA in polar coordinates.

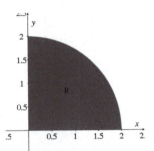

Figure 3.3.1: A region R which has a nice polar description.

A point with Cartesian coordinates (x, y) is said to have *polar coordinates* (r, θ) provided that
$$x = r\cos\theta \quad \text{and} \quad y = r\sin\theta.$$

In general, you can allow r to be negative when discussing polar coordinates, and allowing $r < 0$ leads to simple descriptions of some very cool curves. However, when looking at double integrals, it is standard to assume that r is always chosen to be non-negative. Thus, we shall assume that

$$r \geq 0.$$

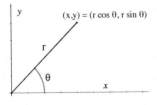

Figure 3.3.2: The polar coordinates r and θ.

Even though we require $r \geq 0$, the periodicity of sine and cosine still implies that every point (x, y) possesses an **infinite** number of polar coordinates. For instance, the point $(x, y) = (1, 1)$ has polar coordinates $(r, \theta) = (\sqrt{2}, \pi/4)$ and, more generally, polar coordinates
$$(r, \theta) = \left(\sqrt{2}, \frac{\pi}{4} + 2\pi n\right),$$

where n is any integer. The origin, $(x, y) = (0, 0)$, has polar coordinates $r = 0$ and $\theta =$ **any** real number.

Note that it is easy to write r^2 in terms of x and y:

$$x^2 + y^2 = r^2\cos^2\theta + r^2\sin^2\theta = r^2.$$

We need to look at how you express infinitesimal pieces of area dA in terms of r and θ. A picture will help, but, of course, we can't really draw it infinitesimally, or even very small; we want you to be able to see what's going on. Recall that we assume that $r \geq 0$.

In Figure 3.3.3, you see an "infinitesimal curved rectangle" (which you shouldn't think of as curving, infinitesimally). That the inner arc length of this curved rectangle is $r\,d\theta$ follows immediately from the definition of the radian measure of an angle, and then, to obtain area, you just multiply by dr, the infinitesimal change in r. Thus, we see that the infinitesimal area dA (frequently called the *area element*) is given by

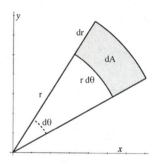

Figure 3.3.3: An infinitesimal chunk of area dA in polar coordinates.

$$dA = r\,dr\,d\theta.$$

You could also derive this replacing dr, $d\theta$, and dA in Figure 3.3.3 by Δr, $\Delta\theta$, and ΔA, then calculating the area inside the big circle sector minus the area of the little sector. We find

$$\Delta A = \frac{\Delta\theta}{2}(r + \Delta r)^2 - \frac{\Delta\theta}{2}r^2 = r\Delta r\Delta\theta + \frac{(\Delta\theta)(\Delta r)^2}{2}.$$

Now, realizing that the triple Δ in the final term becomes negligible compared to double Δ in the other term, in the limit, we once again obtain

$$dA = r\,dr\,d\theta.$$

In the More Depth portion of this section, we shall give a more formal, more general, formula for how expressions for dA are obtained via *changes of coordinates*.

So, if you're given a function $f(x,y)$ and a region R, how do you integrate f over the region R, using polar coordinates? You use

$$x = r\cos\theta, \qquad y = r\sin\theta, \qquad r^2 = x^2 + y^2, \qquad \text{and} \qquad dA = r\,dr\,d\theta$$

to write

$$\iint_R f(x,y)\,dA = \iint_R f(r\cos\theta,\ r\sin\theta)\cdot r\,dr\,d\theta,$$

which you calculate by an iterated integral, using limits of integration that describe the region R in polar coordinates.

Example 3.3.1. Let's look at the example from the beginning of the section.

Integrate $f(x, y) = x^2 + y^2$ over the region R which consists of the quarter disk in the first quadrant, inside the circle of radius 2, centered at the origin.

Solution:

The region R is easily described in polar coordinates as being where $0 \leq r \leq 2$ and $0 \leq \theta \leq \pi/2$.

Thus,

$$\iint_R (x^2 + y^2)\, dA = \int_0^{\pi/2} \int_0^2 r^2 \cdot r\, dr\, d\theta = \int_0^{\pi/2} \left[\frac{r^4}{4} \Big|_0^2 \right] d\theta =$$

$$\int_0^{\pi/2} 4\, d\theta = 2\pi.$$

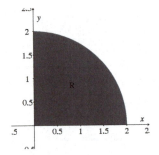

Figure 3.3.4: The region R, where $0 \leq r \leq 2$ and $0 \leq \theta \leq \pi/2$.

Example 3.3.2. Integrate $f(x, y) = x$ over the region R, which is the second quadrant portion of the annulus (a disk with a smaller disk removed) where $2 \leq r \leq 5$.

Solution:

The region is described in polar coordinates as the set of points where $2 \leq r \leq 5$ and $\pi/2 \leq \theta \leq \pi$.

Therefore, we find

$$\iint_R x\, dA = \int_{\pi/2}^{\pi} \int_2^5 (r \cos \theta)\, r\, dr\, d\theta = \int_{\pi/2}^{\pi} \int_2^5 r^2 \cos \theta\, dr\, d\theta =$$

$$\int_{\pi/2}^{\pi} \left[\frac{r^3}{3} \cos \theta \Big|_{r=2}^{r=5} \right] d\theta = \int_{\pi/2}^{\pi} \frac{117}{3} \cos \theta\, d\theta = \frac{117}{3} \sin \theta \Big|_{\pi/2}^{\pi} = -\frac{117}{3}.$$

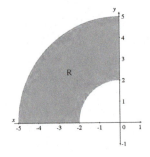

Figure 3.3.5: The region R, where $2 \leq r \leq 5$ and $\pi/2 \leq \theta \leq \pi$.

Note that the answer HAD to come out to be negative; x is less than or equal to 0 in the second quadrant.

Example 3.3.3. Integrate

$$g(x,y) \;=\; \frac{1}{\sqrt{x^2+y^2}}$$

over the trapezoid R bounded by the graphs of $y = x$, $x = 1$, $x = 2$, and the x-axis.

Solution:

The region R is **not** one that obviously looks like one that you'd want to describe in polar coordinates. On the other hand, in polar coordinates, the integrand, g, simply becomes

$$\frac{1}{\sqrt{x^2+y^2}} \;=\; \frac{1}{r},$$

and this is so simple that it's reasonable to try using polar coordinates.

In polar coordinates, the ray where $y = x \geq 0$ is just where $\theta = \pi/4$. Thus,

$$0 \;\leq\; \theta \;\leq\; \frac{\pi}{4}.$$

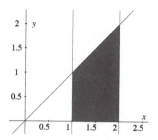

Figure 3.3.6: The trapezoid R.

The tougher question is: for a fixed θ between 0 and $\pi/4$, what do the corresponding r values do in the region R?

Well...clearly the r value starts at the r on the line $x = 1$ and ends at the r value on the line $x = 2$, but what are those r values in terms of θ?

This really isn't so bad. In polar coordinates, the line where $x = 1$ becomes the line where $r\cos\theta = 1$, i.e., $r = \sec\theta$. The line where $x = 2$ is given by $r\cos\theta = 2$ in polar coordinates, i.e., $r = 2\sec\theta$.

Hence, we find

$$\iint_R \frac{1}{\sqrt{x^2+y^2}}\, dA \;=\; \int_0^{\pi/4} \int_{\sec\theta}^{2\sec\theta} \frac{1}{r}\cdot r\,dr\,d\theta \;=\; \int_0^{\pi/4} \sec\theta\,d\theta \;=$$

$$\ln|\sec\theta + \tan\theta|\ \Big|_0^{\pi/4} \;=\; \ln\!\left(\sqrt{2}+1\right) - \ln 1 \;=\; \ln\!\left(\sqrt{2}+1\right).$$

It is unlikely that you've memorized $\int \sec\theta\,d\theta$. To see that $\ln|\sec\theta + \tan\theta|$ is an anti-derivative of $\sec\theta$, you may differentiate $\ln|\sec\theta + \tan\theta|$ and simplify to obtain $\sec\theta$.

Example 3.3.4. In this example, we wish to calculate an improper integral, one which is related to the *normal distribution* from probability and statistics.

We want to calculate the definite integral

$$I \;=\; \int_{-\infty}^{\infty} e^{-x^2}\,dx.$$

An elementary function is a function which is a constant function, a power function (with an arbitrary real exponent), a polynomial function, an exponential function, a logarithmic function, a trigonometric function, or inverse trigonometric function, or any finite combination of such functions using addition, subtraction, multiplication, division, or composition.

You should recall that there is no elementary function which is an anti-derivative of e^{-x^2}; that is, there is no "nice" formula for the indefinite integral

$$\int e^{-x^2}\, dx.$$

Well then, how are we supposed to calculate the value of the improper definite integral I? The answer is...we get **very** tricky.

In our definition of I, the variable x is a dummy variable; we could have called it anything. So, we can also write

$$I = \int_{-\infty}^{\infty} e^{-y^2}\, dy.$$

Then,

$$I^2 = \left(\int_{-\infty}^{\infty} e^{-x^2}\, dx\right)\cdot\left(\int_{-\infty}^{\infty} e^{-y^2}\, dy\right) = \int_{-\infty}^{\infty}\int_{-\infty}^{\infty} e^{-(x^2+y^2)}\, dx\, dy.$$

This is the iterated integral of $f(x,y) = e^{-(x^2+y^2)}$ over the region where $-\infty < x < \infty$ and $-\infty < y < \infty$, i.e., over the region R consisting of the entire xy-plane.

Therefore, we have

$$I^2 = \iint_R e^{-(x^2+y^2)}\, dA,$$

where R is the whole xy-plane. Switching into polar coordinates, the xy-plane is described by $0 \leq r < \infty$ and $0 \leq \theta < 2\pi$, and our integral becomes

$$I^2 = \iint_R e^{-(x^2+y^2)}\, dA = \int_0^{2\pi}\int_0^{\infty} e^{-r^2} r\, dr\, d\theta.$$

Note that the extra r factor saves us; a substitution of $u = -r^2$ allows us to integrate easily:

$$I^2 = \int_0^{2\pi}\left[\int_0^{-\infty} e^u\left(-\frac{1}{2}du\right)\right] d\theta = -\frac{1}{2}\int_0^{2\pi}\left[e^u\,\Big|_0^{-\infty}\right] d\theta = \pi.$$

Thus,

$$I = \int_{-\infty}^{\infty} e^{-x^2}\, dx = \sqrt{\pi}.$$

More Depth:

We want to explain in a general manner why, in polar coordinates,

$$dA = r\, dr\, d\theta.$$

We shall carry out this discussion using "infinitesimal" language; we could be more technical and use small changes and limits, but this would make the discussion very cumbersome.

In fact, the formula $dA = r\,dr\,d\theta$ should be a bit confusing. After all, in Cartesian (rectangular) coordinates,

$$dA = dx\,dy.$$

But, if you take $x = r\cos\theta$ and $y = r\sin\theta$, calculate the differentials

$$dx = \frac{\partial x}{\partial r}\,dr + \frac{\partial x}{\partial \theta}\,d\theta = \cos\theta\,dr - r\sin\theta\,d\theta$$

and

$$dy = \frac{\partial y}{\partial r}\,dr + \frac{\partial y}{\partial \theta}\,d\theta = \sin\theta\,dr + r\cos\theta\,d\theta$$

and multiply dx times dy, you **don't** get $r\,dr\,d\theta$. What's going on here?

The answer is that $dx\,dy$ and $r\,dr\,d\theta$ do not yield the area of the same infinitesimal "chunks" or, what are usually referred to as *elements of area*. The area element associated to infinitesimal changes dr and $d\theta$ is the infinitesimal area of the region, labeled dA in Figure 3.3.7, where one side is obtained by holding r constant and letting θ change by $d\theta$, and an adjacent side is obtained by holding θ constant and letting r change by dr.

The area given by $dx\,dy$ is the area of the rectangle indicated in Figure 3.3.8, where one side is obtained by holding y constant and letting x change by dx, and an adjacent side is obtained by holding x constant and letting y change by dy.

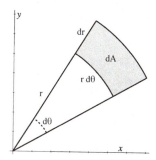

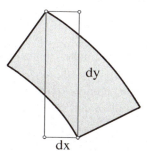

Figure 3.3.7: Area element dA associated with dr and $d\theta$.

Figure 3.3.8: The corresponding dx and dy.

The problem is not that the sides of the polar region curve, because, infinitesimally, the region labeled dA can be thought of as a parallelogram. The problem is that this parallelogram does not have sides parallel to the x- and y-axes.

In Figure 3.3.9, we show a (red) parallelogram with the "same" infinitesimal area as the curved polar region. We also know from Section 1.5 how to calculate the area

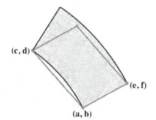

Figure 3.3.9: A parallelogram with the "same" infinitesimal area.

of a parallelogram which has one side given by the vector (p, q) between two adjacent corners, and has an adjacent side given by the vector (r, s) between two adjacent corners; the area of the parallelogram is the absolute value of the determinant of the matrix that has the two vectors in its columns.

Therefore, we want to know the infinitesimal vectors $\overrightarrow{(a,b)(c,d)}$ and $\overrightarrow{(a,b)(e,f)}$, in terms of r, θ, dr, and $d\theta$. We will then take the absolute value of the determinant of the matrix with these vectors in its columns.

If θ is constant, then $dx = \frac{\partial x}{\partial r}\, dr = \cos\theta\, dr$ and $dy = \frac{\partial y}{\partial r}\, dr = \sin\theta\, dr$. If r is constant, then $dx = \frac{\partial x}{\partial \theta}\, d\theta = -r\sin\theta\, d\theta$ and $dy = \frac{\partial y}{\partial \theta}\, d\theta = r\cos\theta\, dr$. Hence, in Figure 3.3.9,

$$(c, d) \;=\; (a, b) + (\cos\theta\, dr,\; \sin\theta\, dr)$$

and

$$(e, f) \;=\; (a, b) + (-r\sin\theta\, d\theta,\; r\cos\theta\, d\theta).$$

Finally, we calculate, assuming that r, $d\theta$ and dr are non-negative:

$$dA \;=\; \left| \det \begin{bmatrix} \cos\theta\, dr & -r\sin\theta\, d\theta \\ \sin\theta\, dr & r\cos\theta\, d\theta \end{bmatrix} \right| \;=\; \left| \det \begin{bmatrix} \cos\theta & -r\sin\theta \\ \sin\theta & r\cos\theta \end{bmatrix} \right| dr\, d\theta \;=\;$$

$$\left| r\cos^2\theta + r\sin^2\theta \right| dr\, d\theta \;=\; r\, dr\, d\theta.$$

Actually, our derivation above of dA, in terms of polar coordinates, works for any local change of coordinates (recall Section 2.6). But, first, let's rephrase our result above.

Consider the function $\Phi : \mathbb{R}^2 \to \mathbb{R}^2$ given by

$$(x, y) \;=\; \boldsymbol{\Phi}(r, \theta) \;=\; (r\cos\theta,\; r\sin\theta).$$

This is the formal polar change of coordinates function. As we discussed in Section 2.6, Φ defines a local change of coordinates near any point $\mathbf{p}$ where $d_{\mathbf{p}}\boldsymbol{\Phi}$ is invertible, i.e., where the determinant of the Jacobian matrix is non-zero. Therefore, polar coordinates provide a local change of coordinates near any point where

$$\det\left[d\boldsymbol{\Phi} \right] \;\neq\; 0.$$

However, for the polar change of coordinates $\boldsymbol{\Phi}$,

$$\left[d\boldsymbol{\Phi} \right] \;=\; \begin{bmatrix} \cos\theta & -r\sin\theta \\ \sin\theta & r\cos\theta \end{bmatrix} \qquad \text{and} \qquad \det\left[d\boldsymbol{\Phi} \right] \;=\; r.$$

The fact that polar coordinates do not yield a local change of coordinates at the origin does not affect any of our integral calculations, since the origin has zero area, i.e., forms a set of measure zero in $\mathbb{R}^2$.

Thus, if $r > 0$ (i.e., if we're not at the origin), then polar coordinates really do provide a local coordinate change of coordinates, and

$$dA \;=\; \left| \det\left[d\boldsymbol{\Phi} \right] \right| dr\, d\theta,$$

assuming that dr and $d\theta$ are non-negative.

What was special in our discussion about polar coordinates? Nothing. The general theorem is:

Theorem 3.3.5. *Suppose that $\boldsymbol{\Phi}$ is a C^1 function from an open neighborhood of a point $\mathbf{p}$ in $\mathbb{R}^2$ into $\mathbb{R}^2$, and suppose that $\det\left[d_{\mathbf{p}}\boldsymbol{\Phi}\right] \neq 0$.*

Then, $(x, y) = \boldsymbol{\Phi}(u, v)$ is a local change of coordinates at $\mathbf{p}$ and, near $\mathbf{p}$, the area element $dA = dx\, dy$ is given by

$$dA = \left| \det\left[d\boldsymbol{\Phi}\right]\right| du\, dv.$$

Thus, if $f = f(x, y)$ is Riemann integrable on a region R, contained in an open subset $\mathcal{U}$ of $\mathbb{R}^2$ and $\boldsymbol{\Phi}$ is a C^1 change of coordinates from an open set $\mathcal{W}$ in $\mathbb{R}^2$ onto $\mathcal{U}$, then $f \circ \boldsymbol{\Phi}$ is Riemann integrable on the region $M = \boldsymbol{\Phi}^{-1}(R)$ and

$$\iint_R f\, dA = \iint_M (f \circ \boldsymbol{\Phi}) \cdot \left| \det\left[d\boldsymbol{\Phi}\right]\right| du\, dv.$$

This follows from Theorem 7.3.8 of Trench, [8].

Remark 3.3.6. It is worth noting that Theorem 3.3.5 is the 2-dimensional version of *integration by substitution*.

Example 3.3.7. Consider the tilted rectangle R in the xy-plane give by $1 \leq x - y \leq 3$ and $2 \leq x + y \leq 5$. See Figure 3.3.10. Calculate the double integral

$$\iint_R xy\, dA.$$

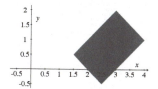

Figure 3.3.10: The region R is a tilted rectangle.

Solution:

To calculate this in the coordinates x and y would be fairly painful; we would need to split the double integral up into three separate iterated integrals. (You should think about why that would be true.)

However, we will let $u = x - y$ and $v = x + y$, so that R is the region where $1 \leq u \leq 3$ and $2 \leq v \leq 5$. In the set up of Theorem 3.3.5, we have

$$(u, v) = \mathbf{\Phi}^{-1}(x, y) = (x - y, x + y),$$

and M is precisely the rectangle in the uv-plane where $1 \leq u \leq 3$ and $2 \leq v \leq 5$.

Solving for x and y in terms of u and v, we find

$$x = \frac{1}{2}(v + u) \quad \text{and} \quad y = \frac{1}{2}(v - u),$$

i.e.,

$$(x, y) = \mathbf{\Phi}(u, v) = \left(\frac{1}{2}(v + u), \frac{1}{2}(v - u) \right).$$

This is a global change of coordinates (in fact, it's a *linear change of coordinates*), because we started with C^1 functions for u and v, and solved to obtain that x and y are C^1 functions of u and v.

We easily calculate

$$xy = \frac{1}{4}(v^2 - u^2) \quad \text{and} \quad dA = dx\, dy = \left| \det \begin{bmatrix} 1/2 & 1/2 \\ -1/2 & 1/2 \end{bmatrix} \right| du\, dv = \frac{1}{2} du\, dv.$$

And so, applying Theorem 3.3.5, we find

$$\iint_R xy\, dA = \int_2^5 \int_1^3 \frac{1}{4}(v^2 - u^2) \cdot \frac{1}{2}\, du\, dv = \frac{1}{8} \int_2^5 \left[uv^2 - \frac{u^3}{3} \Big|_{u=1}^{u=3} \right] dv =$$

$$\frac{1}{8} \int_2^5 \left((3v^2 - 9) - \left(v^2 - \frac{1}{3} \right) \right) dv = \frac{13}{2}.$$

Online answers to select exercises are here.

Basics:

3.3.1   Exercises

In each of the following exercises, a region R in $\mathbb{R}^2$ is described in terms of x and y, or in terms of r and θ. You are also given an integral in terms of x and y, or in terms of r and θ. **(a) Sketch the region R. (b) Calculate the given integral in whichever coordinates seem most convenient, possibly switching from the given coordinate system to the other.**

1. Calculate $\iint_R x\, dA$, where R is the region where $0 \leq \theta \leq \pi/4$ and $1 \leq r \leq 2$.

2. Calculate $\iint_R y\,dA$, where R is the region where $0 \le \theta \le \pi$ and $1 \le r \le 2$.

3. Calculate $\iint_R x^3\,dA$, where R is the region where $0 \le r\cos\theta \le 3$ and $1 \le r\sin\theta \le 2$.

4. Calculate $\iint_R r^2\,dA$, where R is the region where $0 \le r\cos\theta \le 3$ and $1 \le r\sin\theta \le 2$.

5. Calculate $\iint_R r\sin\theta\,dA$, where R is the region where $0 \le x \le 2$ and $0 \le y \le \sqrt{4 - x^2}$.

6. Calculate $\iint_R (x^2 + y^2)\,dA$, where R is the region where $0 \le \theta \le \pi/4$ and $0 \le r \le \sqrt{\theta}$.

7. Calculate $\iint_R \dfrac{2\cos\theta}{r^2 + 1}\,dA$, where R is the region where $\pi/2 \le \theta \le \pi$ and $0 \le r \le 1$.

8. Calculate $\iint_R (x^2 + y^2)^2\,dA$, where R is the region where $0 \le \theta \le \pi$ and $e^{-\theta} \le r \le 1$.

9. Calculate $\iint_R e^{x^2 + y^2}\,dA$, where R is the region where $0 \le \theta < 2\pi$ and $0 \le r \le 3$.

10. Calculate $\iint_R xye^{x^2 + y^2}\,dA$, where R is the region where $0 \le r\cos\theta \le 2\pi$ and $0 \le r\sin\theta \le 3$.

In each of the following exercises, you are given an iterated integral in Cartesian coordinates x and y. Convert the integral into an iterated integral in terms of the polar coordinates r and θ, and evaluate the new iterated integral.

11. $\displaystyle\int_0^2 \int_0^x 1\,dy\,dx$

12. $\displaystyle\int_0^1 \int_0^{\sqrt{1 - y^2}} x\,dx\,dy$

13. $\displaystyle\int_{-1}^0 \int_0^{\sqrt{1 - y^2}} x\,dx\,dy$

14. $\displaystyle\int_0^1 \int_0^{\sqrt{1 - y^2}} y\,dx\,dy$

15. $\displaystyle\int_0^2 \int_0^{\sqrt{4-x^2}} \frac{1}{x^2+y^2+1}\, dy\, dx$

16. $\displaystyle\int_0^\infty \int_0^\infty \frac{1}{(x^2+y^2+1)^2}\, dy\, dx$

17. $\displaystyle\int_{-3}^0 \int_0^{\sqrt{9-x^2}} xe^{(x^2+y^2)^{3/2}}\, dy\, dx$

18. $\displaystyle\int_1^3 \int_{-y}^y \frac{y^2}{x^2+y^2}\, dx\, dy$

19. $\displaystyle\int_1^3 \int_{-y}^y \frac{x}{y}\, dx\, dy$

20. $\displaystyle\int_{-\infty}^\infty \int_1^\infty \frac{1}{(x^2+y^2)^{3/2}}\, dx\, dy$

More Depth:

In each of the following exercises, you are given an integral. Each integral leads to an "obvious" choice of substitutions $u = f(x,y)$ and $v = g(x,y)$ so that $(u,v) = \Phi^{-1}(x,y)$ is a C^1 change of coordinates on some open set containing the given region R. (a) Verify that Φ^{-1} is a C^1 change of coordinates by explicitly producing Φ (and noting that Φ^{-1} and Φ are C^1). (b) Use your change of coordinates to evaluate the given integral.

21. Calculate $\displaystyle\iint_R (y+x)^2(y-x)\, dA$, where R is the region given by $0 \le y+x \le 2$ and $0 \le y - x \le 4$. ▶

22. Calculate $\displaystyle\iint_R (x+y^3)y\, dA$, where R is the region given by $0 \le x+y^3 \le 2$ and $0 \le y \le 4$.

23. Calculate $\displaystyle\iint_R (2x+\sqrt{y})y^3\, dA$, where R is the region given by $3 \le 2x+\sqrt{y} \le 5$ and $1 \le y \le 2$.

24. Calculate $\displaystyle\iint_R \frac{e^x+y}{y^5}\, dA$, where R is the region given by $1 \le e^x + y \le 4$ and $-2 \le y \le -1$.

In each of the following exercises, use the given change of coordinates to evaluate the given integral.

25. $\displaystyle\iint_R (2x-3y)\sin(x+y)\, dA$, where R is the region given by $0 \le 2x - 3y \le 6$ and $0 \le x+y \le \pi$, and $(u,v) = \Phi^{-1}(x,y) = (2x-3y, x+y)$. ▶

26. $\iint_R (xy^3 + y^4) \, dA$, where R is the region given by $1 \le y \le 2$ and $-1 \le x + y \le 1$, and $(u, v) = \mathbf{\Phi}^{-1}(x, y) = (x + y, y^3)$.

27. $\iint_R \left(\dfrac{y^2 + x}{2y} \right) \left(\dfrac{y^2 - x}{2y} \right) \, dA$, where R is the region given by $1 \le \frac{y^2 + x}{2y} \le 2$, and $1 \le \frac{y^2 - x}{2y} \le 3$, and $(u, v) = \mathbf{\Phi}^{-1}(x, y) = \left(\frac{y^2 + x}{2y}, \ \frac{y^2 - x}{2y} \right)$.

28. $\iint_R \sqrt{x + \sqrt{x^2 + y^2}} \, dA$, where R is the region given by $x > 0$, $y > 0$, $2 \le \sqrt{x + \sqrt{x^2 + y^2}} \le 4$ and $1 \le \sqrt{-x + \sqrt{x^2 + y^2}} \le 2$, and $(u, v) = \mathbf{\Phi}^{-1}(x, y) = \left(\sqrt{x + \sqrt{x^2 + y^2}}, \ \sqrt{-x + \sqrt{x^2 + y^2}} \right)$. (Hint: Calculate uv and $u^2 - v^2$.)

3.4   Integration in $\mathbb{R}^3$ and $\mathbb{R}^n$

In the previous two sections, we looked at double integrals, Riemann integrals over regions in $\mathbb{R}^2$. We also want to look at *triple integrals*, Riemann integrals over solid regions in $\mathbb{R}^3$, which have a wide range of physical applications. Of course, as always, integrals should be thought of as continuous sums of infinitesimal contributions.

While integration in $\mathbb{R}^2$ and $\mathbb{R}^3$ will be of primary interest throughout the remainder of this book, it is not substantially more difficult to discuss integration in $\mathbb{R}^n$, but we relegate that discussion to the More Depth portion of this section.

Basics:

As in the previous section, we want to begin with a motivating example, which should make it clear why we want to define integrals over 3-dimensional regions, i.e., why we want to define *triple integrals*.

Example 3.4.1. Suppose that we have a solid object in $\mathbb{R}^3$, i.e., an object which occupies a region T in xyz-space. The density, δ, the mass per unit volume, of the object may vary at different points in the object, and so, we write $\delta = \delta(x, y, z)$ and, at least for now, we'll assume that δ is continuous.

The question is: given the region T in $\mathbb{R}^3$, and the density function, can we determine the mass of the object?

Well...how can we **estimate** the mass of the object? As before, we chop up T into lots of little pieces, but, this time, the pieces are rectangular solids, solid boxes, or, near the boundary, perhaps just partial boxes. We estimate the mass of each little box, either ignoring partial boxes or not, and add all of these little masses together, to approximate the mass of the whole object.

How do we estimate the mass of each little box? If you think about what we did in the previous section, it should be obvious. We take any point $\mathbf{p}$ (a *sample point*) in the little box, evaluate $\delta(\mathbf{p})$, and assume – since the box is small and since δ is continuous – that $\delta(\mathbf{p})$ is a reasonable approximation of the density at every point in the small box; hence, $\delta(\mathbf{p})$ times the volume of the little box should be a good estimate of the mass of the little box.

Adding together all of these little mass contributions gives an **estimate** of the mass of the object. Then, we take the limit as the size (here, length, width, and height) of the little boxes approaches 0. If this limit exists, and does not depend on precisely how the sizes of the little boxes approach zero or on how the sample points are chosen, then we say that this limit is the mass of the object and that the function δ is *Riemann integrable* over the region T.

As was the case when we defined the double integral, in order to define the triple integral, the Riemann integral over a solid region in $\mathbb{R}^3$, we must define partitions of solid regions, sample points, Riemann sums, take limits, and state a number of theorems. We cover all of this, and, more generally, discuss integration in $\mathbb{R}^n$, for any n, in the More Depth portion of this section. For now, as before, we wish to give the quick intuition, state the theorems that we will usually use, and look at some examples.

It is instructive to compare the discussion below, for triple integrals, with the discussion in the previous section for double integrals. The words are practically identical, except that 2-dimensional concepts, like area, are replaced by 3-dimensional analogs, like volume.

If T is a subset of $\mathbb{R}^3$ (which will need to be reasonable in some sense) and $f = f(x, y, z)$ is a real-valued function on T, we would like to define the *triple integral*

$$\iiint_T f(x, y, z) \, dV,$$

which we think of as being **a continuous sum of the values of f multiplied by infinitesimal pieces of volume, for points in the set T.**

What does this mean? You should think of dividing T into lots of little solid boxes, where, near the boundary of T, you might have just pieces of boxes, which we'll call partial boxes. Any little box has volume, which we'll denote by ΔV; of course, $\Delta V = \Delta x \Delta y \Delta z$, where Δx, Δy, and Δz are the length, width, and height of the box. Pick a point $\mathbf{s}$, a sample point, inside the little box, and look at the product $f(\mathbf{s})\Delta V$.

You do this for each little box in T, either including or excluding the partial boxes near the boundary. Then, you add up all of these products $f(\mathbf{s})\Delta V$; this is a *Riemann sum*. Finally, you take the limit of this process, as the lengths, widths, and heights of all of the little boxes approach zero, provided that the limit exists and doesn't depend on any of the choices that were made.

If, in fact, the limit exists and doesn't depend on any of the choices that were made, then we say that f is *Riemann integrable on T*, and the limit is the *Riemann integral of f on (or, over) T*; in $\mathbb{R}^3$, this Riemann integral is frequently referred to as the *triple integral of f on R*, and is denoted by

$$\iiint_T f(x, y, z) \, dV.$$

In our discussion of integration in $\mathbb{R}^2$, we first dealt with integrating over rectangles, and then generalized the discussion to more general regions.

Rather than first considering triple integrals over boxes, we will go ahead and state the fundamental result that we use in calculating over more-general regions. However, even discussing "nice" regions in $\mathbb{R}^3$ is fairly complicated, so we need to spend a little time on this topic.

What we want to do is take a triple integral $\iiint_T f(x, y, z)\, dV$ and write it as an iterated integral, i.e., we want to have an equality of the form

$$\iiint_T f(x, y, z)\, dV \;=\; \int_a^b \int_{u(x)}^{v(x)} \int_{p(x,y)}^{q(x,y)} f(x, y, z)\, dz\, dy\, dx$$

where a and b are constants, u, v, p, and q are continuous functions, and where we would be happy to interchange the roles of any of the coordinates x, y and z. This means that we need have a, b, u, v, p, and q so that

(x, y, z) is in T if and only if $a \leq x \leq b$, $u(x) \leq y \leq v(x)$, and $p(x, y) \leq z \leq q(x, y)$,

possibly permuting the variables x, y and z.

So, we make the following definition,

Definition 3.4.2. *An **iterated region** T in $\mathbb{R}^3$ is a region such that there exist constants a and b, and continuous functions u, v, p, and q such that a point (x, y, z) is in T if and only if*

$$a \leq x \leq b, \ u(x) \leq y \leq v(x), \ \text{and} \ p(x, y) \leq z \leq q(x, y),$$

after, possibly, permuting the variables x, y, and z.

The phrase "after, possibly, permuting the variables x, y, and z" means, for instance, that a region of the form

$$a \leq y \leq b, \ u(y) \leq z \leq v(y), \ \text{and} \ p(y, z) \leq x \leq q(y, z)$$

is also an iterated region.

The following result, which is another version of Fubini's Theorem, should come as no surprise.

Theorem 3.4.3. *Suppose that a and b are constants, u, v, p, and q are continuous functions, and T is the iterated region in* $\mathbb{R}^3$ *given by*

$$a \leq x \leq b, \ u(x) \leq y \leq v(x), \ \text{and} \ p(x,y) \leq z \leq q(x,y).$$

Suppose that $f = f(x,y,z)$ *is a continuous function on T.*

Then, f is Riemann integrable on the region T and the triple integral, the Riemann integral of f over T, can be calculated via an iterated integral:

$$\iiint_T f(x,y,z)\, dV \ = \ \int_a^b \int_{u(x)}^{v(x)} \int_{p(x,y)}^{q(x,y)} f(x,y,z)\, dz\, dy\, dx.$$

In addition, this theorem remains true if the variables x, y, and z are permuted.

This follows from Theorem 3.4.18 in the More Depth portion of this section.

Example 3.4.4. Let's integrate the function $f = f(x,y,z) = y$ over the solid region T which is under the graph of $z = 4 - x^2 - y^2$ and above the first quadrant in the xy-plane. See Figure 3.4.1; we are considering the solid region under the blue surface, and above the purple quarter-disk in the xy-plane. This quarter disk, which we'll call R appears separately in Figure 3.4.2.

In light of Theorem 3.4.3, the initial part of the problem is to represent T as an iterated region, because this determines the limits of integration in the iterated integral.

You first want to decide which coordinate will be the farthest inside the iterated integral. Here, we will choose the z coordinate, since, by definition of the solid region T, for every (x,y,z) in T, $0 \leq z \leq 4 - x^2 - y^2$. We **could** choose x or y for the inside coordinate (and we will later, as an example), but, here, we choose z.

How do you decide on the other two pairs of limits of integration? You **project** the solid region T into the coordinate plane of the remaining two coordinates, here, into the xy-plane. What does this mean? Technically, it means that you look at all of the points (x,y) in $\mathbb{R}^2$ which come from points in T, i.e., points (x,y) in $\mathbb{R}^2$ such that there exists a z such that (x,y,z) is in T. More intuitively, if there were a big spotlight pointing down the z-axis (from very far out), the projection of T into the xy-plane would be the shadow of T (this is where the term "projection" comes from). In this example, the projected region in the xy-plane is precisely T, the purple region.

Now, you first change our triple integral into a double integral over R on the outside, and a single integral over the inside coordinate, which, here, is z. The limits of integration for z are determined by: for each point $\mathbf{p}$ in R, the z coordinates of the corresponding points in S go from what function of $\mathbf{p}$ to what function of $\mathbf{p}$?

Figure 3.4.1: The solid region T, and the projection into the xy-plane.

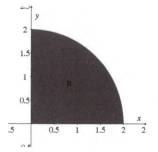

Figure 3.4.2: The projected region R in the xy-plane.

In Figure 3.4.1, we have included a line segment which begins at a point $\mathbf{p}$ in R and extends up to the blue bounding surface. This is the line segment of points in T whose x and y coordinates are simply $\mathbf{p}$. The line segment starts at $z = 0$ and goes up to $z = 4 - x^2 - y^2$; these are the inside limits of integration. In other words,

$$\iiint_T y \, dV = \iint_R \left[\int_0^{4-x^2-y^2} y \, dz \right] dA.$$

In general, the projected region need **not** be the part of the solid region which actually lies in the coordinate plane that you're projecting into. It's a question of where the solid region sticks out the most. See Example 3.4.6.

Now you are reduced to a 2-dimensional problem. You have to write the double integral over R as an iterated integral in terms of x and y.

However, this is easy. The region R is the region is the first quadrant bounded by the curve where the graph of $z = 4 - x^2 - y^2$ hits the plane where $z = 0$; thus, this bounding curve in the xy-plane is given by $0 = 4 - x^2 - y^2$, which, in the first quadrant, means that $y = \sqrt{4 - x^2}$. Therefore, R is the region where $0 \le x \le 2$ and $0 \le y \le \sqrt{4 - x^2}$.

We arrive at

$$\iiint_T y \, dV = \int_0^2 \int_0^{\sqrt{4-x^2}} \int_0^{4-x^2-y^2} y \, dz \, dy \, dx.$$

Now we can calculate

$$\int_0^2 \int_0^{\sqrt{4-x^2}} \int_0^{4-x^2-y^2} y \, dz \, dy \, dx = \int_0^2 \int_0^{\sqrt{4-x^2}} \left[yz \, \Big|_{z=0}^{z=4-x^2-y^2} \right] dy \, dx =$$

$$\int_0^2 \int_0^{\sqrt{4-x^2}} (4y - x^2 y - y^3) \, dy \, dx = \int_0^2 \left(2y^2 - \frac{x^2 y^2}{2} - \frac{y^4}{4} \, \Big|_{y=0}^{y=\sqrt{4-x^2}} \right) dx =$$

$$\int_0^2 \left(2(4 - x^2) - \frac{x^2(4 - x^2)}{2} - \frac{(4 - x^2)^2}{4} \right) dx = \int_0^2 \frac{1}{4} \left(16 - 8x^2 + x^4 \right) dx =$$

$$\frac{1}{4} \left(16x - \frac{8x^3}{3} + \frac{x^5}{5} \right) \Big|_0^2 = \frac{64}{15}.$$

Example 3.4.5. As in the previous example, let's integrate the function

$$f = f(x, y, z) = y$$

over the solid region T which is under the graph of $z = 4 - x^2 - y^2$ and above the first quadrant in the xy-plane. This time, however, we shall project the solid region T into the xz-plane; see Figure 3.4.3.

Figure 3.4.3: The solid region T, and the projection into the xz-plane.

This means that our inside variable will be y, and our projected region R in the xz-plane will be the region in the first quadrant, bounded by the curve where the graph of $z = 4 - x^2 - y^2$ intersects the xz-plane, which is where $y = 0$. Therefore, the projected region R in the xz-plane is bounded by curve given by $z = 4 - x^2$. See Figure 3.4.4.

For each point $\mathbf{p}$ in R, the corresponding points in the solid region T are those whose x and z coordinates are the same as those of $\mathbf{p}$, and the y coordinates go from $y = 0$ to the y coordinate on the blue surface, i.e., $y = \sqrt{4 - x^2 - z}$.

Thus, we find

$$\iiint_T y \, dV = \iint_R \left[\int_0^{\sqrt{4-x^2-z}} y \, dy \right] dA = \int_0^2 \int_0^{4-x^2} \int_0^{\sqrt{4-x^2-z}} y \, dy \, dz \, dx =$$

$$\int_0^2 \int_0^{4-x^2} \left[\frac{y^2}{2} \Big|_{y=0}^{y=\sqrt{4-x^2-z}} \right] dz \, dx = \frac{1}{2} \int_0^2 \int_0^{4-x^2} (4 - x^2 - z) \, dz \, dx =$$

$$\frac{1}{2} \int_0^2 \left[\left(4z - x^2 z - \frac{z^2}{2} \right) \Big|_{z=0}^{z=4-x^2} \right] dx =$$

$$\frac{1}{2} \int_0^2 \left(4(4 - x^2) - x^2(4 - x^2) - \frac{(4 - x^2)^2}{2} \right) dx =$$

$$\frac{1}{4} \int_0^2 \left(16 - 8x^2 + x^4 \right) dx = \frac{64}{15},$$

as we obtained in the previous example.

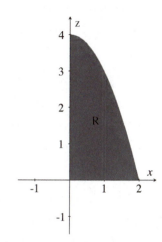

Figure 3.4.4: The projected region R in the xz-plane.

Example 3.4.6. Consider the solid region T bounded by the slanted parabolic cylinder given by $y = 2z - x^2$ and the planes given by $y = 0$ and $z = 2$. See Figure 3.4.5.

Set up an iterated integral for calculating the triple integral of any continuous function $f(x, y, z)$ over the solid region T.

Solution: The plane where $z = 2$ intersects the graph of $y = 2z - x^2$ in the curve where $y = 4 - x^2$, inside the copy of the xy-plane at $z = 2$. Looking at Figure 3.4.5, we see the projection of S into the xy-plane; it's the region R, in the xy-plane, which is bounded by the x-axis and the parabola where $y = 4 - x^2$. See Figure 3.4.6.

For each point $\mathbf{p}$ in R, the points in the region T that correspond to $\mathbf{p}$, the points whose x and y coordinates are those of $\mathbf{p}$, go from the graph of $y = 2z - x^2$ up to the plane where $z = 2$. Therefore,

$$\iiint_T f(x, y, z) \, dV = \iint_R \left[\int_{(y+x^2)/2}^2 f(x, y, z) \, dz \right] dA =$$

Figure 3.4.5: The solid region T and its projection into the xy-plane.

$$\int_{-2}^{2} \int_{0}^{4-x^2} \int_{(y+x^2)/2}^{2} f(x,y,z)\, dz\, dy\, dx.$$

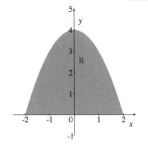

Figure 3.4.6: The projection R, of the solid region T, into the xy-plane.

More Depth:

This does have the amusing/confusing consequence that the length of an interval in $\mathbb{R}^1$ is also it's 1-dimensional volume, and the area of a rectangle in $\mathbb{R}^2$ is its 2-dimensional volume.

We want to give the rigorous definition of the triple integral, the Riemann integral in $\mathbb{R}^3$. Since it's just as easy to define the Riemann integral in $\mathbb{R}^n$, for any n, we will do that.

This discussion will look almost identical to our discussion in Section 3.2 of integration in $\mathbb{R}^2$, and it may be helpful to look back at that case, particularly for definitions of some terms, such as *partition* and *mesh*.

Definition 3.4.7. *Suppose that, for all integers i such that $1 \leq i \leq n$, we have numbers $a_i < b_i$. The **closed n-dimensional box** $\mathcal{B}$ of points $(x_1, \ldots, x_n)$ such that $a_i \leq x_i \leq b_i$, for all i, is denoted by $[a_1, b_1] \times \cdots \times [a_n, b_n] = \Pi_{i=1}^{n}[a_i, b_i]$.*

*The **n-dimensional volume** of a box $\mathcal{B}$, as above, is*

$$\Delta V_{\mathcal{B}} = (b_1 - a_1)(b_2 - a_2) \cdots (b_n - a_n) = \Pi_{i=1}^{n}\Delta x_i.$$

Definition 3.4.8. *A **partition** $\mathcal{P}$ of the box $\Pi_{i=1}^{n}[a_i, b_i]$ consists of a partition of each of the intervals $[a_i, b_i]$ in the product, i.e., for each i, a partition $x_0^i, \ldots, x_j^i, \ldots, x_{m_i}^i$ of $[a_i, b_i]$.*

---

*The j-**subinterval**, I_j^i, of $[a_i, b_i]$, with respect to a partition $\mathcal{P}$, is $[x_{j-1}^i, x_j^i]$. We write I^i for an arbitrary subinterval of $[a_i, b_i]$, with respect to $\mathcal{P}$.*

*A **sub-box** $\widehat{\mathcal{B}}$ of the partition $\mathcal{P}$ is a product of subintervals, with respect to $\mathcal{P}$, with one subinterval from each of the $[a_i, b_i]$, i.e., a box*

$$\widehat{\mathcal{B}} = I^1 \times \cdots \times I^n = \Pi_{i=1}^{n} I^i.$$

*The **mesh** of $\mathcal{P}$, denoted $\|\mathcal{P}\|$, is the maximum of all the meshes of the partitions of the intervals $[a_i, b_i]$.*

*A **sample set** $\mathcal{S}$ for $\mathcal{P}$ is a set of sample points $\mathbf{s}_{\widehat{\mathcal{B}}}$, one for each sub-box $\widehat{\mathcal{B}}$ of the partition $\mathcal{P}$.*

Definition 3.4.9. *Suppose that f is a real-valued function, which is defined on an n-dimensional box $\mathcal{B}$. A **Riemann sum** for f on $\mathcal{B}$ is a summation*

$$\sum_{\widehat{\mathcal{B}}} f\left(\mathbf{s}_{\widehat{\mathcal{B}}}\right) \Delta V_{\widehat{\mathcal{B}}},$$

for some partition of $\mathcal{B}$ and sample points $\mathbf{s}_{\widehat{\mathcal{B}}}$.

Definition 3.4.10. (**The Riemann Integral**) *Suppose that f is defined on the closed box $\mathcal{B}$, and that there exists a real number L such that, for all $\epsilon > 0$, there exists $\delta > 0$ such that, for all partitions $\mathcal{P}$ of $\mathcal{B}$, with mesh less than δ, and for all sample sets $\mathcal{S}$ for $\mathcal{P}$,*

$$\left| \sum_{\widehat{\mathcal{B}}} f(\mathbf{s}_{\widehat{\mathcal{B}}})\Delta V_{\widehat{\mathcal{B}}} \; - \; L \right| < \epsilon.$$

*Then, we say that f is **Riemann integrable** on $\mathcal{B}$ (in $\mathbb{R}^n$), and that the (n-dimensional) **Riemann integral** of f on $\mathcal{B}$ exists and equals L. We write*

$$\int_{\mathcal{B}} f \, dV \; = \; L.$$

Note that, beyond the triple integral, when $n \geq 4$, it is common to stop writing multiple integral signs, and simply let the dimension of the box tell you the dimensionality of the integral.

Occasionally, it is important to emphasize the dimension in which the integral is taking place. In that case, we denote the integral in $\mathbb{R}^n$ by $\int_{\mathcal{B}} f \, dV^n$.

Once again, we need to discuss sets of *measure zero*; as before, see Chapter 11 of Rudin, [7], or Chapter 5 of Fleming, [3], for a complete treatment.

Definition 3.4.11. *A subset A of $\mathbb{R}^n$ has **measure zero** in $\mathbb{R}^n$ provided that the n-dimensional volume of A is defined and that volume is 0.*

Technically, a subset A of $\mathbb{R}^n$ has measure zero in $\mathbb{R}^n$ if and only if, for all $\epsilon > 0$, there exists a finite number, or infinite sequence, of n-dimensional closed boxes whose union contains A and the sum of the volumes of the rectangles is $< \epsilon$.

*A property defined at points in $\mathbb{R}^N$ is said to hold **almost everywhere** if and only if the set of points where the property does not hold has measure zero.*

This follows from an easy
generalization of Example
7.1.5 of Trench, [8].

Theorem 3.4.12. *Any subset of $\mathbb{R}^n$ which is contained in a union of a finite number, or infinite sequence, of graphs of real-valued continuous functions on compacts subsets of $\mathbb{R}^{n-1}$ has measure zero.*

We want to give a name to the types of "nice" regions that we want to consider in $\mathbb{R}^n$.

Definition 3.4.13. *A* **standard region** *in $\mathbb{R}^n$ is a compact set whose boundary is contained in a union of a finite number of graphs of real-valued continuous functions on compacts subsets of $\mathbb{R}^{n-1}$.*

Now that we have an idea of what "measure zero" means, we can give an improved theorem about when a function on a closed box is Riemann integrable.

Theorem 3.4.14. *Suppose that f is bounded on the closed n-dimensional box R, i.e., that there exists M such that, for all $\mathbf{x}$ in R, $|f(\mathbf{x})| \leq M$.*

Then, f is Riemann integrable on R if and only if f is continuous almost everywhere in R, i.e., the set of points in R at which f is not continuous has measure zero.

As in the 2-dimensional case, we want to be able to integrate over regions other than rectangles/boxes. Hence, we once again look at the *extension by zero*.

Suppose that f is a function which is defined on a bounded region R in R^n. Let $\mathcal{B}$ be a closed box containing R. We define a new function, an extension $\hat{f}$ of the function f to a function on the entire enclosing box, which is zero outside of R, i.e., we define $\hat{f}$ to be the function on $\mathcal{B}$ given by

$$\hat{f}(\mathbf{x}) = \begin{cases} f(\mathbf{x}), & \text{if } \mathbf{x} \text{ is in } R; \\ 0, & \text{if } \mathbf{x} \text{ is in } \mathcal{B}, \text{ but } \mathbf{x} \text{ is not in } R. \end{cases}$$

We call such an $\hat{f}$ an *extension of f by zero*.

Definition 3.4.15. *Suppose that R is a bounded region in $\mathbb{R}^n$, and that f is a real-valued function, defined on R. Let $\mathcal{B}$ be any closed box which encloses R, and let $\hat{f}$ be the function on $\mathcal{B}$ which is the extension of f by zero.*

> *Then, we say that f is **Riemann integrable on** R if and only if $\hat{f}$ is Riemann integrable on $\mathcal{B}$ and, in this case, we define the Riemann integral of f on R by*
>
> $$\int_R f \, dV \;=\; \int_{\mathcal{B}} \hat{f} \, dV.$$

It is important that this definition does not depend on the choice of the box $\mathcal{B}$ which encloses the region R.

The big theorem which tells us when a function is integrable is:

> **Theorem 3.4.16.** *Suppose that R is a bounded subset of $\mathbb{R}^n$ and that the boundary of R has measure zero. Then, a function f on R is Riemann integrable if and only if f is bounded on R and f is continuous almost everywhere in the interior of R.*
>
> *In particular, a continuous function on a standard region is Riemann integrable.*

See, for instance, page 185 of Fleming, [3].

The theorem that allows us to split n-dimensional integrals is:

> **Theorem 3.4.17.** *Suppose that R_1 and R_2 are bounded subsets of $\mathbb{R}^n$ that each have a boundary with measure zero, and that R_1 and R_2 intersect in a set of measure zero (which includes not intersecting at all). Suppose that f is a function which is Riemann integrable on R_1 and on R_2.*

> *Then, the union R of R_1 and R_2 is a bounded subset of $\mathbb{R}^n$ whose boundary has measure 0, f is Riemann integrable on R, and*
>
> $$\int_R f \, dV \;=\; \int_{R_1} f \, dV \;+\; \int_{R_2} f \, dV.$$

See part 7 of Theorem 5.4 of Fleming, [3].

The question remains as to how you actually effectively calculate Riemann integrals. The answer, of course, is that you use iterated integrals. Before we can state the relevant result, we need to first define *iterated regions* carefully.

Define the *i-th projection*, $\pi_i : \mathbb{R}^n \to \mathbb{R}^{n-1}$ to be the function which omits the i-th coordinate, i.e.,

$$\pi_i(x_1, \ldots, x_{i-1}, x_i, x_{i+1}, \ldots, x_n) \;=\; (x_1, \ldots, x_{i-1}, x_{i+1}, \ldots, x_n).$$

If R is a subset of $\mathbb{R}^n$, then $\pi_i(R)$ is the image of R under the function π_i, i.e., the set of points in $\mathbb{R}^{n-1}$ that come from omitting the i-coordinate from any point in R

We can now state the primary theorem for calculating high-dimensional integrals.

> **Theorem 3.4.18.** *Suppose that $\widehat{R}$ is a standard region in $\mathbb{R}^{n-1}$, and that p and q are continuous real-valued functions on $\widehat{R}$. Fix an integer i with $1 \leq i \leq n$. Let R be the subset of $\mathbb{R}^n$ such that $\mathbf{x} = (x_1, \ldots, x_n)$ is in R if and only if $\pi_i(\mathbf{x})$ is in $\widehat{R}$ and $p(\pi_i(\mathbf{x})) \leq x_i \leq q(\pi_i(\mathbf{x}))$.*
>
> *Then, R is a standard region and, for any continuous function f on R,*
>
> $$\int_R f \, dV^n \;=\; \int_{\widehat{R}} \left[\int_{p(\pi_i(\mathbf{x}))}^{q(\pi_i(\mathbf{x}))} f \, dx_i \right] dV^{n-1}.$$

See Theorem 5.6 of Fleming, [3]. Note that Fleming is using the Lebesgue integral, not the Riemann integral, but the two notions agree for continuous functions on standard regions.

Note that the inside integral on the right in the above formula yields a continuous function on $\widehat{R}$ by Theorem 3.1.9.

By iterating the result of Theorem 3.4.18, we can use iterated integrals to calculate Riemann integrals of continuous functions on *iterated regions*, as defined below.

> **Definition 3.4.19.** *We define an **iterated region** in $\mathbb{R}^1$ to be a closed, bounded interval $[a, b]$.*
>
> *For $n > 1$, a subset R of $\mathbb{R}^n$ is an **iterated region** if and only if there exists an i such that $\widehat{R} = \pi_i(R)$ is an iterated region in $\mathbb{R}^{n-1}$ and there exist continuous real-valued functions p and q on $\widehat{R}$ such that $\mathbf{x} = (x_1, \ldots, x_n)$ is in R if and only if $\pi_i(\mathbf{x})$ is in $\widehat{R}$ and $p(\pi_i(\mathbf{x})) \leq x_i \leq q(\pi_i(\mathbf{x}))$.*

You should convince yourself that, when $n = 3$, this definition of an iterated region does, in fact, agree with the definition that we gave in Definition 3.4.2.

Finally, we wish to state two important theorems about the Riemann integral. Though these theorems are important, they are not used often in explicit calculations, where we use iterated integrals and the Fundamental Theorem of Calculus.

> **Theorem 3.4.20. (Linearity)** *Suppose that a and b are constants, and that f and g are Riemann integrable on a region R in $\mathbb{R}^n$.*
>
> *Then, $af + bg$ is Riemann integrable on R, and*
>
> $$\int_R (af + bg) \, dV \;=\; a \int_R f \, dV \;+\; b \int_R g \, dV.$$

See parts 1 and 2 of Theorem 5.4 of Fleming, [3].

> **Theorem 3.4.21. (Monotonicity)** *Suppose that f and g are Riemann integrable on a region R in $\mathbb{R}^n$ and that, for all $\mathbf{x}$ in R, $f(\mathbf{x}) \leq g(\mathbf{x})$.*
> *Then,*
> $$\int_R f \, dV \ \leq \ \int_R g \, dV.$$

See part 4 of Theorem 5.4 of Fleming, [3].

3.4.1 Exercises

Online answers to select exercises are here.

Basics:

1. Calculate

$$\iiint_T (x + 3z^2) \, dV,$$

where T is the solid region in $\mathbb{R}^3$ given by $0 \leq x \leq 1$, $0 \leq y \leq x$, and $0 \leq z \leq xy$.

2. Calculate

$$\iiint_T (x + 3z^2) \, dV,$$

where T is the solid region in $\mathbb{R}^3$ given by $0 \leq y \leq 1$, $0 \leq z \leq y$, and $0 \leq x \leq yz$.

3. Calculate

$$\iiint_T 1 \, dV,$$

where T is the solid region in $\mathbb{R}^3$ given by $0 \leq x \leq 1$, $0 \leq y \leq x$, and $0 \leq z \leq xy$.

4. Calculate

$$\iiint_T 1 \, dV,$$

where T is the solid region in $\mathbb{R}^3$ given by $0 \leq y \leq 1$, $0 \leq z \leq y$, and $0 \leq x \leq yz$.

5. Calculate

$$\iiint_T 4z e^{x-y} \, dV,$$

where T is the solid region in $\mathbb{R}^3$ given by $-1 \leq z \leq 1$, $z \leq x \leq 1$, and $z \leq y \leq x$.

6. Calculate

$$\iiint_T (3x^2 + 4y^3 + 5z^4) \, dV,$$

where T is the solid region in $\mathbb{R}^3$ given by $0 \leq x \leq 2$, $0 \leq y \leq x^2$, and $0 \leq z \leq x+y$.

7. Calculate

$$\iiint_T yze^{2x}\, dV,$$

where T is the solid region in $\mathbb{R}^3$ given by $0 \le x \le 1$, $3 \le z \le 5$, and $-4 \le y \le 2$.

8. Calculate

$$\iiint_T x^2 yz\, dV,$$

where T is the solid region in $\mathbb{R}^3$ given by $0 \le x \le 1$, $x^2 \le z \le 5$, and $x^3 \le y \le 2$.

9. Calculate

$$\iiint_T y^2 z \cos(xz)\, dV,$$

where T is the solid region in $\mathbb{R}^3$ given by $-2 \le y \le -1$, $0 \le z \le \pi/2$, and $y \le x \le 1$.

10. Calculate

$$\iiint_T \frac{xyz}{(x^2 + y^2 + z^2)^{3/2}}\, dV,$$

where T is the solid region in $\mathbb{R}^3$ given by $0 \le z \le 2$, $0 \le y \le z$, and $y \le x \le z$.

In each of the following exercises, you are given an iterated integral of the form

$$\int_a^b \int_{u(x)}^{v(x)} \int_{p(x,y)}^{q(x,y)} f(x, y, z)\, dz\, dy\, dx.$$

This iterated integral calculates the value of a triple integral $\iiint_T f\, dV$ over a solid region T in $\mathbb{R}^3$, which is also equal to a double integral

$$\iint_R \left[\int_{p(x,y)}^{q(x,y)} f(x, y, z)\, dz \right] dA$$

over the projected region R in the xy-plane. (a) Calculate the iterated integral. (b) Sketch the projected region R. (c) By hand, or using technology, sketch the solid region T.

11. $\displaystyle\int_1^2 \int_2^4 \int_0^{25 - x^2 - y^2} 1\, dz\, dy\, dx$

12. $\displaystyle\int_1^2 \int_2^{2x} \int_0^{25 - x^2 - y^2} 1\, dz\, dy\, dx$

13. $\displaystyle\int_0^2 \int_0^{\sqrt{4 - x^2}} \int_0^y x\, dz\, dy\, dx$

14. $\displaystyle\int_{-2}^{2}\int_{-1}^{3-x^2}\int_{-4}^{4} xe^{z/4}\, dz\, dy\, dx$

15. $\displaystyle\int_{-2}^{2}\int_{-1}^{3-x^2}\int_{0}^{\sqrt{16-x^2-y^2}} \frac{1}{\sqrt{16-x^2-y^2}}\, dz\, dy\, dx$

16. $\displaystyle\int_{0}^{1}\int_{e^x}^{e}\int_{-y}^{y} z\, dz\, dy\, dx$

Figure 3.4.7: The solid tetrahedron T.

In each of the following exercises, you are given a function $f = f(x, y, z)$, and a solid region T is described and sketched. Calculate the value of $\iiint_T f\, dV$: (a) by first projecting T into the xy-plane, and (b) by first projecting T into the xz-plane.

17. $f(x, y, z) = x$, and T is the solid tetrahedron in the 1st octant, bounded by the three coordinate planes and the plane where $x + 2y + z = 3$. See Figure 3.4.7.

18. $f(x, y, z) = 12yz$, and T is the solid half-cone, bounded by the graph of $z = \sqrt{x^2 + y^2}$ and the plane where $z = 5$. See Figure 3.4.8.

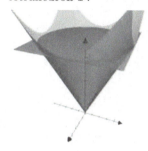

Figure 3.4.8: The solid half-cone T.

19. $f(x, y, z) = 1$, and T is the solid which is bounded by the graph of $y = 4 - x^2 - z^2$ and the plane where $y = 0$. See Figure 3.4.9.

20. $f(x, y, z) = e^y$, and T is the solid which is bounded by the graph of $z = x^2 + 1$ and the planes where $y = 1$, $y = 2$, and $z = 3$. See Figure 3.4.10.

21. Write a brief essay, explaining, non-technically, the definition of the definite integral $\iiint_T f(x, y, z)\, dV$.

22. Consider a definite integral $\iiint_T f(x, y, z)\, dV$.

Figure 3.4.9: The solid T between $y = 4 - x^2 - z^2$ and $y = 0$.

    (a) If $f(x, y, z) = 1$, what does this integral give you?

    (b) If $f(x, y, z)$ is the density of a solid object at the point (x, y, z), what does this integral give you?

More Depth:

In each of the following exercises, you are given a function $f = f(x, y, z)$ and a solid box $T = [a_1, b_1] \times [a_2, b_2] \times [a_3, b_3]$. (a) Calculate $\iiint_T f\, dV$. (b) Partition each of the subintervals $[a_i, b_i]$ into two subintervals of equal length to obtain a partition of T into eight sub-boxes. Calculate the Riemann sum for f which uses this partition and in which the sample points are chosen to be the centers of the sub-boxes.

23. $f(x, y, z) = 8$, and $S = [0, 2] \times [1, 5] \times [2, 8]$.

Figure 3.4.10: The solid T between $z = x^2 + 1$, $y = 1$ $y = 2$, and $z = 3$.

24. $f(x, y, z) = 8xyz$, and $S = [0, 2] \times [1, 5] \times [2, 8]$.

25. $f(x, y, z) = xy + z^2$, and $S = [-2, 2] \times [-8, 8] \times [-4, 4]$.

26. $f(x, y, z) = xe^y \cos z$, and $S = [1, 5] \times [0, 2] \times [-\pi, \pi]$.

In each of the following exercises, you are given a function $f = f(w, x, y, z)$ and an iterated region R in $\mathbb{R}^4$. Calculate $\int_R f \, dV$.

27. $f(w, x, y, z) = 2wx + 4yz$, and R is the region where $0 \leq w \leq 1$, $0 \leq x \leq w$, $0 \leq y \leq x$ and $0 \leq z \leq y$.

28. $f(w, x, y, z) = 1$, and R is the region where $0 \leq w \leq 2$, $w/2 \leq x \leq 1$, $0 \leq y \leq x+w$ and $0 \leq z \leq x^2 y$.

29. $f(w, x, y, z) = \dfrac{y}{1 + z^2}$, and R is the region where $0 \leq w \leq 1$, $0 \leq x \leq 1$, $0 \leq y \leq x$ and $-w \leq z \leq w$.

30. $f(w, x, y, z) = \sqrt{w + 2}$, and R is the region where $-2 \leq w \leq 0$, $-2 \leq x \leq w$, $-2 \leq y \leq x$ and $-2 \leq z \leq y$.

3.5 Volume

One of the most basic applications of multivariable integration is to find volumes of solid regions by using either double or triple integrals. However, as we shall see, the triple integral point of view leads to more general applications later.

Since we defined and looked at examples of double and triple integrals in the previous two sections, we can leap right into finding volumes of solid regions in space.

Example 3.5.1. Find the volume of the solid region T under the graph of the hyperbolic paraboloid given by

$$z = 3 + \frac{x^2}{4} - \frac{y^2}{9}$$

and above the rectangle in the xy-plane where $0 \le x \le 2$ and $0 \le y \le 3$. See Figure 3.5.1.

Solution:

In terms of triple integrals, you should simply think that the volume of T is the continuous sum of all of the infinitesimal volumes dV as you move through the points of T.

Thus,

$$\text{volume of } T \;=\; \iiint_T 1\, dV \;=\; \iiint_T dV.$$

By definition of the solid region T, the projected region in the xy-plane is the rectangle $R = [0, 2] \times [0, 3]$, and for each point $\mathbf{p}$ in R, the points in T that lie over $\mathbf{p}$, i.e., those points (x, y, z) in T whose projection is $\mathbf{p}$, satisfy

$$0 \le z \le 3 + \frac{x^2}{4} - \frac{y^2}{9}.$$

Therefore,

$$\text{volume of } T \;=\; \iiint_T dV \;=\; \iint_R \left[\int_0^{3 + \frac{x^2}{4} - \frac{y^2}{9}} dz \right] dA \;=$$

$$\iint_R \left(3 + \frac{x^2}{4} - \frac{y^2}{9} - 0 \right) dA.$$

Figure 3.5.1: The solid region T.

Notice that this last double integral is precisely what you'd get if you had initially decided to calculate the volume using rectangular solids that have an infinitesimal base in the xy-plane, but a finite height given by the top z-coordinate minus the bottom z-coordinate, i.e., of height

$$\left(3 + \frac{x^2}{4} - \frac{y^2}{9}\right) - 0.$$

Then you would write

$$dV = (\text{height})\, dA = \left(3 + \frac{x^2}{4} - \frac{y^2}{9} - 0\right) dA,$$

and integrate

$$\iint_R \left(3 + \frac{x^2}{4} - \frac{y^2}{9} - 0\right) dA,$$

precisely as above.

By either approach, the resulting double integral over a rectangle is simple:

$$\iint_R \left(3 + \frac{x^2}{4} - \frac{y^2}{9}\right) dA = \int_0^2 \int_0^3 \left(3 + \frac{x^2}{4} - \frac{y^2}{9}\right) dy\, dx =$$

$$\int_0^2 \left[3y + \frac{x^2 y}{4} - \frac{y^3}{27}\, \Big|_{y=0}^{y=3}\right] dx = \int_0^2 \left(8 + \frac{3x^2}{4}\right) dx =$$

$$8x + \frac{x^3}{4}\, \Big|_0^2 = 18.$$

Remark 3.5.2. Does it matter whether you approach volume problems, like the previous one, with a triple integral of 1 over the solid region or a double integral of the height over the projected region? Yes and no.

If the volume can be calculated by a triple integral, via an iterated integral involving 3 partial integrals, then the volume can be calculated just as easily via a double integral of the height over the projected region. So, for the calculation of volume, it doesn't really matter which approach you take.

However, for other applications, like the mass of an object with variable density, it is frequently the case that you are integrating a function that depends on x, y, and z, like $\delta(x, y, z)$ and you need to multiply this by $dV = dz\, dy\, dx$. If you look at dV as being $h(x, y)\, dy\, dx$, where h is the height, then the function that you're integrating needs to depend only on x and y, not on z.

The point is that the triple integral approach to calculating volume is no more difficult than the double integral approach, and is more versatile for future applications.

Example 3.5.3. Find the volume of the solid region T between the planes given by $z = x + y$ and $z = 1 + 2x + 3y$, and above the triangle in the xy-plane which has vertices $(1, 0, 0)$, $(0, 1, 0)$, and $(2, 1, 0)$. The planes are shown in Figure 3.5.2.

Solution:

Let R denote the projection of T into the xy-plane; this is just the specified triangle. See Figure 3.5.3. Note that, for (x, y) in R, the plane where $z = 1 + 2x + 3y$ is always above the plane where $z = x + y$. Also, note that R is bounded by the lines where $y = 1$, $x + y = 1$, and $y = x - 1$.

Thus,

$$\text{volume} \;=\; \iiint_T dV \;=\; \iint_R \left[\int_{x+y}^{1+2x+3y} dz \right] dA.$$

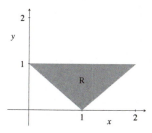

Figure 3.5.2: The planes where $z = x + y$ and $z = 1 + 2x + 3y$.

The region R is where $0 \le y \le 1$ and the x-coordinates always start on the line $x + y = 1$, i.e., start at $x = 1 - y$, and go to the x-coordinate on $y = x - 1$, i.e., go to $x = 1 + y$.

Therefore,

$$\text{volume} \;=\; \iint_R \left[\int_{x+y}^{1+2x+3y} dz \right] dA \;=\; \int_0^1 \int_{1-y}^{1+y} \int_{x+y}^{1+2x+3y} dz \, dx \, dy \;=\;$$

$$\int_0^1 \int_{1-y}^{1+y} \big((1 + 2x + 3y) - (x + y) \big) \, dx \, dy \;=\; \int_0^1 \int_{1-y}^{1+y} (1 + x + 2y) \, dx \, dy \;=\;$$

$$\int_0^1 \left[x + \frac{x^2}{2} + 2yx \, \Big|_{x=1-y}^{x=1+y} \right] dy \;=\;$$

$$\int_0^1 \left[\left((1 + y) + \frac{(1 + y)^2}{2} + 2y(1 + y) \right) - \left((1 - y) + \frac{(1 - y)^2}{2} + 2y(1 - y) \right) \right] dy \;=\;$$

$$\int_0^1 (4y + 4y^2) \, dy \;=\; 4 \left(\frac{y^2}{2} + \frac{y^3}{3} \right) \Big|_0^1 \;=\; \frac{10}{3}.$$

Figure 3.5.3: The projected triangle R.

Example 3.5.4. Consider the solid wedge T bounded by the trigonometric cylinder given by $z = \cos x$ and the planes given by $y = 0$ and $z = y$, for $-\pi/2 \le x \le \pi/2$. See Figure 3.5.4.

Figure 3.5.4: The solid wedge T.

Set up an iterated integral for calculating the triple integral of any continuous function $f(x, y, z)$ over the solid region T, and then let the function be the constant function 1 to calculate the volume.

Solution:

While we could project the region T into a different coordinate plane, the easiest, most obvious, projection is into the xz-plane, where we obtain the region R shown in Figure 3.5.5. Thus, R is the region in the xz-plane given by $-\pi/2 \leq x \leq \pi/2$ and $0 \leq z \leq \cos x$.

For each (x, z) in R, what do the y-coordinates of the corresponding points in T do? They start where $y = 0$ and go to the plane where $y = z$.

Therefore,

$$\iiint_T f(x, y, z)\, dV = \int_{-\pi/2}^{\pi/2} \int_0^{\cos x} \int_0^z f(x, y, z)\, dy\, dz\, dx,$$

and so, the volume of T is

$$\int_{-\pi/2}^{\pi/2} \int_0^{\cos x} \int_0^z dy\, dz\, dx = \int_{-\pi/2}^{\pi/2} \int_0^{\cos x} z\, dz\, dx =$$

$$\int_{-\pi/2}^{\pi/2} \left[\frac{z^2}{2} \Big|_{z=0}^{z=\cos x} \right] dx = \int_{-\pi/2}^{\pi/2} \frac{\cos^2 x}{2}\, dx =$$

$$\frac{1}{2} \int_{-\pi/2}^{\pi/2} \frac{1 + \cos(2x)}{2}\, dx = \frac{1}{4} \left(x + \frac{\sin(2x)}{2} \right) \Big|_{-\pi/2}^{\pi/2} = \frac{\pi}{4}.$$

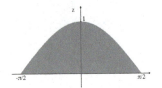

Figure 3.5.5: The projected region R.

Online answers to select exercises are here.

3.5.1   Exercises

Basics:

In each of the following exercises: (a) Sketch, or have technology sketch, the solid region T described by the given inequalities. (b) Find the volume of T.

1. T is the solid region in $\mathbb{R}^3$ given by $0 \leq x \leq 1$, $0 \leq y \leq 1$, and $0 \leq z \leq 4 - x^2 - y^2$.

2. T is the solid region in $\mathbb{R}^3$ given by $0 \leq x \leq 1$, $0 \leq y \leq 1$, and $-1 \leq z \leq x^2 - y^2 + 2$.

3. T is the solid region in $\mathbb{R}^3$ given by $0 \leq y \leq 1$, $0 \leq x \leq y$, and $0 \leq z \leq y^2 - x^2 + 2$.

4. T is the solid region in $\mathbb{R}^3$ given by $0 \leq x \leq 2$, $0 \leq y \leq x^2$, and $0 \leq z \leq x + y$.

5. T is the solid region in $\mathbb{R}^3$ given by $-1 \leq z \leq 1$, $z \leq x \leq 1$, and $-1 \leq y \leq x$.

6. T is the solid region in $\mathbb{R}^3$ given by $0 \leq y \leq 1$, $y^2 \leq z \leq 5$, and $y^3 \leq x \leq 2$.

7. T is the solid region in $\mathbb{R}^3$ given by $\pi/2 \leq y \leq \pi$, $0 \leq x \leq \sin y$, and $0 \leq z \leq y$.

8. T is the solid region in $\mathbb{R}^3$ given by $0 \leq x \leq 1$, $0 \leq y \leq e^x$, and $-2 \leq z \leq 5$.

In each of the following exercises, you are given an iterated integral of the form

$$\int_a^b \int_{u(x)}^{v(x)} \int_{p(x,y)}^{q(x,y)} 1 \, dz \, dy \, dx \;=\; \int_a^b \int_{u(x)}^{v(x)} \int_{p(x,y)}^{q(x,y)} dz \, dy \, dx.$$

This iterated integral calculates the volume of a solid region T in $\mathbb{R}^3$, which is also equal to a double integral

$$\iint_R \left[\int_{p(x,y)}^{q(x,y)} dz \right] dA$$

over the projected region R in the xy-plane. (a) Calculate the volume of T. (b) Sketch the projected region R. (c) By hand, or using technology, sketch the solid region T.

9. $\displaystyle\int_1^2 \int_2^4 \int_0^{25-x^2-y^2} dz \, dy \, dx$

10. $\displaystyle\int_1^2 \int_2^{2x} \int_0^{25-x^2-y^2} dz \, dy \, dx$

11. $\displaystyle\int_0^2 \int_0^{\sqrt{4-x^2}} \int_0^y dz \, dy \, dx$

12. $\displaystyle\int_{-2}^2 \int_{-1}^{3-x^2} \int_{-4}^4 dz \, dy \, dx$

13. $\displaystyle\int_0^4 \int_0^x \int_0^{\sqrt{16-x^2}} dz \, dy \, dx$

14. $\displaystyle\int_0^1 \int_{e^x}^e \int_{-y}^y dz \, dy \, dx$

In each of the following exercises, a solid region T is described and sketched. Calculate the volume of T.

15. T is the solid tetrahedron in the 1st octant, bounded by the three coordinate planes and the plane where $x + 2y + z = 3$. See Figure 3.5.6.

Figure 3.5.6: The solid tetrahedron T.

16. T is the solid which is bounded by the graph of $z = x^2 + 1$ and the planes where $y = 1$ $y = 2$, and $z = 3$. See Figure 3.5.7.

17. T is the solid which is bounded by the graph of $z = x^2 + 1$ and the planes where $z = 2y - 1$, and $y = 2$. See Figure 3.5.8.

18. T is the solid which is below the graph of $z = 1 - x^2 - y^2$ and above the triangle with vertices $(0, 0, 0)$, $(1, 0, 0)$, and $(0, 1, 0)$. See Figure 3.5.9.

Figure 3.5.7: The solid T between $z = x^2 + 1$, $y = 1$ $y = 2$, and $z = 3$.

Figure 3.5.8: The solid T between $z = x^2 + 1$, $z = 2y - 1$, and $y = 2$.

Figure 3.5.9: The solid T under $z = 1 - x^2 - y^2$ and above a triangle.

3.6   Integration with Cylindrical and Spherical Coordinates

In this section, we describe, and give examples of, computing triple integrals in the cylindrical coordinates r, θ, and z, and in spherical coordinates ρ, ϕ, and θ.

In the More Depth portion of this section, we will address how you integrate in $\mathbb{R}^3$ or, more generally, in $\mathbb{R}^n$, using any C^1 change of coordinates.

Just as some double integrals don't look very nice in terms of the Cartesian coordinates x and y, many triple integrals don't look particularly nice in terms of x, y, and z. There are two other standard sets of coordinates that are used in space: *cylindrical coordinates* and *spherical coordinates*.

Cylindrical Coordinates

Cylindrical coordinates are easy, given that we already know about polar coordinates in the xy-plane from Section 3.3. Recall that in the context of multivariable integration, we always assume that $r \geq 0$.

Cylindrical coordinates for $\mathbb{R}^3$ are simply what you get when you use polar coordinates r and θ for the xy-plane, and just let z be z. Therefore, we still have that $r = \sqrt{x^2 + y^2}$, but now r is not the distance from the origin; it's the distance from the z-axis.

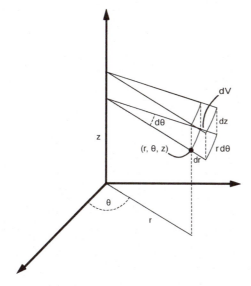

Figure 3.6.1: In cylindrical coordinates, $dV = r\,dr\,d\theta\,dz$.

Our expression for the volume element dV is also easy now; since $dV = dz\, dA$, and $dA = r\, dr\, d\theta$ in polar coordinates, we find that $dV = dz\, r\, dr\, d\theta = r\, dz\, dr\, d\theta$ in cylindrical coordinates.

Thus, to integrate, you use:

Integration in Cylindrical Coordinates:

To perform triple integrals in cylindrical coordinates, and to switch from cylindrical coordinates to Cartesian coordinates, you use:

$$x = r\cos\theta, \qquad y = r\sin\theta, \qquad z = z, \qquad \text{and} \qquad dV = dz\, dA = r\, dz\, dr\, d\theta.$$

Example 3.6.1. Find the volume of the solid region T which is above the half-cone given by $z = \sqrt{x^2 + y^2}$ and below the hemisphere where $z = \sqrt{8 - x^2 - y^2}$.

Solution:

Note that, in cylindrical coordinates, the half-cone is given by $z = \sqrt{r^2} = r$ and the hemisphere is given by $z = \sqrt{8 - r^2}$.

To find the volume, we need to calculate $\iiint_T dV$.

Figure 3.6.2: The "snow cone" T and the projected region R.

The projected region R in the xy-plane, or $r\theta$-plane, is the inside of the circle (thought of as lying in a copy of the xy-plane) along which the two surfaces intersect. To find this circle, we set the two z's equal to each other and find

$$r = \sqrt{8 - r^2}, \qquad \text{or, equivalently,} \qquad r^2 = 8 - r^2.$$

We find

$$2r^2 = 8, \qquad \text{so} \qquad r^2 = 4, \qquad \text{and, hence,} \qquad r = 2.$$

Thus, R is the disk in the xy-plane where $r \leq 2$.

For each point $\mathbf{p}$ in R, the corresponding points which lie over it in the solid region T have z-coordinates which start on the half-cone where $z = r$ and end on the hemisphere where $z = \sqrt{8 - r^2}$.

Therefore,

$$\iiint_T dV = \iint_R \left[\int_r^{\sqrt{8-r^2}} dz \right] dA = \int_0^{2\pi} \int_0^2 \int_r^{\sqrt{8-r^2}} r\, dz\, dr\, d\theta =$$

$$\int_0^{2\pi} \int_0^2 \left[rz \, \Big|_{z=r}^{z=\sqrt{8-r^2}} \right] dr\, d\theta = \int_0^{2\pi} \int_0^2 \left(r\sqrt{8-r^2} - r^2 \right) dr\, d\theta.$$

The inner integral is easy, via the substitution $u = 8 - r^2$ in the first term. We obtain

$$\text{volume} \;=\; \iiint_T dV \;=\; \int_0^{2\pi} \frac{16}{3}\left(\sqrt{2}-1\right) d\theta \;=\; \frac{32\pi}{3}\left(\sqrt{2}-1\right).$$

Example 3.6.2. Let R be the region in the xy-plane, or $r\theta$-plane, which is bounded by the curves given by $r = 1 + \theta^2$ and $r = 1 + \theta + \theta^2$, for $0 \le \theta \le \pi$.

Integrate the function $f(x,y) = 1/\sqrt{x^2 + y^2}$ over the solid region T which lies above the region R and is bounded by the plane where $z = 1$ and the half-cone where $z = 1 + 2\sqrt{x^2 + y^2}$.

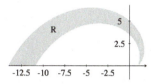

Figure 3.6.3: The spiraling plane region R.

Solution: The problem is given in a mixture of cylindrical and Cartesian coordinates, but the region R is so clearly set up for nice integration in polar coordinates that it should be obvious that you want to use cylindrical coordinates for space.

In Figure 3.6.3 and Figure 3.6.4, we show the region R and the plane and cone. It is difficult to sketch the solid region T, but, fortunately, there's no need to do so. After noting that $f = 1/r$ and that the cone is given by $z = 1 + 2r$, we can go ahead and calculate:

$$\iiint_T \frac{1}{\sqrt{x^2 + y^2}}\, dV \;=\; \iint_R \left[\int_1^{1+2r} \frac{1}{r}\, dz\right] dA \;=$$

$$\int_0^\pi \int_{1+\theta^2}^{1+\theta+\theta^2} \int_1^{1+2r} \frac{1}{r}\, dz\, r\, dr\, d\theta \;=\; \int_0^\pi \int_{1+\theta^2}^{1+\theta+\theta^2} \int_1^{1+2r} dz\, dr\, d\theta \;=$$

$$\int_0^\pi \int_{1+\theta^2}^{1+\theta+\theta^2} 2r\, dr\, d\theta \;=\; \int_0^\pi \left[r^2\, \Big|_{r=1+\theta^2}^{r=1+\theta+\theta^2}\right] d\theta \;=\; \int_0^\pi \left((1+\theta+\theta^2)^2 - (1+\theta^2)^2\right) d\theta \;=$$

$$\int_0^\pi \theta(2 + \theta + 2\theta^2)\, d\theta \;=\; \int_0^\pi (2\theta + \theta^2 + 2\theta^3)\, d\theta \;=\; \frac{\pi^2}{6}\left(6 + 2\pi + 3\pi^2\right).$$

Figure 3.6.4: The plane and cone which form the vertical bounds of the solid region T.

Spherical Coordinates

There is a third common set of coordinates for $\mathbb{R}^3$, other than the Cartesian coordinates x, y, and z, or the cylindrical coordinates r, θ, and z. It is sometimes convenient to use *spherical coordinates* ρ, θ, and ϕ.

Pick a point $\mathbf{p}$ in space, other than the origin, and draw the line segment L from the origin to $\mathbf{p}$. We let ρ denote the length of L, i.e., the distance from the origin to $\mathbf{p}$. We orthogonally project L into the xy-plane, and let θ be the angle, $0 \le \theta < 2\pi$, between the positive x-axis and this projected line segment. If the Cartesian coordinates of $\mathbf{p}$ are (x, y, z), then θ is precisely the polar coordinate angle of the point (x, y). Note that we also have that $\rho^2 = x^2 + y^2 + z^2$. Finally, we let ϕ, where $0 \le \phi \le \pi$, be the angle between the positive z-axis and L. See Figure 3.6.5.

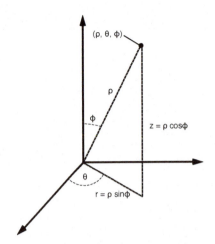

Figure 3.6.5: A point with spherical coordinates (ρ, θ, ϕ).

Figure 3.6.5 makes it clear that the polar coordinate r of the point (x, y) is $\rho \sin \phi$, and that $z = \rho \cos \phi$.

In order to obtain an expression for the infinitesimal volume element dV in spherical coordinates, we need to include the infinitesimal changes in ρ, θ, and ϕ; this makes for the much more complicated diagram in Figure 3.6.6.

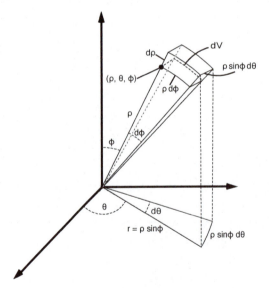

Figure 3.6.6: In spherical coordinates, $dV = \rho^2 \sin \phi \, d\rho \, d\phi \, d\theta$.

<table>
<tr><td>In the More Depth portion, we shall derive the formula for dV in spherical coordinates, or in any coordinates, in a more analytic way.</td></tr>
</table>

In the diagram, we see that the volume element is given, in spherical coordinates, by

$$dV = \rho^2 \sin \phi \, d\rho \, d\phi \, d\theta.$$

Thus, to integrate in spherical coordinates, you use:

Integration in Spherical Coordinates:

To perform triple integrals in spherical coordinates, and to switch from spherical coordinates to cylindrical or Cartesian coordinates, you use:

$$r = \rho \sin \phi;$$

$$x = r \cos \theta = \rho \sin \phi \cos \theta;$$

$$y = r \sin \theta = \rho \sin \phi \sin \theta;$$

$$z = \rho \cos \phi;$$

and

$$dV = \rho^2 \sin \phi \, d\rho \, d\phi \, d\theta,$$

where $\rho \geq 0$, $0 \leq \theta < 2\pi$, and $0 \leq \phi \leq \pi$.

Figure 3.6.7: Graph of $\phi = c$, where $0 < c < \pi/2$.

Figure 3.6.8: Graph of $\phi = c$, where $\pi/2 < c < \pi$.

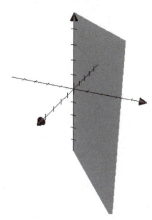

Figure 3.6.9: Graph of $\theta = c$.

Remark 3.6.3. We also make frequent use of the fact that $\rho^2 = x^2 + y^2 + z^2$. This follows from the definition of ρ and the Pythagorean Theorem, but it is a good exercise to take the expressions for x, y and z in spherical coordinates, square them, add them, and use the Fundamental Trigonometric Identity (twice) to verify that you get ρ^2.

Note that $\rho = 0$ corresponds to exactly one point: the origin. What are θ and ϕ for the origin? Anything at all, as long as $0 \leq \theta < 2\pi$, and $0 \leq \phi \leq \pi$.

While $\rho = 0$ describes a single point, if c is a constant, greater than 0, then $\rho = c$ describes a sphere of radius c, centered at the origin.

The equation $\phi = 0$ describes the positive z-axis (plus the origin), while $\phi = \pi$ describes the negative z-axis (plus the origin). The equation $\phi = \pi/2$ describes the xy-plane. If $0 < c < \pi/2$, then $\phi = c$ describes a right circular cone, surrounding the positive z-axis. If $\pi/2 < c < \pi$, then $\phi = c$ describes a right circular cone, surrounding the negative z-axis.

If c is a constant, the equation $\theta = c$ describes a half-plane, which is perpendicular to the xy-plane, and which extends outward from the z-axis at an angle θ with the positive x-axis. In other words, in the xy-plane, you take a ray, which makes an angle of θ with the positive x-axis, and then you move this ray up and down the z-axis to form a plane which gets chopped off on one side by the z-axis.

Example 3.6.4. Integrate 1 over the 1st octant portion of ball of radius R, centered at the origin, to obtain its volume. Verify that you obtain 1/8 of $4\pi R^3/3$, i.e., $\pi R^3/6$.

Solution:

This is really easy in spherical coordinates. Let's call the solid region T. Clearly, the region T is described in spherical coordinates by letting ρ go from 0 to R, θ go from 0 to $\pi/2$, and ϕ go from 0 to $\pi/2$. We calculate

$$\text{volume} \;=\; \iiint_T dV \;=\; \int_0^{\pi/2} \int_0^{\pi/2} \int_0^{R} \rho^2 \sin\phi \, d\rho \, d\phi \, d\theta \;=\;$$

$$\int_0^{\pi/2} \int_0^{\pi/2} \left[\frac{\rho^3 \sin\phi}{3} \, \Big|_{\rho=0}^{\rho=R} \right] d\phi \, d\theta \;=\; \frac{R^3}{3} \int_0^{\pi/2} \int_0^{\pi/2} \sin\phi \, d\phi \, d\theta \;=\;$$

$$\frac{R^3}{3} \int_0^{\pi/2} \left[-\cos\phi \, \Big|_0^{\pi/2} \right] d\theta \;=\; \frac{R^3}{3} \int_0^{\pi/2} 1 \, d\theta \;=\; \frac{\pi R^3}{6}.$$

Figure 3.6.10: An eighth of a ball.

Example 3.6.5. Integrate z over the 1st octant portion of ball of radius R, centered at the origin.

Solution:

This is the solid region T from the previous example. As before, the region T is described in spherical coordinates by letting ρ go from 0 to R, θ go from 0 to $\pi/2$, and ϕ go from 0 to $\pi/2$.

We calculate

$$\iiint_T z \, dV \;=\; \int_0^{\pi/2} \int_0^{\pi/2} \int_0^{R} \rho\cos\phi \cdot \rho^2 \sin\phi \, d\rho \, d\phi \, d\theta \;=\;$$

$$\frac{R^4}{4} \int_0^{\pi/2} \int_0^{\pi/2} \sin\phi\cos\phi \, d\phi \, d\theta \;=\; \frac{R^4}{4} \int_0^{\pi/2} \left[\frac{\sin^2\phi}{2} \, \Big|_0^{\pi/2} \right] d\theta \;=\; \frac{\pi R^4}{16}.$$

Example 3.6.6. Let's redo the problem from Example 3.6.1. We'll find the volume of the solid region T which is above the half-cone given by $z = \sqrt{x^2 + y^2}$ and below the hemisphere where $z = \sqrt{8 - x^2 - y^2}$, but, this time, we'll use spherical coordinates.

It should be clear that ρ goes from 0 to $\sqrt{8}$, and that θ goes all the way around, i.e., from 0 to 2π. But, what about ϕ?

Some people may see quickly that the cone where $z = \sqrt{x^2 + y^2}$ is precisely where $\phi = \pi/4$, and so ϕ goes from 0 to $\pi/4$. However, if it isn't so obvious that this cone is where $\phi = \pi/4$, how do you figure it out?

Figure 3.6.11: The "snow cone" T from Example 3.6.1.

You just switch $z = \sqrt{x^2 + y^2}$ into spherical coordinates, passing through cylindrical coordinates along the way. In cylindrical coordinates, $r = \sqrt{x^2 + y^2}$; so our equation becomes $z = r$. Now, recalling Figure 3.6.5, we use that $z = \rho \cos \phi$, and that $r = \rho \sin \phi$, so that our equation becomes

$$\rho \cos \phi \;=\; \rho \sin \phi,$$

and so, either $\rho = 0$ and we're at the origin, or $\cos \phi = \sin \phi$, which means $\phi = \pi/4$. We don't have to worry about $\rho = 0$ as a separate case, since the equation $\phi = \pi/4$ already includes the possibility that $\rho = 0$ (since the origin has every possible θ and ϕ).

Now that we know that our cone is where $\phi = \pi/4$, we find that the volume of our snow cone is

$$\text{volume} \;=\; \int_0^{2\pi} \int_0^{\pi/4} \int_0^{\sqrt{8}} \rho^2 \sin \phi \, d\rho \, d\phi \, d\theta \;=\; \frac{8\sqrt{8}}{3} \int_0^{2\pi} \int_0^{\pi/4} \sin \phi \, d\phi \, d\theta \;=\;$$

$$\frac{8\sqrt{8}}{3} \int_0^{2\pi} \left[-\cos \phi \, \Big|_0^{\pi/4} \right] d\theta \;=\; \frac{16\sqrt{2}}{3} \cdot 2\pi \cdot \left(-\frac{1}{\sqrt{2}} + 1 \right) \;=\; \frac{32\pi}{3}(\sqrt{2} - 1),$$

as we found in Example 3.6.1.

+ Linear Algebra:

In the More Depth portion of Section 3.3, we gave a theorem which tells us how changes of coordinates affect integration in $\mathbb{R}^2$. That theorem generalizes in an obvious way to integration in $\mathbb{R}^n$ for any $n \geq 1$ and, in particular, applies to the spherical change of coordinates on $\mathbb{R}^3$.

To understand the statement in $\mathbb{R}^n$, you need to know about determinants of $n \times n$ matrices, which is why this section requires a bit of linear algebra beyond what you may know. Of course, if you care only about $n = 1, 2, 3$, then you're fine without knowing about determinants of larger matrices; you already know how to take determinants of square matrices of those sizes.

Theorem 3.6.7. *Suppose that $\mathbf{\Phi}$ is a C^1 function from an open neighborhood of a point $\mathbf{p}$ in $\mathbb{R}^n$ into $\mathbb{R}^n$, and suppose that $\det\left[d_{\mathbf{p}}\mathbf{\Phi}\right] \neq 0$.*

Then, $(x_1, \ldots, x_n) = \mathbf{\Phi}(u_1, \ldots, u_n)$ is a C^1 local change of coordinates at $\mathbf{p}$ and, near $\mathbf{p}$, the n-dimensional volume element $dV^n = dx_1 dx_2 \cdots dx_n$ is given by

$$dV^n \;=\; \Big| \det\left[d\mathbf{\Phi}\right] \Big| \, du_1 du_2 \cdots du_n.$$

This follows from Theorem 7.3.8 of Trench, [8].

Thus, if $f = f(\mathbf{x})$ is Riemann integrable on a region R, contained in an open subset $\mathcal{U}$ of $\mathbb{R}^n$ and Φ is a C^1 change of coordinates from an open set $\mathcal{W}$ in $\mathbb{R}^n$ onto $\mathcal{U}$, then $f \circ \Phi$ is Riemann integrable on the region $T = \Phi^{-1}(R)$ and

$$\int_R f \, dV^n \;=\; \int_T (f \circ \Phi) \cdot \left| \det [d\Phi] \right| \, du_1 du_2 \cdots du_n.$$

Let's apply Theorem 3.6.7 to the case where $n = 3$ and our change of coordinates is given by spherical coordinates, i.e., where

$$x = \rho \sin \phi \cos \theta, \qquad y = \rho \sin \phi \sin \theta, \qquad \text{and} \qquad z = \rho \cos \phi,$$

so that

$$(x, y, z) \;=\; \Phi(\rho, \theta, \phi) \;=\; (\rho \sin \phi \cos \theta, \; \rho \sin \phi \sin \theta, \; \rho \cos \phi).$$

Theorem 3.6.7 tells us that to calculate dV in spherical coordinates, we need to calculate $\left| \det [d\Phi] \right|$. We find

$$\left| \det [d\Phi] \right| = \left| \det \begin{bmatrix} \sin \phi \cos \theta & -\rho \sin \phi \sin \theta & \rho \cos \phi \cos \theta \\ \sin \phi \sin \theta & \rho \sin \phi \cos \theta & \rho \cos \phi \sin \theta \\ \cos \phi & 0 & -\rho \sin \phi \end{bmatrix} \right| =$$

$$\left| -\rho^2 \sin^3 \phi \cos^2 \theta - \rho^2 \sin \phi \cos^2 \phi \sin^2 \theta - \rho^2 \sin \phi \cos^2 \phi \cos^2 \theta - \rho^2 \sin^3 \phi \sin^2 \theta \right| =$$

$$\rho^2 \sin \phi \sin^2 \phi (\cos^2 \theta + \sin^2 \theta) + \rho^2 \sin \phi \cos^2 \phi (\sin^2 \theta + \cos^2 \theta) =$$

$$\rho^2 \sin \phi (\sin^2 \phi + \cos^2 \phi) \;=\; \rho^2 \sin \phi.$$

Therefore, we recover what we already knew:

$$dV \;=\; dx \, dy \, dz \;=\; \rho^2 \sin \phi \, d\rho \, d\phi \, d\theta.$$

Online answers to select exercises are here.

3.6.1   Exercises

Basics:

In each of the following exercises, you are given a function $f = f(r, \theta, z)$, and an iterated description, in cylindrical coordinates, of a solid region T. Calculate $\iiint_T f \, dV$, using cylindrical coordinates.

1. $f(r, \theta, z) = zr \cos \theta$, and T is given by $0 \le z \le 4$, $0 \le r \le z$, $0 \le \theta \le \pi/2$.

2. $f(r, \theta, z) = e^{-z} \sec^2 \theta$, and T is given by $0 \leq \theta \leq \pi/4$, $0 \leq r \leq 2$, and $0 \leq z \leq r$.

3. $f(r, \theta, z) = \dfrac{1}{r^{3/2}}$, and T is given by $0 \leq \theta \leq \pi$, $1 \leq z \leq 2$, and $1 \leq r \leq z^2$.

4. $f(r, \theta, z) = \sin \theta$, and T is given by $0 \leq \theta \leq 2\pi$, $0 \leq r \leq 1 - \cos \theta$, and $0 \leq z \leq r$.

5. $f(r, \theta, z) = e^{-1/r^2}$, and T is given by $0 \leq \theta \leq 2\pi$, $1 \leq r < \infty$, $0 \leq z \leq r^{-4}/2$.

6. $f(r, \theta, z) = \sqrt{z^2 - r^2}$, and T is given by $0 \leq z \leq 2$, $0 \leq \theta \leq \pi/3$, $0 \leq r \leq z \cos \theta$.

In each of the following exercises, you are given a function $f = f(x, y, z)$. You are also given a description of a solid region T in $\mathbb{R}^3$. Calculate $\iiint_T f \, dV$, using cylindrical coordinates. Note that, when $f(x, y, z) = 1$, $\iiint_T f \, dV = \iiint_T dV$ yields the volume of T.

7. $f(x, y, z) = 1$, and T is the solid region inside the sphere given by $x^2 + y^2 + z^2 = 4$ and inside the right circular cylinder given by $x^2 + y^2 = 1$.

8. $f(x, y, z) = x^2 + y^2$, and T is the solid region inside the sphere given by $x^2 + y^2 + z^2 = 4$ and inside the right circular cylinder given by $x^2 + y^2 = 1$.

9. $f(x, y, z) = 1$, and T is the solid region which is bounded below by the half-cone given by $z = \sqrt{x^2 + y^2}$ and is bounded above by the circular paraboloid given by $z = 6 - x^2 - y^2$.

10. $f(x, y, z) = x$, and T is the solid region which is bounded below by the half-cone given by $z = \sqrt{x^2 + y^2}$ and is bounded above by the circular paraboloid given by $z = 6 - x^2 - y^2$.

11. $f(x, y, z) = x^2 + y^2$, and T is the solid region inside the right circular cylinder given by $x^2 + y^2 = 9$, above the plane where $z = 0$, and below the plane where $z = y$.

12. $f(x, y, z) = 5xy$, and T is the solid region in the 1st octant above the half-cone where $z = \sqrt{x^2 + y^2}$ and below the plane where $z = 4$.

In each of the following exercises, you are given a function $f = f(\rho, \theta, \phi)$, and an iterated description, in spherical coordinates, of a solid region T. Calculate $\iiint_T f \, dV$, using spherical coordinates.

13. $f(\rho, \theta, \phi) = \rho^2$, and T is given by $0 \leq \theta \leq 2\pi$, $0 \leq \phi \leq \pi/2$, and $0 \leq \rho \leq 6$.

14. $f(\rho, \theta, \phi) = 1$, and T is given by $0 \leq \theta \leq \pi/2$, $0 \leq \phi \leq \pi/4$, and $0 \leq \rho \leq \cos \phi$.

15. $f(\rho, \theta, \phi) = \theta \cos \phi$, and T is given by $0 \leq \theta \leq \pi/2$, $0 \leq \phi \leq \pi/4$, and $0 \leq \rho \leq \cos \phi$.

16. $f(\rho, \theta, \phi) = \sin\theta$, and T is given by $0 \leq \theta \leq \pi$, $0 \leq \phi \leq \cos^{-1}(3/4)$, and $3\sec\phi \leq \rho \leq 4$.

17. $f(\rho, \theta, \phi) = 1$, and T is given by $0 \leq \theta \leq 2\pi$, $0 \leq \phi \leq \cos^{-1}(3/4)$, and $3\sec\phi \leq \rho \leq 4$.

18. $f(\rho, \theta, \phi) = \csc\phi$, and T is given by $\pi/2 \leq \theta \leq \pi$, $\pi/4 \leq \phi \leq \theta/2$, and $\phi \leq \rho \leq 2$.

In each of the following exercises, you are given a function $f = f(x, y, z)$. You are also given a description of a solid region T in $\mathbb{R}^3$. Calculate $\iiint_T f\, dV$, using spherical coordinates. Note that, when $f(x, y, z) = 1$, $\iiint_T f\, dV = \iiint_T dV$ yields the volume of T.

19. $f(x, y, z) = xz$, and T is the solid region in the 1st octant inside the sphere of radius 1, centered at the origin.

20. $f(x, y, z) = 1$, and T is the solid region inside the sphere given by $x^2 + y^2 + z^2 = 4$ and outside the right circular cylinder given by $x^2 + y^2 = 1$.

21. $f(x, y, z) = 1$, and T is the solid region inside the sphere given by $x^2 + y^2 + z^2 = 16$ and above the plane where $z = 3$.

22. $f(x, y, z) = 1/(x^2 + y^2 + z^2)$, and T is the solid region between the spheres of radius 3 and radius 5, centered at the origin.

23. $f(x, y, z) = 1/(x^2 + y^2 + z^2)$, and T is the solid region between the spheres of radius 3 and radius 5, centered at the origin, and inside the cone where $z = \sqrt{x^2 + y^2}$.

24. $f(x, y, z) = 1$, and T is the solid region between the spheres of radius 3 and radius 5, centered at the origin, and inside the cone where $z = \sqrt{x^2 + y^2}$.

In each of the following exercises, you are given an iterated integral in terms of x, y, and z. Calculate the value of the integral, by using whatever coordinates seem to be the most convenient.

25. $\displaystyle\int_0^5 \int_{-3}^3 \int_{-\sqrt{9-x^2}}^{\sqrt{9-x^2}} 1\, dy\, dx\, dz$

26. $\displaystyle\int_0^5 \int_{-3}^3 \int_{-\sqrt{9-x^2}}^{\sqrt{9-x^2}} z(x^2 + y^2)\, dy\, dx\, dz$

27. $\displaystyle\int_{-2}^2 \int_0^{\sqrt{4-x^2}} \int_{2\sqrt{x^2+y^2}}^{\sqrt{20-x^2-y^2}} dz\, dy\, dx$

28. $\displaystyle\int_{-2}^2 \int_0^{\sqrt{4-x^2}} \int_{2\sqrt{x^2+y^2}}^{\sqrt{20-x^2-y^2}} z\, dz\, dy\, dx$

29. $\displaystyle\int_0^1 \int_0^4 \int_{-2}^2 x \, dz \, dy \, dx$

30. $\displaystyle\int_0^1 \int_0^{\sqrt{1-y^2}} \int_0^{2-x-y} 1 \, dz \, dx \, dy$

In each of the following exercises, you are given a description of a solid region T. Find the volume of T, using whatever coordinates seem to be the most convenient.

31. T is the solid region outside the cylinder where $x^2 + y^2 = 1$, inside the cylinder where $x^2 + y^2 = 4$, and inside the sphere of radius 3, centered at the origin.

32. T is the solid region above the filled-in square with vertices $(0,0,0)$, $(1,0,0)$, $(0,1,0)$, and $(1,1,0)$, and below the circular paraboloid where $z = 9 - x^2 - y^2$.

⊙▶

33. T is the solid region above the xy-plane, and below the circular paraboloid where $z = 9 - x^2 - y^2$.

34. T is the solid region inside the sphere of radius 5, where $0 \le y \le x$.

35. T is the solid region bounded by the half-cones where $z = \sqrt{x^2 + y^2}$ and $z\sqrt{3} = \sqrt{x^2 + y^2}$, and inside the cylinder where $x^2 + y^2 = 1$.

36. T is the solid region bounded by the half-cones where $z = \sqrt{x^2 + y^2}$ and $z\sqrt{3} = \sqrt{x^2 + y^2}$, and inside the sphere where $x^2 + y^2 + z^2 = 1$.

> **+ Linear Algebra:**

In each of the following exercises, you are given C^1 functions $x = \Phi_1(u, v, w)$, $y = \Phi_2(u, v, w)$, and $z = \Phi_3(u, v, w)$. Thus, we can define Φ by

$$(x, y, z) = \Phi(u, v, w) = (\Phi_1(u, v, w), \Phi_2(u, v, w), \Phi_3(u, v, w)).$$

(a) Determine those points $\mathbf{p} = (u_0, v_0, w_0)$ at which Φ is NOT a C^1 local change of coordinates. (b) Near those points $\mathbf{p}$ at which Φ is a C^1 local change of coordinates, find an expression for the volume element dV in terms of u, v, w, du, dv, and dw.

37. $x = 3u - 2v + 5w$, $y = 7v + 4w$, $z = -2w$.

38. $x = u + v + w$, $y = u + 2v + 3w$, $z = -v + 5w$.

39. $x = \tan^{-1} u$, $y = \tan^{-1} v$, $z = \tan^{-1} w$.   ▶

40. $x = u + w$, $y = v^2 + w^2$, $z = vw$.

41. $x = u \cosh v$, $y = u \sinh v$, $z = w$.

42. $x = u \sinh w \cosh v$, $y = u \sinh w \sinh v$, $z = u \cosh w$.

3.7   Average Value

We define and calculate the average values of continuous functions on plane regions in $\mathbb{R}^2$ and solid regions in $\mathbb{R}^3$.

Basics:

   The average value of a finite collection of numbers is easy define: you add up all of the numbers, and divide by the number of numbers. So, the average value of a real-valued function $f = f(x)$, whose domain is a **finite** set of x values, say $x_1 \ldots, x_n$, is given by

$$\frac{f(x_1) + \cdots + f(x_n)}{n} \quad = \quad \frac{\sum_{k=1}^{n} f(x_k)}{n}.$$

   But how should we define the average value of a function $f = f(x, y)$ or $f = f(x, y, z)$ which is defined at an infinite number of points?

   We'll discuss the $f = f(x, y, z)$ case, and make the definition in the case of $f = f(x, y)$ by analogy.

   Suppose that have a function $f = f(x, y, z)$ defined on some bounded solid region T in $\mathbb{R}^3$. We want to use a limit of averages of finite collections of numbers to define the average value of f on T.

   Let's suppose that T has volume V. Chop up T into n little cubical solid boxes, each of volume $\Delta V = V/n$. Of course, as in our discussion of Riemann sums, there might be some partial boxes. This problem can be dealt with as we did before, by extending f by zero to a solid box which contains all of T.

   Now select a sample point $\mathbf{p}_k$ in each little sub-box. Then, the average value of f at the sample points is

$$\frac{f(\mathbf{p}_1) + \cdots + f(\mathbf{p}_n)}{n} \quad = \quad \frac{\sum_{k=1}^{n} f(\mathbf{p}_k)}{n}.$$

We would like to define the average value of f on T to be the limit of these average values over the sample points, if such a limit exists. Can we somehow write this in terms of integrals? Yes.

   Multiply the numerator and denominator by ΔV to obtain

$$\frac{\sum_{k=1}^{n} f(\mathbf{p}_k)}{n} \quad = \quad \frac{\sum_{k=1}^{n} f(\mathbf{p}_k)\Delta V}{n\Delta V} \quad = \quad \frac{\sum_{k=1}^{n} f(\mathbf{p}_k)\Delta V}{V}.$$

Now, the denominator is constant, and equals the total volume of S and, as $n \to \infty$, the numerator simply becomes the triple integral

$$\iiint_T f(x, y, z) \, dV.$$

This is our motivation for making the following definitions.

Definition 3.7.1. *Suppose that R is a bounded region in $\mathbb{R}^2$, which has area $A > 0$, and that $f = f(x, y)$ is a real-valued function, which is Riemann integrable on R.*

Then, the **average value of f on R** *is*

$$\frac{1}{A} \iint_R f(x, y) \, dA.$$

Suppose that T is a bounded region in $\mathbb{R}^3$, which has volume $V > 0$, and that $f = f(x, y, z)$ is a real-valued function, which is Riemann integrable on T.

Then, the **average value of f on T** is

$$\frac{1}{V} \iiint_T f(x, y, z) \, dV.$$

We have assumed here that the *area of R* and the *volume of T* have meaning for us, i.e., that the constant function 1 is Riemann integrable.

These notions of average value can be extended to any dimension in the obvious manner.

Remark 3.7.2. If f is bounded on T, then there exists real numbers m and M such that, for all **p** in T,

$$m \leq f(\mathbf{p}) \leq M.$$

Applying monotonicity (Theorem 3.4.21) twice, we find that

$$mV = \iiint_T m \, dV \leq \iiint_T f \, dV \leq \iiint_T M \, dV = MV.$$

Dividing throughout by $V > 0$, we obtain that the average value of f is between m and M.

In particular, if T is a compact region and f is continuous, it follows from the Extreme Value Theorem, Theorem 1.7.13, that f attains a minimum value, m, on T, and a maximum value, M, on T. What we have shown is that the average value of f is somewhere between the maximum and minimum values. That's good, for, otherwise, our notion of *average value* would be seriously flawed.

Of course, the same argument applies to planar regions in $\mathbb{R}^2$.

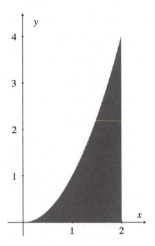

Figure 3.7.1: The region R.

Example 3.7.3. Let $f(x,y)$ be the function which gives the distance squared between the origin and (x,y), i.e., let $f(x,y) = x^2 + y^2$. Find the average value of f on the region R below the parabola given by $y = x^2$ and above the interval $[0,2]$ on the x-axis.

Solution:

We have to calculate two double integrals here: $\iint_R f\, dA$ and another integral to find the area of A. Fortunately, we can use the same limits of integration in both integrals.

We find

$$\iint_R (x^2 + y^2)\, dA = \int_0^2 \int_0^{x^2} (x^2 + y^2)\, dy\, dx = \int_0^2 \left[x^2 y + \frac{y^3}{3} \Big|_{y=0}^{y=x^2} \right] dx =$$

$$\int_0^2 \left(x^4 + \frac{x^6}{3} \right) dx = \frac{x^5}{5} + \frac{x^7}{21} \Big|_0^2 = \frac{1312}{105},$$

and, for the area, we integrate 1:

$$A = \iint_R 1\, dA = \int_0^2 \int_0^{x^2} dy\, dx = \int_0^2 x^2\, dx = \frac{x^3}{3} \Big|_0^2 = \frac{8}{3}.$$

Therefore, the average value of f on the region R is

$$\frac{1}{A} \iint_R (x^2 + y^2)\, dA = \frac{1312/105}{8/3} = \frac{164}{35} \approx 4.6857.$$

Example 3.7.4. Find the average values of $f(x,y,z) = y$ and $g(x,y,z) = z$ on the solid region T below the graph of $z = 4 - x^2 - y^2$ and above the xy-plane.

Solution:

By symmetry, the average y-coordinate had **better be** 0, which would mean that you get 0 before dividing by the volume. Let's check.

Cylindrical coordinates will work very well here. Let R be the projected region in the xy-plane, which is the inside of the circle where $0 = 4 - x^2 - y^2$, i.e., $r = 2$. The surface $z = 4 - x^2 - y^2$ becomes $z = 4 - r^2$, and y becomes $y = r\sin\theta$.

Thus,

$$\iiint_T y\, dV = \iint_R \left[\int_0^{4-r^2} r\sin\theta\, dz \right] dA =$$

$$\int_0^{2\pi} \int_0^2 \int_0^{4-r^2} r\sin\theta\, dz\, r\, dr\, d\theta.$$

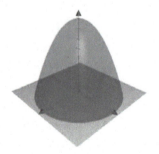

Figure 3.7.2: The region under $z = 4 - x^2 - y^2$ and above $z = 0$.

Moving the θ integral to the inside, we have

$$\iiint_T y \, dV = \int_0^2 \int_0^{4-r^2} r^2 \left[\int_0^{2\pi} \sin\theta \, d\theta \right] dz \, dr.$$

The inside internal is 0, so the value of the triple integral is 0, i.e., the average y-coordinate is 0, as we expected.

However, symmetry won't get us the average z-coordinate. We need to calculate $\iiint_T z \, dV$ and $\iiint_T dV$, and then divide. As above, cylindrical coordinates work well.

We find

$$\iiint_T z \, dV = \int_0^2 \int_0^{4-r^2} \int_0^{2\pi} zr \, d\theta \, dz \, dr = \int_0^2 \int_0^{4-r^2} 2\pi zr \, dz \, dr =$$

$$\pi \int_0^2 \left[rz^2 \Big|_{z=0}^{z=4-r^2} \right] dr = \pi \int_0^2 r(4-r^2)^2 \, dr = -\frac{\pi}{6}(4-r^2)^3 \Big|_0^2 = \frac{32\pi}{3}$$

and

$$V = \iiint_T dV = \int_0^2 \int_0^{4-r^2} \int_0^{2\pi} r \, d\theta \, dz \, dr = \int_0^2 \int_0^{4-r^2} 2\pi r \, dz \, dr =$$

$$2\pi \int_0^2 \left[rz \Big|_{z=0}^{z=4-r^2} \right] dr = 2\pi \int_0^2 \left(4r - r^3 \right) dr = 2\pi \left(2r^2 - \frac{r^4}{4} \right) \Big|_0^2 = 8\pi.$$

Therefore, the average value of z on T is

$$\frac{1}{V} \iiint_S z \, dV = \frac{32\pi/3}{8\pi} = \frac{4}{3},$$

i.e., 1/3 of the way to the top.

The fact that the average z-coordinate is closer to the bottom than the top makes sense; the solid region T is fatter at the bottom, and so has more volume for z-coordinates near the bottom than it does at the top.

3.7.1   Exercises

In each of the following exercises, you are given a function $f = f(x,y)$ and a region R in $\mathbb{R}^2$. Find the average value of f on R. It may be useful to switch into other coordinates.

1. $f(x, y) = x + y$, and R is the region inside the square with vertices $(0, 0)$, $(2, 0)$, $(0, 2)$, and $(2, 2)$.

2. $f(x, y) = x^2 + y^2$, and R is the region inside the square with vertices $(0, 0)$, $(2, 0)$, $(0, 2)$, and $(2, 2)$.

3. $f(x, y) = x^p + y^p$, and R is the region inside the square with vertices $(0, 0)$, $(L, 0)$, $(0, L)$, and (L, L), where $p > 0$ and $L > 0$.

4. $f(x, y) = x^2 + y^2$, and R is the region in the 1st quadrant inside the circle of radius 2, centered at the origin.

5. $f(x, y) = x$, and R is the region in the 1st quadrant inside the circle of radius a, centered at the origin.

6. $f(x, y) = \left(\sqrt{x^2 + y^2} \right)^p$, where $p > 0$, and R is the region in the 1st quadrant inside the circle of radius a, centered at the origin.

7. $f(x, y) = y$, and R is the region above the graph of $y = x^2$, and below the line where $y = 3$.

8. $f(x, y) = x$, and R is the region above the graph of $y = x^p$, and below the line where $y = b$, where $b > 0$ and p is a positive, even, integer.

9. $f(x, y) = y$, and R is the region above the graph of $y = x^p$, and below the line where $y = b$, where $b > 0$ and p is a positive, even, integer.

10. $f(x, y) = y$, and R is the unbounded region below the graph of $y = e^{-x}$ and above the interval $[0, \infty)$ on the x-axis.

In each of the following exercises, you are given a function $f = f(x, y, z)$ and a solid region T in $\mathbb{R}^3$. Find the average value of f on T. It may be useful to switch into other coordinates.

11. $f(x, y, z) = x + y + z$, and T is the solid cube in the 1st octant given by $0 \leq x \leq 2$, $0 \leq y \leq 2$, and $0 \leq z \leq 2$.

12. $f(x, y, z) = \sqrt{x^2 + y^2 + z^2}$ and T is the solid region in the 1st octant inside the sphere of radius 2, centered at the origin.

13. $f(x, y, z) = \left(\sqrt{x^2 + y^2 + z^2} \right)^p$ and S is the solid region in the 1st octant inside the sphere of radius a, centered at the origin, where $p > 0$ and $a > 0$.

14. $f(x, y, z) = x^2 + y^2$, and T is the solid right circular cylinder where $-4 \leq z \leq 4$ and $x^2 + y^2 \leq 4$.

15. $f(x,y,z) = x$, and T is the solid right circular cylinder where $-4 \leq z \leq 4$ and $x^2 + y^2 \leq 4$.

16. $f(x,y,z) = y$, and T is the solid circular half-cylinder where $-4 \leq z \leq 4$, $x^2 + y^2 \leq 4$, and $y \geq 0$.

17. $f(x,y,z) = z$, and T is the solid region above the rectangle, in the xy-plane, where $0 \leq x \leq 1$ and $0 \leq y \leq 5$, and below the hyperbolic paraboloid given by $z = 2 + y^2 - x^2$.

18. $f(x,y,z) = 1/(x^2 + y^2 + z^2)$ and T is the solid region between the spheres, centered at the origin, of radii 3 and 5.

19. Show, explicitly, that the average value that you found for f in Exercise 11 is between the maximum and minimum value of f on the compact region T from the exercise.

20. Show, explicitly, that the average value that you found for f in Exercise 18 is between the maximum and minimum value of f on the compact region T from the exercise.

3.8   Density & Mass

In this section, we discuss how to use integration to calculate the mass of a thin plate or a solid, even when the density of the object varies.

Density is mass per volume. *Area-density* is mass per area (of an ideal 2-dimensional object). Of course, if an object has constant density (respectively, area-density), then, to obtain the mass of the object, you simply multiply the constant density (respectively, area-density) times the volume (respectively, area) of the object. But what do you do when the density or area-density varies from point to point?

The answer, of course, is that you add up all of the infinitesimal mass contributions using integration. Each infinitesimal chunk of mass at a point **p** is given by multiplying the density $\delta(\mathbf{p})$ (respectively, area-density $\delta_{\mathrm{ar}}(\mathbf{p})$) at **p** times the infinitesimal volume dV (respectively, area dA) at **p**. We discussed this in depth in Section 3.2 and Section 3.4, where we used area-density, density, and mass as our motivating examples for defining the Riemann integrals.

Therefore, we make the following definitions:

We remind you that mass, in the metric system, is measured in kilograms, kg. In the English system, mass is measured in slugs. Sometimes, a pound-mass is used, but we shall avoid this.

In the metric system, you multiply the mass, in kilograms, times the acceleration produced by gravity, in (m/s)/s, approx. 9.81, to obtain the weight in Newtons.

In the English system, you multiply the mass, in slugs, times the acceleration produced by gravity, in (ft/s)/s, approx. 32.17, to obtain the weight in pounds.

Definition 3.8.1. *Suppose that an ideal 2-dimensional plate occupies a region R in $\mathbb{R}^2$, and that $\delta_{\mathrm{ar}} = \delta_{\mathrm{ar}}(x, y)$ gives the area-density (mass per unit area) at each point on the plate.*

Then, an infinitesimal piece of mass, an **element of mass**, *dm, is given by*

$$dm \;=\; \delta_{\mathrm{ar}}(x, y)\, dA.$$

Consequently, the total **mass** *of the plate is*

$$\iint_R dm \;=\; \iint_R \delta_{\mathrm{ar}}(x, y)\, dA,$$

provided that the Riemann integral exists.

Suppose that a solid object occupies a region T in $\mathbb{R}^3$, and that $\delta = \delta(x, y, z)$ gives the density (mass per unit volume) at each point in the solid.

Then, an infinitesimal piece of mass, an **element of mass**, *dm, is given by*

$$dm \;=\; \delta(x, y, z)\, dV.$$

> Consequently, the total **mass** of the solid is
>
> $$\iiint_T \delta(x, y, z)\, dV,$$
>
> provided that the Riemann integral exists.

Of course, the density or area-density functions could be essentially any non-negative functions, and so a density and mass problem could be any integration problem where the integrand is non-negative on the region.

Example 3.8.2. Suppose that a thin plate occupies the triangular region R in the xy-plane which has vertices $(0,0)$, $(1,0)$, and $(0,1)$, where all distances are in feet. Suppose that the area-density of the plate, in slugs/ft^2, is given by $\delta_{\text{ar}}(x,y) = 1 + 2x + 4y$. Find the mass of the plate.

Solution:

The triangle is bounded by the lines where $y = 0$, $x = 0$, and $x + y = 1$. Thus, we find

$$\text{mass} = \iint_R \delta_{\text{ar}}\, dA = \int_0^1 \int_0^{1-x} (1 + 2x + 4y)\, dy\, dx =$$

$$\int_0^1 \left[y + 2xy + 2y^2 \ \Big|_{y=0}^{y=1-x} \right] dx = \int_0^1 \left[(1-x) + 2x(1-x) + 2(1-x)^2 \right] dx =$$

$$\int_0^1 3(1-x)\, dx = 3 \left(x - \frac{x^2}{2} \right) \Big|_0^1 = \frac{3}{2} \ \text{slugs}.$$

Example 3.8.3. Let T be the solid region under the plane where $z = y$ and above the rectangle in the xy-plane given by $0 \leq x \leq 2$ and $0 \leq y \leq 1$, where x, y, and z are measured in meters. See Figure 3.8.1.

Now suppose that the density δ at each point in T is given by

$$\delta = \delta(x, y, z) = e^z + xy \ \text{kg/m}^3.$$

Find the mass of T.

Figure 3.8.1: The solid region T.

Solution:

The infinitesimal amount of mass at each point is

$$dm \;=\; \delta\,dV \;=\; (e^z + xy)\,dz\,dy\,dx.$$

The total mass of T is the continuous sum of all of the infinitesimal pieces of mass:

$$\text{mass} \;=\; \int_0^2 \int_0^1 \int_0^y (e^z + xy)\,dz\,dy\,dx \;=\;$$

$$\int_0^2 \int_0^1 (e^z + xyz) \Big|_{z=0}^{z=y} dy\,dx \;=\; \int_0^2 \int_0^1 \left(e^y + xy^2 - 1\right) dy\,dx \;=\;$$

$$\int_0^2 \left(e^y + \frac{xy^3}{3} - y\right) \Big|_{y=0}^{y=1} dx \;=\; \int_0^2 \left(e + \frac{x}{3} - 2\right) dx \;=\;$$

$$(e-2)x + \frac{x^2}{6} \Big|_0^2 \;=\; 2(e-2) + \frac{2}{3} \;\approx\; 2.10 \ \ \text{kg}.$$

Example 3.8.4. Suppose that the density of a ball T of radius R meters varies linearly as a function of the distance, d, from the center. Suppose that the density, in kg/m^3, on the surface of the ball is $A > 0$, and that the density at the center of the ball, in kg/m^3, is $B > 0$.

Determine the mass of the ball.

Solution:

If $B > A$, you might want to think of this as a model for the mass of a planet, where the greater density towards the center is caused by pressure from the weight "above".

We will use spherical coordinates, with the origin at the center of the ball, so that $d = \rho$. Then, what we know is that

$$\delta \;=\; \delta(\rho) \;=\; k\rho + b,$$

for some constants k and b, that $\delta(R) = A$, and $\delta(0) = B$.

Hence, $B = k \cdot 0 + b$, so that $b = B$, and $A = kR + B$, so that $k = (A - B)/R$.

We shall keep in mind that we've solved for k and b in terms of A and B, but continue to use k and b throughout the problem, until the end.

With this density function, using spherical coordinates, our element of mass, dm, becomes

$$dm \;=\; \delta\,dV \;=\; (k\rho + b)\,dV \;=\; k\rho^3 \sin\phi\,d\rho\,d\phi\,d\theta \;+\; b\,dV,$$

and the mass of the ball is

$$\text{mass} \ = \ \iiint_T dm \ = \ k \iiint_T \rho^3 \sin\phi \, d\rho \, d\phi \, d\theta \ + \ b \iiint_T dV \ = $$

$$k \int_0^{2\pi} \int_0^{\pi} \int_0^R \rho^3 \sin\phi \, d\rho \, d\phi \, d\theta \ + \ b \cdot \frac{4\pi R^3}{3}.$$

The iterated integral is easy to calculate; we find

$$\text{mass} \ = \ k \cdot (2\pi)(2)\left(\frac{R^4}{4}\right) \ + \ b \cdot \frac{4\pi R^3}{3} \ = $$

$$\frac{A-B}{R} \cdot \pi R^4 \ + \ \frac{4B\pi R^3}{3} \ = \ \left(A + \frac{B}{3}\right)\pi R^3 \ \text{kg}.$$

As a quick check, note that, if $A = B$, the density of the ball is constantly $\delta = A = B$, and our result becomes what it should:

$$\text{mass} \ = \ \left(\delta + \frac{\delta}{3}\right)\pi R^3 \ = \ \delta \cdot \frac{4\pi R^3}{3} \ = \ \delta \cdot V \ \text{kg},$$

i.e., the constant density times the volume inside a sphere of radius R.

3.8.1   Exercises

Online answers to select exercises are here.

In each of the following exercises, a region R in the xy-plane is given. Assume that the region R is occupied by a thin metal plate, and that x and y are in meters. At each point (x, y) in the plate, you are given the area-density $\delta_{\text{ar}}(x, y)$, in kg/m^2. Calculate the mass of the plate.

Basics:

1. $\delta_{\text{ar}}(x, y) = x + y$, and R is the region inside the square with vertices $(1, 1)$, $(2, 1)$, $(1, 2)$, and $(2, 2)$.

2. $\delta_{\text{ar}}(x, y) = x^2 + y^2$, and R is the region inside the square with vertices $(1, 1)$, $(2, 1)$, $(1, 2)$, and $(2, 2)$.

3. $\delta_{\text{ar}}(x, y) = x^2 + y^2$, and R is the region in the 1st quadrant inside the circle of radius 2, and outside the circle of radius 1, both centered at the origin.

4. $\delta_{\text{ar}}(x, y) = x$, and R is the region in the 1st quadrant inside the circle of radius a, centered at the origin.

5. $\delta_{\mathrm{ar}}(x, y) = y$, and R is the region above the graph of $y = x^2$, and below the line where $y = 3$. ▶

6. $\delta_{\mathrm{ar}}(x, y) = y + 1$, and R is the unbounded region below the graph of $y = e^{-x}$ and above the interval $[0, \infty)$ on the x-axis.

In each of the following exercises, a region T in $\mathbb{R}^3$ is given. Assume that the region T is occupied by a solid, and that x, y, and z are in meters. At each point (x, y, z) in the solid, you are given the density $\delta(x, y, z)$, in kg/m³. Calculate the mass of the solid.

7. $\delta(x, y, z) = x + y + z$, and T is the solid cube in the 1st octant given by $1 \leq x \leq 2$, $1 \leq y \leq 2$, and $1 \leq z \leq 2$.

8. $\delta(x, y, z) = \sqrt{x^2 + y^2 + z^2}$ and T is the solid region in the 1st octant inside the sphere of radius 2, and outside the sphere of radius 1, both centered at the origin.

9. $\delta(x, y, z) = \left(\sqrt{x^2 + y^2 + z^2}\right)^p$ and T is the solid region in the 1st octant inside the sphere of radius b, and outside the sphere of radius a, both centered at the origin, where $p > 0$ and $b > a > 0$.

10. $\delta(x, y, z) = x^2 + y^2$, and T is the solid right circular cylinder where $-4 \leq z \leq 4$ and $x^2 + y^2 \leq 4$.

11. $\delta(x, y, z) = y$, and T is the solid circular half-cylinder where $-4 \leq z \leq 4$, $x^2 + y^2 \leq 4$, and $y \geq 0$. ▶

12. $\delta(x, y, z) = z$, and T is the solid region above the rectangle, in the xy-plane, where $0 \leq x \leq 1$ and $0 \leq y \leq 5$, below the hyperbolic paraboloid given by $z = 2 + y^2 - x^2$, and above the plane where $z = 1$.

13. $\delta(x, y, z) = 1/(x^2 + y^2 + z^2)$ and T is the solid region between the spheres, centered at the origin, of radii 3 and 5.

14. $\delta(x, y, z) = kz$, where $k > 0$, and T is the solid region under the graph of $z = 9 - x^2 - y^2$ and above the plane where $z = 5$.

15. Assume that a region R in the xy-plane is occupied by a thin metal plate and that, at each point (x, y) in the plate, you have the area-density $\delta_{\mathrm{ar}}(x, y)$. Show that the mass of the plate is the area of the plate times the average value of the area-density function on the region R. ▶

16. Assume that a region T in $\mathbb{R}^3$ is occupied by a solid object and that, at each point (x, y, z) in the object, you have the density $\delta(x, y, z)$. Show that the mass of the object is the volume of T times the average value of the density function on the region T.

3.9 Centers of Mass

In this section, we define the *center of mass* of a 2- or 3-dimensional object. In many types of problems, an object of a certain mass can be treated as a point-mass located at the center of mass.

If the density of the object is constant, the center of mass is called the *centroid*, and is determined by the shape of object. As we shall see, the coordinates of the centroid are precisely the average values, as defined in Section 3.7, of x, y, and z over the region.

We need to repeat here some of our discussion about centers of mass from [5].

We frequently discuss objects as though they are located at specific points. When we state that an object of mass m is located at a point P, we are usually thinking of an idealized "point-mass": an imaginary object which occupies a single point in space at any given time. Of course, objects in real life exist at an infinite number of points in space, but we can think of a given solid object as consisting of an infinite number of infinitesimal point-masses.

Now, suppose that you have a solid object, occupying an infinite number of points in space, and a force is acting on the object. We would like to apply Newton's 2nd Law of Motion to determine the acceleration of the object. But, if you're thinking of the object as being composed of an infinite number of point-masses, which are possibly moving separately in a very complicated manner, what does "the acceleration of the object" mean?

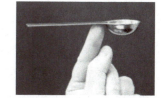

Figure 3.9.1: Balancing a spoon at its center of mass.

The answer is that we can define a point, called the *center of mass* of an object, whose acceleration is given by Newton's 2nd Law. Given an object, or collection of objects, of (total) mass M, the center of mass is a point P in space such that, in many physical problems, the object(s) can be treated as a point-mass, with mass M, located at the point P. If you were trying to balance a rigid wire or metal plate on your finger, the center of mass is where you would place your finger. If two children want to balance on a see-saw, the center of mass is the point where the base (fulcrum) of the see-saw needs to be.

Note that the center of mass of an object need **not** actually be located at a point on the object. Perhaps the easiest example of this is a uniformly dense annulus: think of a thin metal disk with a smaller disk removed from its center; by symmetry, the center of mass is located at the center of the annulus, but that part has been removed. What this means is that there is no place where you could place your finger in order to balance an annulus. Hopefully, this seems intuitively clear.

For a solid object, the center of mass can be determined, in principle, via integration, from the shape of the object and the density function (recall Section 3.8) of the object.

Figure 3.9.2: The center of mass is not located on the annulus itself.

For an object with constant density, the center of mass is also known as the *centroid* of the object; as we shall see, the constant density itself cancels out, and so the centroid is a property of the region. In fact, we shall see that the coordinates of the centroid are the average x-, y-, and z-coordinates in the region (recall Section 3.7).

We recall, in the More Depth portion of this section, the physics discussion that leads to the following definition of the center of mass for a finite collection of point-masses.

Definition 3.9.1. *The **center of mass** of a collection/system of point-masses with masses $m_1, \ldots, m_n$ at positions $\mathbf{r}_1, \ldots, \mathbf{r}_n$, respectively, is the point*

$$\mathbf{r}_{\text{cm}} \;=\; (\overline{x}, \overline{y}, \overline{z}) \;=\; \frac{\displaystyle\sum_{i=1}^{n} m_i \mathbf{r}_i}{\displaystyle\sum_{i=1}^{n} m_i}.$$

Note that the denominator above is simply the total mass of the entire collection of point-masses.

So, how do we go from the case of a finite number of point-masses to the case of a solid object, which has an infinite number of points, with, possibly, varying density δ at each point? Actually, it's easy, and should be what you expect; the individual position vectors $\mathbf{r}_i$ simply become the arbitrary position triple (x, y, z) for the location of each point in the region, m_i becomes the element of mass $dm = \delta\, dV$ (i.e., how much mass is at (x, y, z)), and the finite sum becomes a triple integral. Of course, we do the analogous thing for thin plates and area-density.

In general, you **cannot** cancel the δ's in the numerator and denominator; they're inside the integrals. However, if $\delta > 0$ is constant, you can pull both δ's out of the integrals and cancel them; see Definition 3.9.3.

Definition 3.9.2. *The **center of mass** of a solid object, which occupies a region T in space, and has density function δ is*

$$\mathbf{r}_{\text{cm}} \;=\; (\overline{x}, \overline{y}, \overline{z}) \;=\; \frac{\iiint_T (x, y, z)\, dm}{\iiint_T dm} \;=\; \frac{\iiint_T (x, y, z)\, \delta\, dV}{\iiint_T \delta\, dV},$$

provided that all of the Riemann integrals exist.

Therefore, if we let M denotes the total mass of the object, we have

$$dm \ = \ \delta\, dV, \qquad M \ = \ \iiint_T dm,$$

$$\overline{x} \ = \ \frac{1}{M}\iiint_T x\, dm, \quad \overline{y} \ = \ \frac{1}{M}\iiint_T y\, dm, \quad \text{and} \quad \overline{z} \ = \ \frac{1}{M}\iiint_T z\, dm.$$

For thin plates in the xy-plane, which occupy a region R, we have the analogous formulas for $\overline{x}$ and $\overline{y}$, replacing T with R, replacing triples integrals with double integrals, and using $dm = \delta_{\mathrm{ar}}\, dA$.

When the density is constant, it can be pulled out of the integrals in Definition 3.9.2, and then cancelled. Thus, the center of mass of an object with uniform density is independent of what that constant density is, and is just a property of the region itself; we call the center of mass of a region with uniform density the *centroid*.

Thus, we have:

Definition 3.9.3. *The* **centroid** *of a solid object, which occupies a region T in space (or the centroid of T) is*

$$\mathbf{r}_{\mathrm{centroid}} \ = \ (\overline{x}, \overline{y}, \overline{z}) \ = \ \frac{\iiint_T (x, y, z)\, dV}{\iiint_T dV}.$$

Therefore, if we let V denote the total volume of the object, we have

$$\overline{x} \ = \ \frac{1}{V}\iiint_T x\, dV, \quad \overline{y} \ = \ \frac{1}{V}\iiint_T y\, dV, \quad \text{and} \quad \overline{z} \ = \ \frac{1}{V}\iiint_T z\, dV,$$

i.e., for an object of uniform density, the centroid, which is the center of mass, has coordinates given by the average x, y, and z values in T (recall Section 3.7).

For regions R in the xy-plane, we have the analogous formulas for $\overline{x}$ and $\overline{y}$, replacing T with R, replacing triples integrals with double integrals, and using dA in place of dV.

Note that we have a mild notational issue here. If we have a solid region of variable density, the variables $\overline{x}$, $\overline{y}$, and $\overline{z}$ might denote the coordinates of the center of mass or the coordinates of the centroid, which is where the center of mass **would** be if the density were constant.

This won't cause us a problem. When the density is variable, we'll usually be interested in the center of mass, not the centroid. However, if we are interested in both, then we'll simply explicitly say which we're calculating, and not use the overlined variable notation.

Example 3.9.4. Let $a > 0$ and $b > 0$. Find the centroid of the triangular plate with vertices at $(0,0)$, $(a,0)$, and (a,b).

Solution:

Let R be the region occupied by the plate. The area A of R doesn't require an integral; we know that $A = ab/2$. In addition, the line through the origin and through (a, b) is given by $y = bx/a$. Thus, the x-coordinate of the centroid is given by

$$\overline{x} \;=\; \frac{1}{A} \iint_R x \, dA \;=\; \frac{1}{ab/2} \int_0^a \int_0^{bx/a} x \, dy \, dx \;=\; \frac{2}{ab} \int_0^a (xy) \Big|_{y=0}^{y=bx/a} dx \;=$$

$$\frac{2}{ab} \int_0^a \frac{bx^2}{a} \, dx \;=\; \frac{2}{a^2} \cdot \frac{a^3}{3} \;=\; \frac{2}{3} a.$$

By the symmetry of the problem, we can conclude that the y-coordinate of the centroid is $2/3$ of the way from (a, b) to $(a, 0)$, which would mean that $\overline{y} = b/3$, but let's check:

$$\overline{y} \;=\; \frac{1}{A} \iint_R y \, dA \;=\; \frac{1}{ab/2} \int_0^a \int_0^{bx/a} y \, dy \, dx \;=\; \frac{2}{ab} \int_0^a \left(\frac{y^2}{2} \right) \Big|_{y=0}^{y=bx/a} dx \;=$$

$$\frac{1}{ab} \int_0^a \frac{b^2 x^2}{a^2} \, dx \;=\; \frac{b}{a^3} \cdot \frac{a^3}{3} \;=\; \frac{1}{3} b,$$

as we expected.

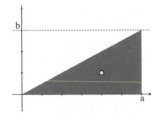

Figure 3.9.3: The center of mass of the right triangle.

Example 3.9.5. Find the centroid of the 1st octant portion of a solid ball of radius R, centered at the origin.

Solution:

By symmetry, we know that $\overline{x} = \overline{y} = \overline{z}$. So, we shall just compute $\overline{z}$. Let B denote the solid region. Note that we know that the volume V of B is given by

$$V \;=\; \frac{1}{8} \cdot \frac{4}{3} \pi R^3 \;=\; \frac{1}{6} \pi R^3.$$

We calculate in spherical coordinates:

$$\overline{z} \;=\; \frac{1}{V} \iiint_B z \, dV \;=\; \frac{6}{\pi R^3} \int_0^{\pi/2} \int_0^{\pi/2} \int_0^R \rho \cos\phi \cdot \rho^2 \sin\phi \, d\rho \, d\phi \, d\theta \;=$$

$$\frac{6}{\pi R^3} \cdot \frac{R^4}{4} \int_0^{\pi/2} \int_0^{\pi/2} \cos\phi \cdot \sin\phi \, d\phi \, d\theta \;=\; \frac{3R}{2\pi} \cdot \frac{\pi}{2} \cdot \left(\frac{\sin^2 \phi}{2} \right) \Big|_0^{\pi/2} \;=\; \frac{3}{8} R.$$

Thus,

$$\overline{x} \;=\; \overline{y} \;=\; \overline{z} \;=\; \frac{3}{8} R.$$

Figure 3.9.4: An eighth of a solid ball.

Example 3.9.6. Consider a solid right circular cylinder T, of radius R and height H, centered around the z-axis, with its base in the xy-plane. Assume that all lengths are measured in meters. Suppose that the density of the cylinder, in kilograms per cubic meter, is given by

$$\delta = 1 + z.$$

Determine the center of mass of the cylinder.

Figure 3.9.5: A solid cylinder with variable density.

Solution:

By the symmetry of T, and using that δ is independent of x and y, we see immediately that

$$\overline{x} = \overline{y} = 0.$$

We need to calculate $\overline{z}$.

Let R denote the projected region in the xy-plane; hence, R is a disk of radius R, centered at the origin. We use $dm = \delta\, dV = (1+z)\, dV$, and calculate:

$$\overline{z} = \frac{\iiint_S z\, dm}{\iiint_S dm} = \frac{\iiint_S z(1+z)\, dV}{\iiint_S (1+z)\, dV} = \frac{\iint_R \left[\int_0^H (z+z^2)\, dz\right] dA}{\iint_R \left[\int_0^H (1+z)\, dz\right] dA} =$$

$$\frac{\pi R^2 \left(\frac{H^2}{2} + \frac{H^3}{3}\right)}{\pi R^2 \left(H + \frac{H^2}{2}\right)} = \frac{3H + 2H^2}{6 + 3H} = \left(\frac{3 + 2H}{6 + 3H}\right) H,$$

which is between 0 and H, as it intuitively should be.

In fact, if we look at the ratio of $\overline{z}$ to H, we see that, when H is close to 0, $\overline{z}$ is close to $H/2$, but, when H is large, $\overline{z}$ is close to $2H/3$. This makes sense, because our density function puts more mass near the top of the cylinder.

More Depth:

Here, we want to look at the physics behind the definition of the center of mass. Since the continuous case, the case handled by integration, follows from the case of a finite number of point-masses, we shall describe the derivation of the center of mass for a finite number of point-masses.

Our treatment follows that in the excellent volumes of The Feyman Lectures on Physics, [2].

Suppose that we have n point-masses, whose masses are constants $m_1, \ldots, m_n$ (not all zero), and the masses are moving in space (or in a plane, or in a line). Let $\mathbf{r}_i = \mathbf{r}_i(t)$ denote the position of the mass m_i, for $i = 1, \ldots, n$, at time t. (In other places, we have used $\mathbf{p}$ to denote position; it is standard in our current situation to use $\mathbf{r}$, for *radial* vector.)

Suppose that we have a force $\mathbf{F}_i = \mathbf{F}_i(t)$ acting on each mass m_i, and that $\mathbf{F}_i$ is the only force acting on m_i. Then, the vector form of Newton's 2nd Law of Motion tells us that the net force acting on an object (of constant mass) is the mass times the acceleration of the object, i.e.,

$$\mathbf{F}_i \;=\; m_i \frac{d^2 \mathbf{r}_i}{dt^2}.$$

Suppose now that we think of all of the masses together as one object, possibly because they're very close together, but possibly not. Then, the total mass of the new "object" is $M = \sum_{i=1}^{n} m_i$, and the net force acting on the object is $\mathbf{F} = \sum_{i=1}^{n} \mathbf{F}_i$. We want to be able to apply Newton's 2nd Law to the collective mass M; that is, we want that $\mathbf{F}$ is equal to the mass M times the acceleration. But what is this "acceleration" the acceleration of? Acceleration is, of course, the second derivative of position, with respect to time, but what position do we take the second derivative of?

Using that the masses m_i are constant, so that M is also constant, we have

$$\mathbf{F} \;=\; \sum_{i=1}^{n} \mathbf{F}_i \;=\; \sum_{i=1}^{n} \frac{d^2 (m_i \mathbf{r}_i)}{dt^2} \;=\; M \frac{d^2 \left[\left(\sum_{i=1}^{n} m_i \mathbf{r}_i \right) / M \right]}{dt^2}.$$

Therefore, if we define a position vector

$$\mathbf{r}_{\mathrm{cm}} \;=\; \mathbf{r}_{\mathrm{cm}}(t) \;=\; \frac{\displaystyle\sum_{i=1}^{n} m_i \mathbf{r}_i}{M} \;=\; \frac{\displaystyle\sum_{i=1}^{n} m_i \mathbf{r}_i}{\displaystyle\sum_{i=1}^{n} m_i},$$

then we obtain that

$$\mathbf{F} \;=\; M \frac{d^2 \mathbf{r}_{\mathrm{cm}}}{dt^2}. \tag{3.2}$$

Remark 3.9.7. Note that if we replaced $\mathbf{r}_{\mathrm{cm}}$ in Formula 3.2 with $\mathbf{r} = \mathbf{r}_{\mathrm{cm}} + t\mathbf{v}_0 + \mathbf{r}_0$, where $\mathbf{v}_0$ and $\mathbf{r}_0$ are constant vectors, then the equation would still hold since the 2nd derivative of $t\mathbf{v}_0 + \mathbf{r}_0$, with respect to t, is zero. However, when there is only one non-zero mass, i.e., when $m_{i_0} \neq 0$ for exactly one index i_0, we naturally want the center of mass to be where that non-zero mass is, namely, at $\mathbf{r}_{i_0}$. Since we want this to be true at all times, and regardless of which mass is non-zero, we are required to choose $\mathbf{v}_0 = \mathbf{0}$ and $\mathbf{r}_0 = \mathbf{0}$, i.e., to define the center of mass as we did.

Richard Feynman (May 11, 1918–February 15, 1988) was an American physicist known for his work on many aspects of physics: the path integral formulation of quantum mechanics, the theory of quantum electrodynamics, and the superfluidity of supercooled liquid helium, particle physics, and his pictorial representation scheme for the mathematical expressions governing the behavior of subatomic particles, which later became known as Feynman diagrams.

Feynman, jointly with Julian Schwinger and Sin-Itiro Tomonaga, received the Nobel Prize in Physics in 1965 for contributions to the development of quantum electrodynamics. During his lifetime, Feynman became one of the best-known scientists in the world. He assisted in the development of the atomic bomb, and was a member of the panel that investigated the Space Shuttle Challenger disaster. In addition to his work in theoretical physics, Feynman has been credited with pioneering the field of quantum computing, and introducing the concept of nanotechnology.

3.9.1 Exercises

Online answers to select exercises are here.

Basics:

In each of the following exercises, you are given a region R in $\mathbb{R}^2$. Find the centroid of R.

1. R is the filled-in triangle with vertices $(0,0)$, $(2,0)$, and $(2,3)$. ▶

2. R is the filled-in triangle with vertices $(0,0)$, $(a,0)$, and (b,c), where a, b, and c are all positive.

3. R is the region under the graph of $y = x^2$ and above the interval $[0,2]$ on the x-axis.

4. R is the half-disk above the x-axis, and below the graph of $y = \sqrt{9 - x^2}$. (Note that one of the coordinates of the centroid can be obtained from symmetry.)

5. R is the region above the graph of $y = x^2$ and below the line where $y = 36$. (Note that one of the coordinates of the centroid can be obtained from symmetry.)

6. R is the region above the graph of $y = e^x$, below the line where $y = e$, where $0 \leq x \leq 1$.

In each of the following exercises, you are given a region R in $\mathbb{R}^2$, and the area-density function $\delta_{\mathrm{ar}}(x,y)$, in kg/m^2, for a thin plate which occupies the region R. Suppose that x and y are in meters. Determine the center of mass of the plate. Note that you may have calculated the masses of some of these plates in the exercises from the previous section.

7. $\delta_{\mathrm{ar}}(x,y) = x + y$, and R is the region inside the square with vertices $(1,1)$, $(2,1)$, $(1,2)$, and $(2,2)$.

8. $\delta_{\mathrm{ar}}(x,y) = x^2 + y^2$, and R is the region inside the square with vertices $(1,1)$, $(2,1)$, $(1,2)$, and $(2,2)$.

9. $\delta_{\mathrm{ar}}(x,y) = x^2 + y^2$, and R is the region in the 1st quadrant inside the circle of radius 2, and outside the circle of radius 1, both centered at the origin.

10. $\delta_{\mathrm{ar}}(x,y) = x$, and R is the region in the 1st quadrant inside the circle of radius a, centered at the origin.

11. $\delta_{\mathrm{ar}}(x,y) = y$, and R is the region above the graph of $y = x^2$, and below the line where $y = 3$. ▶

12. $\delta_{\mathrm{ar}}(x,y) = y + 1$, and R is the unbounded region below the graph of $y = e^{-x}$ and above the interval $[0, \infty)$ on the x-axis.

In each of the following exercises, you are given a region T in $\mathbb{R}^3$. Determine the centroid of T.

13. T is the solid cube in the 1st octant given by $1 \leq x \leq 2$, $1 \leq y \leq 2$, and $1 \leq z \leq 2$.

14. T is the solid region in the 1st octant inside the sphere of radius 2, and outside the sphere of radius 1, both centered at the origin. (Note that the symmetry of the problem implies that the x-, y-, and z-coordinates of the centroid will all be the same.)

15. T is the solid right circular cylinder where $-4 \leq z \leq 4$ and $x^2 + y^2 \leq 4$.

16. T is the solid circular half-cylinder where $-4 \leq z \leq 4$, $x^2 + y^2 \leq 4$, and $y \geq 0$.

17. T is the solid region above the rectangle, in the xy-plane, where $0 \leq x \leq 1$ and $0 \leq y \leq 5$, below the hyperbolic paraboloid given by $z = 2 + y^2 - x^2$, and above the plane where $z = 1$.

18. T is the solid region between the spheres, centered at the origin, of radii 3 and 5.

In each of the following exercises, you are given a region T in $\mathbb{R}^3$, and the density function $\delta(x, y, z)$, in kg/m^3, for a solid object which occupies the region T. Suppose that x, y, and z are in meters. Determine the center of mass of the object. Note that you may have calculated the masses of some of these objects in the exercises from the previous section.

19. $\delta(x, y, z) = x + y + z$, and T is the solid cube in the 1st octant given by $1 \leq x \leq 2$, $1 \leq y \leq 2$, and $1 \leq z \leq 2$.

20. $\delta(x, y, z) = \sqrt{x^2 + y^2 + z^2}$ and T is the solid region in the 1st octant inside the sphere of radius 2, and outside the sphere of radius 1, both centered at the origin. (Note that the symmetry of the problem implies that the x-, y-, and z-coordinates of the center of mass will all be the same.)

21. $\delta(x, y, z) = x^2 + y^2$, and T is the solid right circular cylinder where $-4 \leq z \leq 4$ and $x^2 + y^2 \leq 4$.

22. $\delta(x, y, z) = y$, and T is the solid circular half-cylinder where $-4 \leq z \leq 4$, $x^2 + y^2 \leq 4$, and $y \geq 0$.

23. $\delta(x, y, z) = z$, and T is the solid region above the rectangle, in the xy-plane, where $0 \leq x \leq 1$ and $0 \leq y \leq 5$, below the hyperbolic paraboloid given by $z = 2 + y^2 - x^2$, and above the plane where $z = 1$.

24. $\delta(x, y, z) = 1/(x^2 + y^2 + z^2)$ and T is the solid region between the spheres, centered at the origin, of radii 3 and 5.

25. Suppose that $\mathbf{r}_{cm}(A)$ and $\mathbf{r}_{cm}(B)$ are the centers of mass of two non-intersecting solid objects A and B in $\mathbb{R}^3$, with masses m_A and m_B. Denote the center of mass of the system consisting of both A and B by $\mathbf{r}_{cm}(A \cup B)$. Show that

$$\mathbf{r}_{cm}(A \cup B) = \frac{m_A \mathbf{r}_{cm}(A) + m_B \mathbf{r}_{cm}(B)}{m_A + m_B}.$$

More Depth:

26. The center of mass is sometimes referred to as the *center of gravity*. However, this term is very misleading. It is generally **not** true that the magnitude of the gravitational force between two masses is inversely proportional to the square of the distance between the centers of mass of the objects. The point of this exercise is to see this "problem" occurs even in the simple case of two point-masses m_A and m_B, acting on a third mass M.

Suppose that a mass M is located at the origin, and that point-masses, both with mass m, are located at the points $A = (1,0,0)$ and $B = (0,1,0)$, where a and b are positive. Let $\mathbf{r}_{gf}$ be the point at which a mass of $2m$ would exert the same gravitational force on M as do the two point-masses at the two points A and B. We want to show that $\mathbf{r}_{gf}$ is not equal to $\mathbf{r}_{cm}$, the center of mass of the system consisting of the two point-masses at A and B.

Recall that, according to Newton's Law of Universal Gravitation, if we have a point-mass m, located at any point $\mathbf{r} = (x,y,z) \neq (0,0,0)$ in space, then m exerts a gravitational force $\mathbf{F}$ on M given by

$$\mathbf{F} \;=\; \frac{GMm}{|\mathbf{r}|^2}\left(\frac{\mathbf{r}}{|\mathbf{r}|}\right);$$

that is, a force of magnitude $GMm/|\mathbf{r}|^2$, pointing straight towards m.

(a) Show that $\mathbf{r}_{cm} = 0.5(1,1,0)$.

(b) Use the definition of $\mathbf{r}_{gf}$ to show that

$$\frac{2}{|\mathbf{r}_{gf}|^2}\frac{\mathbf{r}_{gf}}{|\mathbf{r}_{gf}|} \;=\; (1,1,0).$$

(c) Show how you now conclude that $\mathbf{r}_{gf} \neq \mathbf{r}_{cm}$.

3.10 Moments of Inertia

In this section, we define the *moment of inertia* of an object around a given axis in $\mathbb{R}^3$. In a sense, the moment of inertia of an object is an analog of mass for a body which is rotating/revolving about an axis.

Basics:

When an object is moving around an axis, in a circular motion, it is standard to say that the object *rotates* about the axis if the axis passes through the object under consideration. When the axis does not pass through the object, it is standard to say the object *revolves* about the axis.

The moment of inertia is also referred to as the **mass moment of inertia** or the **rotational inertia**.

We are interested in studying solid, rigid, bodies, which are rotating, or revolving, around axes. Because we assume the bodies are rigid, each point in a given body makes one complete revolution around the given axis in the same amount of time. This means that each point has the same *angular velocity*, in radians per second. On the other hand, points that are farther from the axis travel a greater distance than do points closer to the axis, so that the actual speed of points farther from the axis must be greater than the speed of points closer in. Thus, for rotating bodies, it is usually simpler to describe the physics of the objects in terms of angular velocities. In many rotational problems, when velocity is "replaced" with angular velocity, the appropriate "replacement" for mass is a quantity called the *moment of inertia* of the object.

In order to motivate the definition of the moment of inertia, we shall first investigate the case of revolving a point-mass around an axis. Then, of course, to deal with a solid object, we think of the object as being chopped up into an infinite collection of infinitesimal point-masses (or take limits of Riemann sums), and we integrate, to determine the rotational inertia.

Suppose you have a point-mass, of mass m, revolving around a line $\mathfrak{l}$, the **axis of revolution**. We assume that the mass revolves around the axis while staying in a plane which is perpendicular to $\mathfrak{l}$, and moving in a circle of radius r, i.e., the mass stays at a constant distance of r from $\mathfrak{l}$. We can set up a coordinate system so that $\mathfrak{l}$ is the z-axis, and the plane that the point-mass revolves in is the xy-plane. Furthermore, we can select the positive z-axis so that, looking down from the positive z-axis, the point-mass moves counterclockwise.

After making these choices, the distance r is also the radius r from polar or cylindrical coordinates, and the position of the point-mass is given by and

$$\mathbf{p} \ = \ r(\cos\theta,\ \sin\theta,\ 0).$$

The velocity of the point-mass is

$$\mathbf{v} \ = \ \frac{d\mathbf{p}}{dt} \ = \ r\,\frac{d\theta}{dt}\,(-\sin\theta,\ \cos\theta,\ 0). \tag{3.3}$$

The rate of change of θ, $d\theta/dt$, is called the **angular velocity** of the object, and is usually denoted by ω.

Looking at Formula 3.3, and using that the magnitude $|(-\sin\theta, \cos\theta, 0)|$ is 1, we see that the speed $|\mathbf{v}|$ is equal to $r|\omega|$. Therefore, the kinetic energy of the point-mass is

$$K = \frac{1}{2}m|\mathbf{v}|^2 = \frac{1}{2}(mr^2)\omega^2.$$

Thus, if you want to use angular velocity to specify the kinetic energy of the point-mass, of constant mass and constant distance from the axis, then quantity mr^2 is an important constant.

> **Definition 3.10.1.** *Consider a point-mass, in $\mathbb{R}^3$, with mass m, which is a distance r from a line $\mathfrak{l}$. The* **moment of inertia** *of the point-mass about $\mathfrak{l}$ is*
>
> $$I = mr^2.$$

Remark 3.10.2. The moment of inertia I of a point-mass is defined so that the kinetic energy K is given by

$$K = \frac{1}{2}I\omega^2,$$

where ω is the angular velocity of the point-axis around the axis of revolution.

It follows from Conservation of Energy that, to give a point-mass a fixed angular velocity, you must add a larger amount of energy for a point-mass which has a larger moment of inertia. Hence, moment of inertia is a measure of the resistance of a point-mass to being made to revolve around an axis.

Now suppose that a solid object is rotating about an axis, with angular velocity ω. This means that each point in the solid is revolving around the axis with the same angular velocity ω. The total kinetic energy K of the object will once again be

$$K = \frac{1}{2}I\omega^2,$$

provided that we define I to be the continuous sum of all of the infinitesimal moments of inertia from each point in the body.

Thus, we define:

Definition 3.10.3. *Consider a solid, which occupies a region T in $\mathbb{R}^3$, and a line in $\mathbb{R}^3$ (an axis of revolution/rotation) $\mathfrak{l}$. For each point (x, y, z) in T, let $r = r(x, y, z)$ denote the distance from (x, y, z) to the axis $\mathfrak{l}$, and let $\delta = \delta(x, y, z)$ denote the density of the object at (x, y, z).*

Then, we define the **moment of inertia** *of the object about $\mathfrak{l}$ to be*

$$I = \iiint_T r^2 \, dm = \iiint_T r^2 \delta \, dV.$$

For thin plates in a plane in $\mathbb{R}^3$, which occupy a region R, we have an analogous formula for the moment of inertia I, replacing T with R, replacing triples integrals with double integrals, and using $dm = \delta_{\mathrm{ar}} \, dA$.

Remark 3.10.4. The standard case is when the density of the object is uniform, i.e., the density function is just a constant. In this case, our formula for the moment of inertia can be written in another nice form, using that $\delta = M/V$, where M is the total mass of the object and V is its volume.

Then, Definition 3.10.3 becomes

$$I = \iiint_T r^2 \delta \, dV = \iiint_T r^2 \cdot \frac{M}{V} \, dV = M \cdot \frac{1}{V} \iiint_T r^2 \, dV,$$

that is, the moment of inertia equals the total mass of the object times the average value of r^2 (recall Section 3.7).

Of course, in the case of a thin plate occupying a region R, we have

$$I = M \cdot \frac{1}{A} \iint_R r^2 \, dA;$$

hence, once again, the moment of inertia equals the total mass of the object times the average value of r^2

You can find tables of formulas for moments of inertia for various standard objects/shapes; but let's see where three of these come from. Part of the point of working through these standard examples is so that, if you are confronted by a non-standard object/shape, you will know how to calculate the moment of inertia from the definition.

Example 3.10.5. Consider a thin circular plate, of radius a and of uniform area-density $\delta_{\mathrm{ar}} = M/A$, where M is the total mass of the plate and A is its area. Find the moment of inertia of the plate about an axis perpendicular to the plate, passing through its center.

Solution:

We set up the z-axis to be the axis of rotation, so that r, the distance to the axis is also the r from cylindrical coordinates. Then, the plate occupies the disk R in the xy-plane where $r \leq a$. We calculate

$$I \;=\; \frac{M}{A} \iint_R r^2 \, dA \;=\; \frac{M}{A} \int_0^{2\pi} \int_0^a r^2 \cdot r \, dr \, d\theta \;=\;$$

$$\frac{M}{\pi a^2} \cdot 2\pi \cdot \frac{a^4}{4} \;=\; \frac{1}{2} M a^2.$$

Figure 3.10.1: A disk of radius a, with a perpendicular axis through its center.

Example 3.10.6. Consider a ball B (a filled-in sphere) of radius R, with constant density $\delta = M/V$, where M is the total mass and V is the volume. Find the moment of inertia about an axis through the center.

Solution:

Once again, we set up the z-axis to be the axis of rotation, so that r, the distance to the axis is also the r from cylindrical coordinates.

We use cylindrical coordinates, and find

$$I \;=\; \frac{M}{V} \iiint_B r^2 \, dV \;=\; \frac{M}{4\pi R^3/3} \int_0^{2\pi} \int_0^R \int_{-\sqrt{R^2-r^2}}^{\sqrt{R^2-r^2}} r^2 \cdot r \, dz \, dr \, d\theta \;=\;$$

$$\frac{3M}{4\pi R^3} \int_0^{2\pi} \int_0^R 2r^3 \sqrt{R^2 - r^2} \, dr \, d\theta.$$

Making the substitution $u = R^2 - r^2$, so that $2r \, dr = -du$, we obtain

$$I \;=\; \frac{3M}{4\pi R^3} \int_0^{2\pi} \int_{R^2}^0 (R^2 - u) u^{1/2} (-du) \, d\theta \;=\;$$

Figure 3.10.2: A ball of radius R, with an axis through its center.

$$\frac{3M}{4\pi R^3} \int_0^{2\pi} \int_0^{R^2} (R^2 u^{1/2} - u^{3/2}) \, du \, d\theta \;=\; \frac{3M}{4\pi R^3} \int_0^{2\pi} \left(\frac{2R^2 u^{3/2}}{3} - \frac{2u^{5/2}}{5} \right) \Big|_0^{R^2} d\theta \;=\;$$

$$\frac{3M}{4\pi R^3} \cdot 2\pi \cdot \left(\frac{2R^5}{3} - \frac{2R^5}{5} \right) \;=\; \frac{2}{5} M R^2.$$

Figure 3.10.3: A cylinder of radius a and length L, with an axis at one end.

Example 3.10.7. Consider a solid right circular cylinder C, of radius a, length (or height) L, and of constant density $\delta = M/V$. Find the moment of inertia about an axis at one end of cylinder, parallel to one of the circular ends, lying along a diameter; see Figure 3.10.3.

Solution:

Here, we have a mild notational problem. We would like to view C as being vertical, as in Figure 3.10.3, so that we can use cylindrical coordinates to describe the solid region. Hence, we pick the y-axis to be the axis of rotation. The problem is that the distance from a point in C to the axis of rotation is no longer the r from cylindrical coordinates; it is the distance to the y-axis, i.e., $\sqrt{x^2 + z^2}$. Thus, the r^2 that we normally use in the integral for the moment of inertia become $x^2 + z^2$.

Aside from this, the problem is fairly easy. We find

$$I \ = \ \iiint_C (x^2 + z^2)\, dm \ = \ \iiint_C (x^2 + z^2)\, \delta\, dV \ =$$

$$\frac{M}{V} \int_0^{2\pi} \int_0^a \int_0^L (r^2 \cos^2\theta + z^2) \cdot r\, dz\, dr\, d\theta \ =$$

$$\frac{M}{\pi a^2 L} \int_0^{2\pi} \int_0^a \left(r^3 L \cos^2\theta + \frac{rL^3}{3} \right) dr\, d\theta \ = \ \frac{M}{\pi a^2 L} \int_0^{2\pi} \left(\frac{a^4 L \cos^2\theta}{4} + \frac{a^2 L^3}{6} \right) d\theta \ =$$

$$\frac{M}{\pi a^2 L} \left[\frac{\pi a^2 L^3}{3} + \frac{a^4 L}{4} \int_0^{2\pi} \frac{1 + \cos(2\theta)}{2}\, d\theta \right] \ = \ \frac{M}{\pi a^2 L} \left[\frac{\pi a^2 L^3}{3} + \frac{\pi a^4 L}{4} \right] \ =$$

$$M \left[\frac{L^2}{3} + \frac{a^2}{4} \right] \ = \ \frac{1}{3} M L^2 + \frac{1}{4} M a^2.$$

Online answers to select exercises are here.

3.10.1   Exercises

Basics:

1. Let T be the solid region between the two cylinders given by $x^2 + y^2 = 1$ and $x^2 + y^2 = 4$, and between the two planes where $z = 0$ and $z = 3$. Suppose that an object of mass M, and constant density, occupies the region T.

(a) Calculate the moment of inertia of the object about the z-axis.

(b) Calculate the moment of inertia of the object about the y-axis.

2. Let T be the solid region between the two cylinders given by $x^2 + y^2 = a^2$ and $x^2 + y^2 = b^2$, and between the two planes where $z = 0$ and $z = L$, where $b > a > 0$ and $L > 0$. Suppose that an object of mass M, and constant density, occupies the region T.

(a) Calculate the moment of inertia of the object about the z-axis.

(b) Calculate the moment of inertia of the object about the y-axis.

3. Let T be the solid half-cone above the graph of $z = \sqrt{x^2 + y^2}$ and below the plane where $z = 5$. Suppose that an object of mass M, and constant density, occupies the region S.

(a) Calculate the moment of inertia of the object about the z-axis.

(b) Calculate the moment of inertia of the object about the y-axis.

4. Consider a thin circular plate, of radius a, which lies in the xy-plane, with its center at the origin. Suppose that the plate has variable area-density $\delta_{\text{ar}} = kr$, where $k > 0$ and r is the distance to the origin. Find the moment of inertia of the plate about the z-axis.

5. Consider a ball, of radius R, centered at the origin. Suppose that the ball has variable density $\delta = k\sqrt{R^2 - r^2}$, where $k > 0$ and r is the distance to the z-axis. Find the moment of inertia of the ball about the z-axis.

6. (**Parallel Axis Theorem**) Suppose that a solid object, of mass M, but, possibly, variable density, occupies a region T in $\mathbb{R}^3$, and let A denote a line through the center of mass of the object. Denote the moment of inertia of the object about A by I_{cm}. Now let L be a line parallel to A, let D denote the distance between A and L, and let I_L denote the moment of inertia of the object about L. You will show in steps that:
$$I_L = I_{\text{cm}} + MD^2,$$
assuming that all of the relevant integrals are defined.

Set up a coordinate system with the center of mass of the object at the origin, the line A being the z-axis, and the line L being given by $x = D$ and $y = 0$ (and z is anything).

(a) In terms of x and y, what is the distance r_{cm} from a point in T to the line A?

(b) In terms of x and y, what is the distance r_L from a point in T to L?

(c) Show that

$$\iiint_T r_L^2 \, dm \;=\; I_{\mathrm{cm}} \;+\; \iiint_T (-2Dx + D^2) \, dm.$$

(d) Explain why the origin being at the center of mass implies that $\iiint_T -2Dx \, dm$ is 0.

(e) Conclude that
$$I_L \;=\; I_{\mathrm{cm}} \;+\; MD^2.$$

7. Consider a ball B of radius R, with constant density $\delta = M/V$, where M is the total mass and V is the volume. Use the Parallel Axis Theorem to find the moment of inertia of the ball about an axis which is a distance D from the center of the ball. ▶

8. Let T be the solid half-cone above the graph of $z = \sqrt{x^2 + y^2}$ and below the plane where $z = 5$. Suppose that an object of mass M, and constant density, occupies the region T. Use the Parallel Axis Theorem to find the moment of inertia of the object about a line L, which is parallel to the z-axis, at a distance D.

3.11 Surfaces and Area

In this section, we return to looking at parameterized surfaces, as we did in Section 2.8. But, now, rather than being interested in local objects, like the tangent plane, we are interested in using a parameterization to calculate the area of an entire surface.

This will be of great importance to us in Section 4.5, when we want to calculate the flux of a vector field through a surface.

Basics:

Suppose that we are given a surface M in $\mathbb{R}^3$. We might be given M as the graph of a C^1 function of two variables, or as a level surface of a C^1 function of three variables, or as the range of a C^1 parameterization $\mathbf{r}(u, v)$.

The question is: how do you find the area of the surface, i.e., how do you find the *surface area*?

The answer, of course, is that you use integration to add up all of the infinitesimal pieces of area. But this leads us to two new questions: how do you find the infinitesimal surface area, i.e., an *element of area*, and what do you use for limits of integration in your iterated integrals?

It turns out that these questions are relatively easy to deal with for parameterized surfaces. In Section 2.8, we looked at local regular parameterizations, and we will need to put conditions on our parameterizations here, too.

Definition 3.11.1. *A* **basic parameterization** *of a surface consists of:*

1. *a subset D of $\mathbb{R}^2$, the* **domain of the parameterization***, such that the interior* int(D) *of D is non-empty and connected, D is contained in the closure of* int(D)*, and such that the boundary of D is empty or consists of a finite number of points and/or curves (of class C^1);*

2. *a continuously differentiable function $\mathbf{r} = \mathbf{r}(u, v)$, the* **parameterization***, from D into $\mathbb{R}^3$ (where, if D is not open, this means that $\mathbf{r}$ is the restriction of a continuously differentiable function defined on a bigger open set) such that:*

3. *the restriction of $\mathbf{r}$ to* int(D) *is one-to-one, and*

4. *the restriction of $\mathbf{r}$ to* int(D) *is* **regular***, i.e., at all points (u, v) in* int(D)*, the cross product $\mathbf{r}_u \times \mathbf{r}_v$ of the partial derivatives of $\mathbf{r}$ is non-zero (recall Remark 2.8.5).*

We could weaken the conditions somewhat if we wanted. We really need for the boundary of D to have measure zero in $\mathbb{R}^2$, and for $\mathbf{r}$ to be one-to-one and regular when restricted to the complement of a set of measure zero; recall Definition 3.2.14. However, in practice, we always use a basic surface parameterization as given in the definition.

> *The **basic surface** defined by a basic surface parameterization* $\mathbf{r} : D \to \mathbb{R}^3$ *is the image of* $\mathbf{r}$.

Alright. Given this definition of a basic surface parameterization, it should be pretty clear that we're going to define the area of the parameterized surface by an integral, provided it exists. We let

$$\text{surface area} \; = \; \iint_D dS,$$

We reserve dA to denote an element of area in in $\mathbb{R}^2$, i.e., we use dA in place of dS when the surface is flat, and parallel to one of the coordinate planes.

where dS is an element of (surface) area, that is, dS is an infinitesimal amount of area on the surface. But, this just pushes back the question to: how do you calculate dS in terms of the parameterization $\mathbf{r}$?

Fix a point (u_0, v_0) in the domain of $\mathbf{r}$. Then, if you look at the image of $\alpha(u) = \mathbf{r}(u, v_0)$, i.e., if you fix $v = v_0$ and let u vary, you obtain a curve on the surface M; this is the image via $\mathbf{r}$ of (part of) the line where $v = v_0$. Similarly, if you look at the image of $\beta(u) = \mathbf{r}(u_0, v)$, i.e., if you fix $u = u_0$ and let v vary, you obtain another curve on the surface M; this is the image via $\mathbf{r}$ of (part of) the line where $u = u_0$. In this way, the uv-coordinate grid gets sent by $\mathbf{r}$ to a curved "grid" on M. What we'd like to do is find an expression for the area of the infinitesimal "curved parallelograms" in this resulting curved grid on M.

Let's look back at Figure 2.8.5, but now imagine the curved "grid lines", where either u or v is constant as being infinitesimally close together; we give the figure again in Figure 3.11.1.

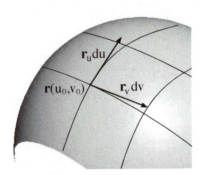

Figure 3.11.1: A parameterized surface, with curves where u and v are constant, and the infinitesimal tangent vectors $\mathbf{r}_u \, du$ and $\mathbf{r}_v \, dv$.

As usual, we think of du and dv as infinitesimal (positive) changes in u and v, respectively. Then, adjacent sides of one of the infinitesimal curved parallelograms on M are spanned by the vector from $\mathbf{r}(u_0, v_0)$ to $\mathbf{r}(u_0 + du, v_0)$ and the vector $\mathbf{r}(u_0, v_0)$ to $\mathbf{r}(u_0, v_0 + dv)$. But these vectors are precisely $\mathbf{r}_u(u_0, v_0) \, du$ and $\mathbf{r}_v(u_0, v_0) \, dv$. Now, we know from Section 1.5 that the area of a parallelogram, in terms of the vectors spanning adjacent sides, is given by the magnitude of the cross product.

Thus, we have:

Theorem 3.11.2. *Given a basic parameterization* $\mathbf{r} : D \to \mathbb{R}^3$, $\mathbf{r} = \mathbf{r}(u,v)$, *for a surface, an element of area on the surface is given by*

$$dS = |\mathbf{r}_u \times \mathbf{r}_v| \, du \, dv,$$

and the surface area of the surface parameterized by $\mathbf{r}$ *is*

$$\text{surface area} = \iint_D dS = \iint_D |\mathbf{r}_u \times \mathbf{r}_v| \, du \, dv,$$

provided that this integral exists.

Of course, a surface could be complicated enough that you need to break the surface up into multiple pieces, and then find basic parameterizations for each piece. This is fine, provided that there are a finite number of pieces and that the overlaps have no area, e.g., when the overlaps consist of a finite number of points and/or curves of class C^1.

In the study of differential geometry, "area" is typically defined for surfaces in terms of the *area form*, or the *2-dimensional volume form*. This requires you to have a *orientable C^1 manifold*. While an entire basic surface may not be an orientable C^1 manifold, the image of the interior of the domain of the parameterization is. This is enough for us here, but we shall have to deal with orientations in some detail in Section 4.5.

An important special case of a parameterized surface, and the area element, is given by the graph of a function $z = f(x,y)$. The standard parameterization of this surface is given by

$$\mathbf{r}(u,v) = (u, \ v, \ f(u,v)),$$

but there's no good reason to introduce new variables as parameters; we can simply use x and y as parameters, and use

$$\mathbf{r}(x,y) = (x, \ y, \ f(x,y)).$$

If f is continuously differentiable, then so is $\mathbf{r}$, it is automatically one-to-one, and you easily calculate

$$\mathbf{r}_x \times \mathbf{r}_y = (1, \ 0, \ f_x) \times (0, \ 1, \ f_y) = (-f_x, \ -f_y, \ 1).$$

Therefore, $\mathbf{r}$ is also automatically regular everywhere, and its area element is given by

$$dS = |\mathbf{r}_x \times \mathbf{r}_y| \, dy \, dx = \sqrt{f_x^2 + f_y^2 + 1} \, dy \, dx.$$

Hence, we have:

Theorem 3.11.3. *Given a continuously differentiable function* $f = f(x,y)$, *defined on a subset D of $\mathbb{R}^2$ which has non-empty interior and whose boundary (if there is one) consists of a finite number of points and/or curves (of class C^1), the graph M of $f : D \to \mathbb{R}$ is a surface with a basic parameterization* $\mathbf{r} : D \to \mathbb{R}^3$ *given by*

$$\mathbf{r}(x,y) = \big(x, \ y, \ f(x,y)\big).$$

Furthermore,

$$\mathbf{r}_x \times \mathbf{r}_y \;=\; (-f_x,\; -f_y,\; 1)$$

and

$$dS \;=\; \sqrt{f_x^2 + f_y^2 + 1}\, dy\, dx.$$

Thus, the surface area of M is given by

$$\iint_D \sqrt{f_x^2 + f_y^2 + 1}\, dy\, dx,$$

provided this integral exists.

Example 3.11.4. Let's begin with an example where we already know the answer; let's find the surface area of a sphere of radius R. Without loss of generality, we'll assume that the sphere is centered at the origin.

We have at least two reasonable choices for how to parameterize this; we can use spherical coordinates to parameterize the entire sphere, or we can use $z = \sqrt{R^2 - x^2 - y^2}$ to parameterize the top half of the sphere, and then double the answer. We'll do both, as an example, and to show that we obtain the same answer either way.

Using spherical coordinates:

The sphere of radius R, centered at the origin, is parameterized by the θ and ϕ from spherical coordinates (recall Section 3.6), with ρ being fixed at R. Therefore, we use

$$x \;=\; R \sin\phi \cos\theta, \qquad y \;=\; R \sin\phi \sin\theta, \qquad \text{and} \qquad z \;=\; R \cos\phi,$$

i.e.,

$$\mathbf{r}(\theta, \phi) \;=\; (R \sin\phi \cos\theta,\; R \sin\phi \sin\theta,\; R \cos\phi) \;=\; R(\sin\phi \cos\theta,\; \sin\phi \sin\theta,\; \cos\phi),$$

where $0 \le \theta \le 2\pi$ and $0 \le \phi \le \pi$.

We find

$$\mathbf{r}_\theta \;=\; R(-\sin\phi \sin\theta,\; \sin\phi \cos\theta,\; 0),$$

$$\mathbf{r}_\phi \;=\; R(\cos\phi \cos\theta,\; \cos\phi \sin\theta,\; -\sin\phi),$$

and we easily calculate

$$\mathbf{r}_\theta \times \mathbf{r}_\phi \;=\; R^2(-\sin^2\phi \cos\theta,\; -\sin^2\phi \sin\theta,\; -\sin\phi \cos\phi \sin^2\theta - \sin\phi \cos\phi \cos^2\theta) \;=$$

$$-R^2 \sin\phi (\sin\phi \cos\theta,\; \sin\phi \sin\theta,\; \cos\phi).$$

Thus,

$$|\mathbf{r}_\theta \times \mathbf{r}_\phi| \;=\; R^2 \sin\phi \sqrt{\sin^2\phi\cos^2\theta + \sin^2\phi\sin^2\theta + \cos^2\phi} \;=\; R^2 \sin\phi,$$

and

$$dS \;=\; R^2 \sin\phi \, d\phi \, d\theta.$$

Hence,

$$\text{surface area} \;=\; \int_0^{2\pi}\int_0^\pi R^2 \sin\phi \, d\phi \, d\theta \;=\; 2\pi R^2 \int_0^\pi \sin\phi \, d\phi \;=\; 4\pi R^2,$$

which, of course, is what we knew we'd get.

> Note that $\mathbf{r}_\theta \times \mathbf{r}_\phi = \mathbf{0}$ when $\phi = 0$ or $\phi = \pi$, so that $\mathbf{r}$ is not regular at such points. However, $\phi = 0$ and $\phi = \pi$ correspond to points on the boundary of the domain of $\mathbf{r}$, and Definition 3.11.1 allows $\mathbf{r}$ not to be regular on the boundary.

Using Cartesian coordinates:

Now, let's calculate the area again, by doubling the area of the graph of

$$z \;=\; f(x, y) = \; \sqrt{R^2 - x^2 - y^2},$$

where the domain is the disk where $x^2 + y^2 \leq R^2$.

We find

$$f_x \;=\; \frac{-x}{\sqrt{R^2 - x^2 - y^2}} \qquad \text{and} \qquad f_y \;=\; \frac{-y}{\sqrt{R^2 - x^2 - y^2}}.$$

Note that there is a technical problem here: f is not differentiable on the boundary circle where $x^2 + y^2 = R^2$, and yet we wish to use Theorem 3.11.3.

We will get around this problem by finding the surface area of the graph over the disk D_a where $x^2 + y^2 \leq a^2$, with $0 < a < R$, and taking the limit as $a \to R$ from the left.

Using Theorem 3.11.3, we find that the area of the graph of $z \;=\; f(x, y)$, over D_a, is given by

$$\iint_{D_a} \sqrt{\frac{x^2}{R^2 - x^2 - y^2} + \frac{y^2}{R^2 - x^2 - y^2} + 1} \;\; dy \, dx.$$

Switching to polar coordinates in the xy-plane, we find that this integral is

$$\int_0^{2\pi}\int_0^a \frac{R}{\sqrt{R^2 - r^2}} \; r \, dr \, d\theta.$$

We leave it as an exercise for you to make the substitution $u = R^2 - r^2$ and find that this integral equals

$$2\pi R^2 \;-\; 2\pi R \sqrt{R^2 - a^2}.$$

Therefore, we recover that the surface area of the entire sphere is

$$2 \cdot \lim_{a \to R^-} \left(2\pi R^2 - 2\pi R\sqrt{R^2 - a^2} \right) = 4\pi R^2.$$

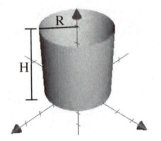

Figure 3.11.2: A right circular cylinder.

Example 3.11.5. Let's look at another example where we already know the answer. Let's take a right circular cylinder (with no top or bottom), of radius R and height H, and show that it has the surface area that we expect: $2\pi RH$.

We will center our cylinder around the z-axis, with its base in the xy-plane; see Figure 3.11.2.

How do you come up with a parameterization for the cylinder? You think of how to specify a location on the cylinder with two real numbers. An easy way would be to specify the height of the point, its z-coordinate, and to specify how far you go around the cylinder, i.e., specify an angle θ in cylindrical coordinates.

Thus, we have

$$x = R\cos\theta, \qquad y = R\sin\theta, \qquad \text{and} \qquad z = z,$$

i.e.,

$$\mathbf{r}(\theta, z) = (R\cos\theta, \ R\sin\theta, \ z),$$

where $0 \le \theta \le 2\pi$ and $0 \le z \le H$.

We calculate

$$\mathbf{r}_\theta = R(-\sin\theta, \ \cos\theta, \ 0),$$

$$\mathbf{r}_z = (0, \ 0, \ 1),$$

and

$$\mathbf{r}_\theta \times \mathbf{r}_z = R(\cos\theta, \ \sin\theta, 0).$$

Therefore,

$$|\mathbf{r}_\theta \times \mathbf{r}_z| = R,$$

and

$$\text{surface area} = \int_0^{2\pi} \int_0^H R\, dz\, d\theta = 2\pi RH,$$

as we expected.

Example 3.11.6. Now let's look at an example where you probably don't know the answer. Consider the surface M which is a (one-sided) right circular cone of radius R and height H.

We position the cone as shown in Figure 3.11.3. If we let r denote the radius of the circle in each z cross section, then, using similar triangles, we find that $r/R = z/H$, i.e.,

$$r = \frac{R}{H} z.$$

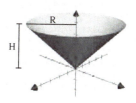

Figure 3.11.3: A right circular cone.

Now, as in the previous example, we specify a position on the cone by its height, z, and the angle θ indicating how far around you go. As we already have an r in the problem, we will denote our parameterization by $\mathbf{p}$. We have

$$\mathbf{p}(\theta, z) = (r\cos\theta,\ r\sin\theta,\ z) = \left(\frac{R}{H} z\cos\theta,\ \frac{R}{H} z\sin\theta,\ z\right) =$$

$$z\left(\frac{R}{H}\cos\theta,\ \frac{R}{H}\sin\theta,\ 1\right),$$

where $0 \le \theta \le 2\pi$ and $0 \le z \le H$.

Now, we calculate

$$\mathbf{p}_\theta = z\left(-\frac{R}{H}\sin\theta,\ \frac{R}{H}\cos\theta,\ 0\right),$$

$$\mathbf{p}_z = \left(\frac{R}{H}\cos\theta,\ \frac{R}{H}\sin\theta,\ 1\right),$$

and

$$\mathbf{p}_\theta \times \mathbf{p}_z = z\left(\frac{R}{H}\cos\theta,\ \frac{R}{H}\sin\theta,\ -\frac{R^2}{H^2}\right) = \frac{Rz}{H}\left(\cos\theta,\ \sin\theta,\ -\frac{R}{H}\right).$$

Hence,

$$|\mathbf{p}_\theta \times \mathbf{p}_z| = \frac{Rz}{H}\sqrt{1 + \frac{R^2}{H^2}},$$

and

Note that $\mathbf{p}_\theta \times \mathbf{p}_z = \mathbf{0}$ when $z = 0$, so that $\mathbf{p}$ is not regular at such points. However, $z = 0$ corresponds to points on the boundary of the domain of $\mathbf{p}$.

$$\text{surface area} = \int_0^{2\pi}\int_0^H \frac{Rz}{H}\sqrt{1 + \frac{R^2}{H^2}}\, dz\, d\theta = \frac{R}{H}\sqrt{1 + \frac{R^2}{H^2}} \cdot 2\pi \cdot \frac{H^2}{2} =$$

$$\pi R\sqrt{H^2 + R^2}.$$

The quantity $\sqrt{H^2 + R^2}$ is the length of the hypotenuse of the right triangle with legs of length H and R, and is called the *slant height*. If we let L denote this slant height, our formula for the surface area of at the cone becomes very simple; the surface area is πRL.

More Depth:

Later, in Section 4.5, Section 4.6, and Section 4.7, we will want to deal with surfaces that are more general than the images of basic parameterizations. What we want is to look at surface which break up into a finite number of pieces, where each piece is a basic regular surface.

Thus, we define

Definition 3.11.7. *A **piecewise-regular surface** M is one which can be written as a finite union of basic surfaces,*

$$M = M_1 \cup M_2 \cup \cdots \cup M_k,$$

where, for any two of the basic surfaces M_i and M_j, where $i \neq j$, the intersection $M_i \cap M_j$ is empty or contained in a finite union of points and piecewise-regular curves.

Note that this definition allows for surfaces in which the pieces cross through each other. Later, we will need to add conditions to prevent this. Typically, we will need for our surface to be a *submanifold* or *submanifold with boundary*, in the topological sense. See Definition 4.5.8.

Since we have allowed the pieces in piecewise-regular surfaces to intersect each other in, at most, sets which have no area, we immediately conclude:

Theorem 3.11.8. *The surface area of a piecewise-regular surface is the sum of the surface areas of the basic surface pieces.*

Figure 3.11.4: A right circular cone, with a top.

Example 3.11.9. Let's return to the right circular cone of radius R and height H, from Example 3.11.6, except that we now close the surface by putting a top on it. Call this closed cone M.

Then, M is certainly a piecewise-regular surface; it's the union of the cone, which we know is a basic surface from Example 3.11.6, and the disk of radius R that closes the top, and the cone and the disk intersect along a circle.

Hence, the surface area of M is what you would have known it was, even without reading Theorem 3.11.8: the sum of the surface area of the cone and the area of the disk. Back in Example 3.11.6, we found the surface area of the cone to be $\pi R \sqrt{H^2 + R^2}$, and so

$$\text{surface area of } M = \pi R \sqrt{H^2 + R^2} + \pi R^2.$$

3.11.1 Exercises

Online answers to select exercises are here.

In each of the following exercises, you are given a function $z = f(x, y)$, and a region R in the xy-plane. (a) Calculate the area element dS of the graph of f, in terms of x, y, dx, and dy. (b) Calculate the (total) area of the graph of f (restricted to the domain R).

Basics:

1. $z = x + y$, and R is given by $0 \leq x \leq 1$ and $0 \leq y \leq 1$.

2. $z = ax + by + c$, where a, b, and c are constants, and R is given by $0 \leq x \leq 1$ and $0 \leq y \leq 1$.

3. $z = (2/3)(x^{3/2} + y^{3/2})$, and R is given by $4 \leq x \leq 12$ and $4 \leq y \leq 12$.

4. $z = x^2$, and R is given by $0 \leq x \leq 1$ and $0 \leq y \leq x$.

5. $z = x^2 + y^2$, and R is the disk of radius 5, centered at the origin in the xy-plane.

6. $z = 2x^2 + 2y^2$, and R is the disk of radius 5, centered at the origin in the xy-plane.

7. $z = x^2 - y^2$, and R is the disk of radius 5, centered at the origin in the xy-plane.

8. $z = \sqrt{36 - x^2 - y^2}$, and R is the disk of radius 5, centered at the origin in the xy-plane.

In each of the following exercises, you are given a region D in the uv-plane, and a basic parameterization $\mathbf{r} : D \to \mathbb{R}^3$. (a) Calculate the area element dS, in terms of u, v, du, and dv. (b) Set up the appropriate double integral for the area of the corresponding surface (the image of $\mathbf{r}$). (c) If instructed to do so, evaluate your integral from part (b) to calculate the surface area.

9. D is a disk of radius 3, centered at the origin, in the uv-plane, and $\mathbf{r}(u, v) = (au, bv, 7)$, where a and b are non-zero constants. Calculate the surface area.

10. D is given by $0 \leq u \leq 1$ and $0 \leq v \leq 1$, and $\mathbf{r}(u, v) = (u, v, u + v)$. Calculate the surface area.

11. D is given by $0 \leq u \leq 2$ and $0 \leq v \leq 3$, and $\mathbf{r}(u, v) = (u^2, v^2, \sqrt{2}\, uv)$. Calculate the surface area.

12. D is given by $2 \leq u \leq 2\sqrt{3}$ and $2 \leq v \leq 2\sqrt{3}$, and

$$\mathbf{r}(u, v) = \left(\frac{u^2}{2}, \frac{v^2}{2}, \frac{u^3 + v^3}{3} \right).$$

Calculate the surface area.

13. D is given by $0 \leq u \leq 2\pi$ and $0 \leq v \leq 1$, and $\mathbf{r}(u,v) = (v \cos u, v \sin u, v^3)$.

14. D is given by $0 \leq u \leq 1$ and $0 \leq v \leq u$, and $\mathbf{r}(u,v) = (v, e^u, u^3)$.

In each of the following exercises, a surface M is described. Parameterize M, calculate dS, and give an integral for the surface area of M. Do NOT calculate the value of this integral.

15. M is the portion of the plane where $2x + 3y - 6z = 4$, and $0 \leq x \leq 1$ and $x \leq y \leq 1$.

16. M is the portion of the plane where $2x + 3y - 6z = 4$, and $0 \leq x \leq 1$ and $x \leq z \leq 1$.

17. M is the portion of the sphere where $x^2 + y^2 + z^2 = 36$, which is above the 1st quadrant in the xy-plane.

18. M is the ellipsoid given by $\dfrac{x^2}{9} + \dfrac{y^2}{4} + z^2 = 1$.

More Depth:

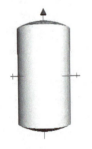

Figure 3.11.5

19. Calculate the surface area of the piecewise-regular surface $M_1 \cup M_2$, where M_1 is the portion of the sphere where $x^2 + y^2 + z^2 = 4$ and $z \leq 1$, and M_2 is the disk that fills in the hole in the cut-off sphere. See Figure 3.11.5.

20. Calculate the surface area of the piecewise-regular surface which is the boundary of the solid region inside the sphere of radius 2, centered at the origin, and inside the cylinder given by $x^2 + y^2 = 1$. See Figure 3.11.6.

21. Calculate the surface area of the piecewise-regular surface which is the boundary of the solid region above the half-cone where $z = \sqrt{x^2 + y^2}$ and below the half-cone where $z = 5 - \sqrt{x^2 + y^2}$. See Figure 3.11.7.

22. As we saw in Exercise 26 in Section 3.9, it is generally not true that the gravitational force exerted by a solid object of mass M is the same as that of a point-mass of M, located at the center of mass of the object.

However, the gravitational force exerted by a sphere of uniform area-density is exactly the same as that of a point-mass located at the center, containing all of the mass of the sphere. By considering a solid ball as being made up of spherical shells, we conclude that a solid ball of total mass m, in which the density is constant at each fixed distance from the center, will produce a gravitational force equal to that of a point-mass of m at the center of the ball.

In this exercise, you will verify the above claim about spheres of uniform area-density.

Figure 3.11.6

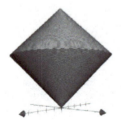

Figure 3.11.7

Suppose that we have a point-mass of M at the origin, and a sphere X of radius R, centered at $(0, 0, a)$, where $a > R > 0$. Suppose that the sphere has mass m and a uniform area-density; thus, $\delta_{\mathrm{ar}} = m/(4\pi R^2)$. Writing dS for an infinitesimal piece of surface area, the corresponding infinitesimal mass is

$$dm = \delta_{\mathrm{ar}} \, dS = \frac{m}{4\pi R^2} \, dS.$$

According to Newton's Law of Universal Gravitation, an infinitesimal mass dm, located at any point $\mathbf{r} = (x, y, z) \neq (0, 0, 0)$ in space, exerts an infinitesimal gravitational force $d\mathbf{F}$ on M given by

$$d\mathbf{F} = \frac{GM \, dm}{|\mathbf{r}|^2} \left(\frac{\mathbf{r}}{|\mathbf{r}|} \right).$$

We want to integrate $d\mathbf{F}$ over the entire sphere and see that we end up the force that would be exerted by a point-mass of m, located at $(0, 0, a)$. Thus, we want to show that

$$\frac{GMm}{a^2}(0, 0, 1) = \iint_X \frac{GM}{|\mathbf{r}|^2} \left(\frac{\mathbf{r}}{|\mathbf{r}|} \right) dm = \frac{GMm}{4\pi R^2} \iint_X \frac{\mathbf{r}}{|\mathbf{r}|^3} \, dS.$$

The symmetry of the sphere around the z-axis allows us to conclude that the x and y components of this double integral are, in fact, 0. Thus, it remains for you to show that the z component satisfies the equality above, i.e., after canceling and rearranging, that

$$\iint_X \frac{z}{(x^2 + y^2 + z^2)^{3/2}} \, dS = \frac{4\pi R^2}{a^2}.$$

Accomplish this by parameterizing X, using z as one parameter, and using the angle θ as your other parameter, where θ indicates the position on the circles obtained by taking z-cross sections of the sphere.

23. The famous *Möbius Strip*, or *Möbius Band*, is the basic surface given by the parameterization

$$\mathbf{r}(u, v) = \left(\left(R - u \sin \left(\frac{v}{2} \right) \right) \cos v, \ \left(R - u \sin \left(\frac{v}{2} \right) \right) \sin v, \ u \cos \left(\frac{v}{2} \right) \right),$$

where L and R are constants such that $0 < L < R$, and $-L \leq u \leq L$ and $0 \leq v \leq 2\pi$. Other parameterizations can also be used.

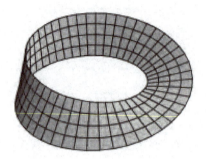

Figure 3.11.8: One view of the Möbius Strip

This describes a line segment of length $2L$, parallel to the z-axis, at a distance R from the z-axis, which is revolved around the z-axis, as the line rotates about its center, to flip over $180°$ as the line segment makes a full revolution about the z-axis. You can imagine taking a thin strip of paper, twisting one end $180°$, and taping or gluing it to the other end.

(a) Verify that $\mathbf{r}$ is **not** one-to-one, by showing that, in fact, the line segment does "flip" as v goes from 0 to 2π. Show this by verifying that

$$\mathbf{r}(u, 0) = \mathbf{r}(-u, 2\pi),$$

for $-L \leq u \leq L$.

(b) Calculate $\mathbf{r}_u \times \mathbf{r}_v$.

(c) Calculate $|\mathbf{r}_u \times \mathbf{r}_v|$, and give the element of area dS on the Möbius Strip.

(d) Show that $(\mathbf{r}_u \times \mathbf{r}_v)(u, 0) = -(\mathbf{r}_u \times \mathbf{r}_v)(-u, 2\pi)$, even though, by (a), the parameterization gives you the same point at $(u, 0)$ and $(-u, 2\pi)$.

Part (d) is a reflection of the fact that the Möbius Strip is not *orientable*; this means that it is not possible to make a continuous, consistent, selection of "one side" of the Möbius Strip. We shall discuss orientability in some depth in Section 4.5.

Chapter 4

Integration and Vector Fields

Throughout the earlier sections of this book, we have dealt with individual vectors or, perhaps, as in the case of the tangent plane, a collection of vectors, which were all based at the same point.

However, in many physical settings, there are vector quantities associated with each point in a region in $\mathbb{R}^2$ or $\mathbb{R}^3$. Gravitational and electromagnetic force fields are important examples. The collection of velocity vectors of a flowing fluid is another important example.

When we have a collection of vectors, defined at each point in some region, we call the collection a *vector field*. When we want to calculate how much work is done on an object as it moves through a force field, or want to calculate the rate at which fluid is flowing into, or out of, a given surface, given the velocity vector field of the fluid, we have to calculate integrals that involve vector fields.

Hence, the subject matter in this chapter is frequently referred to simply as "integrating vector fields".

4.1   Vector Fields

Suppose that we have a region R in $\mathbb{R}^n$. A *vector field* on R consists of a collection of vectors $\mathbf{F}(\mathbf{p})$ in $\mathbb{R}^n$, one for each point $\mathbf{p}$ in R. In this context, it is important that $\mathbf{F}(\mathbf{p})$ is thought of as a vector based at $\mathbf{p}$.

Vector fields arise in many important physical situations, including the gravitational vector field, electromagnetic vector fields, and the velocity vector fields of flowing fluids.

<hr>

Basics:

Before we give the definition of a vector field, we should motivate the definition with an important example.

<hr>

The value of G is approximately 6.67384×10^{-11} m^3/(kg·s^2).

Example 4.1.1. Suppose that a point-mass M is located at a point (x_0, y_0, z_0) in space. Then, according to Newton's Law of Universal Gravitation, at each point $(x, y, z) \neq (x_0, y_0, z_0)$ in space, the mass M exerts a gravitational force $\mathbf{F} = \mathbf{F}(x, y, z)$ on any point-mass m, located at (x, y, z); that force has magnitude GMm/d^2, where G is the universal gravitational constant, and d is the distance between the point-masses. In addition, the vector $\mathbf{F}$ points from (x, y, z) towards the point (x_0, y_0, z_0).

It follows that, if we denote the displacement vector by

$$\mathbf{d}(x, y, z) = (x_0, y_0, z_0) - (x, y, z),$$

then the force vector $\mathbf{F}$ at (x, y, z), the force exerted by M on a mass M at (x, y, z), is given by

$$\mathbf{F} \ = \ \mathbf{F}(x, y, z) \ = \ \frac{GMm}{|\mathbf{d}|^2} \cdot \frac{\mathbf{d}}{|\mathbf{d}|} \ = \ \frac{GMm\mathbf{d}}{|\mathbf{d}|^3}.$$

We actually looked at this back in Example 1.2.18, and you may wonder what the difference is now. In Example 1.2.18, the location of m was fixed, and so we had just one force vector. Here, our perspective is different; now we think of a collection of force vectors, created by M, by placing m at every possible point in space (other than at the point (x_0, y_0, z_0)).

<hr>

What we saw above was that gravitational force naturally associates to each point $\mathbf{p} \neq (x_0, y_0, z_0)$, in $\mathbb{R}^3$, a vector $\mathbf{F}(\mathbf{p})$ in $\mathbb{R}^3$, which should be thought of as based at $\mathbf{p}$. This is what is known as a *vector field*.

Definition 4.1.2. *Suppose that $\mathcal{U}$ is a subset of $\mathbb{R}^n$. Then, a **vector field on** $\mathcal{U}$ (or an n-**dimensional vector field on** $\mathcal{U}$, or a **vector field in** $\mathbb{R}^n$) is a function*

$$\mathbf{F} : \mathcal{U} \to \mathbb{R}^n,$$

where, for each $\mathbf{p}$ in $\mathcal{U}$, $\mathbf{F}(\mathbf{p})$ is considered as a vector based at $\mathbf{p}$.

Typically, we're not interested in arbitrary functions $\mathbf{F}$ when looking at vector fields; we want $\mathbf{F}$ to be continuous, or C^1, or C^∞.

Example 4.1.3. You can try to get an idea of what a vector field $\mathbf{F}$ in $\mathbb{R}^2$ or $\mathbb{R}^3$ "looks like" by picking some points $\mathbf{p}$ in the domain of the vector field and drawing each vector $\mathbf{F}(\mathbf{p})$, based at $\mathbf{p}$. You can draw a few vectors by hand, and there many computer applications for drawing vector fields.

However, even if you use computer software, it can be difficult to visualize 3-dimensional vector fields. For instance, in Figure 4.1.1, we had software generate some vectors in the 3-dimensional gravitational vector field

$$\mathbf{F} = \frac{(1-x,\ 1-y,\ 1-z)}{\left((1-x)^2 + (1-y)^2 + (1-z)^2\right)^{3/2}},$$

which is defined on all of $\mathbb{R}^3$ other than at the point $(1, 1, 1)$.

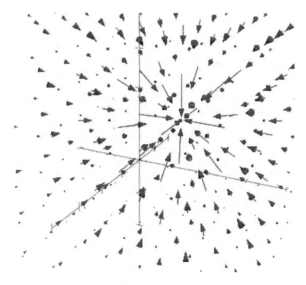

Figure 4.1.1: Gravitational force field exerted by a mass at $(1, 1, 1)$.

We refer to this as a gravitational vector field because, as we discussed in Example 4.1.1, this vector field would represent the gravitational force, exerted by a point-mass M at $(1,1,1)$, on a point-mass m at any point $(x, y, z) \neq (1, 1, 1)$, if M and m are chosen so that the constant GMm is equal to 1.

Note how the vectors get larger in magnitude as you get closer to the point $(1, 1, 1)$.

Example 4.1.4. Vector fields in $\mathbb{R}^2$ are much easier to sketch by hand than vector fields in $\mathbb{R}^3$, and they're easier to examine after a computer has sketched them.

Consider, for example, the vector field on all of $\mathbb{R}^2$ given by

$$\mathbf{v} \; = \; \mathbf{v}(x, y) \; = \; (-y, \, x).$$

How would you sketch this by hand? The easy answer is: pick some points, calculate the corresponding vectors, and then draw the arrows representing the vectors, based at the points you picked.

So, for instance, pick $(x, y) = (0, 0)$. Since $\mathbf{v}(0, 0) = (0, 0)$, we draw the zero vector at the origin, i.e., we just draw a point.

Now we'll pick $(x, y) = (1, 0)$. We find $\mathbf{v}(1, 0) = (0, 1)$. Thus, based at the point $(1, 0)$ we draw the vector $(0, 1)$; this means that you draw an arrow from $(1, 0)$ to $(1, 0) + (0, 1) = (1, 1)$.

So far, our sketch of the vector field $\mathbf{v}$ looks like Figure 4.1.2.

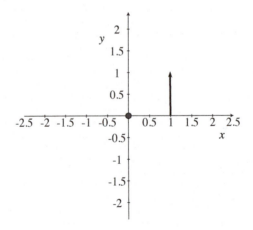

Figure 4.1.2: Two vectors in our vector field.

We'll calculate three more by hand. We find

$$\mathbf{v}(0, 1) = (-1, 0), \qquad \mathbf{v}(-1, 0) = (0, -1), \qquad \text{and} \qquad \mathbf{v}(0, -1) = (1, 0).$$

We add the corresponding vectors to our sketch in Figure 4.1.3.

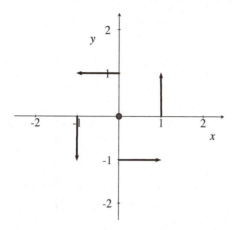

Figure 4.1.3: A few more vectors in our vector field.

Of course, this drawing of arrows by hand gets very tedious very quickly. We can use software to graph MANY vectors in our vector field.

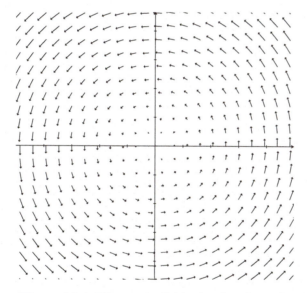

Figure 4.1.4: The vector field $\mathbf{v}(x, y) = (-y, x)$.

You can see that the vectors are circling counterclockwise around the origin, with decreasing magnitude as they get close to the origin, where $\mathbf{v}(0,0) = \mathbf{0}$.

In fact, it's easy to show that the vector field is always perpendicular to the position vector from the origin, and that the magnitude equals the distance from the origin:

$$(x, y) \cdot \mathbf{v}(x, y) \;=\; (x, y) \cdot (-y, x) \;=\; -xy + xy = 0$$

and

$$|\mathbf{v}(x,y)| \;=\; |(-y,x)| \;=\; \sqrt{(-y)^2 + x^2} \;=\; |(x,y)|.$$

How might a vector field such as that in Example 4.1.4 arise physically? You could have a disk, which is spinning counterclockwise around the origin, where the vector field $\mathbf{v}$ gives the velocity vector at each point on the disk at an instant in time.

Naturally, we refer to a vector field which gives the velocity vectors of a moving object or a flowing fluid, a *velocity vector field* or, simply as a *velocity field*.

Consider, for instance, a flowing stream. If the water supply stays constant, then the velocity field of the flowing water will vary with the position, but, at each point, should roughly be constant in time, since the velocity should be determined by the shape of a stream bed, and the obstructions in it.

A reasonable question to ask is: if there's an object in the stream, flowing along with the velocities prescribed by a velocity field $\mathbf{v}$, what path will it follow, given its starting location?

What we're asking for is to find a parameterized curve $\mathbf{p} = \mathbf{p}(t)$, where t denotes time, which gives the position of the object at each time, given that we want $\mathbf{p}(0)$ to be some specified starting point, and, at any time t, we want the velocity of the object $\mathbf{p}'(t)$ to be the velocity specified by the velocity vector field $\mathbf{v}$, which, at the position $\mathbf{p}(t)$ of the object, is $\mathbf{v}(\mathbf{p}(t))$. Thus, we want, for all t (or for some interval of t values),

$$\frac{d\mathbf{p}}{dt} \;=\; \mathbf{v}(\mathbf{p}(t)).$$

In fact, whether a vector field is a velocity vector field or not, parameterized curves such as those above are given a name:

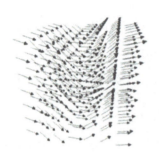

Figure 4.1.5: Velocity field of a flowing stream.

Flows are also known as *flow lines*, *flow curves*, and *integral curves* of the vector field.

If $\mathbf{F}$ is C^1 and $\mathbf{q}$ is a point in the interior of D, then the Picard-Lindelöf Theorem guarantees that on some open interval I containing 0 in $\mathbb{R}$, there exists a unique flow $\mathbf{p} : I \to D$ of $\mathbf{F}$ such that $\mathbf{p}(0) = \mathbf{q}$.

> **Definition 4.1.5.** *Suppose that $\mathbf{F}$ is a vector field on a region D in $\mathbb{R}^n$, and that $\mathbf{p} = \mathbf{p}(t)$ is a differentiable function from a non-empty open interval I in $\mathbb{R}$ into D. Then, $\mathbf{p}$ is a **flow of $\mathbf{F}$** provided that, for all t in I,*
>
> $$\frac{d\mathbf{p}}{dt} \;=\; \mathbf{F}(\mathbf{p}(t)).$$

Example 4.1.6. In Example 4.1.4, we had a vector field $\mathbf{v}(x,y) = (-y,x)$ which appeared to flow around in circles. Let's show that $\mathbf{p}(t) = r(\cos t, \sin t)$, where r is a constant, is a flow of the vector field.

We calculate

$$\frac{d\mathbf{p}}{dt} = r(-\sin t, \cos t)$$

and

$$\mathbf{v}(\mathbf{p}(t)) = \mathbf{v}(r\cos t, r\sin t) = (-r\sin t, r\cos t) = = r(-\sin t, \cos t).$$

Thus, $\mathbf{p}(t) = r(\cos t, \sin t)$ is a flow of $\mathbf{v}$.

> To produce **all** of the flows, you would need to simultaneously solve $x' = -y$ and $y' = x$, where the primes denote derivatives with respect to t.

> In this easy example, we can do this explicitly. We need $y'' = -y$, and the general solution to this, as we saw in Corollary 2.7.15 of [4], is $y = a\cos t + b\sin t$. It follows that $x = -a\sin t + b\cos t$.

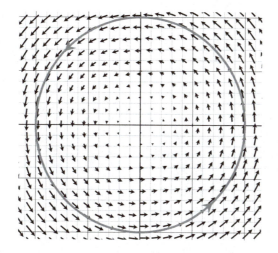

Figure 4.1.6: The vector field $\mathbf{v}$, together with one flow.

Remark 4.1.7. To sketch a flow by hand, you basically just "connect the dots", while "following the arrows". As you connect the dots by following the arrows, you make sure that every time you pass through an arrow in the vector field (or come really close to one), that the arrow is tangent to your curve (or close to it) and points in the direction that you're moving your pen or pencil.

Of course, you'd have to pay attention to the length/magnitude of the vectors in order to have any idea how fast you'd be moving along the flow, i.e., to know the speed of the actual parameterized flow.

An important way of obtaining vector fields is by taking a real-valued function $f = f(x, y)$, $f = f(x, y, z)$, or, more generally, $f = f(\mathbf{x})$ for $\mathbf{x}$ in $\mathbb{R}^n$ and then taking the gradient vector at each point. This is just a mild change in how we think about the gradient vector; before, we've considered the gradient vector of a function at various points, but now, we are considering the gradient vectors of a function at **all** points at the same time.

Definition 4.1.8. *Suppose that $\mathcal{U}$ is an open subset of $\mathbb{R}^n$, and that all of the partial derivatives of $f : \mathcal{U} \to \mathbb{R}$ exist. Suppose that D is a subset of $\mathcal{U}$ (which could be all of $\mathcal{U}$).*

*Then, the **gradient vector field of** f, or, simply, the **gradient field of** f on D is the vector field $\mathbf{F}$ on D given by*

$$\mathbf{F}(\mathbf{x}) = \vec{\nabla} f(\mathbf{x}),$$

for all $\mathbf{x}$ in D.

---

*A vector field $\mathbf{F}$, which is the gradient vector field of some function f, is called a **conservative vector field**. Given a conservative vector field $\mathbf{F}$, any function f such that $\mathbf{F} = \vec{\nabla} f$ is called a **potential function** for $\mathbf{F}$.*

Be warned: In physics, it is standard to say that a potential function f for a vector field $\mathbf{F}$ is a function such that $\mathbf{F} = -\vec{\nabla} f$. We shall see why the minus sign is desirable in physics in Section 4.3.

We shall give several other characterizations of conservative fields in Section 4.3.

Remark 4.1.9. We actually first looked at the gradient vector field and its relationship to level curves and surfaces back in Example 2.7.9. As we discussed there (in different terminology), the gradient vector field $\vec{\nabla} f$ is always perpendicular to the level sets of the potential function f.

Note that, in Example 2.7.9, we scaled the vector field, so that all of the vectors had the same length; this frequently helps you see and understand the vector field (or, at least, its directions) better.

How is this scaling accomplished? It's easy. Given a vector field $\mathbf{F}$ (whether it's a gradient field or not), form a new vector field

$$\mathbf{V} = k \frac{\mathbf{F}}{|\mathbf{F}|},$$

where $k > 0$ is a constant (frequently chosen to be 1), and the new vector field $\mathbf{V}$ is defined on the same region as $\mathbf{F}$, except that the points where $\mathbf{F} = \mathbf{0}$ are not in the domain of $\mathbf{V}$. At a given point, the scaled vector field points in the same direction as unscaled field, but has magnitude k everywhere.

We usually say that the vector field $\mathbf{F}$ has been *normalized* to produce the vector field $\mathbf{V}$.

Example 4.1.10. Let's find the gradient field of $f(x,y) = x^3 - 5xy^2 + 2y^4$ on $\mathbb{R}^2$.

This is easy; we calculate

$$\vec{\nabla} f(x,y) \;=\; (3x^2 - 5y^2, \; -10xy + 8y^3).$$

In Figure 4.1.7, we give the vector field $\vec{\nabla} f$, normalized (as discussed in Remark 4.1.9), and the level curve where $f = 1$.

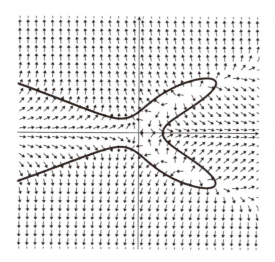

Figure 4.1.7: The normalized vector field $\mathbf{F} = \vec{\nabla} f / |\vec{\nabla} f|$, with a level curve of f.

Note that the level curve is absolutely **not** a flow of the vector field; for a curve to be a flow, the vectors must be tangent to the curve. The vectors in the gradient field $\vec{\nabla} f$ are **perpendicular** to the level curve.

Before we leave this section, we want to discuss some basic notation using the *gradient operator* $\vec{\nabla}$ and use it to define two important notions related to a vector field on a region in $\mathbb{R}^3$: *divergence* and *curl*.

Fix a dimension n. It is extremely convenient to think of $\vec{\nabla}$ as an *operator* given by

$$\vec{\nabla} \;=\; \left(\frac{\partial}{\partial x_1}, \; \frac{\partial}{\partial x_2}, \; \cdots, \; \frac{\partial}{\partial x_n} \right),$$

where calling it an "operator" means that we use it in various operations in which we use multiplicative notation to indicate that you apply the partial derivatives.

For instance, what looks like scalar multiplying f times the "vector" $\vec{\nabla}$ yields the gradient vector field:

$$\left(\frac{\partial}{\partial x_1},\ \frac{\partial}{\partial x_2},\ \cdots,\ \frac{\partial}{\partial x_n} \right) f \ =\ \left(\frac{\partial f}{\partial x_1},\ \frac{\partial f}{\partial x_2},\ \cdots,\ \frac{\partial f}{\partial x_n} \right) \ =\ \vec{\nabla} f.$$

Now, suppose we have a vector field

$$\mathbf{F} \ =\ \mathbf{F}(x_1,\ldots,x_n) \ =\ \big(F_1(x_1,\ldots,x_n),\ F_2(x_1,\ldots,x_n),\ \ldots,\ F_n(x_1,\ldots,x_n)\big),$$

on a region in $\mathbb{R}^n$.

How should we interpret the dot product $\vec{\nabla} \cdot \mathbf{F}$? Essentially, you do exactly what you do normally; you take the sum of the products of the entries in the corresponding components. It's just that this time, the "product" with $\partial/\partial x_i$ means take the partial derivative. Thus,

$$\vec{\nabla} \cdot \mathbf{F} \ =\ \frac{\partial}{\partial x_1} \cdot F_1 \ +\ \frac{\partial}{\partial x_2} \cdot F_2 \ +\ \cdots \ +\ \frac{\partial}{\partial x_n} \cdot F_n \ =\ \frac{\partial F_1}{\partial x_1} \ +\ \frac{\partial F_2}{\partial x_2} \ +\ \cdots \ +\ \frac{\partial F_n}{\partial x_n},$$

assuming that all of these partial derivatives exist.

Our main interest in dotting $\vec{\nabla}$ with a vector field will be in the 3-dimensional case, where the vector field is of class C^1, but we go ahead and make the definition in all dimensions.

If the region $\mathcal{U}$ is not open, then $\mathbf{F}$ being C^1 means that $\mathbf{F}$ is the restriction to $\mathcal{U}$ of a C^1 vector field on an open set containing $\mathcal{U}$.

The divergence of a vector field will be of great importance to us later, when we look at the Divergence Theorem in Section 4.6.

As we shall see, the physical interpretation of divergence is that it represents *flux density* at each point.

Definition 4.1.11. *Suppose that*

$$\mathbf{F} \ =\ \mathbf{F}(\mathbf{x}) \ =\ \big(F_1(\mathbf{x}), F_2(\mathbf{x}), \ldots, F_n(\mathbf{x})\big)$$

is a C^1 vector field on a region $\mathcal{U}$ in $\mathbb{R}^n$.

*Then, the **divergence** of $\mathbf{F}$ is*

$$\operatorname{div} \mathbf{F} \ =\ \vec{\nabla} \cdot \mathbf{F} \ =\ \frac{\partial F_1}{\partial x_1} \ +\ \frac{\partial F_2}{\partial x_2} \ +\ \cdots \ +\ \frac{\partial F_n}{\partial x_n}.$$

Example 4.1.12. Calculate the divergence of the vector field

$$\mathbf{F} \ =\ (3x^2yz + ye^{\sin z},\ 5xy^3 - x\tan^{-1} z,\ -7yz + x^2y^4).$$

Solution:

While we put some ugly functions into the definition of the vector field, you should notice that, as far as divergence is concerned, this is a very simple vector field:

$$\vec{\nabla} \cdot \mathbf{F} = \frac{\partial}{\partial x}\left(3x^2yz + ye^{\sin z}\right) + \frac{\partial}{\partial y}\left(5xy^3 - x\tan^{-1}z\right) + \frac{\partial}{\partial z}\left(-7yz + x^2y^4\right).$$

So,

$$\vec{\nabla} \cdot \mathbf{F} = 6xyz + 15xy^2 - 7y.$$

It is extremely important to notice that the divergence of a vector field yields a scalar function, **not** a vector field, just as the dot product of two vectors is not another vector, but is, instead, a scalar.

Now we come to a more-complicated operation. Suppose again that

$$\mathbf{F} = \mathbf{F}(x,y,z) = \big(P(x,y,z),\ Q(x,y,z),\ R(x,y,z)\big)$$

is a C^1 vector field on a region T in $\mathbb{R}^3$.

How do we calculate the cross product $\vec{\nabla} \times \mathbf{F}$? You do what you do for vectors:

$$\vec{\nabla} \times \mathbf{F} = \begin{vmatrix} \mathbf{i} & \mathbf{j} & \mathbf{k} \\ \dfrac{\partial}{\partial x} & \dfrac{\partial}{\partial y} & \dfrac{\partial}{\partial z} \\ P & Q & R \end{vmatrix} =$$

$$\begin{vmatrix} \dfrac{\partial}{\partial y} & \dfrac{\partial}{\partial z} \\ Q & R \end{vmatrix}\mathbf{i} - \begin{vmatrix} \dfrac{\partial}{\partial x} & \dfrac{\partial}{\partial z} \\ P & R \end{vmatrix}\mathbf{j} + \begin{vmatrix} \dfrac{\partial}{\partial x} & \dfrac{\partial}{\partial y} \\ P & Q \end{vmatrix}\mathbf{k} =$$

$$\left(\frac{\partial R}{\partial y} - \frac{\partial Q}{\partial z}\right)\mathbf{i} - \left(\frac{\partial R}{\partial x} - \frac{\partial P}{\partial z}\right)\mathbf{j} + \left(\frac{\partial Q}{\partial x} - \frac{\partial P}{\partial y}\right)\mathbf{k}.$$

Should you memorize this final result? Absolutely not. Just calculate the cross product with the gradient operator every time.

The curl of a vector field will be of great importance to us later, when we look at Stokes' Theorem in Section 4.7.

As we shall see, the direction of the curl at a point **p** represents an axis, through **p**, about which the vector field is rotating, in a direction given by the right-hand rule, while the magnitude of the curl measures the amount of rotation, or *rotational density*, around the axis.

Definition 4.1.13. *Suppose that*

$$\mathbf{F} \ = \ \mathbf{F}(x,y,z) \ = \ \big(P(x,y,z), \ Q(x,y,z), \ R(x,y,z)\big)$$

is a C^1 vector field on a region T in $\mathbb{R}^3$.

   Then, the **curl** *of* **F** *is*

$$\operatorname{curl} \mathbf{F} \ = \ \vec{\nabla} \times \mathbf{F} \ =$$

$$\left(\frac{\partial R}{\partial y} - \frac{\partial Q}{\partial z}\right) \mathbf{i} \ - \ \left(\frac{\partial R}{\partial x} - \frac{\partial P}{\partial z}\right) \mathbf{j} \ + \ \left(\frac{\partial Q}{\partial x} - \frac{\partial P}{\partial y}\right) \mathbf{k}.$$

Example 4.1.14. Calculate the curl of the vector field $\mathbf{F} = (x + y, \ ye^x, \ z \sin x)$.

Solution: We calculate

$$\vec{\nabla} \times \mathbf{F} \ = \ \begin{vmatrix} \mathbf{i} & \mathbf{j} & \mathbf{k} \\ \dfrac{\partial}{\partial x} & \dfrac{\partial}{\partial y} & \dfrac{\partial}{\partial z} \\ x + y & ye^x & z \sin x \end{vmatrix} \ =$$

$$\begin{vmatrix} \dfrac{\partial}{\partial y} & \dfrac{\partial}{\partial z} \\ ye^x & z \sin x \end{vmatrix} \mathbf{i} \ - \ \begin{vmatrix} \dfrac{\partial}{\partial x} & \dfrac{\partial}{\partial z} \\ x + y & z \sin x \end{vmatrix} \mathbf{j} \ + \ \begin{vmatrix} \dfrac{\partial}{\partial x} & \dfrac{\partial}{\partial y} \\ x + y & ye^x \end{vmatrix} \mathbf{k} \ =$$

$$(0, \ -z \cos x, \ ye^x - 1).$$

Remark 4.1.15. A 2-dimensional vector field $\mathbf{F}(x,y) = (P(x,y), \ Q(x,y))$ on a region D in $\mathbb{R}^2$ can be viewed as a vector field in $\mathbb{R}^3$ by considering the points in D to have 0 for their z-coordinates and thinking of **F** as being

$$\mathbf{F}(x,y,z) \ = \ (P(x,y), \ Q(x,y), \ 0),$$

with domain consisting of the points $(x, y, 0)$ where (x, y) is in D. Technically, we should use a different letter to denote this "new" vector field $\mathbf{F}$, but usually we just use the original vector field variable.

If you now calculate the curl of $\mathbf{F}(x, y, z)$, find easily that

$$\vec{\nabla} \times F \;=\; \left(\frac{\partial Q}{\partial x} - \frac{\partial P}{\partial y} \right) \mathbf{k}.$$

For this reason, the scalar function

$$\frac{\partial Q}{\partial x} - \frac{\partial P}{\partial y} \;=\; Q_x - P_y$$

is referred to as the 2-**dimensional curl** of the vector field

$$\mathbf{F}(x, y) = (P(x, y), \, Q(x, y)).$$

More Depth:

It is easy to show, by direct calculation, that the divergence and curl are linear operations, that is:

Theorem 4.1.16. *Suppose that a and b are constants, and that $\mathbf{F}$ and $\mathbf{G}$ are C^1 vector fields on a region T in $\mathbb{R}^3$. Then,*

$$\vec{\nabla} \cdot (a\mathbf{F} + b\mathbf{G}) \;=\; a(\vec{\nabla} \cdot \mathbf{F}) + b(\vec{\nabla} \cdot \mathbf{G})$$

and

$$\vec{\nabla} \times (a\mathbf{F} + b\mathbf{G}) \;=\; a(\vec{\nabla} \times \mathbf{F}) + b(\vec{\nabla} \times \mathbf{G}).$$

Later, it will be important to us that the curl of a vector field has no divergence, and gradient vector fields have no curl, that is:

Theorem 4.1.17. *Suppose that*

$$\mathbf{F} \;=\; \mathbf{F}(x, y, z) \;=\; \big(P(x, y, z), \; Q(x, y, z), \; R(x, y, z) \big)$$

is a C^2 vector field on a region T in $\mathbb{R}^3$, and that f is a real-valued C^2 function on the region T.

Then,

$$\vec{\nabla} \cdot (\vec{\nabla} \times \mathbf{F}) \;=\; \operatorname{div}(\operatorname{curl} \mathbf{F}) \;=\; 0 \quad \text{and} \quad \vec{\nabla} \times (\vec{\nabla} f) \;=\; \operatorname{curl}(\vec{\nabla} f) \;=\; \mathbf{0}.$$

There is converse of Theorem 4.1.17, assuming that T is simply-connected (recall Definition 1.7.16).

If T is simply-connected and $\vec{\nabla} \times \mathbf{E} = \mathbf{0}$, then there exists an f such that $\mathbf{E} = \vec{\nabla} f$; we shall discuss this more in Section 4.3. Also, if T is simply-connected and $\vec{\nabla} \cdot E = 0$, then there exists a vector field $\mathbf{F}$ such that $\mathbf{E} = \vec{\nabla} \times \mathbf{F}$.

Proof. These follow by direct calculation.

$$\vec{\nabla} \cdot (\vec{\nabla} \times F) = \vec{\nabla} \cdot \left(\left(\frac{\partial R}{\partial y} - \frac{\partial Q}{\partial z} \right) \mathbf{i} - \left(\frac{\partial R}{\partial x} - \frac{\partial P}{\partial z} \right) \mathbf{j} + \left(\frac{\partial Q}{\partial x} - \frac{\partial P}{\partial y} \right) \mathbf{k} \right) =$$

$$\frac{\partial^2 R}{\partial x \partial y} - \frac{\partial^2 Q}{\partial x \partial z} - \frac{\partial^2 R}{\partial x \partial y} + \frac{\partial^2 P}{\partial y \partial z} + \frac{\partial^2 Q}{\partial x \partial z} - \frac{\partial^2 P}{\partial y \partial z} = 0,$$

and

$$\vec{\nabla} \times (\vec{\nabla} f) = \vec{\nabla} \times \left(\frac{\partial f}{\partial x}, \frac{\partial f}{\partial y}, \frac{\partial f}{\partial z} \right) =$$

$$\left(\frac{\partial^2 f}{\partial y \partial z} - \frac{\partial^2 f}{\partial y \partial z} \right) \mathbf{i} - \left(\frac{\partial^2 f}{\partial x \partial z} - \frac{\partial^2 f}{\partial x \partial z} \right) \mathbf{j} + \left(\frac{\partial^2 f}{\partial x \partial y} - \frac{\partial^2 f}{\partial x \partial y} \right) \mathbf{k} = \mathbf{0}.$$

$\square$

Remark 4.1.18. In Definition 4.1.5 and Example 4.1.4, we defined and and gave an example of flows of vector fields. We motivated the definition of a flow by discussing the case of velocity vector fields. However, regardless of the physical origin or interpretation of a vector field $\mathbf{F}$ on a region $\mathcal{U}$ in $\mathbb{R}^n$, a flow of $\mathbf{F}$ is defined as a function $\mathbf{p} = \mathbf{p}(t)$, from an open interval I in $\mathbb{R}$ into $\mathcal{U}$ such that, for all t in I,

$$\frac{d\mathbf{p}}{dt} = \mathbf{F}(\mathbf{p}(t)).$$

However, if $\mathbf{F}$ is a force field, it is a **mistake** to think of a flow of the vector field as a parameterization of the path that an object would follow if the object were acted upon by the force field.

Consider, for instance, the case of a point-mass, with mass 1 kg, which is initially at rest at the origin, and is acted upon by the force field $\mathbf{F} = (12y, 2)$ Newtons.

If $\mathbf{r}(t) = (x(t), y(t))$ is the position of the point-mass, in meters, then we are given that $\mathbf{r}(0) = \mathbf{0}$ meters, $\mathbf{r}'(0) = \mathbf{0}$ meters/second, and, from Newton's 2nd Law of Motion,

$$\mathbf{F}(\mathbf{r}(t)) = m\mathbf{a},$$

and so, as $m = 1$ kg,

$$(12y, 2) = (x'', y'').$$

Solving the initial value problem $y'' = 2$, $y'(0) = 0$, and $y(0) = 0$, we find $y(t) = t^2$. Then, $x'' = 12y = 12t^2$. Since $x'(0) = 0$ and $x(0) = 0$, we find $x(t) = t^4$. Thus,

$$\mathbf{r}(t) = (t^4, t^2),$$

and we see that the image of $\mathbf{r}$ lies along $x = y^2$ (but you only obtain the part where $y \geq 0$).

On the other hand, the flow $\mathbf{p}(t) = (x(t), y(t))$ of $\mathbf{F}$, which is at the origin at $t = 0$, satisfies

$$\mathbf{p}'(t) = \mathbf{F}(\mathbf{p}(t)),$$

i.e.,

$$(x', y') = (12y, 2).$$

Solving the initial value problem $y' = 2$, and $y(0) = 0$, we find $y(t) = 2t$. Then, $x' = 12y = 24t$. Since $x(0) = 0$, we find $x(t) = 12t^2$. Thus,

$$\mathbf{p}(t) = (12t^2, 2t),$$

and we see that the image of $\mathbf{p}$ lies along $x = 3y^2$.

We sketch the flow of the vector field and the path of the particle, moving under the influence of the force field, in Figure 4.1.8.

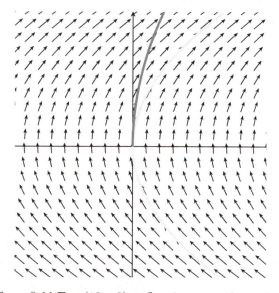

Figure 4.1.8: The force field $\mathbf{F} = (12y, 2)$, a flow in green, the path of a particle in red.

Example 4.1.19. Finally in this section, we'd like to calculate the divergence of another vector field; one that will be important to us later.

Fix a point (a, b, c) in $\mathbb{R}^3$, and let k be a constant. Consider the vector field defined on all of $\mathbb{R}^3$ except at (a, b, c):

$$\mathbf{F} = \frac{k}{\left((x-a)^2 + (y-b)^2 + (z-c)^2\right)^{3/2}} (x-a, y-b, z-c) = (P, Q, R).$$

We want to show that $\vec{\nabla} \cdot \mathbf{F} = 0$.

This is a straightforward, but slightly messy calculation. However, we can use the symmetry of the vector field to reduce the amount of work required.

We calculate

$$\frac{\partial P}{\partial x} = \frac{\partial}{\partial x} \left[\frac{k(x-a)}{\left((x-a)^2 + (y-b)^2 + (z-c)^2\right)^{3/2}} \right] =$$

$$k \cdot \frac{\left((x-a)^2 + (y-b)^2 + (z-c)^2\right)^{3/2} - 3(x-a)^2\left((x-a)^2 + (y-b)^2 + (z-c)^2\right)^{1/2}}{\left((x-a)^2 + (y-b)^2 + (z-c)^2\right)^3}$$

$$k \cdot \frac{(x-a)^2 + (y-b)^2 + (z-c)^2 - 3(x-a)^2}{\left((x-a)^2 + (y-b)^2 + (z-c)^2\right)^{5/2}} =$$

$$k \cdot \frac{(y-b)^2 + (z-c)^2 - 2(x-a)^2}{\left((x-a)^2 + (y-b)^2 + (z-c)^2\right)^{5/2}}.$$

By symmetry, we immediately conclude that

$$\frac{\partial Q}{\partial y} = \frac{\partial}{\partial y} \left[\frac{k(y-b)}{\left((x-a)^2 + (y-b)^2 + (z-c)^2\right)^{3/2}} \right] =$$

$$k \cdot \frac{(x-a)^2 + (z-c)^2 - 2(y-b)^2}{\left((x-a)^2 + (y-b)^2 + (z-c)^2\right)^{5/2}}$$

and

$$\frac{\partial R}{\partial z} = \frac{\partial}{\partial z} \left[\frac{k(z-c)}{\left((x-a)^2 + (y-b)^2 + (z-c)^2\right)^{3/2}} \right] =$$

$$k \cdot \frac{(x-a)^2 + (y-b)^2 - 2(z-c)^2}{\left((x-a)^2 + (y-b)^2 + (z-c)^2\right)^{5/2}}.$$

Now, it's easy to see that the numerators of $\partial P / \partial x$, $\partial Q / \partial y$, and $\partial R / \partial z$ cancel out, i.e., that

$$\vec{\nabla} \cdot \mathbf{F} = \frac{\partial P}{\partial x} + \frac{\partial Q}{\partial y} + \frac{\partial R}{\partial z} = 0.$$

4.1.1   Exercises

Online answers to select exercises are here.

Basics:

1. Suppose that a point-mass of $M = 5$ kg is located at the point $(1, 2, -1)$, where all coordinates are in meters. Let $\mathbf{F} = \mathbf{F}(x, y, z)$ be the gravitational force field that M exerts on a mass of 2 kg at any point in $\mathbb{R}^3$, other than at the point $(1, 2, -1)$.

   What are the values of $\mathbf{F}(0, 0, 0)$, $\mathbf{F}(1, 2, 0)$, and $\mathbf{F}(2, 4, 5)$? ▶

2. Suppose that a point-mass of $M = 100$ kg is located at the point $(2, 5, 3)$, where all coordinates are in meters. Let $\mathbf{F} = \mathbf{F}(x, y, z)$ be the gravitational force field that M exerts on a mass of 10 kg at any point in $\mathbb{R}^3$, other than at the point $(2, 5, 3)$. What are the values of $\mathbf{F}(0, 0, 0)$, $\mathbf{F}(1, 2, 0)$, and $\mathbf{F}(2, 4, 5)$?

In each of the following exercises, you are given a vector field $\mathbf{v} = \mathbf{v}(x, y)$ on $\mathbb{R}^2$. Sketch at least 6 vectors in the given vector field.

3. $\mathbf{v}(x, y) = (1, x)$.   4. $\mathbf{v}(x, y) = (y, -x)$.

5. $\mathbf{v}(x, y) = (-x, -y)$.   6. $\mathbf{v}(x, y) = (0, -x)$. ▶

7. $\mathbf{v}(x, y) = (x, x)$.   8. $\mathbf{v}(x, y) = (y - x^2, y)$.

You are given four functions $\mathbf{F} = \mathbf{F}(x, y)$ defining vector fields on $\mathbb{R}^2$ and, where $\mathbf{F} \neq 0$, sketches of the normalized vector fields $\mathbf{F}/|\mathbf{F}|$. Match the appropriate vector fields and sketches. (Ignore the red dots for now.)

9. $\mathbf{F}(x, y) = (y \cos x, \sin x)$.   10. $\mathbf{F}(x, y) = (-2x, 2y)$.

11. $\mathbf{F}(x, y) = (-2x, 1)$.   12. $\mathbf{F}(x, y) = (e^{-y}, -xe^{-y} + 1)$.

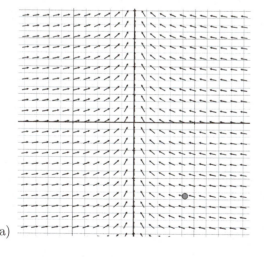

(a)

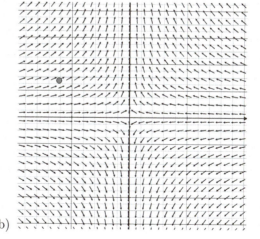

(b)

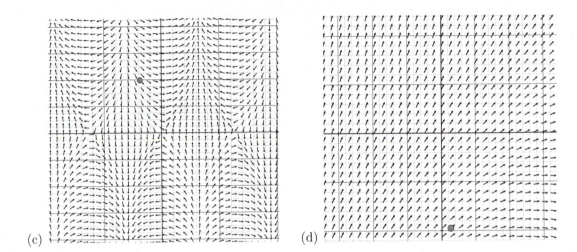

(c)   (d)

13. Consider the vector field **F** from Exercise 9, and its matching normalized graph.

    (a) Sketch a flow, which starts at the red dot, on the graph of the normalized vector field.

    (b) Assume that **F** is conservative, so that there exists a real-valued potential function f such that $\mathbf{F} = \vec{\nabla} f$. On the graph of the normalized vector field, sketch a level curve of f which passes through the red dot.

    (c) Produce a potential function f for **F**. If one is not obvious, you may use the method of Example 3.1.7.

14. Consider the vector field **F** from Exercise 10, and its matching normalized graph.

    (a) Sketch a flow, which starts at the red dot, on the graph of the normalized vector field.

    (b) Assume that **F** is conservative, so that there exists a real-valued potential function f such that $\mathbf{F} = \vec{\nabla} f$. On the graph of the normalized vector field, sketch a level curve of f which passes through the red dot.

    (c) Produce a potential function f for **F**. If one is not obvious, you may use the method of Example 3.1.7.

15. Consider the vector field **F** from Exercise 11, and its matching normalized graph.

    (a) Sketch a flow, which starts at the red dot, on the graph of the normalized vector field.

    (b) Assume that **F** is conservative, so that there exists a real-valued potential function f such that $\mathbf{F} = \vec{\nabla} f$. On the graph of the normalized vector field, sketch a level curve of f which passes through the red dot.

(c) Produce a potential function f for $\mathbf{F}$. If one is not obvious, you may use the method of Example 3.1.7.

16. Consider the vector field $\mathbf{F}$ from Exercise 12, and its matching normalized graph.

(a) Sketch a flow, which starts at the red dot, on the graph of the normalized vector field.

(b) Assume that $\mathbf{F}$ is conservative, so that there exists a real-valued potential function f such that $\mathbf{F} = \vec{\nabla} f$. On the graph of the normalized vector field, sketch a level curve of f which passes through the red dot.

(c) Produce a potential function f for $\mathbf{F}$. If one is not obvious, you may use the method of Example 3.1.7.

In each of the following exercises, you are given a real-valued function $f(x, y)$. (a) Find the gradient vector field $\mathbf{F} = \vec{\nabla} f$. (b) Sketch vectors in the gradient vector field at the points $(0,0)$, $(1,0)$, $(1,1)$, and $(0,1)$.

17. $f(x, y) = 3x - 2y + 7$.

18. $f(x, y) = x^2 + y^2$.

19. $f(x, y) = x + \tan^{-1} y$.

20. $f(x, y) = x \tan^{-1} y$.

In each of the following exercises, you are given a vector field $\mathbf{F}(x, y) = (P(x, y), Q(x, y))$. (a) Calculate the divergence of F. (b) Calculate the 2-dimensional curl of F.

21. $\mathbf{F}(x, y) = (2x, 5y)$.

22. $\mathbf{F}(x, y) = (5y, 2x)$.

23. $\mathbf{F}(x, y) = \left(\dfrac{x^2 + y^2}{2}, xy \right)$.

24. $\mathbf{F}(x, y) = \left(xy, \dfrac{x^2 + y^2}{2} \right)$.

25. $\mathbf{F}(x, y) = (y \cos x, y \sin x)$.

26. $\mathbf{F}(x, y) = (x \ln y, x + y)$.

In each of the following exercises, you are given a vector field of $\mathbf{F}(x, y, z) = (P(x, y, z), Q(x, y, z), R(x, y, z))$. (a) Calculate the divergence of F. (b) Calculate the curl of F.

27. $\mathbf{F}(x, y, z) = (2x,\ 5y,\ -3z)$.

28. $\mathbf{F}(x, y, z) = (-3z,\ 2x,\ 5y)$.

29. $\mathbf{F}(x, y, z) = (y\cos x,\ y\sin x,\ z)$.   ▶

30. $\mathbf{F}(x, y, z) = (x + y + z,\ xyz,\ 2x + 3y + 4z)$.

More Depth:

In each of the following exercises, you are given a force field $\mathbf{F} = \mathbf{F}(x, y)$ Newtons, where the coordinates are in meters. **(a)** Determine the flow $\mathbf{p}(t)$, of $\mathbf{F}$, such that $\mathbf{p}(0) = (0, 0)$. **(b)** Determine the position $\mathbf{r}(t)$ of an particle which has a mass of 1 kg, and is at rest at the origin at time $t = 0$. **(c)** Sketch the vector field, the flow, and the path of the particle.

31. $\mathbf{F}(x, y) = (6, 12x)$.

32. $\mathbf{F}(x, y) = (y^2, 60)$.

33. Let $\mathbf{F}(x, y, z) = (x + y + z,\ xyz,\ xy + yz + xz)$. Calculate the curl $\vec{\nabla} \times \mathbf{F}$ of $\mathbf{F}$, and then verify, explicitly, that the divergence of the curl is zero, i.e., that $\vec{\nabla} \cdot (\vec{\nabla} \times \mathbf{F}) = 0$.

34. Let $f(x, y, z) = xe^y + z\sin x$. Calculate $\vec{\nabla}f$ and verify, explicitly, that the curl of the gradient is zero, i.e., that $\vec{\nabla} \times (\vec{\nabla}f) = \mathbf{0}$.

4.2 Line Integrals

Given a vector field $\mathbf{F}$ and an oriented piecewise-smooth curve C, we define $\int_C \mathbf{F} \cdot d\mathbf{r}$, the *line integral* of $\mathbf{F}$ along C. In the important case where $\mathbf{F}$ is a force field, the line integral represents the amount of work done by the force field on an object moving along the curve C.

Basics:

Suppose that $\mathbf{F}$ is a constant force, which acts on an object, as the object moves with a displacement vector $\mathbf{d}$. Then, as we saw in Theorem 1.6.21, the work done on the object by the force, i.e., the energy imparted to the object by the force, is given by simply taking the dot product:

$$\text{work done by constant force } = \mathbf{F} \cdot \mathbf{d},$$

regardless of the path along which the object actually moved.

But, how do you calculate the work done on the object if the object is moving along a curve C and through a **non-constant** vector force field $\mathbf{F} = \mathbf{F}(\mathbf{x})$?

With all of our experience with integrals, the answer to this question should be obvious: if the force field is on a region in $\mathbb{R}^3$, then to calculate the total work done, you chop up the curve C into a collection of infinitesimal displacement vectors

$$d\mathbf{r} = (dx, dy, dz),$$

calculate the infinitesimal work done

$$\mathbf{F}(x, y, x) \cdot d\mathbf{r}$$

by the vector field at each point (x, y, z) on C, and add up all of the infinitesimal contributions via an integral (of course, formally, you take limits of Riemann sums, using "small" displacement vectors).

The resulting integral is denoted by

$$\int_C \mathbf{F}(\mathbf{x}) \cdot d\mathbf{r} \;=\; \int_C \mathbf{F} \cdot d\mathbf{r},$$

and is called the *line integral of* $\mathbf{F}$ *along* C. As we shall see in examples, **the line integral is not, in general, independent of the path.**

However, as we discuss in the next section, it is extremely important that, **for certain types of vector fields**, the line integral is, in fact, independent of the path, as long as one starts and ends at the specified points.

Line integrals are sometimes referred to as *curve integrals*. This would seem to be the more reasonable terminology, since C is often not a line. Nonetheless, "line integral" is by far the more common term, and we shall stick with it.

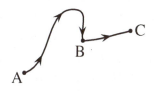

Figure 4.2.1: An oriented, piecewise-regular curve from $\mathbf{A}$ to $\mathbf{C}$.

We will use the notion of an oriented, piecewise-regular curve C and a piecewise-regular parameterization $\mathbf{r}(t)$ from Definition 1.6.19. If we have a vector field $\mathbf{F}$ and such a parameterization $\mathbf{r}(t)$ of C, then, for each t, the value of $\mathbf{F}$ at the point given by $\mathbf{r}(t)$ is $\mathbf{F}(\mathbf{r}(t))$ and $d\mathbf{r} = (d\mathbf{r}/dt)dt = \mathbf{r}'(t)\,dt$. Hence, $\mathbf{F} \cdot d\mathbf{r}$ becomes $\mathbf{F}(\mathbf{r}(t)) \cdot \mathbf{r}'(t)\,dt$, and we make the following definition.

> **Definition 4.2.1.** *Suppose that $\mathbf{F}$ is a continuous vector field on an open region $\mathcal{U}$ in $\mathbb{R}^n$, and that $\mathbf{r} : [a,b] \to \mathcal{U}$ is a piecewise-regular parameterization of a curve. Let C be the oriented, piecewise-regular curve, from $\mathbf{r}(a)$ to $\mathbf{r}(b)$, defined by $\mathbf{r}$.*
>
> *Then, the* **line integral** *of $\mathbf{F}$ along C is defined by*
>
> $$\int_C \mathbf{F} \cdot d\mathbf{r} \;=\; \int_a^b \mathbf{F}(\mathbf{r}(t)) \cdot \mathbf{r}'(t)\,dt,$$
>
> *and this integral is independent of which piecewise-regular parameterization is chosen for C.*
>
> *If $\mathbf{F}$ is a force field, then $\int_C \mathbf{F} \cdot d\mathbf{r}$ is the* **work done by** $\mathbf{F}$ *along the oriented curve C.*

For the proof of the independence statement, see Section 6.2 of [3]. It is shown that every piecewise-regular parameterization yields the same integral as the parameterization by arc length. Recall Remark 1.6.16.

Be careful when reading this independence statement. The line integral typically changes if you pick a different curve C from $\mathbf{r}(a)$ to $\mathbf{r}(b)$. But, if you fix the oriented, piecewise-regular curve C, then it doesn't matter what piecewise-regular parameterization you use for C.

Example 4.2.2. Consider the force field $\mathbf{F}(x,y) = (y^2 - x^3, x^5)$ Newtons, where x and y are in meters. Let C_1 be the oriented curve from $(0,0)$ to $(1,1)$, along the graph of $y = x$. Let C_2 be the oriented curve from $(0,0)$ to $(1,1)$, along the graph of $y = x^2$. Calculate the work done by $\mathbf{F}$ along C_1 and C_2, and show that they are not equal.

Solution:

We parameterize C_1 by $\mathbf{r}(t) = (t,t)$, for $0 \leq t \leq 1$. We parameterize C_2 by $\hat{\mathbf{r}}(t) = (t, t^2)$, for $0 \leq t \leq 1$.

We calculate, in joules,

$$\int_{C_1} \mathbf{F} \cdot d\mathbf{r} \;=\; \int_0^1 \mathbf{F}(\mathbf{r}(t)) \cdot \mathbf{r}'(t)\,dt \;=\; \int_0^1 (t^2 - t^3, t^5) \cdot (1,1)\,dt \;=\; \int_0^1 (t^2 - t^3 + t^5)\,dt \;=\; \frac{1}{4},$$

and

$$\int_{C_2} \mathbf{F} \cdot d\mathbf{r} \;=\; \int_0^1 \mathbf{F}(\hat{\mathbf{r}}(t)) \cdot \hat{\mathbf{r}}'(t)\,dt \;=\; \int_0^1 (t^4 - t^3, t^5) \cdot (1, 2t)\,dt \;=\; \int_0^1 (t^4 - t^3 + 2t^6)\,dt \;=\; \frac{33}{140}.$$

Since $1/4 = 35/140 \neq 33/140$, the line integral does not depend **just** on $\mathbf{F}$ and the starting and final points; it also depends on the particular oriented curve between the starting and ending points.

Remark 4.2.3. There is another common notation for the line integral that looks very different.

Suppose the vector field is given, in terms of component functions, by

$$\mathbf{F}(x,y) = \big(P(x,y), Q(x,y)\big) \quad \text{or} \quad \mathbf{F}(x,y,z) = \big(P(x,y,z), Q(x,y,z), R(x,y,z)\big).$$

If you also write $\mathbf{r}$ in component form,

$$\mathbf{r} = (x,\, y) \quad \text{or} \quad \mathbf{r} = (x,\, y,\, z),$$

then

$$d\mathbf{r} = (dx,\, dy) \quad \text{or} \quad d\mathbf{r} = (dx,\, dy,\, dz),$$

and

$$\int_C \mathbf{F} \cdot d\mathbf{r} = \int_C P\, dx + Q\, dy \quad \text{or} \quad \int_C P\, dx + Q\, dy + R\, dz.$$

This different notation is typically used when you are not going to introduce a new parameter name, but, rather, use one of the variables x, y, or z as a parameter, i.e., when you can easily write the other variables in terms of one specific one along the given curve.

Example 4.2.4. Let's look again at Example 4.2.2, in which we considered the vector field $\mathbf{F}(x,y) = (y^2 - x^3, x^5)$, and the oriented curves C_1 from $(0,0)$ to $(1,1)$, along the graph of $y = x$, and C_2 from $(0,0)$ to $(1,1)$, along the graph of $y = x^2$. We calculated the work done by $\mathbf{F}$ along C_1 and C_2 by calculating the two appropriate line integrals, using a parameter t both times.

We will calculate both of these line integrals again, but this time we will use the notation in Remark 4.2.3, without introducing a new parameter variable.

Along the graph of $y = x$, we have $dy = dx$, and we can write the line integral along C_1 in terms of x and dx, as x goes from 0 to 1:

$$\int_{C_1} \mathbf{F} \cdot d\mathbf{r} = \int_{C_1} (y^2 - x^3)\, dx + x^5\, dy = \int_0^1 (x^2 - x^3)\, dx + x^5\, dx =$$

$$\left. \frac{x^3}{3} - \frac{x^4}{4} + \frac{x^6}{6} \right|_0^1 = \frac{1}{4}.$$

Along the graph of $y = x^2$, we have $dy = 2x\,dx$, and we can write the line integral along C_2 in terms of x and dx, as x goes from 0 to 1:

$$\int_{C_2} \mathbf{F} \cdot d\mathbf{r} \;=\; \int_{C_2} (y^2 - x^3)\,dx + x^5\,dy \;=\; \int_0^1 ((x^2)^2 - x^3)\,dx + x^5 \cdot 2x\,dx \;=$$

$$\int_0^1 (x^4 - x^3 + 2x^6)\,dx \;=\; \frac{33}{140}.$$

Of course, the two values of the line integrals are the same as what we obtained in Example 4.2.2.

Example 4.2.5. The notation

$$\int_C \mathbf{F} \cdot d\mathbf{r} \;=\; \int_C P\,dx + Q\,dy$$

is particularly useful when the oriented curve C is a line segment which parallel to one of the coordinate axes, i.e., when C is an oriented curve where either x or y has a constant value.

Suppose, for instance, we consider the vector field

$$\mathbf{F}(x, y) = \left(x^2 2^y y + y^3,\; x \cos x + e^y \sin y\right)$$

and want to calculate the line integral of $\mathbf{F}$ along the oriented line segment C from $(0, 1)$ to $(3, 1)$.

The curve C is where $y = 1$ and x goes from 0 to 3. Along this curve, since y is fixed, $dy = 0$. Thus, we have

$$\int_C \mathbf{F} \cdot d\mathbf{r} \;=\; \int_C (x^2 2^y y + y^3)\,dx \;+\; (x \cos x + e^y \sin y)\,dy \;=$$

$$\int_C (x^2 2^1 1 + 1^3)\,dx,$$

where we used that, along C, $y = 1$ and $dy = 0$.

Therefore, since x goes from 0 to 3 along C,

$$\int_C \mathbf{F} \cdot d\mathbf{r} \;=\; \int_0^3 (2x^2 + 1)\,dx \;=\; \frac{2x^3}{3} + x \,\Big|_0^3 \;=\; 21.$$

Before we look at more examples, it is convenient to establish some notation for combining oriented curves and for changing the orientation on a curve.

Definition 4.2.6. *Suppose that we have an oriented piecewise-regular curve C_1 from a point $\mathbf{a}$ to a point $\mathbf{b}$ and an oriented piecewise-regular curve C_2 from $\mathbf{b}$ to a point $\mathbf{c}$.*

*Then, $C_1 + C_2$ denotes the **concatenation** of C_1 with C_2, which is the oriented piecewise-regular curve formed by first taking C_1 from $\mathbf{a}$ to $\mathbf{b}$, and then continuing along C_2 from $\mathbf{b}$ to $\mathbf{c}$.*

*The oriented piecewise-regular curve $-C_1$ denotes the curve C_1, with the **opposite orientation**, and so is an oriented piecewise-regular curve from $\mathbf{b}$ to $\mathbf{a}$.*

"Subtracting" a curve means adding the negation; thus,

$$-C_2 - C_1 = -C_2 + (-C_1) = -(C_1 + C_2),$$

is the oriented piecewise-regular curve from $\mathbf{c}$ to $\mathbf{a}$, which is first $-C_2$ from $\mathbf{c}$ to $\mathbf{b}$, and then is $-C_1$ from $\mathbf{b}$ to $\mathbf{a}$.

Even though we use additive notation, concatenation is **not** commutative. In fact, unless $\mathbf{a} = \mathbf{c}$, $C_2 + C_1$ would be undefined here, since C_2 ends at $\mathbf{c}$ and C_1 starts at $\mathbf{a}$. On the other hand, concatenation is associative.

Later, we shall also use this summation and negation notation even when the curves are disjoint. In this case, the term "concatenation" is not usually used. However, the order in which the curves are written doesn't matter in this case; you have to use separate parameterizations for the different pieces. What's important is that the sum of curves still corresponds to a sum of line integrals and negating a curve negates the line integral.

It is an easy, but important, result that concatenating curves or reversing orientation affects line integrals in the obvious ways; we leave the proof as an exercise.

Theorem 4.2.7. *Suppose that $\mathbf{F}$ is a continuous vector field on an open region $\mathcal{U}$ in $\mathbb{R}^n$. Suppose that we have an oriented piecewise-regular curve C_1 in $\mathcal{U}$ from a point $\mathbf{a}$ to a point $\mathbf{b}$ and an oriented piecewise-regular curve C_2 in $\mathcal{U}$ from $\mathbf{b}$ to a point $\mathbf{c}$.*

Then,

$$\int_{C_1 + C_2} \mathbf{F} \cdot d\mathbf{r} = \int_{C_1} \mathbf{F} \cdot d\mathbf{r} + \int_{C_2} \mathbf{F} \cdot d\mathbf{r}$$

and

$$\int_{-C_1} \mathbf{F} \cdot d\mathbf{r} = -\int_{C_1} \mathbf{F} \cdot d\mathbf{r}.$$

Now, let's look at a lengthy example.

Example 4.2.8. Consider the force field $\mathbf{F} = (x - y^2, -x^2)$, in Newtons, where x and y are measured in meters. Find the work done by $\mathbf{F}$ along:

a. the oriented line segment C_1 from $(0,0)$ to $(2,-2)$;

b. the oriented line segment C_2 from $(2,-2)$ to $(0,-2)$;

c. the oriented line segment C_3 from $(0,-2)$ to $(0,0)$; and

d. the entire oriented, piecewise-regular closed curve $C = C_1 + C_2 + C_3$ formed by concatenating the three oriented line segments above.

> Recall that a *closed* curve is one which starts and ends at the same point.

Solution: In Figure 4.2.2, we have sketched the vector field and the relevant line segments; we have labeled the points $(0,0)$, $(2,-2)$, and $(0,-2)$, respectively, as $\mathbf{a}$, $\mathbf{b}$, and $\mathbf{c}$.

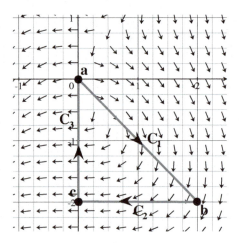

Figure 4.2.2: The vector field $\mathbf{F} = (x - y^2, -x^2)$.

The work is given by the relevant line integrals.

a.

To find the work done by $\mathbf{F}$ along C_1, we need to parameterize the line segment C_1 from $\mathbf{a} = (0,0)$ to $\mathbf{b} = (2,-2)$. The most obvious parameterization is

> Recall that this was our standard parameterization for a line segment from Section 1.4.

$$\mathbf{r}(t) = (0,0) + t\big((2,-2) - (0,0)\big) = (2t, -2t) = 2(t, -t),$$

where $0 \leq t \leq 1$. This means that we substitute $x = 2t$ and $y = -2t$ into our vector field $\mathbf{F} = (x - y^2, -x^2)$; we obtain

$$\mathbf{F}(\mathbf{r}(t)) = \big(2t - (-2t)^2, -(2t)^2\big) = (2t - 4t^2, -4t^2) = 2(t - 2t^2, -2t^2).$$

We also have $\mathbf{r}'(t) = 2(1, -1)$.

Thus,

$$\text{work done by } \mathbf{F} \text{ along } C_1 \; = \; \int_{C_1} \mathbf{F} \cdot d\mathbf{r} \; =$$

$$\int_0^1 \mathbf{F}(\mathbf{r}(t)) \cdot \mathbf{r}'(t)\, dt \; = \; \int_0^1 2(t - 2t^2, \; -2t^2) \cdot 2(1, -1)\, dt \; =$$

$$4 \int_0^1 (t - 2t^2 + 2t^2)\, dt \; = \; 2 \text{ joules.}$$

We could have used a different parameterization to obtain the same result. We could, for instance, parameterize C_1 by a parameterization that "goes at half the speed", or "takes twice as long" as our previous parameterization.

Consider the parameterization of the oriented curve C_1 by

$$\mathbf{p}(t) \; = \; (t, -t),$$

where $0 \leq t \leq 2$. This means that we substitute $x = t$ and $y = -t$ into our vector field $\mathbf{F} = (x - y^2, \; -x^2)$; we obtain

$$\mathbf{F}(\mathbf{p}(t)) \; = \; \left(t - (-t)^2, \; -t^2\right) \; = \; (t - t^2, \; -t^2),$$

and we also have $\mathbf{p}'(t) = (1, -1)$.

Thus, we find

$$\text{work done by } \mathbf{F} \text{ along } C_1 \; = \; \int_{C_1} \mathbf{F} \cdot d\mathbf{p} \; =$$

$$\int_0^2 \mathbf{F}(\mathbf{p}(t)) \cdot \mathbf{p}'(t)\, dt \; = \; \int_0^2 (t - t^2, \; -t^2) \cdot (1, -1)\, dt \; =$$

$$\int_0^2 t\, dt \; = \; 2 \text{ joules.}$$

Of course, this is the same as what we obtained with our previous parameterization.

Could you have known ahead of time that the work along C_1 would be positive? Well...yes, if you look at Figure 4.2.2. What you can see in Figure 4.2.2 is that, for any parametrization $\mathbf{r}$ of the oriented curve C_1, the individual dot products

$$\mathbf{F} \cdot d\mathbf{r} \; = \; \mathbf{F}(\mathbf{r}(t)) \cdot \mathbf{r}'(t)\, dt$$

are all positive, because the angle between the vector $\mathbf{F}(\mathbf{r}(t))$ and $\mathbf{r}'(t)$ is clearly less than $90°$ ($\pi/2$ radians), and we are using positive infinitesimal dt.

However, for curves through a vector field for which the angle between $\mathbf{F}$ and $d\mathbf{r}$ is sometimes less than 90° and sometimes greater than 90°, you can't just look at a sketch and decide easily whether or not the line integral is positive or negative.

---

b.

To calculate the work done by $\mathbf{F} = (x - y^2, -x^2)$ along the oriented line segment C_2 from $(2, -2)$ to $(0, -2)$, we parameterize C_2 by

$$\mathbf{r}(t) = (2, -2) + t\big((0, -2) - (2, -2)\big) = (2 - 2t, -2) = 2(1 - t, -1),$$

We could also parameterize in the "wrong" direction, and let t go from a bigger value to a smaller one. You should verify that you get same result if use $\mathbf{r}(t) = (t, -2)$, but have t start at 2 and go to 0.

as t goes from 0 to 1, and calculate the line integral

$$\int_{C_2} \mathbf{F} \cdot d\mathbf{r} = \int_0^1 \mathbf{F}(\mathbf{r}(t)) \cdot \mathbf{r}'(t) \, dt$$

$$\int_0^1 (2 - 2t - 4, -(2 - 2t)^2) \cdot 2(-1, 0) \, dt = \int_0^1 (-2t - 2, -4(1 - t)^2) \cdot 2(-1, 0) \, dt$$

$$\int_0^1 4(t + 1) \, dt = 4\left(\frac{t^2}{2} + t\right)\Big|_0^1 = 6 \text{ joules.}$$

As in part (a), looking at the vector field $\mathbf{F}$ and the oriented line C_2 from $\mathbf{b}$ to $\mathbf{c}$ (which we show again in Figure 4.2.3), you can see that the angle between $\mathbf{F}$ and $d\mathbf{r}$ is always less than 90°; thus, the individual dot products $\mathbf{F}(\mathbf{r}(t)) \cdot \mathbf{r}'(t)$ are all positive, and so the integral is positive (since we also go from a smaller t value to a larger one).

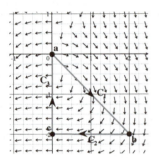

Figure 4.2.3: The vector field $\mathbf{F} = (x - y^2, -x^2)$.

We could have calculated this line integral using our previously discussed alternative notation, and letting x go from a smaller value to larger one. We calculate

$$\int_{C_2} \mathbf{F} \cdot d\mathbf{r} = \int_{C_2} (x - y^2) \, dx + (-x^2) \, dy = \int_2^0 (x - 4) \, dx + 0 = 6 \text{ joules,}$$

where we used that, along C_2, y is constantly -2, so $dy = 0$, and x goes from 2 to 0.

---

c.

To calculate the work done by $\mathbf{F} = (x - y^2, -x^2)$ along the oriented line segment C_3 from $(0, -2)$ to $(0, 0)$, we parameterize C_3 by

$$\mathbf{r}(t) = (0, t),$$

as t goes from -2 to 0, and calculate the line integral

$$\int_{C_3} \mathbf{F} \cdot d\mathbf{r} = \int_{-2}^0 \mathbf{F}(\mathbf{r}(t)) \cdot \mathbf{r}'(t) \, dt = \int_{-2}^0 (-t^2, 0) \cdot (0, 1) \, dt = \int_{-2}^0 0 \, dt = 0.$$

Note that you can easily see that C_3 is perpendicular to the vector field $\mathbf{F}$.

d.

Finally, the work done by $\mathbf{F}$ around the entire oriented closed curve C is simply the sum of the line integrals along C_1, C_2, and C_3.

Thus, the total work done by $\mathbf{F}$ along C is

$$\int_C \mathbf{F} \cdot d\mathbf{r} \ = \ \int_{C_1+C_2+C_3} \mathbf{F} \cdot d\mathbf{r} \ = \ \int_{C_1} \mathbf{F} \cdot d\mathbf{r} \ + \ \int_{C_2} \mathbf{F} \cdot d\mathbf{r} \ + \ \int_{C_3} \mathbf{F} \cdot d\mathbf{r} \ = $$

$$2 \ + \ 6 \ + \ 0 \ = \ 8 \text{ joules.}$$

Remark 4.2.9. In the previous example, we calculated a line integral over a closed curve; we started at $(0,0)$ and we ended at $(0,0)$. It is worth noting that, for a line integral

$$\int_C \mathbf{F} \cdot d\mathbf{r},$$

where C is a closed, oriented, piecewise-regular curve, it is **irrelevant** what point is chosen as the the starting and ending point.

Consider a closed, oriented, piecewise-regular curve C, and two possible starting/ending points $\mathbf{a}$ and $\mathbf{b}$ on C. This divides C into two oriented pieces C_1 and C_2, where C_1 starts at $\mathbf{a}$ and ends at $\mathbf{b}$, while C_2 starts at $\mathbf{b}$ and ends at $\mathbf{a}$; we have $C = C_1 + C_2$. See Figure 4.2.4. Written as $C_1 + C_2$, the curve C is considered as starting and ending at the point marked $\mathbf{a}$.

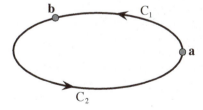

Figure 4.2.4: The closed curve $C = C_1 + C_2 = C_2 + C_1$.

However, it's also true that $C = C_2 + C_1$ and, written this way, the curve C is considered as starting and ending at the point marked $\mathbf{b}$.

That the line integral along C, starting and ending at $\mathbf{a}$, is the same as the line integral along C, starting and ending at $\mathbf{b}$, is just a reflection of the fact that

$$\int_C \mathbf{F} \cdot d\mathbf{r} \;=\; \int_{C_1} \mathbf{F} \cdot d\mathbf{r} \;+\; \int_{C_2} \mathbf{F} \cdot d\mathbf{r} \;=\; \int_{C_2} \mathbf{F} \cdot d\mathbf{r} \;+\; \int_{C_1} \mathbf{F} \cdot d\mathbf{r}.$$

Thus, in problems where you are supposed to calculate line integrals along closed curves, there is no need for a starting/ending point to be explicitly given.

Example 4.2.10. Now let's look at an example of a line integral of a vector field in $\mathbb{R}^3$.

Consider the vector field on $\mathbb{R}^3$ given by

$$\mathbf{F} \;=\; \left(yz + e^x, \;\; xz + \cos y, \;\; xy + \frac{1}{1+z^2} \right).$$

We want to calculate the line integral of $\mathbf{F}$ along the two different curves from $(1,0,0)$ to $(0,0,1)$.

We let C_1 denote the oriented line segment from $(1,0,0)$ to $(0,0,0)$, i.e., an oriented portion of the x-axis. We let C_2 be the oriented line segment from $(0,0,0)$ to $(0,0,1)$, i.e., an oriented portion of the z-axis. Then, $C_1 + C_2$ is an oriented, piecewise-regular curve from $(1,0,0)$ to $(0,0,1)$.

Now let C_3 denote the counterclockwise (in the xz-plane) oriented quarter of the circle of radius 1, centered at the origin, in the xz-plane; this curve also starts at $(1,0,0)$ and ends at $(0,0,1)$.

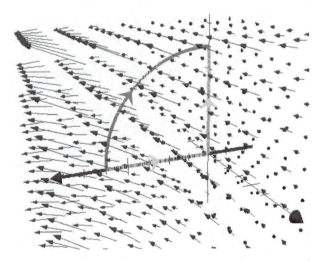

Figure 4.2.5: The vector field $\mathbf{F} \;=\; \left(yz + e^x, \;\; xz + \cos y, \;\; xy + \frac{1}{1+z^2} \right)$ and two curves from $(1,0,0)$ to $(0,0,1)$.

We parameterize C_1 by $\mathbf{r}(t) = (1-t,\, 0,\, 0)$, where $0 \le t \le 1$, and calculate

$$\int_{C_1} \mathbf{F} \cdot d\mathbf{r} = \int_0^1 \mathbf{F}(\mathbf{r}(t)) \cdot \mathbf{r}'(t)\, dt = \int_0^1 (e^{1-t},\, 1,\, 1) \cdot (-1,\, 0,\, 0)\, dt =$$

$$\int_0^1 -e^{1-t}\, dt = e^{1-t}\, \big|_0^1 = 1 - e.$$

We parameterize C_2 by $\mathbf{r}(t) = (0,\, 0,\, t)$, where $0 \le t \le 1$, and calculate

$$\int_{C_2} \mathbf{F} \cdot d\mathbf{r} = \int_0^1 \mathbf{F}(\mathbf{r}(t)) \cdot \mathbf{r}'(t)\, dt = \int_0^1 \left(1,\, 1,\, \frac{1}{1+t^2}\right) \cdot (0,\, 0,\, 1)\, dt =$$

$$\int_0^1 \frac{1}{1+t^2}\, dt = \tan^{-1} t\, \big|_0^1 = \frac{\pi}{4}.$$

Thus,

$$\int_{C_1+C_2} \mathbf{F} \cdot d\mathbf{r} = \int_{C_1} \mathbf{F} \cdot d\mathbf{r} + \int_{C_2} \mathbf{F} \cdot d\mathbf{r} = 1 - e + \frac{\pi}{4}.$$

Now we'll calculate the line integral of $\mathbf{F}$ along C_3. We parameterize C_3 by

$$\mathbf{r}(t) = (\cos t,\, 0,\, \sin t),$$

for $0 \le t \le \pi/2$.

We calculate

$$\int_{C_3} \mathbf{F} \cdot d\mathbf{r} = \int_0^{\pi/2} \mathbf{F}(\mathbf{r}(t)) \cdot \mathbf{r}'(t)\, dt =$$

$$\int_0^{\pi/2} \left(e^{\cos t},\, \cos t \sin t + 1,\, \frac{1}{1+\sin^2 t}\right) \cdot (-\sin t,\, 0,\, \cos t)\, dt$$

$$\int_0^{\pi/2} -e^{\cos t} \sin t\, dt + \int_0^{\pi/2} \frac{1}{1+\sin^2 t} \cos t\, dt.$$

Substituting $u = \cos t$ into the first integral and $w = \sin t$ into the second, we find

$$\int_{C_3} \mathbf{F} \cdot d\mathbf{r} = e^{\cos t} + \tan^{-1}(\sin t)\, \Big|_0^{\pi/2} =$$

$$e^0 + \tan^{-1}(1) - e^1 - \tan^{-1}(0) = 1 + \frac{\pi}{4} - e.$$

Note that the line integrals of $\mathbf{F}$ along $C_1 + C_2$ and C_3 were the same. You may wonder if every oriented piecewise-regular curve from $(1, 0, 0)$ to $(0, 0, 1)$ would yield the same line integral. The answer is: yes.

In fact, as we shall discuss in the next section, $\mathbf{F}$ is what is known as a *conservative vector field*, which means that for every starting point $\mathbf{a}$ and every ending point $\mathbf{b}$, the line integral of $\mathbf{F}$ along any oriented piecewise-regular curve from $\mathbf{a}$ to $\mathbf{b}$ does not depend on the specific curve chosen, but only on $\mathbf{F}$, $\mathbf{a}$, and $\mathbf{b}$.

More Depth:

We want to investigate the relationship between the work done by a force field on an object and the kinetic energy of the object.

Suppose that an object is moving with velocity $\mathbf{v} = \mathbf{v}(t)$, where t denotes time. For each time t, we denote the speed $|\mathbf{v}(t)|$ by v_t. Thus, the kinetic energy of the object of mass m, at time t, is

$$\text{kinetic energy} \;=\; \frac{1}{2}\,mv_t^2.$$

We have the following theorem relating work done by a force field and the change in kinetic energy of an object, in the special case where the force field describes the **net** force (or total force, or the sum of all forces) acting on the object.

Theorem 4.2.11. *Suppose that $\mathbf{F}_{\text{net}}$ is a continuous force field on an open subset $\mathcal{U}$ of $\mathbb{R}^3$ (or any $\mathbb{R}^n$). Suppose that a point-mass, of mass m, is moving through $\mathcal{U}$, and $\mathbf{F}_{\text{net}}$ is the net force which acts on the point-mass. Let $\mathbf{r}(t)$ denote the position of the point-mass at time t, for $t_0 \leq t \leq t_1$. Let v_0 and v_1 denote the speeds of the object at times t_0 and t_1, respectively. Let C be the oriented curve parameterized by $\mathbf{r}(t)$.*

Then, the work done by $\mathbf{F}_{\text{net}}$ on the point-mass m, along the curve C, is equal to the change in kinetic energy of the point-mass, i.e.,

$$\int_C \mathbf{F}_{\text{net}} \cdot d\mathbf{r} \;=\; \frac{1}{2}\,mv_1^2 - \frac{1}{2}\,mv_0^2.$$

Proof. This proof uses two basic facts/results. First, letting $\mathbf{a}$ denote acceleration, Newton's 2nd Law of Motion tells us that

$$\mathbf{F}_{\text{net}} \;=\; m\mathbf{a}.$$

Second, since $v_t^2 = \mathbf{v} \cdot \mathbf{v}$,

$$\frac{d}{dt}(v_t^2) \;=\; \frac{d}{dt}(\mathbf{v} \cdot \mathbf{v}) \;=\; \mathbf{v} \cdot \frac{d\mathbf{v}}{dt} + \frac{d\mathbf{v}}{dt} \cdot \mathbf{v} \;=\; 2\,\mathbf{a} \cdot \mathbf{v},$$

so that

$$\frac{d}{dt}\left(\frac{1}{2}mv_t^2\right) = m(\mathbf{a}\cdot\mathbf{v}).$$

Therefore,

$$\int_C \mathbf{F}_{\text{net}}\cdot d\mathbf{r} = \int_{t_0}^{t_1} \mathbf{F}_{\text{net}}(\mathbf{r}(t))\cdot\mathbf{r}'(t)\,dt = \int_{t_0}^{t_1} m\,\mathbf{a}\cdot\mathbf{v}\,dt = \int_{t_0}^{t_1} \frac{d}{dt}\left(\frac{1}{2}mv_t^2\right)dt =$$

$$\frac{1}{2}mv_1^2 - \frac{1}{2}mv_0^2.$$

$\square$

Remark 4.2.12. It is **extremely** important, in Theorem 4.2.11, that the force field gives the **net** force acting on the object.

In many problems in which an object is moving through a force field, there are other forces acting on the object, in addition to the force from the force field. If we write $\mathbf{F}_{\text{ext}}$ for the sum of all forces which are external to the force field, then the net force acting on the object is $\mathbf{F} + \mathbf{F}_{\text{ext}}$, and the argument in Theorem 4.2.11 tells us that

$$\int_C \mathbf{F}\cdot d\mathbf{r} + \int_C \mathbf{F}_{\text{ext}}\cdot d\mathbf{r} = \frac{1}{2}mv_1^2 - \frac{1}{2}mv_0^2.$$

Example 4.2.13. A force field $\mathbf{F}$, in Newtons, acts on a particle of mass 2 kg, as the particle moves in the xy-plane in such a way that its position is given by $\mathbf{r}(t) = (t, t^2)$ meters, at times t seconds, for $1 \leq t \leq 5$.

    a. Assume that $\mathbf{F}$ is the net force acting on the mass. Determine the work done by $\mathbf{F}$ on the particle between times $t = 1$ and $t = 5$ seconds.

    b. Assume that there are external forces $\mathbf{F}_{\text{ext}}(x, y) = (y, -x)$, in Newtons, which are also acting on the particle. Determine the work done by $\mathbf{F}$ on the particle between times $t = 1$ and $t = 5$ seconds.

Solution: Let C denoted the curve parameterized by $\mathbf{r}$.

a.   By Theorem 4.2.11, what we need to do is calculate the change in kinetic energy, which means that we need the speeds v_1 and v_5. This is easy; we calculate the velocity

$$\mathbf{v}(t) = \mathbf{r}'(t) = (1, 2t)\ \text{m/s},$$

so that the speed v_t is given by

$$v_t = \sqrt{1^2 + (2t)^2} = \sqrt{1 + 4t^2} \text{ m/s.}$$

Therefore, the work done by $\mathbf{F}$ is

$$\int_C \mathbf{F} \cdot d\mathbf{r} = \frac{1}{2}(2)\Big((1 + 4(5)^2) - (1 + 4(1)^2)\Big) = 96 \text{ joules.}$$

b.   By Remark 4.2.12 and our result in part (a), we know that the work done by $\mathbf{F}$ is

$$\int_C \mathbf{F} \cdot d\mathbf{r} = 96 - \int_C \mathbf{F}_{\text{ext}} \cdot d\mathbf{r}.$$

We calculate

$$\int_C \mathbf{F}_{\text{ext}} \cdot d\mathbf{r} = \int_1^5 \mathbf{F}_{\text{ext}}(\mathbf{r}(t)) \cdot \mathbf{r}'(t) \, dt = \int_1^5 (t^2, -t) \cdot (1, 2t) \, dt = \int_1^5 (t^2 - 2t^2) \, dt =$$

$$\int_1^5 -t^2 \, dt = -\frac{t^3}{3} \Big|_1^5 = -\frac{124}{3} \text{ joules.}$$

Therefore, the work done by $\mathbf{F}$ is

$$\int_C \mathbf{F} \cdot d\mathbf{r} = 96 - \left(-\frac{124}{3}\right) = \frac{412}{3} \text{ joules.}$$

Online answers to select exercises are here.

4.2.1   Exercises

Basics:

In each of the following exercises, you are given a vector field $\mathbf{F}$, and a parameterization $\mathbf{r} = \mathbf{r}(t)$ of a curve. Find the line integral of $\mathbf{F}$ along the curve.

1. $\mathbf{F}(x, y) = (y, x + y)$, and $\mathbf{r}(t) = (t, t^2)$, $0 \le t \le 2$.

2. $\mathbf{F}(x, y) = (y, x + y)$, and $\mathbf{r}(t) = (2t, 4t^2)$, $0 \le t \le 1$.

3. $\mathbf{F}(x, y) = (1, -3)$, and $\mathbf{r}(t) = 5(\cos t, \sin t)$, $0 \le t \le \pi/2$.

4. $\mathbf{F}(x, y) = (1, -3)$, and $\mathbf{r}(t) = 5(\cos t, \sin t)$, $0 \le t \le 2\pi$.

5. $\mathbf{F}(x, y) = (ye^x + y^2, e^x)$, and $\mathbf{r}(t) = (t, e^t)$, $-1 \le t \le 1$.

6. $\mathbf{F}(x, y) = (x^2, xe^{y^2} + \ln x)$, and $\mathbf{r}(t) = (t^2, 4)$, $2 \le t \le 3$.

7. $\mathbf{F}(x, y, z) = -(x,\ y,\ z)$, and $\mathbf{r}(t) = (t, t^2, t^3)$, $0 \le t \le 1$.

8. $\mathbf{F}(x, y, z) = (xy,\ yz,\ zx)$, and $\mathbf{r}(t) = (t, 2t+1, 3t+2)$, $-1 \le t \le 2$.

9. $\mathbf{F}(x, y, z) = (z, -x, y)$, and $\mathbf{r}(t) = (\cos t, \sin t, \sin t)$, $0 \le t \le 2\pi$. ▶

10. $\mathbf{F}(x, y, z) = (ye^x,\ \tan^{-1}(\ln(xy+z^2)),\ x^2+y^2+z^2)$, and $\mathbf{r}(t) = (t, 1, -t^2)$, $1 \le t \le 2$.

In each of the following exercises, you are given a force field F in Newtons, and an oriented curve C, where all coordinates are in meters. Find the work done by F along C.

11. $\mathbf{F}(x, y) = (-y, x)$, and C is the oriented line segment from $(0, 0)$ to $(2, 3)$.

12. $\mathbf{F}(x, y) = (-y, x)$, and C is the circle of radius 2, centered at the origin, oriented clockwise.

13. $\mathbf{F}(x, y) = \left(\dfrac{-y}{\sqrt{x^2 + y^2}}, \dfrac{x}{\sqrt{x^2 + y^2}} \right)$, and C is the semi-circle of radius 3, centered

    at the origin, oriented counterclockwise, from $(3, 0)$ to $(-3, 0)$. ▶

14. $\mathbf{F}(x, y) = \left(\dfrac{x}{\sqrt{x^2 + y^2}}, \dfrac{y}{\sqrt{x^2 + y^2}} \right)$, and C is the oriented line segment from $(1, 1)$
    to $(1, 5)$.

15. $\mathbf{F}(x, y, z) = (2, -3, 5)$, and C is the oriented line segment from $(1, 0, -1)$ to $(5, 7, -4)$.

16. $\mathbf{F}(x, y, z) = (xz, y + z, x^2)$, and C is the oriented curve in $\mathbb{R}^3$ where $z = \sqrt{x}$ and $y = x^3$, which goes from $(1, 1, 1)$ to $(4, 64, 2)$.

17. $\mathbf{F}(x, y, z) = \dfrac{1}{(x^2 + y^2 + z^2)^{3/2}}(x, y, z)$, and C is the counterclockwise circle of radius 2 in the xy-plane.

18. $\mathbf{F}(x, y, z) = (x, y, z)$, and C is the oriented quarter-circle of radius 5, centered at the origin, in the plane where $z = y$, which starts at the point $(5, 0, 0)$ and ends at $\left(0, \dfrac{5}{\sqrt{2}}, \dfrac{5}{\sqrt{2}} \right)$.

In each of the following exercises, you are given a vector field F, an oriented curve C, and a sketch of the vector field and curve. (a) Use the sketch to determine if $\int_C \mathbf{F} \cdot d\mathbf{r}$ is positive, negative, or (approximately) zero. (b) Calculate $\int_C \mathbf{F} \cdot d\mathbf{r}$, and verify that your answer to part (a) was correct. Note that the sketch of the vector field may be scaled.

19. $\mathbf{F} = (1, 1)$, and C is the oriented line segment from $(0.5, 0)$ to $(0.5, 1)$.

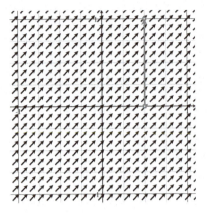

20. $\mathbf{F} = (1, 1)$, and C is the oriented line segment from $(1, 0)$ to $(0, 1)$.

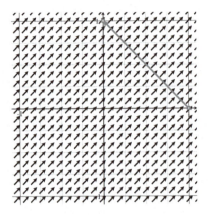

21. $\mathbf{F} = (1, 1)$, and C is the counterclockwise circle of radius 0.5, centered at the origin.

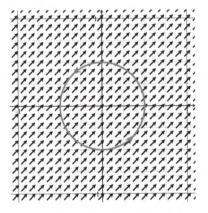

22. $\mathbf{F} = (1, 1)$, and C is the oriented portion of the parabola where $y = -x^2$, from $(0, 0)$ to $(-1, -1)$.

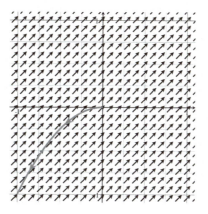

23. $\mathbf{F} = (-x, -y)$, and C is the oriented line segment from $(0.5, 0)$ to $(0.5, 1)$.

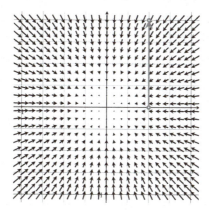

24. $\mathbf{F} = (-x, -y)$, and C is the oriented portion of the parabola where $y = -x^2$, from $(0, 0)$ to $(-1, -1)$.

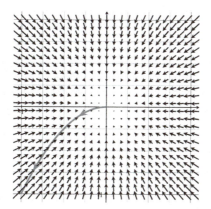

In each of the following exercises, you are given a force field $\mathbf{F}$ in Newtons, and an oriented piecewise-regular curve C, where all coordinates are in meters. Find the work done by $\mathbf{F}$ along C.

25. $\mathbf{F}(x, y) = (y, -x)$, and $C = C_1 + C_2$, where C_1 is the oriented line segment from $(1, 0)$ to $(0, 1)$, and C_2 is the oriented quarter-circle of radius 1, centered at the origin, from $(0, 1)$ to $(-1, 0)$.

26. $\mathbf{F}(x, y) = (y, -x)$, and $C = C_1 + C_2$, where C_1 is the oriented line segment from $(1, 0)$ to $(0, -1)$, and C_2 is the oriented quarter-circle of radius 1, centered at the origin, from $(0, -1)$ to $(-1, 0)$.

27. $\mathbf{F}(x, y, z) = (1, -z, y)$, and $C = C_1 + C_2$, where C_1 is the portion of the parabola where $y = x^2$ and $z = 0$ from $(0, 0, 0)$ to $(1, 1, 0)$, and C_2 is the line segment from $(1, 1, 0)$ to $(2, 3, 4)$.

28. $\mathbf{F}(x, y, z) = \left(-\dfrac{y}{z}, \dfrac{x}{z}, xye^{z^2}\right)$, and $C = C_1 + C_2 + C_3$, where C_1, C_2, and C_3 are all in the plane where $z = 1$, C_1 is the oriented quarter-circle from $(2, 0, 1)$ to $(0, 2, 1)$, C_2 is the line segment from $(0, 2, 1)$ to $(0, 0, 1)$, and C_3 is the line segment from $(0, 0, 1)$ to $(2, 0, 1)$ (which closes the curve C).

More Depth:

In each of the following exercises, the net force acting on a particle of mass 3 kg is given by the vector field $\mathbf{F}$, in Newtons. You are given the position $\mathbf{r}(t)$, in meters, of the particle at time t seconds, between times $t = a$ and $t = b$. Calculate the work by $\mathbf{F}$ on the particle between times a and b.

29. $\mathbf{r}(t) = (4t, 3t - 2, -5t + 7)$, and $3 \le t \le 6$.

30. $\mathbf{r}(t) = (t, t^2, t^3)$, and $0 \le t \le 1$.

31. $\mathbf{r}(t) = (t, \cos t, \sin t)$, and $0 \le t \le \pi$.

32. $\mathbf{r}(t) = (e^t, e^{3t}, e^{5t})$, and $-1 \le t \le 0$.

In each of the following exercises, a force field $\mathbf{F}$ and an external force field $\mathbf{F}_{\text{ext}}$ are acting on a particle of mass 3 kg, where both force fields are measured in Newtons and, together, $\mathbf{F}$ and $\mathbf{F}_{\text{ext}}$ include all of the forces acting on the particle. You are given $\mathbf{F}_{\text{ext}}$ and the position $\mathbf{r}(t)$, of the particle, in meters, at time t seconds, between times $t = a$ and $t = b$. Calculate the work by $\mathbf{F}$ on the particle between times a and b.

33. $\mathbf{F}_{\text{ext}}(x, y, z) = -(x, y, z)$, $\mathbf{r}(t) = (4t, 3t - 2, -5t + 7)$, and $3 \le t \le 6$.

34. $\mathbf{F}_{\text{ext}}(x, y, z) = -(x, y, z)$, $\mathbf{r}(t) = (t, t^2, t^3)$, and $0 \le t \le 1$.

35. $\mathbf{F}_{\text{ext}}(x, y, z) = (2xy^3, 3x^2y^2, z^4)$, $\mathbf{r}(t) = (t, t^2, t^3)$, and $0 \le t \le 1$.

36. $\mathbf{F}_{\text{ext}}(x, y, z) = (2xy^3, 3x^2y^2, z^4)$, $\mathbf{r}(t) = (t, \cos t, \sin t)$, and $0 \le t \le \pi$. (Hint: what is $(t^2 \cos^3 t)'$?)

4.3 Conservative Vector Fields

Recall from Definition 4.1.8 that a vector field **F** which is the gradient vector field of some function f is called a *conservative vector field*.

Conservative vector fields are very special, and can also be characterized as vector fields for which line integrals depend on the vector field and the starting and ending points of the curve, but not on the particular curve itself. This is also equivalent to the line integral along any **closed** oriented piecewise-regular curve being zero.

For *simply-connected* open subsets of $\mathbb{R}^2$ and $\mathbb{R}^3$, conservative vector fields can be further characterized as those vector fields for which the curl is zero.

If **F** is conservative and f is such that $\mathbf{F} = \vec{\nabla} f$, then f is called a *potential function* for **F**.

The *Fundamental Theorem of Line Integrals* tells us how line integrals of conservative vector fields are related to potential functions for the vector field.

Suppose that f is a function on an open subset $\mathcal{U}$ of $\mathbb{R}^n$, and that all of the partial derivatives of f exist. Then, we may consider the gradient vector field $\mathbf{F} = \vec{\nabla} f$.

Basics:

However, when you start with the vector field **F**, instead of starting with f, and wish to assert that **F** is the gradient vector field of **some** f, it is more common to use our terminology from Definition 4.1.8, which we repeat here.

> **Definition 4.3.1.** *Suppose that* **F** *is a continuous vector field on an open subset* $\mathcal{U}$ *of* $\mathbb{R}^n$.
>
> *Then,* **F** *is a* **conservative vector field** *if and only if there exists a differentiable function* f *on* $\mathcal{U}$ *such that* $\mathbf{F} = \vec{\nabla} f$.
>
> *When such a function* f *exists, it is called a* **potential function** *for the conservative vector field* **F**.

Remark 4.3.2. Suppose that **F** is a conservative vector field on a connected open subset $\mathcal{U}$ of $\mathbb{R}^n$, and that we have two potential functions, f and g, for **F**.

Then, $\vec{\nabla} f = \mathbf{F} = \vec{\nabla} g$ and, as we saw back in Theorem 2.1.29, this implies that f and g differ by a constant, i.e., there exists a constant C such that $f = g + C$. Thus, on a connected open set, if you know one potential function, you know them all; just add an arbitrary constant.

We could also have concluded this by using the Fundamental Theorem. If f and g are potential functions for **F**, and **a** is in $\mathcal{U}$, then for any **x** is $\mathcal{U}$ and for any oriented, piecewise-regular curve from **a** to **x**, $\int_C \mathbf{F} \cdot d\mathbf{r} = f(\mathbf{x}) - f(\mathbf{a}) = g(\mathbf{x}) - g(\mathbf{a})$, so that $f(\mathbf{x}) = g(\mathbf{x}) + C$, where $C = f(\mathbf{a}) - g(\mathbf{a})$.

What's so special about a conservative vector field?

Suppose that $\mathbf{F}$ is a conservative vector field, and that f is a potential function for $\mathbf{F}$. To be precise, we mean that f is a differentiable function on an open subset $\mathcal{U}$ of $\mathbb{R}^n$, and that the vector field $\mathbf{F} = \vec{\nabla} f$ is a continuous vector field on $\mathcal{U}$.

Consider a piecewise-regular parameterization $\mathbf{r} : [t_0, t_1] \to \mathcal{U}$ of an oriented curve C that goes from $\mathbf{a} = \mathbf{r}(t_0)$ to $\mathbf{b} = \mathbf{r}(t_1)$. Then, the Chain Rule, Theorem 2.4.9, tells us that

$$\frac{d}{dt}\big(f(\mathbf{r}(t))\big) \;=\; \vec{\nabla} f(\mathbf{r}(t)) \cdot \mathbf{r}'(t) \;=\; \mathbf{F}(\mathbf{r}(t)) \cdot \mathbf{r}'(t).$$

But, this means that the line integral is easy to calculate:

$$\int_C \mathbf{F} \cdot d\mathbf{r} \;=\; \int_{t_0}^{t_1} \mathbf{F}(\mathbf{r}(t)) \cdot \mathbf{r}'(t)\, dt \;=\; \int_{t_0}^{t_1} \frac{d}{dt}\big(f(\mathbf{r}(t))\big)\, dt \;=$$

$$f(\mathbf{r}(t)) \Big|_{t_0}^{t_1} \;=\; f(\mathbf{r}(t_1)) - f(\mathbf{r}(t_0)) \;=\; f(\mathbf{b}) - f(\mathbf{a}).$$

We have just proved:

Theorem 4.3.3. (Fundamental Theorem of Line Integrals)
   Suppose that $\mathbf{F}$ is a conservative vector field, and that f is a potential function for $\mathbf{F}$, i.e., suppose that $\mathbf{F} = \vec{\nabla} f$. Let $\mathbf{a}$ and $\mathbf{b}$ be points in $\mathcal{U}$, and let C be an oriented, piecewise-regular curve in $\mathcal{U}$ from $\mathbf{a}$ to $\mathbf{b}$.
   Then,

$$\int_C \mathbf{F} \cdot d\mathbf{r} \;=\; f(\mathbf{b}) - f(\mathbf{a}).$$

In particular, line integrals of conservative vector fields do not depend on the choice of curves from a given starting point to a given ending point.

   Furthermore, if C is closed curve, so that $\mathbf{a} = \mathbf{b}$, then the line integral is 0. Thus, line integrals of conservative fields along closed curves are zero.

Example 4.3.4. Consider the function

$$f(x, y, z) \;=\; xyz + e^x + \sin y + \tan^{-1} z$$

on $\mathbb{R}^3$.

Then,

$$\vec{\nabla} f \;=\; \left(yz + e^x, \; xz + \cos y, \; xy + \frac{1}{1 + z^2} \right).$$

If we let $\mathbf{F}$ denote this gradient field, then this is the same as the vector field $\mathbf{F}$ from Example 4.2.10.

Thus, the Fundamental Theorem of Line Integrals implies that, for all $\mathbf{a}$ and $\mathbf{b}$ in $\mathbb{R}^3$, and for all oriented piecewise-regular curves C from $\mathbf{a}$ to $\mathbf{b}$,

$$\int_C \mathbf{F} \cdot d\mathbf{r} = f(\mathbf{b}) - f(\mathbf{a}).$$

In particular, in Example 4.2.10, we found that we obtained the same value for the line integrals along two very different curves from $(1, 0, 0)$ to $(0, 0, 1)$. The Fundamental Theorem explains why: regardless of what of we choose for our oriented, piecewise-regular curve C from $(1, 0, 0)$ to $(0, 0, 1)$, what we obtain for the line integral is

$$\int_C \mathbf{F} \cdot d\mathbf{r} = f(0, 0, 1) - f(1, 0, 0) =$$

$$0 \cdot 0 \cdot 1 + e^0 + \sin 0 + \tan^{-1}(1) - \left(1 \cdot 0 \cdot 0 + e^1 + \sin 0 + \tan^{-1}(0)\right) =$$

$$1 + \frac{\pi}{4} - e.$$

which, of course, is what we found both times in Example 4.2.10.

Example 4.3.5. Let x and y be measured in meters, and consider the force field $\mathbf{F} = (3x^2, 2y)$ Newtons.

Find the work done by $\mathbf{F}$, acting on an object which moves along the oriented curve $C = C_1 + C_2 + C_3$, where C_1 is the oriented line segment from $(2, 0)$ to $(2, 3)$, C_2 is the oriented line segment from $(2, 3)$ to $(0, 3)$, C_3 is the oriented quarter of the circle of radius 3, centered at the origin, from $(0, 3)$ to $(-3, 0)$, and $C = C_1 + C_2 + C_3$.

Solution:

We could parameterize C_1, C_2, and C_3 separately, calculate the three corresponding line integrals, and then add the results. However, if we can find an $f = f(x, y)$ so that $\mathbf{F} = \vec{\nabla} f$, then we could apply the Fundamental Theorem of Line Integrals and simply calculate

$$\int_C \mathbf{F} \cdot d\mathbf{r} = f(-3, 0) - f(2, 0);$$

the intermediate points and curves would be irrelevant.

Is there an f such that $\mathbf{F} = \vec{\nabla} f$ and, if such an f exists, how do we find one?

In general, this would be a slightly difficult question, and we shall look at harder examples later. However, here, it's easy to produce an f such that

$$\mathbf{F} \;=\; (3x^2, 2y) \;=\; \vec{\nabla} f,$$

i.e., such that

$$\frac{\partial f}{\partial x} \;=\; 3x^2 \quad \text{and} \quad \frac{\partial f}{\partial y} \;=\; 2y.$$

You can quickly see that

$$f(x,y) \;=\; x^3 + y^2$$

is such a function.

> You might think "Ah. I never have to calculate line integrals from the definition. I can just apply the Fundamental Theorem of Line Integrals". This is **not** the case. The Fundamental Theorem applies only to **conservative** vector fields.

Thus, rather than parameterizing three curves and calculating three line integrals, we calculate the work simply by calculating

$$\int_C \mathbf{F} \cdot d\mathbf{r} \;=\; f(-3,0) - f(2,0) \;=\; (-3)^3 + 0^2 - \left(2^3 + 0^2\right) \;=\; -35 \;\; \text{joules.}$$

We know from Remark 4.3.2 that, given a conservative vector field on a connected set, there's essentially only one potential function for the vector field; all of the other potential functions come from adding a constant to a particular one. But, given a conservative vector field, how do you find one potential function?

Example 4.3.6. Consider the vector field $\mathbf{F}$ on $\mathbb{R}^2$ given by

$$\mathbf{F} \;=\; (e^y + 3x^2, \, xe^y + \cos y).$$

   a. Show that $\mathbf{F}$ is conservative, by producing a potential function f for $\mathbf{F}$.

> Actually, even though it was phrased differently, we looked at another example of finding a potential function back in Example 3.1.7.

   b. Calculate the line integral $\int_C \mathbf{F} \cdot d\mathbf{r}$, where C is the oriented piece of the parabola given by $y = x^2$ from $(0,0)$ to $(1,1)$.

Solution:

(a) We want to find a function (or all functions) f such that $\mathbf{F} = \vec{\nabla} f$, that is, such that

$$(e^y + 3x^2, \, xe^y + \cos y) \;=\; \left(\frac{\partial f}{\partial x}, \frac{\partial f}{\partial y} \right).$$

Thus, we need to simultaneously solve the two equations

$$\frac{\partial f}{\partial x} \;=\; e^y + 3x^2 \quad \text{and} \quad \frac{\partial f}{\partial y} \;=\; xe^y + \cos y. \tag{4.1}$$

As in Example 3.1.7, we anti-differentiate the first equation to obtain

$$f \;=\; \int (e^y + 3x^2)\, dx \;=\; xe^y + x^3 + A(y),$$

where $A(y)$ is a function which may depend on y, but does not depend on x.

At this point, we know f, except that we have to determine the function $A(y)$. What you **don't** do here is anti-differentiate the $\partial f / \partial y$ equation. Instead, it's much easier to take the partial derivative, with respect to y, of what we've obtained thus far for f; this will give us $\partial f / \partial y$, and we require it to equal what we were given for $\partial f / \partial y$ in Formula 4.1.

Hence, we take

$$f \;=\; \int (e^y + 3x^2)\, dx \;=\; xe^y + x^3 + A(y), \qquad (4.2)$$

and take its partial derivative with respect to y to obtain:

$$\frac{\partial f}{\partial y} \;=\; xe^y + A'(y),$$

where we can use the prime notation since $A(y)$ is a function of one variable.

What Formula 4.1 tells us is that

$$\frac{\partial f}{\partial y} \;=\; xe^y + \cos y.$$

Therefore, we need

$$xe^y + A'(y) \;=\; xe^y + \cos y.$$

This means we want

$$A'(y) \;=\; \cos y.$$

So,

$$A(y) \;=\; \sin y + C,$$

where C is a constant.

> If, in fact, $\mathbf{F}$ were not a conservative vector field, this is where you would see it; you would find that all of the x terms did not cancel out, and obtain that $A'(y)$ depends on x. This would be a contradiction.

Combining this with Formula 4.2, we obtain that every potential function for f is of the form

$$f \;=\; xe^y + x^3 + \sin y + C.$$

(b)

Using the potential function $f = xe^y + x^3 + \sin y$ and applying the Fundmental Theorem, Theorem 4.3.3, we find

$$\int_C \mathbf{F} \cdot d\mathbf{r} \;=\; f(1,1) - f(0,0) \;=\; e + 1 + \sin(1).$$

Looking at the previous example leads to a new question: is there an "easy" way to tell if a vector field is, or is not, conservative, other than by trying to find a potential function and succeeding or failing?

The answer is "yes", and we will state this important result as part of a theorem which gives various characterizations of conservative vector fields. First, we wish to motivate the theorem.

Suppose that $\mathbf{F}$ is conservative and $\mathbf{F} = \vec{\nabla} f$ on some open subset of $\mathbb{R}^n$. Assume that $\mathbf{F}$ is continuously differentiable (so that f is of class C^2). Then, the Fundamental Theorem of Line Integrals implies that, for fixed points $\mathbf{a}$ and $\mathbf{b}$, the value of a line integral of $\mathbf{F}$ along an oriented piecewise-regular curve from $\mathbf{a}$ to $\mathbf{b}$ is independent of the curve; it also implies that line integrals of $\mathbf{F}$ along closed curves are all zero.

Furthermore, if we have the vector field

$$\mathbf{F}(x,y) \ = \ (P(x,y), Q(x,y)) \ = \ (f_x, f_y)$$

on an open subset of $\mathbb{R}^2$, then

$$P_y \ = \ (f_x)_y \ = \ f_{xy} \ = \ f_{yx} \ = \ (f_y)_x \ = \ Q_x,$$

that is, the 2-dimensional curl is zero, i.e.,

$$Q_x - P_y \ = \ 0.$$

If, instead, $\mathbf{F}$ is a vector field on an open subset of $\mathbb{R}^3$, then, by Theorem 4.1.17,

$$\operatorname{curl} \mathbf{F} \ = \ \vec{\nabla} \times \mathbf{F} \ = \ \vec{\nabla} \times \left(\vec{\nabla} f \right) \ = \ \mathbf{0}.$$

Thus, in $\mathbb{R}^2$ or $\mathbb{R}^3$, if the curl of $\mathbf{F}$ is not zero, then $\mathbf{F}$ is not conservative.

The theorem that we are about to present also states that the converses of the above statements are true, if we assume that the regions are simply-connected; recall Definition 1.7.16. We shall discuss this further in the More Depth portion of this section, but we shall state the basic equivalence theorem now.

Recall from Definition 1.7.16 that a simply-connected region $\mathcal{U}$ is one such that, for every continuously parameterized closed curve C in $\mathcal{U}$, C can be continuously deformed/collapsed to a point, in such a way that the deformation remains inside $\mathcal{U}$ through the process.

Also recall that, in $\mathbb{R}^2$, you should think of a simply-connected region as a connected region with no holes. However, you may remove a finite number of points, or open or closed balls, from $\mathbb{R}^3$ and the resulting space will still be simply-connected.

Theorem 4.3.7. *Suppose that $\mathbf{F}$ is a continuous vector field on a connected open subset $\mathcal{U}$ of $\mathbb{R}^n$. Then, the following are equivalent:*

1. *$\mathbf{F}$ is a conservative vector field, i.e., there exists a differentiable function f on $\mathcal{U}$ such that $\mathbf{F} = \vec{\nabla} f$;*

2. *for each given pair of points $\mathbf{a}$ and $\mathbf{b}$ in $\mathcal{U}$, for all oriented, piecewise-regular curves C in $\mathcal{U}$ from $\mathbf{a}$ to $\mathbf{b}$, the line integral $\int_C \mathbf{F} \cdot d\mathbf{r}$ exists, and its value is independent of the curve C;*

3. *for every closed oriented, piecewise-regular curve C in $\mathcal{U}$, $\int_C \mathbf{F} \cdot d\mathbf{r} = 0$.*

> *Furthermore, if $\mathcal{U}$ is a simply-connected open subset of $\mathbb{R}^2$ or $\mathbb{R}^3$ and $\mathbf{F}$ is continuously differentiable, then all of these conditions are also equivalent to*
>
> $$\operatorname{curl} \mathbf{F} = \mathbf{0}.$$
>
> *When $\mathcal{U}$ is a subset of $\mathbb{R}^2$ and $\mathbf{F}(x,y) = \big(P(x,y), Q(x,y)\big)$, we mean that $Q_x - P_y$, the 2-dimensional curl of $\mathbf{F}$, equals 0, i.e.,*
>
> $$\frac{\partial Q}{\partial x} = \frac{\partial P}{\partial y}.$$

Vector fields with zero curl are called *irrotational*. Thus, on simply-connected open sets in $\mathbb{R}^2$ or $\mathbb{R}^3$, a vector field is irrotational if and only if it is conservative.

If $\mathbf{F}$ is conservative, then Theorem 4.1.17 tells us that $\vec{\nabla} \times \mathbf{F} = \mathbf{0}$. It's the converse statement that requires simple-connectedness.

Example 4.3.8. The vector field

$$\mathbf{F} = (x - y^2, -x^2) = \big(P(x,y), Q(x,y)\big)$$

in Example 4.2.8 is not a conservative vector field. We can use Theorem 4.3.7 to see this in two different ways.

One way to see this is to see that the line integral around the closed curve C in Example 4.2.8 was equal to 8, not 0.

The second way to see that $\mathbf{F}$ is not conservative is to calculate its (2-dimensional) curl $Q_x - P_y$ and see that it's not zero, i.e., see that $Q_x \neq P_y$. This is the quick way, which doesn't require us to actually compute a line integral.

We find

$$Q_x = \frac{\partial}{\partial x}(-x^2) = -2x,$$

while

$$P_y = \frac{\partial}{\partial y}(x - y^2) = -2y.$$

Since $-2x \neq -2y$, $\mathbf{F}$ is not conservative.

Example 4.3.9. You are given two vector fields $\mathbf{F}$ and $\mathbf{G}$. Decide if one, both, or neither vector field is conservative, and produce a potential function for whatever conservative fields you find.

$$\mathbf{F} = \left(2xy + \frac{2y}{1+y^2},\ y^2 + \frac{1}{1+x^2}\right) \quad \text{and} \quad \mathbf{G} = \left(y^2 + \frac{1}{1+x^2},\ 2xy + \frac{2y}{1+y^2}\right).$$

Solution:

Both vector fields are defined on all of $\mathbb{R}^2$, which is simply-connected.

To check if $\mathbf{F}$ is conservative, we check: is it true that

$$\frac{\partial}{\partial x}\left(y^2 + \frac{1}{1+x^2}\right) = \frac{\partial}{\partial y}\left(2xy + \frac{2y}{1+y^2}\right)?$$

No; we find

$$\frac{\partial}{\partial x}\left(y^2 + \frac{1}{1+x^2}\right) = \frac{-2x}{(1+x^2)^2},$$

while

$$\frac{\partial}{\partial y}\left(2xy + \frac{2y}{1+y^2}\right) = 2x + \frac{2(1-y^2)}{(1+y^2)^2}.$$

Therefore, the 2-dimensional curl of $\mathbf{F}$ is not zero, and so $\mathbf{F}$ is **not** conservative.

To check if $\mathbf{G}$ is conservative, we check: is it true that

$$\frac{\partial}{\partial x}\left(2xy + \frac{2y}{1+y^2}\right) = \frac{\partial}{\partial y}\left(y^2 + \frac{1}{1+x^2}\right)?$$

Yes, because we easily calculate that each side equals $2y$. Therefore, $\mathbf{G}$ is conservative, and we need to find a potential function for it.

We want a function $g = g(x, y)$ such that

$$\frac{\partial g}{\partial x} = y^2 + \frac{1}{1+x^2} \quad \text{and} \quad \frac{\partial g}{\partial y} = 2xy + \frac{2y}{1+y^2}.$$

Thus,

$$g = \int\left(y^2 + \frac{1}{1+x^2}\right) dx = xy^2 + \tan^{-1} x + A(y),$$

and we need

$$\frac{\partial g}{\partial y} = 2xy + A'(y) = 2xy + \frac{2y}{1+y^2}.$$

It follows that we must have $A'(y) = \dfrac{2y}{1+y^2}$, and so

$$A(y) = \int\left(\frac{2y}{1+y^2}\right) dy = \ln(1+y^2) + C.$$

Thus, every potential function g for $\mathbf{G}$ is of the form

$$g = xy^2 + \tan^{-1} x + \ln(1+y^2) + C.$$

Example 4.3.10. Is it possible to look at a sketch of a vector field **F** and tell if the vector field is conservative or not? Sometimes.

If you can spot an oriented closed curve C, parameterized by some $\mathbf{r}(t)$, such that the dot products $\mathbf{F}(\mathbf{r}(t)) \cdot \mathbf{r}'(t)$ are always positive or always negative, then $\int_C \mathbf{F} \cdot d\mathbf{r}$ will be correspondingly positive or negative; by Theorem 4.3.7, this would imply that **F** is not conservative.

For instance, in Figure 4.3.1, we have sketched the vector field $\mathbf{F}(x, y) = (-y, x)$, together with a curve along which $\mathbf{F}(\mathbf{r}(t)) \cdot \mathbf{r}'(t)$ is clearly positive for all t.

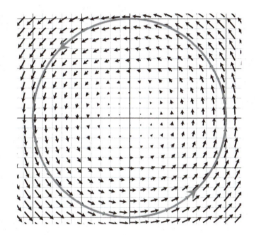

Figure 4.3.1: The vector field $\mathbf{F} = (-y, x)$ and a closed curve along which the line integral is clearly positive.

Consequently, **F** is definitely not conservative.

Seeing that $\mathbf{F}(x, y) = (P(x, y), Q(x, y))$ is conservative is typically more difficult. Essentially the only way to see this in the sketch of the vector field is to notice that, for every fixed x value, the x components of the vectors in **F** are fixed and that, for every fixed y value, the y components of the vectors in **F** are fixed. This implies that $P(x, y)$ is really just a function $P(x)$ of x, and $Q(x, y)$ is really just a function $Q(y)$ of y.

Assuming these $P(x)$ and $Q(y)$ are continuous functions, they have anti-derivatives; call one choice of anti-derivatives $\hat{P}(x)$ and $\hat{Q}(y)$. Then, $f(x, y) = \hat{P}(x) + \hat{Q}(y)$ is a potential function for **F**, and so **F** is conservative.

For instance, the vector field in Figure 4.3.2 appears to be conservative; for each fixed x-coordinate, the x component of the vector field is constant and, for each fixed y-coordinate, the y component is constant.

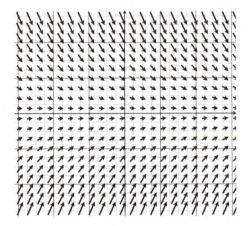

Figure 4.3.2: The vector field **F** which appears to be conservative.

Of course, vector fields of the form $\mathbf{F} = (P(x), Q(y))$ are very special, even among conservative vector fields; but they are essentially the only types of conservative fields that can be detected just by looking at a sketch at **F**.

Physicists define a potential function for a conservative vector field **F** to be a function f such that $\mathbf{F} = -\vec{\nabla} f$. This introduction of the minus sign in the definition means that, for physicists, the potential energy is the value of the potential function, not negative the value.

If $\mathbf{F} = \mathbf{F}(\mathbf{x})$ is a conservative **force** field, and the units of **x** are length units, then any potential function f for **F** has energy units. In this setting, the negation of the value of the potential function, i.e., the value of $-f$, is referred to as the *potential energy* of **F** or the potential energy of an object being acted on by **F**.

Of course, potential functions are unique (on a connected open set) only up to the addition of a constant. So, either you talk about the **change** in the potential energy, so that the constant is irrelevant, or you fix the value of the constant by specifying where you want the potential energy to be zero.

This definition must be modified in the appropriate way if you want the potential energy to be 0 "at infinity", i.e., if you want the limit of the potential energy to be 0 as the distance from the origin approaches ∞.

> **Definition 4.3.11.** *Suppose that* **F** *is a continuous, conservative force field on a connected open subset* $\mathcal{U}$ *of* $\mathbb{R}^n$, *that* **p** *is a point in* $\mathcal{U}$, *and that* f *is a potential function for* **F** *such that* $f(\mathbf{p}) = 0$.
>
> *Then, for all* **x** *in* $\mathcal{U}$, $-f(\mathbf{x})$ *is the* **potential energy of** **F** **at** **x** *(normalized so that the potential energy is 0 at* **p***).*

Example 4.3.12. Near the surface of the Earth, the gravitational force field which acts on a mass m acts essentially "straight down", i.e., it is approximated well by the force field

$$\mathbf{F} = -mg\,\mathbf{k} = (0, 0 - mg) \text{ Newtons},$$

where m is in kilograms, and g is 9.78 (m/s)/s.

It is easy to find a potential function f for this vector field, namely

$$f(x, y, z) \;=\; -mgz \;\; \text{Newton-meters},$$

where z is the height above the surface of the (approximately flat) Earth.

Thus, what **every** potential function for $\mathbf{F}$ looks like is

$$g(x, y, z) \;=\; -mgz + C \;\; \text{Newton-meters},$$

where C is an arbitrary constant.

In our current setting, it is common to look for a potential function which is 0 when $z = 0$; this means we use simply

$$f(x, y, z) \;=\; -mgz$$

as our potential function. Or, as is more standard, we write h for the height above the surface of the Earth, and the gravitational potential is given by

$$f(h) \;=\; -mgh.$$

As you may already know, the value of $-f$, i.e., of mgh, is known as the *potential energy* of the object or of the gravitational force field.

More Depth:

Example 4.3.13. Suppose that x, y, and z are measured in meters. Consider the force field

$$\mathbf{F} \;=\; \left(z^2 e^x + e^y, \; xe^y + 2y, \; 2ze^x + 3z^2\right) \;\; \text{Newtons}.$$

Calculate the work done by $\mathbf{F}$ on a particle as the particle moves along straight lines from $(1, 0, 0)$ to $(0, 2, 0)$, then goes from $(0, 2, 0)$ to $(0, 0, 3)$, and finally goes from $(0, 0, 3)$ to $(4, 0, 0)$.

Solution:

Of course, you **could** parameterize each oriented line segment, calculate three separate line integrals, and add them together. However, your hope should be that $\mathbf{F}$ is a conservative vector field; if it is, then you find a potential function f, and calculate line integral via the Fundamental Theorem. You'd get just $f(4, 0, 0) - f(1, 0, 0)$.

We don't want to go to the trouble of trying to produce a potential function, before we quickly check whether or not we'll succeed. As $\mathbf{F}$ is defined on all of $\mathbb{R}^3$, which is simply-connected, Theorem 4.3.7 tells us that we want to check $\vec{\nabla} \times \mathbf{F}$ and see whether it's zero or not.

We calculate

$$\vec{\nabla} \times \mathbf{F} = \begin{vmatrix} \mathbf{i} & \mathbf{j} & \mathbf{k} \\ \dfrac{\partial}{\partial x} & \dfrac{\partial}{\partial y} & \dfrac{\partial}{\partial z} \\ z^2 e^x + e^y & x e^y + 2y & 2z e^x + 3z^2 \end{vmatrix} =$$

$$\begin{vmatrix} \dfrac{\partial}{\partial y} & \dfrac{\partial}{\partial z} \\ x e^y + 2y & 2z e^x + 3z^2 \end{vmatrix} \mathbf{i} - \begin{vmatrix} \dfrac{\partial}{\partial x} & \dfrac{\partial}{\partial z} \\ z^2 e^x + e^y & 2z e^x + 3z^2 \end{vmatrix} \mathbf{j} + \begin{vmatrix} \dfrac{\partial}{\partial x} & \dfrac{\partial}{\partial y} \\ z^2 e^x + e^y & x e^y + 2y \end{vmatrix} \mathbf{k} =$$

$$(0 - 0)\mathbf{i} - (2z e^x - 2z e^x)\mathbf{j} + (e^y - e^y)\mathbf{k} = \mathbf{0}.$$

Thus, the curl is zero; so $\mathbf{F}$ is conservative, and a potential function exists.

Now we need to find a potential function. We need to simultaneously solve

$$\frac{\partial f}{\partial x} = z^2 e^x + e^y, \qquad \frac{\partial f}{\partial y} = x e^y + 2y, \qquad \text{and} \qquad \frac{\partial f}{\partial z} = 2z e^x + 3z^2.$$

We integrate the first equation with respect to x:

$$f = \int (z^2 e^x + e^y)\, dx = z^2 e^x + x e^y + A(y, z).$$

We then take the partial derivative of this with respect to y and require it to equal what we're told $\partial f / \partial y$ should equal:

$$\frac{\partial f}{\partial y} = x e^y + \frac{\partial A}{\partial y} = x e^y + 2y.$$

Therefore, $\dfrac{\partial A}{\partial y} = 2y$, and so,

$$A(y, z) = \int 2y\, dy = y^2 + B(z).$$

This means that now we know that

$$f = z^2 e^x + x e^y + A(y, z) = z^2 e^x + x e^y + y^2 + B(z).$$

Now, we take the partial derivative of this with respect to z and require it to equal what it should:

$$\frac{\partial f}{\partial z} \;=\; 2ze^x + B'(z) \;=\; 2ze^x + 3z^2.$$

This tells us that $B'(z) = 3z^2$, so that $B(z) = z^3 + C$.

Finally, we conclude that every potential function for $\mathbf{F}$ is given by

$$f \;=\; z^2 e^x + xe^y + y^2 + z^3 + C.$$

Now that we have our potential function(s), calculating the work is easy; we find

$$f(4,0,0) - f(1,0,0) \;=\; 4 - 1 \;=\; 3 \;\text{ joules.}$$

Example 4.3.14. Consider the vector field

$$\mathbf{F} = \left(-\frac{y}{x^2 + y^2}, \; \frac{x}{x^2 + y^2} \right),$$

which is defined on $\mathbb{R}^2$ minus the origin. This region is **not** simply-connected.

We leave it as an exercise for you to check that the 2-dimensional curl of $\mathbf{F}$ is zero, but that the line integral along the counterclockwise oriented circle of radius 1, centered at the origin, equals 2π. In particular, since the line integral along a closed curve is not zero, $\mathbf{F}$ is **not** conservative.

Thus, $\mathbf{F}$ has curl equal to zero and, yet, is not conservative; this example shows that the assumption of simple-connectedness is essential in Theorem 4.3.7.

Example 4.3.15. Two very important force fields are the gravitational force field associated with a mass, and the electric force field associated to an electric charge.

Suppose that we have a point-mass M located at the origin in $\mathbb{R}^3$. Let $\mathbf{r}(x,y,z) = (x,y,z)$ be the radial vector, and consider a point-mass of m, located at an arbitrary point $(x,y,z) \neq (0,0,0)$. We let $r = |\mathbf{r}|$.

Then, as we discussed in Example 4.1.1, Newton's Law of Universal Gravitation tells us that there is the gravitational force field exerted by M on m:

$$\mathbf{F}_{\text{grav}} \;=\; -\left(\frac{GMm}{r^3} \right) \mathbf{r},$$

where G is the universal gravitational constant. In Cartesian coordinates, this is

$$\mathbf{F}_{\text{grav}} \;=\; \frac{k}{\left(x^2 + y^2 + z^2\right)^{3/2}}\,(x, y, z),$$

where k is the constant $-GMm$.

_____

Now suppose that we have an electric charge q_1 located at the origin in $\mathbb{R}^3$. Let $\mathbf{r}(x, y, z) = (x, y, z)$ be the radial vector, and consider a second charge q_2, located at an arbitrary point $(x, y, z) \neq (0, 0, 0)$. We let $r = |\mathbf{r}|$.

Then, Coulomb's Law, tells us that the electric force exerted by q_1 on q_2 is given by

$$\mathbf{F}_{\text{elec}} \;=\; k_e \frac{q_1 q_2}{r^3}\,\mathbf{r},$$

where k_e is the *Coulomb constant*. In Cartesian coordinates, this is again of the form

$$\mathbf{F}_{\text{elec}} \;=\; \frac{k}{\left(x^2 + y^2 + z^2\right)^{3/2}}\,(x, y, z),$$

where now k is the constant $k_e q_1 q_2$.

_____

Thus, both the gravitational force field and the electric force field have the same form:

$$\mathbf{F} \;=\; \frac{k}{\left(x^2 + y^2 + z^2\right)^{3/2}}\,(x, y, z),$$

and are defined on the simply-connected space consisting of $\mathbb{R}^3$ minus the origin.

You can check that the curl of $\mathbf{F}$ is zero. But if you just look for a potential function, it's not hard to see one. To get the $(x^2 + y^2 + z^2)^{-3/2}$ after taking the gradient, it should occur to you that you should start with $(x^2 + y^2 + z^2)^{-1/2}$. Once you think of this, it's easy to see that the potential functions for $\mathbf{F}$ are

$$f(x, y, z) \;=\; -k(x^2 + y^2 + z^2)^{-1/2} + C \;=\; -\frac{k}{\sqrt{x^2 + y^2 + z^2}} + C.$$

In addition, if we decide that we want the gravitational or electric potential to be 0 when we're infinitely far from the origin, i.e., when $x^2 + y^2 + z^2 \to \infty$, then we must pick $C = 0$, so that the potential is simply

$$f(x, y, z) \;=\; -\frac{k}{\sqrt{x^2 + y^2 + z^2}}.$$

Remark 4.3.16. You may wonder how we can reconcile our gravitational potential calculation in Example 4.3.15 with our earlier calculation in Example 4.3.12, in which we assumed that the gravitational field is approximately constant near the surface of the Earth.

In Example 4.3.12, we concluded that the gravitational potential function was $-mgz$, where z is distance above the surface of the Earth (having set the potential equal to 0 at the surface). This doesn't look anything like the potential function(s)

$$f(x, y, z) = -k(x^2 + y^2 + z^2)^{-1/2} + C = -\frac{k}{\sqrt{x^2 + y^2 + z^2}} + C$$

from Example 4.3.15. What's going on?

First, we assume that the Earth can reasonably be approximated as a ball of radius R, and we put the origin at its center. Let the mass of the Earth be M. As we saw in Exercise 22 in Section 3.11, at a distance $r \geq R$, the gravitational force that the Earth exerts on a mass m is the same as that exerted by a point-mass with mass M, located at the origin, i.e., for $r \geq R$,

$$\mathbf{F}_{\text{grav}} = -\left(\frac{GMm}{r^3}\right)\mathbf{r},$$

where $\mathbf{r} = \mathbf{r}(x, y, z) = (x, y, z)$ is the position of the mass m.

> We are also assuming here that the density of the Earth is constant at a fixed distance from the center.

The magnitude of this force is GMm/r^2, which, at the surface of the Earth, i.e., when $r = R$, is supposed to be mg. Thus, g is determined by the equation

$$\frac{GMm}{R^2} = mg, \quad \text{i.e.,} \quad GMm = mgR^2.$$

Therefore, our potential calculation from Example 4.3.15 tells us that the gravitational potential is

$$f(x, y, z) = \frac{mgR^2}{\sqrt{x^2 + y^2 + z^2}} + C = \frac{mgR^2}{r} + C.$$

If we want the gravitational potential to be zero at the surface of the Earth, that means it needs to be zero when $r = R$; thus, we need

$$0 = \frac{mgR^2}{R} + C,$$

and so, we must have $C = -mgR$.

Hence, our (mathematical) gravitational potential, in terms of r, is

$$f(r) = \frac{mgR^2}{r} - mgR,$$

where $r \geq R$ is the distance from the center of the Earth. At a height of h above the surface of the Earth, $r = R + h$, and we find that the gravitational potential is

$$\frac{mgR^2}{R+h} - mgR = mgR\left(\frac{R}{R+h} - \frac{R+h}{R+h}\right) = -mgh\left(\frac{R}{R+h}\right).$$

Finally, we see that, when h is small compared to R, then $R/(R+h)$ is approximately 1, and the (mathematical) potential is approximately $-mgh$, as we found in Example 4.3.12.

Remark 4.3.17. In Remark 4.2.12, we discussed the situation in which an object is moving through a force field, and there are additional "external" forces acting on the object. If we write $\mathbf{F}_{\text{ext}}$ for the sum of the forces which are external to the force field $\mathbf{F}$, we saw that, for a curve C from $\mathbf{a}$ to $\mathbf{b}$, the work done by the force field along C plus the work done by the external force along C equals the change in kinetic energy, i.e.,

$$\int_C \mathbf{F} \cdot d\mathbf{r} + \int_C \mathbf{F}_{\text{ext}} \cdot d\mathbf{r} = \frac{1}{2}mv_1^2 - \frac{1}{2}mv_0^2.$$

If $\mathbf{F}$ is a conservative force field, with potential function f, then, $-f$ is the potential energy of the force field, and the Fundamental Theorem of Line Integrals tells us that we have

$$f(\mathbf{b}) - f(\mathbf{a}) + \int_C \mathbf{F}_{\text{ext}} \cdot d\mathbf{r} = \frac{1}{2}mv_1^2 - \frac{1}{2}mv_0^2,$$

that is,

$$\int_C \mathbf{F}_{\text{ext}} \cdot d\mathbf{r} = \frac{1}{2}mv_1^2 - \frac{1}{2}mv_0^2 + (-f(\mathbf{b})) - (-f(\mathbf{a})).$$

In words, this says that the external work, i.e., the external energy supplied, equals the change in kinetic energy plus the change in potential energy; this is usually referred to as *conservation of energy*.

We would like to discuss the equivalent characterizations of a conservative vector field given in Theorem 4.3.7.

The Fundamental Theorem of Line Integrals tells us that (1) of Theorem 4.3.7 implies (2) and (3), and we saw back in Theorem 4.1.17 that the curl of a conservative vector field is zero. The proof that the curl of a vector field on a simply-connected open set being zero implies any of the other properties of the vector field is beyond the scope of this textbook; see Theorem 8.4 of [3].

What we would like to discuss is why (2) and (3) of Theorem 4.3.7 are equivalent, and why (2) implies (1). Fix a continuous vector field $\mathbf{F}$.

Suppose that the line integral $\int_C \mathbf{F} \cdot d\mathbf{r}$ depends on the starting and ending points of C, but is independent of the actual curve C. Then, the line integral around a closed curve, from a point $\mathbf{a}$ to itself, is independent of the closed curve, and we may pick the parameterized curve which never moves, i.e., the parameterized curve given by $\mathbf{r}(t) = \mathbf{a}$. But the line integral around this constant "curve" is clearly 0, and so the line integral around any closed curve must be 0.

Suppose that the line integral around every (oriented, piecewise-regular) closed curve is 0. Let C_1 and C_2 be two, possibly different, curves from a point $\mathbf{a}$ to a point $\mathbf{b}$. Then, $C_1 - C_2$ is a closed curve, and so

$$0 = \int_{C_1 - C_2} \mathbf{F} \cdot d\mathbf{r} = \int_{C_1} \mathbf{F} \cdot d\mathbf{r} - \int_{C_2} \mathbf{F} \cdot d\mathbf{r}.$$

Therefore,

$$\int_{C_1} \mathbf{F} \cdot d\mathbf{r} = \int_{C_2} \mathbf{F} \cdot d\mathbf{r},$$

i.e., the line integral is independent of the curve from $\mathbf{a}$ to $\mathbf{b}$.

Finally, assume that line integrals of $\mathbf{F}$ along a curve are independent of the curve, except for the starting and ending points. Recall that we are assuming that $\mathbf{F}$ is a continuous vector field on a connected open subset $\mathcal{U}$ of $\mathbb{R}^n$. Let us write $\mathbf{F}$ in terms of its component functions:

$$\mathbf{F}(\mathbf{x}) = \big(F_1(\mathbf{x}), F_2(\mathbf{x}), \ldots, F_n(\mathbf{x})\big).$$

We would like to produce a function f so that $\mathbf{F} = \vec{\nabla} f$.

Fix a point $\mathbf{a}$ in $\mathcal{U}$. Since $\mathcal{U}$ is a connected open set, for every $\mathbf{x}$ in $\mathcal{U}$, there exists an oriented, piecewise-regular curve $C_{\mathbf{x}}$ from $\mathbf{a}$ to $\mathbf{x}$. Define a function $f : \mathcal{U} \to \mathbb{R}$ by

$$f(\mathbf{x}) = \int_{C_{\mathbf{x}}} \mathbf{F} \cdot d\mathbf{r},$$

and note that, by our assumption, this function is independent of which oriented, piecewise-regular curve from $\mathbf{a}$ to $\mathbf{x}$ we select for $C_{\mathbf{x}}$. we claim that $\mathbf{F} = \vec{\nabla} f$. We shall show that $F_1 = \partial f / \partial x_1$; the same argument applies to the other components.

We need to show that

$$\lim_{h \to 0} \frac{f(\mathbf{x} + h\mathbf{i}) - f(\mathbf{x})}{h} = F_1(\mathbf{x}),$$

where, for h close enough to 0, $\mathbf{x} + h\mathbf{i}$ is in $\mathcal{U}$ because $\mathcal{U}$ is an open set. Regardless of what curve is used for $C_{\mathbf{x}}$, for $C_{\mathbf{x}+h\mathbf{i}}$, we use $C_{\mathbf{x}} + C_h$, where C_h is parameterized by

$$\mathbf{r}(t) = \mathbf{x} + t\mathbf{i},$$

for $0 \leq t \leq h$.

Then,

$$\lim_{h \to 0} \frac{f(\mathbf{x} + h\mathbf{i}) - f(\mathbf{x})}{h} \;=\; \lim_{h \to 0} \frac{\int_{C_{\mathbf{x}} + C_h} \mathbf{F} \cdot d\mathbf{r} \;-\; \int_{C_{\mathbf{x}}} \mathbf{F} \cdot d\mathbf{r}}{h} \;=\;$$

$$\lim_{h \to 0} \frac{\int_{C_h} \mathbf{F} \cdot d\mathbf{r}}{h} \;=\; \lim_{h \to 0} \frac{\int_0^h \mathbf{F}(\mathbf{r}(t)) \cdot \mathbf{r}'(t)\, dt}{h} \;=\; \lim_{h \to 0} \frac{\int_0^h \mathbf{F}(\mathbf{r}(t)) \cdot \mathbf{i}\, dt}{h} \;=\;$$

$$\lim_{h \to 0} \frac{\int_0^h F_1(\mathbf{r}(t))\, dt}{h} \;=\; F_1(\mathbf{r}(0)) \;=\; F_1(\mathbf{x}),$$

where the next-to-last equality follows from the (single variable) Fundamental Theorem of Calculus.

Online answers to select exercises are here.

4.3.1   Exercises

Basics:

In each of the following exercises, you are given a real-valued function f. Let $\mathbf{F} = \vec{\nabla} f$, and use the Fundamental Theorem of Line Integrals to calculate $\int_C \mathbf{F} \cdot d\mathbf{r}$ for the given curve C.

1. $f(x, y) = xe^{\cos y} + y \ln(1 + x^2)$, and C is the circle of radius 7, centered at the origin, and oriented counterclockwise.

2. $f(x, y) = xe^{\cos y} + y \ln(1 + x^2)$, and C is the oriented curve consisting of the line segment from $(0, 0)$ to $(3, 4)$, then the line segment from $(3, 4)$ to $(-1, 7)$, and then the line segment from $(-1, 7)$ to $(1, \pi)$.

3. $f(x, y) = \tan^{-1}(x^3 + y^2)$, and C is the top half of the circle of radius 1, centered at the origin, oriented clockwise.

4. $f(x, y) = \tan^{-1}(x^3 + y^2)$, and C is the circle of radius 1, centered at $(3, 4)$, oriented clockwise.

5. $f(x, y) = x^3 + xy^2$, and C consists of the top half of the circle of radius 1, centered at the origin, oriented clockwise, followed by the line segment from $(1, 0)$ to $(1, 1)$.

6. $f(x, y) = x^3 + xy^2$, and C consists of the top half of the circle of radius 1, centered at the origin, oriented clockwise, followed by the line segment from $(1, 0)$ to $(1, 1)$, followed by the line segment from $(1, 1)$ to $(-1, 0)$.

In each of the following exercises, you are given a vector field $\mathbf{F} = \mathbf{F}(x, y)$ on a simply-connected region in $\mathbb{R}^2$. Calculate the 2-dimensional curl of $\mathbf{F}$ and, if $\mathbf{F}$ is conservative, find a potential function f for $\mathbf{F}$.

7. $\mathbf{F}(x, y) = (x^3, y^5)$.

8. $\mathbf{F}(x, y) = (y^5, x^3)$.

9. $\mathbf{F}(x, y) = (y^2 + 2xy^3, 2xy + 3x^2y^2)$. ▶

10. $\mathbf{F}(x, y) = (2xe^y + x^2, x^2e^y + \tan^{-1} y)$.

11. $\mathbf{F}(x, y) = (y \sin x, \cos x)$.

12. $\mathbf{F}(x, y) = (y \sin x, -\cos x)$.

13. $\mathbf{F}(x, y) = \left(y + \dfrac{1}{1 + x^2}, \ x + \dfrac{1}{1 + y^2} \right)$.

14. $\mathbf{F}(x, y) = (x^2 + y^2, xy)$.

In each of the following exercises, you are given a vector field $\mathbf{F} = \mathbf{F}(x, y)$ from the previous exercises. Assume that $\mathbf{F}$ represents a force field, in Newtons, and that x and y are in meters. Calculate the work done by $\mathbf{F}$ along the given oriented curve C.

15. $\mathbf{F}(x, y) = (x^3, y^5)$, and C is the circle of radius 5, centered at the origin, and oriented counterclockwise.

16. $\mathbf{F}(x, y) = (y^5, x^3)$, and C is the oriented line segment from $(0, 0)$ to $(-5, 2)$.

17. $\mathbf{F}(x, y) = (y^2 + 2xy^3, 2xy + 3x^2y^2)$, and C consists of the top half of the circle of radius 1, centered at the origin, oriented clockwise, followed by the line segment from $(1, 0)$ to $(1, 1)$, followed by the line segment from $(1, 1)$ to $(-1, 0)$. ▶

18. $\mathbf{F}(x, y) = (2xe^y + x^2, x^2e^y + \tan^{-1} y)$, and C is the ellipse given by $2(x - 1)^2 + 3(y + 2)^2 = 1$, oriented clockwise.

19. $\mathbf{F}(x, y) = (y \sin x, \cos x)$, and C is the oriented curve which starts at $(0, 0)$ and moves along the parabola where $y = x^2$ to the point $(2, 4)$. ▶

20. $\mathbf{F}(x, y) = (y \sin x, -\cos x)$, and C is the oriented curve consisting of the line segment from $(0, 0)$ to $(3, 4)$, then the line segment from $(3, 4)$ to $(-1, 7)$, and then the line segment from $(-1, 7)$ to $(\pi, 1)$.

21. $\mathbf{F}(x, y) = \left(y + \dfrac{1}{1 + x^2},\ x + \dfrac{1}{1 + y^2}\right)$, and C is the oriented curve parameterized by $\mathbf{r}(t) = (t^3, t^5) + (1 - t)(te^t + t\sin t,\ e^{-t}\tan^{-1} t)$, for $0 \leq t \leq 1$.

22. $\mathbf{F}(x, y) = (x^2 + y^2, xy)$, and C is the circle of radius 3, centered at the origin, and oriented counterclockwise.

In each of the following exercises, you are given a sketch of a vector field $\mathbf{F} = \mathbf{F}(x, y)$. Decide whether or not the vector field is (or, appears to be) conservative.

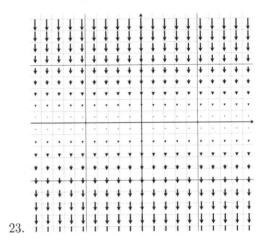

23.

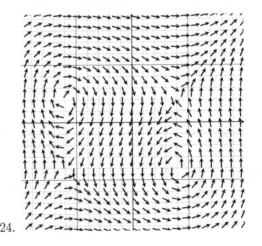

24.

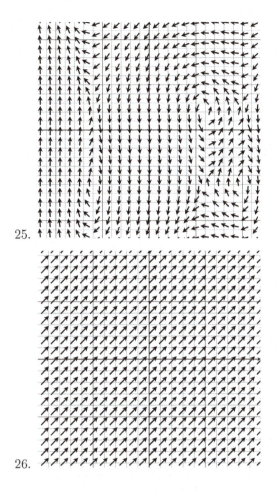

25.

26.

In each of the following exercises, you are given a vector field $\mathbf{F} = \mathbf{F}(x, y, z)$ on a simply-connected region in $\mathbb{R}^3$. Calculate the curl of $\mathbf{F}$ and, if $\mathbf{F}$ is conservative, find a potential function f for $\mathbf{F}$.

More Depth:

27. $\mathbf{F}(x, y, z) = (x, y, z)$.

28. $\mathbf{F}(x, y, z) = (y, x, z)$.

29. $\mathbf{F}(x, y, z) = (z, x, y)$.

30. $\mathbf{F}(x, y, z) = (2x + y + z, \ x, \ x + 2z)$.

31. $\mathbf{F}(x, y, z) = (x, \ x + 2z, \ 2x + y + z)$.

32. $\mathbf{F}(x, y, z) = (yz + e^y + x, \ xz + xe^y + e^z + y^2, \ xy + ye^z + z^3)$.

33. $\mathbf{F}(x, y, z) = (\sin y, \ x \cos y + \cos z, \ -y \sin z + e^z)$.

34. $\mathbf{F}(x, y, z) = (x, \ x + y, \ x + y + z)$.

In each of the following exercises, you are given a vector field $\mathbf{F} = \mathbf{F}(x, y, z)$ from the previous exercises. Assume that F represents a force field, in Newtons, and that x, y, and z are in meters. Calculate the work done by F along the given oriented curve C.

35. $\mathbf{F}(x, y, z) = (x, y, z)$, and C is the oriented curve consisting of the line segment from $(0, 0, 0)$ to $(1, 1, 0)$, then the line segment from $(1, 1, 0)$ to $(5, 1, 7)$, and then the line segment from $(5, 1, 7)$ to $(2, -3, 5)$.

36. $\mathbf{F}(x, y, z) = (y, x, z)$, and C is the oriented curve consisting of the line segment from $(0, 0, 0)$ to $(1, 1, 0)$, then the line segment from $(1, 1, 0)$ to $(5, 1, 7)$, and then the line segment from $(5, 1, 7)$ to $(2, -3, 5)$.

37. $\mathbf{F}(x, y, z) = (z, x, y)$, and C is the oriented curve consisting of the line segment from $(0, 0, 0)$ to $(2, -3, 5)$.

38. $\mathbf{F}(x, y, z) = (2x + y + z, \ x, \ x + 2z)$, and C is the oriented curve that consists of the quarter circle of radius 2, centered at the origin, in the xy-plane, from $(2, 0, 0)$ to $(0, 2, 0)$, followed by the line segment from $(0, 2, 0)$ to $(3, 1, -2)$.

39. $\mathbf{F}(x, y, z) = (x, \ x + 2z, \ 2x + y + z,)$, and C is the oriented curve parameterized by $\mathbf{r}(t) = (e^t, e^{2t} + 1, -e^t)$ for $0 \leq t \leq 1$.

40. $\mathbf{F}(x, y, z) = (yz + e^y + x, \ xz + xe^y + e^z + y^2, \ xy + ye^z + z^3)$, and C is the oriented curve that consists of the quarter circle of radius 3, centered at the origin, in the xz-plane, from $(3, 0, 0)$ to $(0, 0, 3)$, followed by the line segment from $(0, 0, 3)$ to $(1, 1, 1)$.

41. $\mathbf{F}(x, y, z) = (\sin y, \ x \cos y + \cos z, \ -y \sin z + e^z)$, and C is the oriented curve that consists of the line segment from $(0, 0, 0)$ to $(3, 0, 0)$, followed by the quarter circle of radius 3, centered at the origin, in the xz-plane, from $(3, 0, 0)$ to $(0, 0, 3)$, followed by the line segment from $(0, 0, 3)$ to (π, π, π). $\quad \blacktriangleright$

42. $\mathbf{F}(x, y, z) = (x, \ x + y, \ x + y + z)$, and C is the oriented helix parameterized by $\mathbf{r}(t) = (\cos t, \sin t, t)$, for $0 \leq t \leq \pi$.

43. Write a brief essay in which you explain the various equivalent characterizations of a conservative vector field.

44. Verify that the 2-dimensional curl of the vector field

$$\mathbf{F} = \left(-\frac{y}{x^2 + y^2}, \ \frac{x}{x^2 + y^2} \right),$$

from Example 4.3.14 is zero. Then, explain, in your own words, why Example 4.3.14 is important.

45. Suppose that $\mathbf{F}$ and $\mathbf{G}$ are continuously differentiable vector fields on all of $\mathbb{R}^3$, and that, for every closed oriented, piecewise-regular curve C in $\mathbb{R}^3$,

$$\int_C \mathbf{F} \cdot d\mathbf{r} \;=\; \int_C \mathbf{G} \cdot d\mathbf{r}.$$

Show that the curl of $\mathbf{F}$ is equal to the curl of $\mathbf{G}$.

46. Suppose that $\mathbf{F} = (P, Q, R, S)$ is a continuously differentiable conservative vector field on all of $\mathbb{R}^4$. What relationships must the partial derivatives of P, Q, R, and S satisfy?

4.4   Green's Theorem

For a 2-dimensional conservative vector field, the line integral around closed curves is zero and the 2-dimensional curl is zero. In this section, we will see that, even when the vector field is not conservative, there is a relationship between line integrals around closed curves and the 2-dimensional curl.

That relationship is provided by *Green's Theorem*, which says, with some technical assumptions, that the line integral of a vector field around a closed curve C is equal to the integral of the 2-dimensional curl over the region enclosed by C.

Basics:

More technically, a **simple closed curve** in $\mathbb{R}^n$, where $n \geq 2$ is the image of a continuous function $\mathbf{p} : [0, 1] \to \mathbb{R}^n$ such that $\mathbf{p}(0) = \mathbf{p}(1)$ and such that the restriction of $\mathbf{p}$ to the interval $[0, 1)$ is one-to-one.

It will be important to us, in the More Depth portion, that this counterclockwise orientation is the orientation so that the region R is on your left as you move around ∂R in the given direction.

We give the main idea of the proof in the More Depth portion of this section. For a complete proof, see [3], p. 360.

Suppose that $\mathcal{U}$ is an open subset of $\mathbb{R}^2$, and that $\mathbf{F} = (P(x, y), Q(x, y))$ is a continuously differentiable conservative vector field on $\mathcal{U}$. Let C be a closed, piecewise-regular curve in $\mathcal{U}$. Then, as we discussed in the previous section,

$$\int_C \mathbf{F} \cdot d\mathbf{r} = 0 \qquad \text{and} \qquad \frac{\partial Q}{\partial x} - \frac{\partial P}{\partial y} = 0.$$

Is there a more general relationship between line integrals around closed curves and the 2-dimensional curl $Q_x - P_y$? Yes.

First, we need the notion of a *simple, closed curve*. A simple, closed curve in $\mathbb{R}^2$ is a closed curve which does not intersect itself (other than to be closed).

Second, there is a theorem, the *Jordan Curve Theorem*, which says that a simple, closed curve C in $\mathbb{R}^2$ separates $\mathbb{R}^2$ into an outside piece, which is unbounded, and an inside piece which is bounded by C.

Typically, our assumptions will include "let R be the compact region which is bounded by the simple, closed, oriented, piecewise-regular curve C"; it is common to let ∂R denote the boundary curve C, and say that ∂R is given the *positive orientation* if it is oriented counterclockwise.

Now we can state:

Theorem 4.4.1. (Green's Theorem, Version 1) *Let R be the region in $\mathbb{R}^2$ which is bounded by a simple, closed, oriented, piecewise-regular curve, ∂R, which is oriented counterclockwise. Let $\mathbf{F} = (P(x, y), Q(x, y))$ be a continuously differentiable vector field on an open subset of $\mathbb{R}^2$ which contains R.*

   Then,

$$\int_{\partial R} \mathbf{F} \cdot d\mathbf{r} = \iint_R (Q_x - P_y) \, dA.$$

Example 4.4.2. Suppose that x and y are given in meters, and you have the force field

$$\mathbf{F} = \left(4y + \ln(1 + x^2),\ 6x + e^y \sin y\right) \quad \text{Newtons.}$$

Calculate the work done by $\mathbf{F}$ on an object that starts at $(0,0)$, travels along a line to $(1,0)$, then travels along a line to $(0,2)$, and finally, travels along a line back to $(0,0)$.

Solution:

The work is given by the line integral around the given triangle. We **could** parameterize each line segment, set-up the three line integrals, try to evaluate the integrals, and then add the results together. However, Green's Theorem gives us a much easier method.

The triangle that the object moves along is the boundary ∂R of the filled triangle R in Figure 4.4.1. The 2-dimensional curl of $\mathbf{F} = (P, Q)$, in Newtons per meter, is

$$Q_x - P_y = 6 - 4 = 2.$$

Therefore, Green's Theorem tells us that

$$\text{work} = \int_{\partial R} \mathbf{F} \cdot d\mathbf{r} = \iint_R 2\, dA = 2\big(\text{area inside the triangle}\big) = 2 \text{ joules.}$$

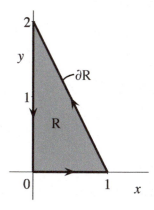

Figure 4.4.1: The region R and its oriented boundary ∂R.

Example 4.4.3. Consider the vector field

$$\mathbf{V} = \left(6x^2 y + 2y^3 + \tan^{-1} x,\ x^3 + 3xy^2 + e^y\right).$$

Calculate the line integral of $\mathbf{V}$ along the curve which starts at $(-3, 0)$, then goes clockwise along the circle of radius 3, centered at the origin, until reaching $(3, 0)$, and then goes along the line from $(3, 0)$ back to $(-3, 0)$.

Solution:

Let R denote the region given bounded by the given closed curve. Note that, as an oriented curve, the boundary of R has been given the negative orientation, i.e., it is oriented clockwise.

Thus, writing ∂R for the positively oriented boundary of R, what we have been asked to calculate is

$$\int_{-\partial R} \mathbf{V} \cdot d\mathbf{r} = -\int_{\partial R} \mathbf{V} \cdot d\mathbf{r},$$

and we can calculate $\int_{\partial R} \mathbf{V} \cdot d\mathbf{r}$ using Green's Theorem.

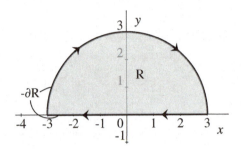

Figure 4.4.2: The region R and its negatively oriented boundary $-\partial R$.

Letting $P = 6x^2y + 2y^3 + \tan^{-1} x$ and $Q = x^3 + 3xy^2 + e^y$, we find

$$Q_x - P_y \;=\; 3x^2 + 3y^2 - (6x^2 + 6y^2) \;=\; -3(x^2 + y^2).$$

Therefore, by Green's Theorem, we have

$$\int_{-\partial R} \mathbf{V} \cdot d\mathbf{r} \;=\; -\int_{\partial R} \mathbf{V} \cdot d\mathbf{r} \;=\; -\iint_R -3(x^2 + y^2)\, dA,$$

which is clearly nicest to integrate using polar coordinates.

We calculate

$$\int_{-\partial R} \mathbf{V} \cdot d\mathbf{r} \;=\; -\iint_R -3(x^2 + y^2)\, dA \;=\; -\int_0^\pi \int_0^3 -3r^2 \cdot r\, dr\, d\theta \;=\;$$

$$3 \int_0^\pi \left[\frac{r^4}{4} \;\Big|_0^3 \right] d\theta \;=\; \frac{243\pi}{2}.$$

Remark 4.4.4. In the two previous examples, we explicitly gave you a starting/ending point on the closed curve. This was just for convenience. We remind you of Remark 4.2.9; for a line integral around a closed curve, it is irrelevant which point on the closed curve is used as the starting/ending point.

More Depth:

Before we state a more general version of Green's Theorem, let's look at an illuminating example.

Example 4.4.5. Suppose that we have a continuously differentiable vector field $\mathbf{F} = (P(x,y)\, Q(x,y))$ on a region enclosed by a simple, closed piecewise-regular curve from which we removed the interior of another, enclosed, simple, closed piecewise-regular curve. Does Green's Theorem tell us anything about this situation?

For instance, suppose that we integrate $Q_x - P_y$ not over the entire filled triangle T from Example 4.4.2, but, instead, over the region R which consists of T minus the disk D centered at $(0.4, 0.4)$ of radius 0.2. See Figure 4.4.3.

Then, we have

$$\iint_R (Q_x - P_y)\, dA = \iint_T (Q_x - P_y)\, dA - \iint_D (Q_x - P_y)\, dA = \int_{\partial T} \mathbf{F} \cdot d\mathbf{r} - \int_{\partial D} \mathbf{F} \cdot d\mathbf{r},$$

where we used Green's Theorem twice to write the last equality, and the orientations on ∂T and ∂D are both counterclockwise.

Of course, we can write this final sum as a single line integral:

$$\iint_R (Q_x - P_y)\, dA = \int_{\partial T} \mathbf{F} \cdot d\mathbf{r} - \int_{\partial D} \mathbf{F} \cdot d\mathbf{r} = \int_{\partial T - \partial D} \mathbf{F} \cdot d\mathbf{r}.$$

Note that $-\partial D$ will be the **clockwise** oriented circle.

Thus, we see that we obtain a more general Green's Theorem if we agree that the positive orientation for the boundary of a region R which has holes in it is the orientation in which the closed boundary curves on the outside of the region are oriented counterclockwise, but the closed boundary curves on the inside are oriented clockwise.

This can be incorporated in a single easy rule: the positive orientation for each piece of ∂R, the orientation that you need for Green's Theorem, is the orientation such that, if you move in the orientation direction, the interior of the region is always on your **left**.

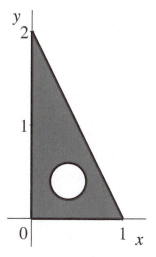

Figure 4.4.3: The region R consisting of the filled triangle T minus the disk D.

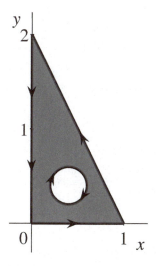

Figure 4.4.4: The region R with the positive orientation on the boundary.

Theorem 4.4.6. (Green's Theorem, Version 2) *Let R be a compact region in $\mathbb{R}^2$, which is the closure of its interior, and whose boundary, ∂R, consists of a finite number of simple, closed, oriented, piecewise-regular curves, which are oriented so that the interior of R is always on the left as you move in the direction of the orientation.*

Let $\mathbf{F} = (P(x,y), Q(x,y))$ be a continuously differentiable vector field on an open subset of $\mathbb{R}^2$ which contains R.

Then,

$$\int_{\partial R} \mathbf{F} \cdot d\mathbf{r} = \iint_R (Q_x - P_y)\, dA.$$

There is a subtlety present in Theorem 4.4.6 that was not present in Example 4.4.5: the vector field does not need to exist everywhere in the missing parts, i.e., the vector field may be undefined in the holes in the region. This can be useful in various problems.

Example 4.4.7. Consider the vector field from Example 4.3.14:

$$\mathbf{F} \;=\; (P,\,Q) \;=\; \left(-\frac{y}{x^2+y^2},\; \frac{x}{x^2+y^2}\right).$$

The 2-dimensional curl $Q_x - P_y$ of this vector field is 0; however, the line integral along the counterclockwise oriented circle of radius 1, centered at the origin, equals 2π. Hence, $\mathbf{F}$ is **not** a conservative vector field.

This doesn't contradict Theorem 4.3.7, since the region on $\mathbf{F}$ is defined is $\mathbb{R}^2$ minus the origin, which is not simply-connected.

The interesting thing is that, even though $\mathbf{F}$ is not defined at the origin, we can use our improved version of Green's Theorem in Theorem 4.4.6 to conclude that the line integral around any counterclockwise oriented, simple, closed, piecewise-regular curve C is 0 if origin is not inside C, and is 2π if the origin is inside C. How do you see this?

First, if the origin is not contained inside C, then $\mathbf{F}$ is defined and C^1 on an open neighborhood containing C and its interior, and $Q_x - P_y = 0$. Then, our original version of Green's Theorem from Theorem 4.4.1, or even Theorem 4.3.7 on equivalent conditions for a conservative field, tells us that

> To use Theorem 4.3.7 here, you need to know that the interior of a simple closed curve is simply-connected; this is true, but isn't easy. It is a consequence of the Jordan-Schönflies Theorem.

$$\int_C \mathbf{F} \cdot d\mathbf{r} \;=\; 0.$$

Now, let C_1 denote the unit circle, centered at the origin, oriented counterclockwise. A quick calculation yields that

$$\int_{C_1} \mathbf{F} \cdot d\mathbf{r} \;=\; 2\pi.$$

Let C_2 denote a circle of any positive radius less than 1, centered at the origin, oriented counterclockwise, and let R denote the compact region between C_1 and C_2. Then, the positively oriented boundary ∂R of R is $C_1 - C_2$.

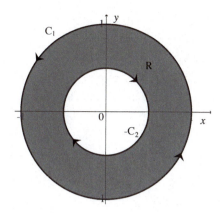

Figure 4.4.5: The region R and its positively oriented boundary ∂R.

There is an open neighborhood of R on which the vector field $\mathbf{F}$ is defined and is C^1; the open neighborhood just can't contain the origin. Thus, we can apply Theorem 4.4.6 to conclude that

$$\int_{C_1 - C_2} \mathbf{F} \cdot d\mathbf{r} \;=\; \int_{\partial R} \mathbf{F} \cdot d\mathbf{r} \;=\; \iint_R (Q_x - P_y)\,dA \;=\; \iint_R 0\,dA \;=\; 0.$$

Therefore,

$$\int_{C_1} \mathbf{F} \cdot d\mathbf{r} \;-\; \int_{C_2} \mathbf{F} \cdot d\mathbf{r} \;=\; 0,$$

and so

$$\int_{C_2} \mathbf{F} \cdot d\mathbf{r} \;=\; \int_{C_1} \mathbf{F} \cdot d\mathbf{r} \;=\; 2\pi.$$

Now, suppose that C is any counterclockwise oriented, simple, closed, piecewise-regular curve which has 0 in the interior of the region it bounds. Let C_2 be a small enough circle, centered at the origin, oriented counterclockwise, so that C_2 is contained in the interior of the region bounded by C. Let R be the region between C and C_2. Then, just as in our argument above with circle Theorem 4.4.6 tells us that

$$\int_C \mathbf{F} \cdot d\mathbf{r} \;-\; \int_{C_2} \mathbf{F} \cdot d\mathbf{r} \;=\; \int_{C - C_2} \mathbf{F} \cdot d\mathbf{r} \;=\; \int_{\partial R} \mathbf{F} \cdot d\mathbf{r} \;=\; \iint_R 0\,dA \;=\; 0,$$

i.e.,

$$\int_C \mathbf{F} \cdot d\mathbf{r} \;=\; \int_{C_2} \mathbf{F} \cdot d\mathbf{r} \;=\; 2\pi.$$

Finally, we would like to give a sketch of the proof of Green's Theorem, or, at least, the big idea.

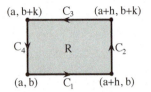

Figure 4.4.6: The filled rectangle R, with its positively oriented boundary.

We will first prove Green's Theorem for rectangular regions. To deal with more-general regions, we chop them up into lots of little rectangular regions (with some small missing or extra parts). Thus, we'll have to explain why Green's Theorem continues to hold when you put rectangular regions together.

Consider the filled rectangle R in Figure 4.4.6, and suppose that we have a C^1 vector field $\mathbf{F} = (P, Q)$ defined on an open set containing R.

Let's first calculate the line integral of $\mathbf{F}$ along $C_1 + C_3$. Since y is constant on each of C_1 and C_3, dy is zero on both of these pieces. Hence, we find

$$\int_{C_1+C_3} \mathbf{F} \cdot d\mathbf{r} = \int_{C_1+C_3} P\,dx + Q\,dy = \int_{C_1} P\,dx + \int_{C_3} P\,dx =$$

$$\int_a^{a+h} P(x,b)\,dx + \int_{a+h}^a P(x,b+k)\,dx = -\int_a^{a+h} \Big[P(x,b+k) - P(x,b) \Big]\,dx =$$

$$-\int_a^{a+h} \int_b^{b+k} \frac{\partial P}{\partial y}\,dy\,dx = \iint_R -\frac{\partial P}{\partial y}\,dA.$$

The calculation along $C_2 + C_4$ is similar, but now $dx = 0$. We find

$$\int_{C_2+C_4} \mathbf{F} \cdot d\mathbf{r} = \int_{C_2+C_4} P\,dx + Q\,dy = \int_{C_2} Q\,dy + \int_{C_4} Q\,dy =$$

$$\int_b^{b+k} P(a+h,y)\,dy + \int_{b+k}^b P(a,y)\,dy = \int_b^{b+k} \Big[P(a+h,y) - P(a,y) \Big]\,dy =$$

$$\int_b^{b+k} \int_a^{a+h} \frac{\partial P}{\partial x}\,dx\,dy = \iint_R \frac{\partial P}{\partial x}\,dA.$$

Using that $\partial R = C_1 + C_2 + C_3 + C_4$, and adding our two calculations above, we find

$$\int_{\partial R} \mathbf{F} \cdot d\mathbf{r} = \iint_R \left(\frac{\partial Q}{\partial x} - \frac{\partial P}{\partial y} \right)\,dA,$$

which is Green's Theorem for our rectangular region.

As we wrote, regions of more-general shapes are handled by chopping them up into small rectangles (with missing or extra small parts). Our question now is: why, when you combine rectangles does Green's Theorem still hold for the new region?

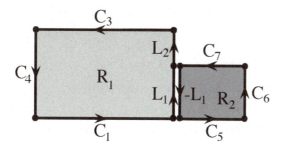

Figure 4.4.7: Two rectangular regions, with their positively oriented boundaries.

Consider the two rectangles R_1 and R_2 in Figure 4.4.7, with positively oriented boundaries. You are supposed to imagine that there is no gap between R_1 and R_2; we have drawn them slightly separated so that we can clearly indicate the positive orientations for each boundary. Note that on the overlapping line segment on their boundaries, which is labelled L_1 and $-L_1$, the positive orientations for the boundaries of R_1 and R_2 produce **opposite** orientations on the overlapping line segment. This is crucial.

Now, consider the region R, which is the union of R_1 and R_2 (remembering that there is no gap between them). Then, the positively oriented boundary of R is $C_1 + C_5 + C_6 + C_7 + L_2 + C_3 + C_4$. Since we derived Green's Theorem for rectangles, we calculate

$$\iint_R \left(\frac{\partial Q}{\partial x} - \frac{\partial P}{\partial y} \right) dA = \iint_{R_1} \left(\frac{\partial Q}{\partial x} - \frac{\partial P}{\partial y} \right) dA + \iint_{R_2} \left(\frac{\partial Q}{\partial x} - \frac{\partial P}{\partial y} \right) dA =$$

$$\int_{\partial R_1} \mathbf{F} \cdot d\mathbf{r} + \int_{\partial R_2} \mathbf{F} \cdot d\mathbf{r} = \int_{C_1 + L_1 + L_2 + C_3 + C_4} \mathbf{F} \cdot d\mathbf{r} + \int_{C_5 + C_6 + C_7 - L_1} \mathbf{F} \cdot d\mathbf{r} =$$

$$\int_{C_1 + L_1 + L_2 + C_3 + C_4 + C_5 + C_6 + C_7 - L_1} \mathbf{F} \cdot d\mathbf{r} = \int_{C_1 + L_2 + C_3 + C_4 + C_5 + C_6 + C_7} \mathbf{F} \cdot d\mathbf{r} =$$

$$\int_{\partial R} \mathbf{F} \cdot d\mathbf{r}.$$

Therefore, Green's Theorem holds for the combined region.

4.4.1 Exercises

Online answers to select exercises are here.

Basics:

In each of the following exercises, you are given a vector field $\mathbf{F} = \mathbf{F}(x, y)$, and an oriented, closed curve C in the xy-plane. Calculate the line integral

$\int_C \mathbf{F} \cdot d\mathbf{r}$ **in two different ways: (a) using the definition of the line integral, and (b) using Green's Theorem.**

1. $\mathbf{F}(x, y) = (3, -5)$, and C is the square with corners at $(0,0)$, $(1,0)$, $(1,1)$, and $(0,1)$, oriented counterclockwise.

2. $\mathbf{F}(x, y) = (ax+by+c, px+qy+r)$, where a, b, c, p, q, and r are constants, and C is the square with corners at $(0,0)$, $(1,0)$, $(1,1)$, and $(0,1)$, oriented counterclockwise.

3. $\mathbf{F}(x, y) = (y^2 + e^x, 2xy + 5x + e^y)$, and C is the circle of radius 2, centered at the origin, oriented clockwise.

4. $\mathbf{F}(x, y) = (y^2 + e^x, 2xy + 5x + e^y)$, and C is the circle of radius 2, centered at $(3,5)$, oriented clockwise.

5. $\mathbf{F}(x, y) = (x^2y + \sin x, xy + \cos y)$, and C is the triangle with corners at $(0,0)$, $(2,0)$, and $(0,3)$, oriented counterclockwise.

6. $\mathbf{F}(x, y) = (x^2y + \sin x, xy + \cos y)$, and C is the closed, oriented curve which starts at $(0,0)$, moves along the parabola where $y = x^2$, to $(2,4)$, then moves along the line segment from $(2,4)$ to $(2,0)$, and finally moves along the line segment from $(2,0)$ to $(0,0)$.

In each of the following exercises, you are given a force field $\mathbf{F} = \mathbf{F}(x, y)$, in Newtons, and a oriented, closed curve C in the xy-plane, where x and y are in meters. Use Green's Theorem to calculate the work done by $\mathbf{F}$ along C.

7. $\mathbf{F}(x, y) = (e^x, \ln(1 + y^2))$, and C is the square with corners at $(0,0)$, $(1,0)$, $(1,1)$, and $(0,1)$, oriented counterclockwise.

8. $\mathbf{F}(x, y) = (5y, -3x)$, and C is the triangle with corners at $(1,5)$, $(3,5)$, and $(1,11)$, oriented counterclockwise.

9. $\mathbf{F}(x, y) = (x^5 - y^3, x^3 - y^5)$, and C is the curve which starts at $(0,0)$, moves along a line segment to $(1/\sqrt{2}, 1/\sqrt{2})$, moves counterclockwise along the circle of radius 1, centered at the origin, to the point $(0,1)$, and then moves along a line segment, back to $(0,0)$.

10. $\mathbf{F}(x, y) = \left(\dfrac{3xy}{1 + x^2}, \ln(1 + x^2) \right)$, and C is the rectangle with corners at $(1,0)$, $(5,0)$, $(5,2)$, and $(1,2)$, oriented clockwise.

11. $\mathbf{F}(x, y) = (a\cos x + b\cos y, p\sin x + q\sin y)$, where a, b, p, and q are constants, and C is the square with corners at $(0,0)$, $(\pi,0)$, $(0,\pi)$, and (π,π), oriented counterclockwise.

12. $\mathbf{F}(x,y) = (x^4 - y^4, x^5 + y^5)$, and C is the closed, oriented curve which starts at $(0,0)$, moves along the curve where $y = x^3$, to $(1,1)$, then moves along the line segment from $(1,1)$ to $(1,0)$, and finally moves along the line segment from $(1,0)$ to $(0,0)$.

Suppose you wish to calculate a line integral $\int_{C_1} \mathbf{F} \cdot d\mathbf{r}$, where C_1 is not closed. It may be possible to simplify the calculation by using Green's Theorem. You accomplish this by selecting a curve C_2 which closes C_1, i.e., such that $C_1 + C_2$ is closed, and such that the line integral $\int_{C_2} \mathbf{F} \cdot d\mathbf{r}$ is relatively simple to calculate from the definition of a line integral. Then, you use Green's Theorem to calculate $\int_{C_1+C_2} \mathbf{F} \cdot d\mathbf{r}$. Finally, you use that

$$\int_{C_1} \mathbf{F} \cdot d\mathbf{r} = \int_{C_1+C_2} \mathbf{F} \cdot d\mathbf{r} - \int_{C_2} \mathbf{F} \cdot d\mathbf{r}.$$

In each of the following exercises, you are given a vector field $\mathbf{F}$ and an oriented non-closed curve C_1. (a) Select a "good" closing curve C_2. (b) Calculate, from the definition of the line integral, $\int_{C_2} \mathbf{F} \cdot d\mathbf{r}$. (c) Use Green's Theorem to calculate $\int_{C_1+C_2} \mathbf{F} \cdot d\mathbf{r}$. (d) Calculate $\int_{C_1} \mathbf{F} \cdot d\mathbf{r}$ from your answers to (b) and (c).

13. $\mathbf{F}(x,y) = (3, -5)$, and C_1 is the oriented curve which starts at $(1,0)$, moves along the line segment to $(1,1)$, then moves along the line segment to $(0,1)$, and then moves along the line segment to $(0,0)$.

14. $\mathbf{F}(x,y) = (ax + by + c, px + qy + r)$, where a, b, c, p, q, and r are constants, and C_1 is the oriented curve which starts at $(1,0)$, moves along the line segment to $(1,1)$, then moves along the line segment to $(0,1)$, and then moves along the line segment to $(0,0)$. ▶

15. $\mathbf{F}(x,y) = (x^2 y + \sin x, xy + \cos y)$, and C_1 is the oriented curve which starts at $(0,0)$, moves along a line segment to $(2,0)$, and then moves along a line segment to $(0,3)$.

16. $\mathbf{F}(x,y) = (x^5 - y^3, x^3 - y^5)$, and C_1 is the curve which starts at $(0,0)$, moves along a line segment to $(1/\sqrt{2}, 1/\sqrt{2})$, moves counterclockwise along the circle of radius 1, centered at the origin, to the point $(0,1)$.

More Depth:

In each of the following exercises, you are given a continuously differentiable vector field $\mathbf{F}(x,y) = (P(x,y), Q(x,y))$ and a compact plane region R, which is the closure of its interior, and whose boundary, ∂R, consists of a finite number of simple, closed, oriented, piecewise-regular curves. When we write that R is inside a closed curve, or that we have removed a region inside a closed curve, we always mean that the boundary curves are included in R, so that R is compact. **(a) Sketch the region R, and the simple oriented closed curves which comprise ∂R, indicating the positive orientations on the curves. (b) State the equality given by Green's Theorem for your particular F, R, and ∂R. Do not evaluate any integrals.**

17. $\mathbf{F} = (x^5, y^3)$, and R is the disk inside the circle of radius 4, centered at the origin, from which we have removed the disks of radius 1 centered at $(-2, 0)$ and $(2, 0)$.

    ▶

18. $\mathbf{F} = (y^3, x^5)$, and R is the disk inside the circle of radius 4, centered at the origin, from which we have removed the disks of radius 1 centered at $(-2, 0)$ and $(2, 0)$.

19. $\mathbf{F} = \left(-\dfrac{y}{x^2 + y^2}, \dfrac{x}{x^2 + y^2} \right)$, and R is the region inside the ellipse where $\dfrac{x^2}{100} + \dfrac{y^2}{36} = 1$, from which we remove the region inside the ellipse where $\dfrac{x^2}{9} + \dfrac{y^2}{25} = 1$.

20. $\mathbf{F} = (x \tan^{-1} y, \, x \tan^{-1} y)$, and R is the filled-in square with corners $(-4, -4)$, $(-4, 4)$, $(4, -4)$, and $(4, 4)$, from which we have removed the disk of radius 2, centered at the origin.

21. Give examples of two different continuously differentiable vector fields $\mathbf{F}$ on $\mathbb{R}^2$ such that, for every simple, closed, oriented, piecewise-regular curve, ∂R, which bounds a region R, $\int_{\partial R} \mathbf{F} \cdot d\mathbf{r}$ equals the area of R.

22. Derive Green's Theorem in the special case where R is the filled-in square with corners at $(0, 0)$, $(1, 0)$, $(1, 1)$, and $(0, 1)$.

23. Let R be the region in $\mathbb{R}^2$ which is bounded by a simple, closed, oriented, piecewise-regular curve, ∂R, which is oriented counterclockwise. Let A denote the area of R. Show that the x- and y-coordinates of the centroid of R (recall Definition 3.9.3) are given by

$$\overline{x} \; = \; \frac{1}{2A} \int_{\partial R} x^2 \, dy \qquad \text{and} \qquad \overline{y} \; = \; -\frac{1}{2A} \int_{\partial R} y^2 \, dx.$$

24. Use the formulas in the previous exercise to find the centroid of the triangular region with vertices $(0, 0)$, $(a, 0)$, and $(0, b)$, where a and b are positive constants.

4.5 Flux through a Surface

A line integral of a vector field uses the components of the vector field which are **tangent** to a given curve. A *flux integral* uses the components of a vector field which are **perpendicular** to a given surface. If the vector field is the velocity vector field of a moving fluid, the flux integral through a surface measures the rate at which the volume of the fluid which moves through the surface, with respect to time.

For force fields, we still think of flux integrals as measuring the force "flowing" through the given surface. Magnetic and electric flux are vitally important concepts in electromagnetics.

Suppose that **V** is the velocity vector field, measured in meters per second, of a flowing fluid in some region $\mathcal{U}$ in $\mathbb{R}^3$. Consider a fixed surface M in $\mathcal{U}$.

Basics:

Our question is: how can we calculate the volume of fluid which flows through the surface per unit time? This rate of flow through the surface is called the *flux through the surface*.

If we assume that the surface M is a 2-dimensional C^1 submanifold of $\mathbb{R}^3$ (recall Definition 2.6.9), then, at each point **p**, M has two unit normal vectors: vectors of length 1, perpendicular to the tangent plane to M at **p**. We want to pick an *orientation* for M by making a continuous choice of a unit normal vector **n** at each point of M; if such a continuous choice is possible, we say that M is *orientable*. The vectors **n** in our orientation are said to point in the *positive direction*.

We shall discuss orientations more carefully in the More Depth portion of this section. What we have given here is basically an orientation for a C^1 submanifold of $\mathbb{R}^3$. We actually need to discuss orienting *topological submanifolds*, so that our surfaces may have sharp points.

For each point **p** on M, we want to look at the contribution from $\mathbf{V} = \mathbf{V}(\mathbf{p})$ to the total amount fluid moving through the surface per time unit, where a positive contribution is in the direction of $\mathbf{n} = \mathbf{n}(\mathbf{p})$.

As we saw back in Section 1.3, we can decompose **V**, uniquely as a sum of vectors, one of which is a scalar multiple of **n** and one of which is perpendicular to **n**; specifically, since **n** is a unit vector, we have

$$\mathbf{V} = (\mathbf{V} \cdot \mathbf{n})\,\mathbf{n} + \mathbf{T},$$

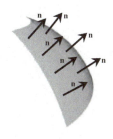

where **T** is perpendicular to **n**. As we know from Section 2.7 and Section 2.8, the vectors which are perpendicular to **n** are precisely the tangent vectors to M (at the point **p**), and so **T** is tangent to M. This tangent component does not contribute to the amount of fluid moving through M, only the component $(\mathbf{V} \cdot \mathbf{n})\,\mathbf{n}$ contributes to the flux.

Figure 4.5.1: The positive direction, determined by a choice of unit normal vectors **n**.

The vector **T** is known as the *orthogonal projection of* **V** *into the tangent plane.*

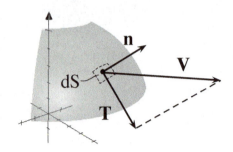

Figure 4.5.2: A surface M, a patch of surface area dS, a unit normal vector, $\mathbf{n}$, a vector in the vector field $\mathbf{V}$, and the orthogonal projection $\mathbf{T}$ of $\mathbf{V}$ into the tangent plane of the surface.

Thus, the scalar $\mathbf{V} \cdot \mathbf{n}$ measures the flux through M; this is the rate, in meters per second, at which fluid is flowing through M, at $\mathbf{p}$, with a positive number meaning that the fluid flows in the direction of $\mathbf{n}$, and a negative number meaning that fluid is flowing in the opposite direction from $\mathbf{n}$.

But, $\mathbf{V} \cdot \mathbf{n}$ measures the fluid flow through M through the single point $\mathbf{p}$ in meters per second. How do we get the infinitesimal **volume** of fluid through M at $\mathbf{p}$? We multiply $\mathbf{V} \cdot \mathbf{n}$ by an infinitesimal piece dS of surface area on M at $\mathbf{p}$, and get the flow in cubic meters per second. Then, of course, we take an integral to add up all of the infinitesimal contributions.

Therefore, the **flux through the oriented surface** M is given by

$$\iint_M \mathbf{V} \cdot \mathbf{n} \, dS,$$

provided this integral exists.

But, this leaves us with another question: how do you actually **calculate**

$$\iint_M \mathbf{V} \cdot \mathbf{n} \, dS?$$

The answer is that there are **two cases**: the easy case, where M is contained in a plane which is parallel to one of the coordinate planes, and the harder case where we use a parameterization of M.

The easy case:

Suppose that we have a vector field

$$\mathbf{V}(x, y, z) \; = \; \big(P(x, y, z), \; Q(x, y, z), \; R(x, y, z)\big).$$

and our surface M is contained in a plane parallel to one of the coordinate planes, i.e., M is contained in a plane defined by setting one of x, y, or z equal to a constant.

To be more specific in this discussion, let's suppose that M is contained in the plane where $y = 7$; this plane is parallel to the xz-plane.

In this case, the unit normal $\mathbf{n}(x, y, z)$ is, in fact, constant and is $\pm \mathbf{j} = \pm(0, 1, 0)$. Of course, we are supposed to start with an **oriented** surface M and, in the description of M, you must be told in some way which of these two orientations you have. Let's suppose, for this discussion, that $\mathbf{n} = -(0, 1, 0)$.

Notice also that dS, the element of surface area, is just surface area in a copy of the xz-plane, i.e., dS is simply $dx\,dz = dz\,dx$. We write dA, instead of dS, for either $dx\,dz$ or $dz\,dx$ (or maybe even polar coordinates in the xz-plane).

Therefore, our flux integral becomes

$$\iint_M \mathbf{V} \cdot \mathbf{n} \, dS \; = \; \iint_M \big(P(x, y, z), \; Q(x, y, z), \; R(x, y, z) \big) \cdot -(0, 1, 0) \, dA \; =$$

$$\iint_M -Q(x, y, z) \, dA \; = \; \iint_M -Q(x, 7, z) \, dA \; = \; \iint_M -Q(x, 7, z) \, dz\,dx,$$

where, in the last equality, we used that, in this example, $y = 7$ on all of M. Thus, we end up with simply a double/iterated integral in terms of x and z.

If you understand the above example, you should see what changes you need to make in the cases where x or z is constant on M, or where a different orientation for M is chosen.

The parameterized surface case:

Recall the definition of a *basic parameterization* $\mathbf{r} : D \to \mathbb{R}^3$ from Definition 3.11.1 and the *basic surface M*, which is its image. A basic surface may fail to be a C^1 submanifold, but the places where it may fail are contained in the union of a finite number of points and curves; thus, a basic surface may fail to be a submanifold, but that failure would take place along a set with no area, and so will not affect the flux integral.

Recall that a basic parameterization $\mathbf{r} = \mathbf{r}(u, v)$ is required to be regular and one-to-one when restricted to $\text{int}(D)$. This means that, on $\text{int}(D)$, $\mathbf{r}_u \times \mathbf{r}_v \neq \mathbf{0}$ and is always normal to M. Therefore, there is an orientation of the image of $\text{int}(D)$ given by picking the *positive* side to be specified by the unit normal vectors

$$\mathbf{n} \; = \; \frac{\mathbf{r}_u \times \mathbf{r}_v}{|\mathbf{r}_u \times \mathbf{r}_v|},$$

where we use that $\mathbf{r}$ is one-to-one on $\text{int}(D)$ to ensure that we are assigning a single $\mathbf{n}$ to each point in the image of $\text{int}(D)$. Of course, we could negate all of these unit normal vectors to pick the opposite orientation as the positive one.

Therefore, we define:

The term "almost-orientation" is very cumbersome. However, the distinction between an almost-orientation, which is an orientation "almost everywhere", and an orientation of the entire image is very important, and we need to distinguish between these concepts in the terminology.

If $\mathbf{r}$ is regular and one-to-one on **all** of D, not just on the interior, then the image M is orientable, not merely almost-orientable, and the two orientations are given by $\mathbf{n} = \pm(\mathbf{r}_u \times \mathbf{r}_v)/|\mathbf{r}_u \times \mathbf{r}_v|$.

The most basic, and most famous, non-orientable surface is the *Möbius Strip*, which we looked at in Exercise 23 in Section 3.11. We shall look at it again in the exercises for this section. The Möbius Strip is, however, a basic surface, and so is almost-orientable.

Flux integrals are typically defined only for truly oriented surfaces, not for almost-oriented surfaces. There is no problem making the definition as we have done; however, important results about flux integrals over orientable surfaces may fail for surfaces which are merely almost-orientable. Stokes' Theorem, Theorem 4.7.2, is true for oriented surfaces, but can fail for almost-orientable surfaces. We shall see this for the Möbius Strip in Exercise 18 of Section 4.7.

Definition 4.5.1. *Suppose that M is a basic surface, parameterized by the basic parameterization $\mathbf{r} : D \to \mathbb{R}^3$, $\mathbf{r} = \mathbf{r}(u, v)$.*

*An **almost-orientation** of M is a choice of either*

$$\mathbf{n} = \frac{\mathbf{r}_u \times \mathbf{r}_v}{|\mathbf{r}_u \times \mathbf{r}_v|} \quad \text{or} \quad \mathbf{n} = \frac{\mathbf{r}_v \times \mathbf{r}_u}{|\mathbf{r}_v \times \mathbf{r}_u|} = -\frac{\mathbf{r}_u \times \mathbf{r}_v}{|\mathbf{r}_u \times \mathbf{r}_v|}$$

*as the **positive** directions at all of the points in the image of the interior of D.*

Now, as we saw in Theorem 3.11.2, the infinitesimal area, the area element, is given by $dS = |\mathbf{r}_u \times \mathbf{r}_v| \, du \, dv$. Thus, canceling the $|\mathbf{r}_u \times \mathbf{r}_v|$ terms, and we obtain

$$\mathbf{n} \, dS = \frac{\mathbf{r}_u \times \mathbf{r}_v}{|\mathbf{r}_u \times \mathbf{r}_v|} \cdot |\mathbf{r}_u \times \mathbf{r}_v| \, du \, dv = (\mathbf{r}_u \times \mathbf{r}_v) \, du \, dv,$$

provided that we use the orientation given by $\mathbf{r}_u \times \mathbf{r}_v$; if we use the opposite orientation, then $\mathbf{n} \, dS = (\mathbf{r}_v \times \mathbf{r}_u) \, du \, dv$.

Therefore, we make the following definition.

Definition 4.5.2. *Suppose that M is a basic surface, parameterized by the basic parameterization $\mathbf{r} : D \to \mathbb{R}^3$, $\mathbf{r} = \mathbf{r}(u, v)$, with an almost-orientation, on the image of the interior of D, given by*

$$\mathbf{n} = (\mathbf{r}_u \times \mathbf{r}_v)/|\mathbf{r}_u \times \mathbf{r}_v|.$$

Suppose that $\mathbf{V}$ is a continuous vector field on M.

*Then, the **flux integral of $\mathbf{V}$ over** M, or the **flux of V through** M, is*

$$\iint_M \mathbf{V} \cdot \mathbf{n} \, dS = \iint_D \mathbf{V}(\mathbf{r}(u, v)) \cdot (\mathbf{r}_u \times \mathbf{r}_v) \, du \, dv,$$

provided that this double integral exists.

Example 4.5.3. Let's look back at Example 3.11.5, where we had a right circular cylinder (with no top or bottom), of radius R and height H, centered around the z-axis, with its base in the xy-plane. We choose the outward almost-orientation (which, here, is actually an orientation), and let M denote the oriented surface. Suppose that x, y and z are in meters.

Let $\mathbf{V}$ be the velocity vector field of a fluid given by

$$\mathbf{V}(x, y, z) = (x + y, -x + y, e^x y^3 z^2) \text{ m/s}.$$

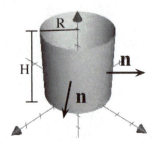

Figure 4.5.3: A right circular cylinder, oriented outward.

Let's find the flux of $\mathbf{V}$ through M.

In Example 3.11.5, we had the parameterization

$$x = R\cos\theta, \qquad y = R\sin\theta, \qquad \text{and} \qquad z = z,$$

i.e.,

$$\mathbf{r}(\theta, z) = (R\cos\theta, \ R\sin\theta, \ z),$$

where $0 \leq \theta \leq 2\pi$ and $0 \leq z \leq H$. We shall use this parameterization again; thus, θ and z are playing the roles of u and v, respectively, from Definition 4.5.2.

We calculated

$$\mathbf{r}_\theta = R(-\sin\theta, \ \cos\theta, \ 0), \quad \mathbf{r}_z = (0, \ 0, \ 1), \quad \text{and} \quad \mathbf{r}_\theta \times \mathbf{r}_z = R(\cos\theta, \ \sin\theta, 0).$$

Note that this does point outward, in agreement with our specified orientation.

Therefore,

$$\iint_M \mathbf{V} \cdot \mathbf{n}\, dS = \int_0^H \int_0^{2\pi} \mathbf{V}(R\cos\theta, R\sin\theta, z) \cdot R(\cos\theta, \ \sin\theta, 0)\, d\theta\, dz =$$

$$\int_0^H \int_0^{2\pi} (R\cos\theta + R\sin\theta, \ -R\cos\theta + R\sin\theta, \ *) \cdot R(\cos\theta, \ \sin\theta, 0)\, d\theta\, dz,$$

where we have placed an asterisk in the third position of the vector field because it doesn't matter what it is; in the dot product, it will be multiplied by 0.

Thus, we find

$$\iint_M \mathbf{V} \cdot \mathbf{n}\, dS = R^2 \int_0^H \int_0^{2\pi} (\cos^2\theta + \sin\theta\cos\theta - \sin\theta\cos\theta + \sin^2\theta)\, d\theta\, dz =$$

$$R^2 \int_0^H \int_0^{2\pi} 1\, d\theta\, dz = 2\pi R^2 H \ \text{ m}^3/\text{s}.$$

Since we picked the outward orientation, the fact that the flux is positive means that the rate at which fluid flows out of the cylinder is greater than the rate at which fluid flows in.

Example 4.5.4. Consider the vector field which goes "straight down":

$$\mathbf{V} = (0, 0, -1) = -\mathbf{k}.$$

Let's find the flux of $\mathbf{V}$ through the right circular half-cone M, whose vertex is at the origin, which is centered around the positive z-axis, which has height $H = 4$ and

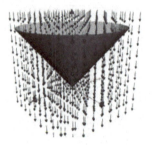

Figure 4.5.4: A right circular cone, and a downward vector field.

radius $R = 2$; we almost-orient M so that "down" is the positive direction, i.e., so that the positive unit normal vectors at all points other than the vertex have a negative z component.

We parameterized such a cone back in Example 3.11.6. We used the parameterization

$$\mathbf{p}(\theta, z) = z\left(\frac{R}{H}\cos\theta, \ \frac{R}{H}\sin\theta, \ 1\right) = z\left(\frac{1}{2}\cos\theta, \ \frac{1}{2}\sin\theta, \ 1\right),$$

where $0 \leq \theta \leq 2\pi$ and $0 \leq z \leq 4$.

We also calculated

$$\mathbf{p}_\theta \times \mathbf{p}_z = \frac{z}{2}\left(\cos\theta, \ \sin\theta, \ -\frac{1}{2}\right);$$

note that this has a negative z component, and so, corresponds to the given orientation.

Therefore, the flux of $\mathbf{V}$ through M is given by

$$\iint_M \mathbf{V} \cdot \mathbf{n}\, dS = \int_0^{2\pi}\int_0^4 \mathbf{V}(\mathbf{p}(\theta, z)) \cdot (\mathbf{p}_\theta \times \mathbf{p}_z)\, dz\, d\theta =$$

$$\int_0^{2\pi}\int_0^4 (0,\, 0,\, -1) \cdot \left(*,\, *,\, -\frac{z}{4}\right)\, dz\, d\theta,$$

where we have once again used $*$'s for entries that don't matter, since they get multiplied by 0's in the dot product.

Hence, we find that

$$\iint_M \mathbf{V} \cdot \mathbf{n}\, dS = \int_0^{2\pi}\int_0^4 \frac{z}{4}\, dz\, d\theta = 4\pi.$$

We motivated our discussion of flux by using velocity vector fields of fluids. However, it is also common to talk about the flux of a force field through a surface. In particular, the flux of the gravitational and electric force fields are of great interest.

As we discussed in Section 4.3, the gravitational force field and the electric force field have the same form:

$$\mathbf{F} = \frac{k}{\left(x^2 + y^2 + z^2\right)^{3/2}}(x, y, z) = \frac{k}{\rho^3}\boldsymbol{\rho},$$

where k is a constant, and $\boldsymbol{\rho}(x, y, z) = (x, y, z)$. As you may have guessed, we have switched from using $\mathbf{r}$ and r for the radial vector and its magnitude to using $\boldsymbol{\rho}$ and ρ because we are going to integrate using spherical coordinates.

Example 4.5.5. Consider a force field

$$\mathbf{F} = \frac{k}{(x^2 + y^2 + z^2)^{3/2}}(x, y, z) = \frac{k}{\rho^3}\boldsymbol{\rho} \quad \text{Newtons}$$

where k is a constant, and x, y, and z are measured in meters.

Find the flux of $\mathbf{F}$ through the outward-oriented sphere M of radius R, centered at the origin.

Solution:

We use spherical coordinates to parameterize the sphere. We looked at this back in Example 3.11.4.

The sphere of radius R, centered at the origin, is parameterized by the θ and ϕ from spherical coordinates with ρ being fixed at R, i.e.,

$$\mathbf{r}(\theta, \phi) = (R\sin\phi\cos\theta,\ R\sin\phi\sin\theta,\ R\cos\phi) = R(\sin\phi\cos\theta,\ \sin\phi\sin\theta,\ \cos\phi),$$

where $0 \leq \theta \leq 2\pi$ and $0 \leq \phi \leq \pi$.

We calculated in Example 3.11.4 that

$$dS = |\mathbf{r}_\theta \times \mathbf{r}_\phi|\, d\phi\, d\theta = R^2\sin\phi\, d\phi\, d\theta.$$

Note that, because our surface M is a sphere of radius R, centered at the origin, we immediately know what the outward-pointing unit normals are:

$$\mathbf{n} = \frac{\boldsymbol{\rho}}{\rho} = \frac{\boldsymbol{\rho}}{R}.$$

Because the expressions for dS and $\mathbf{n}$ are so simple, we do not actually need to use $\mathbf{r}_\theta \times \mathbf{r}_\phi$ in our flux integral. Instead, noting that $\boldsymbol{\rho} \cdot \boldsymbol{\rho} = |\boldsymbol{\rho}|^2 = R^2$, we calculate

$$\iint_M \mathbf{F} \cdot \mathbf{n}\, dS = \int_0^{2\pi}\int_0^\pi \left(\frac{k}{R^3}\boldsymbol{\rho} \cdot \frac{\boldsymbol{\rho}}{R}\right) R^2\sin\phi\, d\phi\, d\theta =$$

$$k\int_0^{2\pi}\int_0^\pi \sin\phi\, d\phi\, d\theta = 4\pi k \quad \text{N-m}^2.$$

Note that this result is independent of the radius R.

Suppose we are looking at the case of a point-charge of Q coulombs at the origin, and the electric force field $\mathbf{F}_e$ that it produces by considering point-charges of 1 coulomb at every point in space, other than the origin. In this case,

$$\mathbf{F}_e = \frac{1}{4\pi\epsilon_0}\frac{Q}{\rho^3}\boldsymbol{\rho} \quad \text{Newtons},$$

where ϵ_0 is the *electric constant*, whose value is approximately 8.854×10^{-12} farads/meter.

Thus, the constant k that we were using throughout this example would be $k = Q/(4\pi\epsilon_0)$, and our calculation tells us that the electric flux through a sphere, of any radius, produced by a charge of Q at its center, is Q/ϵ_0.

This is a special form of *Gauss' Law*, which we shall need when we derive the general form in Section 4.6.

We are about to look at an example of a flux integral through a non-basic surface; however, the surface will be composed of a finite number of almost-oriented basic surface pieces. Thus, the surface will be a piecewise-regular surface, as defined in Definition 3.11.7, where we choose an almost-orientation on each basic surface piece.

Naturally, the flux of a vector field though such surface is defined to be the sum of the fluxes through each of the almost-oriented basic surface pieces.

We shall define a *closed surface* more carefully in the More Depth portion.

If the piecewise-regular surface M bounds a solid region, i.e., if M is *closed*, it is standard to use, as the default positive almost-orientation, the unit normal vectors which point outward from the solid region at each regular point.

In order to deal rigorously with orientations on surfaces bounding a solid region in $\mathbb{R}^3$, we need to look at *topological submanifolds of* $\mathbb{R}^3$ and actual *orientations*, not merely almost-orientations. We shall discuss these concepts in the More Depth portion of this section.

However, we will first go ahead and look at an example.

Example 4.5.6. As in Example 4.5.4, let's consider the vector field

$$\mathbf{V} = (0, 0, -1) = -\mathbf{k},$$

and once again consider the right circular (half) cone whose vertex is at the origin, which is centered around the positive z-axis, and which has height 4 and radius 2. However, this time, let's close the cone by placing a disk of radius 2 on top, and let M denote the resulting piecewise-regular surface.

Figure 4.5.5: A right circular cone, with a top, and a downward vector field.

We'll call the cone portion of this piecewise-regular surface M_c and the top disk M_d. Since M bounds a solid region, we use the standard almost-orientation on each piece of M; this is the almost orientation that points outward from the solid bounded region. This means that positive unit normal vectors on M_c point downward (at an angle), but positive unit normal vectors on M_d point straight up.

The flux of $\mathbf{V}$ through M is the sum of the flux through M_c and M_d, i.e.,

$$\iint_M \mathbf{V} \cdot \mathbf{n}\, dS \;=\; \iint_{M_c} \mathbf{V} \cdot \mathbf{n}\, dS \;+\; \iint_{M_d} \mathbf{V} \cdot \mathbf{n}\, dS.$$

We calculated the flux $\iint_{M_c} \mathbf{V} \cdot \mathbf{n}\, dS$ in Example 4.5.4; we found

$$\iint_{M_c} \mathbf{V} \cdot \mathbf{n}\, dS \;=\; 4\pi.$$

Since M_d is contained in the plane where $z = 4$, we are in the easy case for flux integrals. We have that $\mathbf{n}$ on M_d is simply $\mathbf{k} = (0, 0, 1)$ and $dS = dA$, where dA denotes an element of area in the xy-plane. Therefore,

$$\iint_{M_d} \mathbf{V} \cdot \mathbf{n}\, dS \;=\; \iint_{M_d} (0, 0, -1) \cdot (0, 0, 1)\, dA \;=\; -\iint_{M_d} dA \;=$$

$$-(\text{area of the disk of radius } 2) \;=\; -4\pi.$$

Hence,

$$\iint_M \mathbf{V} \cdot \mathbf{n}\, dS \;=\; \iint_{M_c} \mathbf{V} \cdot \mathbf{n}\, dS \;+\; \iint_{M_d} \mathbf{V} \cdot \mathbf{n}\, dS \;=\; 4\pi - 4\pi \;=\; 0.$$

Is it a coincidence that this total flux over the closed, oriented surface is 0? No. As we shall see in the next section, it's a reflection of the fact that the divergence of the vector field $\mathbf{V}$ is 0.

Remark 4.5.7. In the example above, we found that the flux of a particular vector field through a particular closed surface was 0.

If you think of the vector field as the velocity vector field of a flowing fluid, you might wonder how the flux through a closed surface could be anything **but** 0. After all, if more fluid flows in than out, then the fluid shouldn't fit into the same space. If more fluid flows out than flows in, you should end up with some sort of vacuum inside your surface.

However, it is possible that fluid compresses or decompresses, or that there is a source of fluid production somewhere inside the closed surface, or that there is a drain, a sink, inside the closed surface.

Because of such physical interpretations, if a closed surface M is oriented outward, and the flux of a vector field through M is positive, we say that M contains a *source* of the vector field; if the flux is negative we say that M contains a *sink* of the vector field.

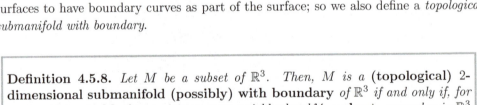

Figure 4.5.6: Part of an edge of a right circular cone, with a top.

We defined piecewise-regular surfaces in Definition 3.11.7. To discuss flux, it is standard to require that the surfaces being integrated over are **oriented** piecewise-regular surfaces, and the orientation part makes things quite a bit more complicated than our use of the weaker concept of almost-orientations.

We call a connected surface *orientable* if it has two well-defined sides; then we declare one side to be what we'll call the *positive side* and the other side the *negative side*. But how do we make this rigorous, mathematically, when the surface has sharp edges, and so doesn't have a choice of just two unit normal vectors along an edge?

We want, somehow, to make sense of the unit normal vectors at the regular points being on the same side as we approach each edge. This means we want to be able to locally discuss two different sides of the surface. If you think about the xy-plane, sitting in $\mathbb{R}^3$, you can pick the positive side to be the side containing the positive z-axis, and the negative side to be the side containing the negative z-axis. We would like to be able to do such a thing for our orientable, piecewise-regular surfaces.

What we do is require our surface to locally be like the xy-plane in $\mathbb{R}^3$ after we make a "local continuous change of coordinates"; this means that we use a continuous bijection with a continuous inverse, rather than using a differentiable change of coordinates. After we use our change of coordinates to make our surface locally look like the xy-plane in $\mathbb{R}^3$, we can use the notions of positive and negatives sides there in order to locally pick a positive and a negative side of our surface. Note that we won't get positive and negative unit normal vectors everywhere, just a notion of the positive and negative sides.

A surface in $\mathbb{R}^3$ which locally, after a continuous change of coordinates, becomes the xy-plane in $\mathbb{R}^3$, is called a *topological submanifold* of $\mathbb{R}^3$. We also want to allow our surfaces to have boundary curves as part of the surface; so we also define a *topological submanifold with boundary*.

Definition 4.5.8. *Let M be a subset of $\mathbb{R}^3$. Then, M is a* **(topological) 2-dimensional submanifold (possibly) with boundary** *of $\mathbb{R}^3$ if and only if, for every point $\mathbf{p}$ in M, there exists an open neighborhood $\mathcal{U}$, a **chart**, around $\mathbf{p}$ in $\mathbb{R}^3$ and an open neighborhood $\mathcal{V}$ of the origin in $\mathbb{R}^3$, and a continuous bijective function $h : \mathcal{U} \to \mathcal{V}$, whose inverse is continuous, such that $h(\mathbf{p}) = \mathbf{0}$, and such that the image of the points of M in $\mathcal{U}$ equals the intersection of $\mathcal{V}$ with either (i) the xy-plane, or (ii) with the xy-plane where $y \geq 0$.*

*In case (ii), $\mathbf{p}$ is called a **boundary point** in M. If there are no boundary points in M, then M is called, simply, a **submanifold** of $\mathbb{R}^3$.*

As we discussed, a submanifold, possibly with boundary, can be locally oriented, i.e., a choice of a positive side can be made, by using the local continuous changes of coordinates. We want a **global** orientation, and not all surfaces have them. We should point out that, if a surface has different connected pieces, then you could pick different notions of the positive and negative side for each piece; so it's enough to define orientations on connected surfaces.

Definition 4.5.9. *Suppose that M is a connected 2-dimensional submanifold, possibly with boundary, in $\mathbb{R}^3$.*

*If it is possible to pick a collection of open charts, whose union contains M, and choose local orientations in each chart so that, whenever two charts overlap, their positive sides agree, then we say that M **is orientable**. Otherwise, M is* **non-orientable**.

An **orientation** *of an orientable M is a choice of the positive side (which also determines the negative side) of M in $\mathbb{R}^3$. When we say that M is* **oriented**, *we mean that M is orientable and that an orientation has been chosen, i.e., a choice of the positive side has been made.*

Despite the fact that it may seem as though we have been very rigorous, we have not; orientability is a very delicate subject. See, for instance, [1] for details.

Definition 4.5.10. *Suppose that M is a connected, oriented, 2-dimensional submanifold, possibly with boundary, in $\mathbb{R}^3$ which is also a piecewise-regular surface.*

We say that M is a **connected, oriented, piecewise-regular surface** *provided that, at each regular point of M, the positive unit normal vector $\mathbf{n}$ is chosen to point in the direction of the positive side of M as specified by the (topological) orientation.*

The following theorem tells us that orientability is automatic for compact surfaces without boundary.

Theorem 4.5.11. *Suppose that M is a compact, connected, 2-dimensional submanifold (without boundary) of $\mathbb{R}^3$. Then, M is orientable, and there is a well-defined connected, bounded region which is inside M, and a well-defined connected, unbounded region which is outside M.*

The proof of this theorem is far beyond the scope of this textbook. See [1], Chapter VI.

Given the theorem above, we can make the following definition.

> **Definition 4.5.12.** *Suppose that M is a compact, connected, 2-dimensional sub-manifold (without boundary) of $\mathbb{R}^3$.*
>
> *Then, we say that M is a **closed surface**, and the **positive** side of M is the outside, as described in Theorem 4.5.11; so, if M is piecewise-regular, the positive unit normal $\mathbf{n}$, at regular points, is chosen to point out of the solid region which is surrounded by M.*

Online answers to select exercises are here.

Basics:

4.5.1   Exercises

In each of the following exercises, you are given a velocity vector field $\mathbf{V} = \mathbf{V}(x, y, z)$, in meters per second, of a flowing fluid, where x, y, and z are in meters. You are also given a basic parameterization $\mathbf{r} = \mathbf{r}(u, v)$ of a surface M in $\mathbb{R}^3$. Almost-orient M by using $\mathbf{n} = (\mathbf{r}_u \times \mathbf{r}_v)/|\mathbf{r}_u \times \mathbf{r}_v|$ as the positive unit normal vectors. (a) Sketch the surface M, and include some vectors $\mathbf{n}$ in the almost-orientation. (b) Calculate the rate, in m^3/s, at which fluid is flowing through M, i.e., find the flux of $\mathbf{V}$ through the almost-oriented surface M.

1. $\mathbf{V}(x, y, z) = (1, 1, 1)$, $\mathbf{r}(u, v) = (u \cos v, u \sin v, 0)$, where $0 \leq u \leq 2$ and $0 \leq v \leq 2\pi$. ▶

2. $\mathbf{V}(x, y, z) = (z e^y, z \tan^{-1} x, x^2 + y^2 + z^2)$, $\mathbf{r}(u, v) = (u \cos v, u \sin v, 0)$, where $0 \leq u \leq 2$ and $0 \leq v \leq 2\pi$.

3. $\mathbf{V}(x, y, z) = (y, -x, z)$, $\mathbf{r}(u, v) = (u \cos v, u \sin v, 3)$, where $0 \leq u \leq 2$ and $0 \leq v \leq 2\pi$.

4. $\mathbf{V}(x, y, z) = (y, z, -x)$, $\mathbf{r}(u, v) = (u \cos v, u \sin v, 3)$, where $0 \leq u \leq 2$ and $0 \leq v \leq 2\pi$.

5. $\mathbf{V}(x, y, z) = (x, y, z)$, $\mathbf{r}(u, v) = (5 \sin u \cos v, 5 \sin u \sin v, 5 \cos u)$, where $0 \leq u \leq \pi/4$ and $0 \leq v \leq \pi$.

6. $\mathbf{V}(x, y, z) = (y, -x, 1)$, $\mathbf{r}(u, v) = (5 \sin u \cos v, 5 \sin u \sin v, 5 \cos u)$, where $0 \leq u \leq \pi/4$ and $0 \leq v \leq \pi$.

7. $\mathbf{V}(x, y, z) = (2y, xz, -x)$, $\mathbf{r}(u, v) = (u, v, 4 - u^2 - v^2)$, where $0 \leq u \leq 1$ and $0 \leq v \leq 2$. ▶

8. $\mathbf{V}(x, y, z) = (2y, xz, -x)$, $\mathbf{r}(u, v) = (3 \cos u, 3 \sin u, v)$, where $0 \leq u \leq 2\pi$ and $-2 \leq v \leq 2$.

In each of the following exercises, you are given a velocity vector field $\mathbf{V} = \mathbf{V}(x, y, z)$, in meters per second, of a flowing fluid, where x, y, and z are in meters. You are also given a basic surface M in $\mathbb{R}^3$, and an almost-orientation on M. (a) Sketch the surface M, and include some vectors n in the almost-orientation. (b) Parameterize M. (c) Calculate the rate, in m^3/s, at which fluid is flowing through M, i.e., find the flux of $\mathbf{V}$ through the almost-oriented surface M.

9. $\mathbf{V}(x, y, z) = (x, y, z)$, and M is the disk of radius 5, centered at $(0, 7, 0)$, in the plane where $y = 7$, oriented so that the positive direction is in the direction of the positive y-axis.

10. $\mathbf{V}(x, y, z) = (x, y, z)$, and M is the filled-in square with vertices $(-5, 7, -5)$, $(-5, 7, 5)$, $(5, 7, -5)$, and $(5, 7, 5)$, oriented so that the positive direction is in the direction of the positive y-axis.

11. $\mathbf{V}(x, y, z) = (-y, x, z)$, and M is the right circular cylinder (with no top or bottom), centered around the z-axis, of radius 5, between $z = -7$ and $z = 7$, oriented outward. ▶

12. $\mathbf{V}(x, y, z) = (3, 5, 7)$, and M is the right circular half-cone, whose vertex is at the origin, which is centered around the positive z-axis, which has height $H = 12$ and radius $R = 6$, and is almost-oriented "up", i.e., so that the positive unit normal vectors at all points other than the vertex have a positive z component.

13. $\mathbf{V}(x, y, z) = (0, 0, 1)$, and M is the bottom hemisphere (including the bounding circle) of the sphere of radius 3, centered at the origin, oriented downward.

14. $\mathbf{V}(x, y, z) = (0, 0, -z)$, and M is the bottom hemisphere (including the bounding circle) of the sphere of radius 3, centered at the origin, oriented downward.

In each of the following exercises, you are given a force field $\mathbf{F} = \mathbf{F}(x, y, z)$, in Newtons, where x, y, and z are in meters. You are also given a closed surface M. (a) Calculate the flux of $\mathbf{F}$ through M, in N-m^2, where you use the default outward-pointing almost-orientation for M. (b) Decide whether there is a source, a sink, or neither inside the closed surface M.

15. $\mathbf{F}(x, y, z) = (x, y, z)$, and M is the cube, centered at the origin, whose faces are parallel to the coordinate planes, where each side has length 2.

16. $\mathbf{F}(x, y, z) = (z, x, y)$, and M is the cube, centered at the origin, whose faces are parallel to the coordinate planes, where each side has length 2.

17. $\mathbf{F}(x, y, z) = (-y, x, z)$, and M is the right circular cylinder, with a top and bottom, centered around the z-axis, of radius 5, between $z = -7$ and $z = 7$, oriented outward.

18. $\mathbf{F}(x, y, z) = (0, 0, -z)$, and M is the bottom hemisphere of the sphere of radius 3, centered at the origin, with a disk on the top, closing the surface.

19. $\mathbf{F}(x, y, z) = (0, 0, -1)$, and M is the bottom hemisphere of the sphere of radius 3, centered at the origin, with a disk on the top, closing the surface.

20. $\mathbf{F}(x, y, z) = (2y, xz, -x)$, and M consists of the portion of the graph of $z = 4 - x^2 - y^2$, where $z \geq 0$, together with the closing disk in the xy-plane.

More Depth:

21. In Example 4.5.4, we used the parameterization

$$\mathbf{p}(\theta, z) = z\left(\frac{R}{H}\cos\theta, \ \frac{R}{H}\sin\theta, \ 1\right) = z\left(\frac{1}{2}\cos\theta, \ \frac{1}{2}\sin\theta, \ 1\right),$$

where $0 \leq \theta \leq 2\pi$ and $0 \leq z \leq 4$, for the half-cone M. Why does this parameterization yield an almost-orientation, but not an orientation; that is, why don't we get a continuous choice of unit normal vectors **everywhere** by letting

$$\mathbf{n} = \frac{\mathbf{p}_\theta \times \mathbf{p}_z}{|\mathbf{p}_\theta \times \mathbf{p}_z|}?$$

22. Is the half-cone from the previous exercise orientable? Explain, intuitively.

23. Describe, intuitively, what a submanifold, or submanifold with boundary, of $\mathbb{R}^3$ is.

24. Describe, intuitively, what an orientation of a submanifold, or submanifold with boundary, of $\mathbb{R}^3$ is.

25. Explain what part (d) of Exercise 23 from Section 3.11 has to do with producing an almost-orientation and/or an orientation for the Möbius Strip.

26. Explain why the result from part (d) of Exercise 23 from Section 3.11 does **not** prove that the Möbius Strip is non-orientable (even though it is, in fact, non-orientable).

4.6 The Divergence Theorem

Green's Theorem equates the line integral of a vector field along a closed curve, a 1-dimensional geometric object, to the integral of a quantity involving the partial derivatives the components of the vector field, integrated over the 2-dimensional region bounded by the curve.

The *Divergence Theorem* has this same general form, but in one higher dimension on both sides. The Divergence Theorem equates the flux integral of a vector field over a closed surface to the integral of a quantity involving the partial derivatives of the components of the vector field, integrated over the 3-dimensional region bounded by the surface.

The Divergence Theorem is vital for deriving a number of important physical laws such as *Gauss' Law* for electric flux, and *Archimedes' Principle* in hydrostatics.

Basics:

Throughout this section, we need to deal with compact, connected, piecewise-regular closed surfaces M in $\mathbb{R}^3$ (recall Definition 4.5.10), which by Theorem 4.5.11, are orientable and bound a connected solid region; for such surfaces M, we declare the positive direction $\mathbf{n}$ at regular points to be outward from the bounded solid region.

It is very cumbersome to write "compact, connected, piecewise-regular surfaces in $\mathbb{R}^3$" over and over again, so we'll establish some more concise terminology, before stating the Divergence Theorem.

> **Definition 4.6.1.** *A* **ccpr-surface** *M is a compact, connected, piecewise-regular 2-dimensional topological submanifold, without boundary, in $\mathbb{R}^3$, and so M is a closed surface which bounds a connected solid region.*
> *If T is the connected solid region bounded by M, then we write $M = \partial T$.*

Recall now that the divergence of a vector field $\mathbf{F}(x, y, z) = (P(x, y, z), Q(x, y, z), R(x, y, z))$ is

$$\operatorname{div} \mathbf{F} = \vec{\nabla} \cdot \mathbf{F} = \frac{\partial P}{\partial x} + \frac{\partial Q}{\partial y} + \frac{\partial R}{\partial z}.$$

The fundamental result in this section is:

We shall give the idea of the proof in the More Depth portion. For a rigorous proof, see [3], section 8.4.

Theorem 4.6.2. (Divergence Theorem, Version 1) *Suppose that M is a ccpr-surface with its positive, outward orientation. Let T be the compact solid region bounded by M, so that $M = \partial T$. Let $\mathbf{F}$ be a continuously differentiable vector field on an open set in $\mathbb{R}^3$ which contains T.*
Then,

$$\iint_{\partial T} \mathbf{F} \cdot \mathbf{n}\, dS \;=\; \iiint_{T} (\vec{\nabla} \cdot \mathbf{F})\, dV.$$

Example 4.6.3. Let M be any ccpr-surface in $\mathbb{R}^3$, and let $\mathbf{F}$ be any constant vector field on R^3.

Then, the divergence of $\mathbf{F}$ is certainly 0, and so the Divergence Theorem tells us that the flux of $\mathbf{F}$ through M is zero, i.e.,

$$\iint_{M} \mathbf{F} \cdot \mathbf{n}\, dS \;=\; 0.$$

We saw an example of this in Example 4.5.6.

Example 4.6.4. Let T be the region where $z \geq 0$, and $x^2 + y^2 + z^2 \leq 16$, i.e., a solid half-ball of radius 4, where all distances are measured in meters. So, ∂T consists of the top hemisphere, H, together with a bottom disk, D, of radius 4. Give ∂T its positive, outward orientation. Suppose that a fluid is flowing with a velocity vector field given by

$$\mathbf{V} \;=\; (3x + ye^z,\, 5y - x\tan^{-1} z,\, 7z - 4) \;\; \text{m/s}.$$

Find the volume of fluid flowing through ∂T per second, i.e., find the flux of $\mathbf{V}$ through ∂T.

Solution:

We **could** parameterize H and D separately, and calculate the flux from the definition. This would be fairly painful.

Instead, we can use the Divergence Theorem. We calculate the divergence of $\mathbf{V}$, which will have units of $(\text{seconds})^{-1}$:

$$\vec{\nabla} \cdot \mathbf{V} \;=\; \frac{\partial}{\partial x}\left(3x + ye^z\right) \;+\; \frac{\partial}{\partial y}\left(5y - x\tan^{-1} z\right) \;+\; \frac{\partial}{\partial z}\left(7z - 4\right) \;=\; 3 + 5 + 7 \;=\; 15.$$

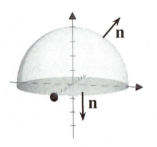

Figure 4.6.1: The half-ball T, and its outward oriented boundary.

Therefore, the Divergence Theorem tells us that the rate at which fluid flows through M is

$$\iint_{\partial T} \mathbf{V} \cdot \mathbf{n} \, dS \;=\; \iiint_T (\overset{\rightarrow}{\nabla} \cdot \mathbf{V}) \, dV \;=\; \iiint_T 15 \, dV \;=$$

$$15 \cdot \frac{1}{2} \left(\frac{4\pi}{3} \cdot 4^3 \right) \;=\; 640\pi \ \text{m}^3/\text{s}.$$

Example 4.6.5. Can we use the Divergence Theorem to calculate flux through a **non-closed** surface? Yes – we just have to be clever; we close the surface, but then subtract the flux that we threw in by closing the surface. Of course, this could be a difficult process, but there are examples where it's fairly simple.

For instance, let's return to Example 4.6.4, in which we had the vector field

$$\mathbf{V} \;=\; (3x + ye^z, \; 5y - x \tan^{-1} z, \; 7z - 4),$$

which has the constant 15 for its divergence.

How can we use the Divergence Theorem to find the flux of $\mathbf{V}$ through just the top hemisphere H in Example 4.6.4, i.e., through H given by $z \geq 0$ and $x^2 + y^2 + z^2 = 16$, oriented upward?

The surface H is a ccpr-surface with boundary, but it's not closed, and so doesn't bound a solid region. However, if we let D be the downward-oriented disk where $z = 0$ and $x^2 + y^2 \leq 16$, as in Example 4.6.4, then the union $M = H \cup D$, of H and D, **is** a ccpr-surface without boundary, which bounds the solid region T where $z \geq 0$, and $x^2 + y^2 + z^2 \leq 16$. Now, the flux is additive on pieces of the surface which intersect along curves, i.e.,

$$\iint_M \mathbf{V} \cdot \mathbf{n} \, dS \;=\; \iint_{H \cup D} \mathbf{V} \cdot \mathbf{n} \, dS \;=\; \iint_H \mathbf{V} \cdot \mathbf{n} \, dS \;+\; \iint_D \mathbf{V} \cdot \mathbf{n} \, dS.$$

Therefore, since we know from our earlier use of the Divergence Theorem in Example 4.6.4 that $\iint_M \mathbf{V} \cdot \mathbf{n} \, dS = 640\pi$, we conclude that

$$\iint_H \mathbf{V} \cdot \mathbf{n} \, dS \;=\; 640\pi \;-\; \iint_D \mathbf{V} \cdot \mathbf{n} \, dS.$$

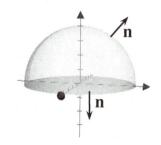

Figure 4.6.2: The closed surface M with its outward orientation.

Of course, for this to be a helpful approach to calculating $\iint_H \mathbf{V} \cdot \mathbf{n} \, dS$, we need for the flux integral $\iint_D \mathbf{V} \cdot \mathbf{n} \, dS$ to be easier to calculate from the definition than $\iint_H \mathbf{V} \cdot \mathbf{n} \, dS$; in our example, this is certainly the case.

The surface D has a constant outward-pointing unit normal $\mathbf{n}$, namely $-\mathbf{k} = (0, 0, -1)$, and is contained in the plane where $z = 0$. The element of surface area dS is simply our flat area element dA. Thus,

$$\iint_D \mathbf{V} \cdot \mathbf{n} \, dS \;=\; \iint_D (\ast, \ast, -4) \cdot (0, 0, -1) \, dA,$$

where we have written $*$'s in the first two components of the vector field since they will be multiplied by 0 in the dot product, so we don't care what they are. The third component of the vector field is just -4, not $7z - 4$, because $z = 0$ on the surface D.

Thus, we find

$$\iint_D \mathbf{V} \cdot \mathbf{n}\, dS \;=\; 4 \cdot \pi(4)^2 \;=\; 64\pi \ \ \mathrm{m}^3/\mathrm{s},$$

and, finally,

$$\iint_H \mathbf{V} \cdot \mathbf{n}\, dS \;=\; 640\pi \;-\; \iint_D \mathbf{V} \cdot \mathbf{n}\, dS \;=\; 640\pi \;-\; 64\pi \;=\; 576\pi \ \ \mathrm{m}^3/\mathrm{s}.$$

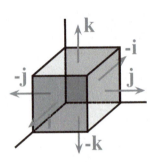

Figure 4.6.3: The closed surface M with its outward orientation.

Example 4.6.6. Consider the vector field

$$\mathbf{F} \;=\; (x^2 y z,\, 2x y^2 z,\, 3x y z^2).$$

Find the flux of $\mathbf{F}$ through the outward-oriented faces of the cube, four of whose vertices are at $(0,0,0)$, $(2,0,0)$, $(0,2,0)$, and $(0,0,2)$. Do this in two different ways: first, by using the definition of the flux and, second, by using the Divergence Theorem.

Solution:

To calculate using the definition, we will have to calculate 6 different flux integrals, one for each of the six sides of the cube. However, as each side is flat, we have $dS = dA$, and the unit normals $\mathbf{n}$ will all be $\pm\mathbf{i}$, $\pm\mathbf{j}$, or $\pm\mathbf{k}$. Also, one of the coordinates will be constant on each face. We will write $*$'s in components of the vector field if they are non-zero, but get multiplied by zero in the dot product. We will denote the faces by M_i, where $1 \le i \le 6$.

On M_1, where $x = 0$, $0 \le y \le 2$, $0 \le z \le 2$:

$$\iint_{M_1} \mathbf{F} \cdot \mathbf{n}\, dS \;=\; \int_0^2 \int_0^2 (0,0,0) \cdot (-\mathbf{i})\, dy\, dz \;=\; 0.$$

On M_2, where $x = 2$, $0 \le y \le 2$, $0 \le z \le 2$:

$$\iint_{M_2} \mathbf{F} \cdot \mathbf{n}\, dS \;=\; \int_0^2 \int_0^2 (4yz, *, *) \cdot \mathbf{i}\, dy\, dz \;=\; \int_0^2 \int_0^2 4yz\, dy\, dz \;=\; 16.$$

On M_3, where $y = 0$, $0 \leq x \leq 2$, $0 \leq z \leq 2$:

$$\iint_{M_3} \mathbf{F} \cdot \mathbf{n}\, dS \;=\; \int_0^2 \int_0^2 (0,0,0) \cdot (-\mathbf{j})\, dx\, dz \;=\; 0.$$

On M_4, where $y = 2$, $0 \leq x \leq 2$, $0 \leq z \leq 2$:

$$\iint_{M_4} \mathbf{F} \cdot \mathbf{n}\, dS \;=\; \int_0^2 \int_0^2 (*, 8xz, *) \cdot j\, dx\, dz \;=\; \int_0^2 \int_0^2 8xz\, dx\, dz \;=\; 32.$$

On M_5, where $z = 0$, $0 \leq x \leq 2$, $0 \leq y \leq 2$:

$$\iint_{M_5} \mathbf{F} \cdot \mathbf{n}\, dS \;=\; \int_0^2 \int_0^2 (0,0,0) \cdot (-\mathbf{k})\, dy\, dx \;=\; 0.$$

On M_6, where $z = 2$, $0 \leq x \leq 2$, $0 \leq y \leq 2$:

$$\iint_{M_6} \mathbf{F} \cdot \mathbf{n}\, dS \;=\; \int_0^2 \int_0^2 (*, *, 12xy) \cdot \mathbf{k}\, dy\, dx \;=\; \int_0^2 \int_0^2 12xy\, dy\, dx \;=\; 48.$$

Therefore, the flux over the entire surface M of the cube is the sum of our previous 6 calculations, i.e.,

$$\iint_M \mathbf{F} \cdot \mathbf{n}\, dS \;=\; 16 + 32 + 48 \;=\; 96.$$

Now, Let T denote the solid cube, so that ∂T is our surface M from above. We want to use the Divergence Theorem to calculate the value of $\iint_{\partial T} \mathbf{F} \cdot \mathbf{n}\, dS$ and see that we again get 96.

We calculate the divergence of $\mathbf{F} = (x^2yz, 2xy^2z, 3xyz^2)$:

$$\vec{\nabla} \cdot \mathbf{F} \;=\; \frac{\partial}{\partial x}\left(x^2yz\right) + \frac{\partial}{\partial y}\left(2xy^2z\right) + \frac{\partial}{\partial z}\left(3xyz^2\right) \;=\; 2xyz + 4xyz + 6xyz = 12xyz.$$

Therefore, the Divergence Theorem tells us that

$$\iint_{\partial T} \mathbf{F} \cdot \mathbf{n}\, dS \;=\; \iiint_T \left(\vec{\nabla} \cdot \mathbf{F}\right) dV \;=\; \int_0^2 \int_0^2 \int_0^2 12xyz\, dz\, dy\, dx \;=$$

$$\int_0^2 \int_0^2 6xyz^2 \left.\right|_{z=0}^{z=2} dy\,dx \;=\; \int_0^2 \int_0^2 24xy\,dy\,dx \;=\; \int_0^2 12xy^2 \left.\right|_{y=0}^{y=2} dx \;=$$

$$\int_0^2 48x\,dx \;=\; 24x^2 \left.\right|_0^2 \;=\; 96,$$

which, of course, is what we knew we'd get.

More Depth:

When we looked at Green's Theorem, back in Section 4.4, we first had a version of theorem in which the plane region had a single boundary component. We then gave a more general version in which the boundary could consist of a finite number of separate closed curves.

We have an analogous generalization of the Divergence Theorem:

Theorem 4.6.7. (Divergence Theorem, Version 2) *Suppose that T is a non-empty compact, connected region in $\mathbb{R}^3$, which is the closure of its interior, such that the boundary ∂T of T consists of a finite number of non-intersecting ccpr-surfaces $M_1, \ldots, M_k$, all oriented outward from T. Let $\mathbf{F}$ be a continuously-differentiable vector field on an open set in $\mathbb{R}^3$ which contains T.*

Then,

$$\iint_{\partial T} \mathbf{F} \cdot \mathbf{n}\,dS \;=\; \sum_{i=1}^k \iint_{M_i} \mathbf{F} \cdot \mathbf{n}\,dS \;=\; \iiint_T (\vec{\nabla} \cdot \mathbf{F})\,dV.$$

Gauss' Law

We can use our new version of the Divergence Theorem to prove *Gauss' Law* for electric flux.

Let $\mathbf{r} = \mathbf{r}(x, y, z) = (x, y, z)$. Suppose that we have point-charges of charge $q_1, \ldots, q_n$, located at points $\mathbf{p}_1, \ldots, \mathbf{p}_n$ in $\mathbb{R}^3$. As we discussed in Example 4.5.5, individually, each charge q_i would produce an electric field $\mathbf{F}_i$ on a unit charge in space, given by

$$\mathbf{F}_i \;=\; \frac{1}{4\pi\epsilon_0} \frac{q_i}{|\mathbf{r} - \mathbf{p}_i|^3} (\mathbf{r} - \mathbf{p}_i),$$

where ϵ_0 is the *electric constant*, and where we have centered the field at $\mathbf{p}_i$.

We also showed that the flux of $\mathbf{F}_i$ through a sphere of arbitrary positive radius, centered at $\mathbf{p}_i$ and oriented outward, is q_i/ϵ_0. In addition, recall that in Example 4.1.19, we showed that the divergence of each of these $\mathbf{F}_i$'s is zero, i.e., $\vec{\nabla} \cdot \mathbf{F}_i = 0$.

Now, suppose that M that is a ccpr-surface in $\mathbb{R}^3$, and that the points $\mathbf{p}_1$, ..., $\mathbf{p}_k$ are in the interior of the solid region enclosed by M, while $\mathbf{p}_{k+1}$, ..., $\mathbf{p}_n$ are in the unbounded region outside of M. Let Q be the sum of the electric charges inside of M, i.e., let $Q = \sum_{i=1}^{k} q_i$, and let $\mathbf{F}$ denote the total electric field from all of the point-charges, i.e., let

$$\mathbf{F} = \sum_{i=1}^{n} \mathbf{F}_i.$$

We want to calculate the flux integral $\iint_M \mathbf{F} \cdot \mathbf{n} \, dS$; we will do this via Theorem 4.6.7.

For each point $\mathbf{p}_i$ contained inside of M, let S_i denote a sphere, centered at $\mathbf{p}_i$, of small enough radius so that S_i is contained inside M and so that S_i contains none of the point-charges other than q_i. Let T denote the solid region which is trapped between M and the S_i, for $1 \leq i \leq k$. Thus,

$$\partial T = M \cup S_1 \cup \cdots \cup S_k,$$

where we orient each piece of the boundary outward from T; note that this means that the orientation on each of the spheres S_i is **into** the sphere.

As there are no point-charges located in T, and since $\vec{\nabla} \cdot \mathbf{F}_i = 0$ at all points where $\mathbf{F}_i$ is defined, we find $\vec{\nabla} \cdot \mathbf{F} = \sum_{i=1}^{n} \vec{\nabla} \cdot \mathbf{F}_i = 0$. Therefore, Theorem 4.6.7 tells us that

$$\iint_{\partial T} \mathbf{F} \cdot \mathbf{n} \, dS = \iiint_T 0 \, dV = 0.$$

Now, we have

$$0 = \iint_{\partial T} \mathbf{F} \cdot \mathbf{n} \, dS = \iint_M \mathbf{F} \cdot \mathbf{n} \, dS + \sum_{i=1}^{k} \iint_{S_i} \mathbf{F} \cdot \mathbf{n} \, dS.$$

Hence,

$$\iint_M \mathbf{F} \cdot \mathbf{n} \, dS = -\sum_{i=1}^{k} \iint_{S_i} \mathbf{F} \cdot \mathbf{n} \, dS = \sum_{i=1}^{k} \sum_{j=1}^{n} \iint_{-S_i} \mathbf{F}_j \cdot \mathbf{n} \, dS,$$

where $-S_i$ is the sphere S_i with its orientation reversed, i.e., oriented outward from the center of the sphere. Now, if $j \neq i$, then $\iint_{-S_i} \mathbf{F}_j \cdot \mathbf{n} \, dS = 0$, since S_i is closed surface which bounds a solid region on which $\vec{\nabla} \cdot \mathbf{F}_j = 0$.

Thus,

$$\iint_M \mathbf{F} \cdot \mathbf{n} \, dS = \sum_{i=1}^{k} \iint_{-S_i} \mathbf{F}_i \cdot \mathbf{n} \, dS = \sum_{i=1}^{k} (q_i/\epsilon_0) = Q/\epsilon_0.$$

Thus, we have shown:

Theorem 4.6.8. (Gauss' Law) *The flux of* **F** *through* M *is equal to* Q/ϵ_0; *thus, the electric flux through a ccpr-surface is proportional to the total charge enclosed by the surface.*

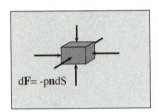

Figure 4.6.4: The force produced by pressure, acting on a submerged object.

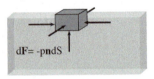

Figure 4.6.5: The force produced by pressure, acting on a partially submerged object.

Archimedes' Principle

We wish to study the buoyancy force, the force upward, exerted by water on an object which is submerged, or partially submerged, in the water.

We will let ρ denote the density of water, which we assume is constant (because we assume that water is incompressible). We let g denote the magnitude of the acceleration produced by gravity, which we also assume is constant (because we will deal with relatively small distances).

We let z denote the vertical coordinate (as usual), with $z = 0$, corresponding to the surface of the water, and positive z being upward. When dealing with a partially submerged object, we assume that air pressure produces a negligible force on the object.

In the water, $z \leq 0$, and the (hydrostatic) pressure p at a point (x, y, z) is a scalar quantity, given by $p = -\rho g z$, with units of force per area. However, this scalar pressure function produces a force on any submerged surface; that force is perpendicular to the surface, points into the surface, and on an infinitesimal area dS is given by

$$d\mathbf{F} = -p\,\mathbf{n}\,dS = \rho g z\,\mathbf{n}\,dS,$$

where **n** is the outward-pointing normal to the surface, and the minus sign is present because the force is directed inward.

We would like to integrate this force due to hydrostatic pressure over the entire surface of a submerged object, in order to obtain the buoyancy force (the net force from the water) which acts on the object.

For this, we need a version of the Divergence Theorem, which applies to scalar functions, instead of vector fields, and in which the integrals produce vector fields, instead of scalars. In fact, it is a little misleading to call this a "divergence theorem", since the divergence of a vector field never appears; nonetheless, the proof of this theorem proceeds exactly like that of the Divergence Theorem, and so the name is used to reflect that fact.

Theorem 4.6.9. (Divergence Theorem, Vector Version) *Suppose that T is a non-empty compact, connected region in $\mathbb{R}^3$, which is the closure of its interior, such that the boundary ∂T of T consists of a finite number of non-intersecting ccpr-surfaces $M_1, \ldots, M_k$, all oriented outward from T. Let f be a real-valued continuously-differentiable function on an open set in $\mathbb{R}^3$ which contains T.*

Then,

$$\iint_{\partial T} f\,\mathbf{n}\,dS \;=\; \sum_{i=1}^{k} \iint_{M_i} f\,\mathbf{n}\,dS \;=\; \iiint_T \vec{\nabla} f\, dV.$$

Suppose that we apply this theorem to our hydrostatic pressure function p, and to a solid object T, whose entire boundary surface is either in contact with the water, or is at, or above, the water level (so that the pressure function can be extended to 0 on the part of the surface that's not in contact with the water).

Then, we have $d\mathbf{F} = \rho g z \mathbf{n}\, dS$ and, applying Theorem 4.6.9, we find

$$\text{buoyancy force} \;=\; \iint_{\partial T} d\mathbf{F} \;=\; \iint_{\partial T} \rho g z \mathbf{n}\, dS \;=\; \rho g \iiint_T \vec{\nabla} z\, dV \;=\;$$

$$\rho g \iiint_T (0,\, 0,\, 1)\, dV \;=\; \rho g V \mathbf{k} \;=\; w\mathbf{k},$$

where V is the volume of T which is below water level, and w is the weight of the water that would occupy the volume V.

This is usually stated simply as:

Theorem 4.6.10. (Archimedes' Principle) *The buoyancy force on a submerged object acts straight up, with a magnitude equal to the weight of the displaced water.*

We wish to give the idea of the proof of the Divergence Theorem, in any of its forms. As with Green's Theorem, the idea is to chop the region (now, a solid region) into rectangular pieces (now, rectangular solids), understand why the theorem is true on these special regions, and then use that the orientations make boundary contributions cancel out when you put your special rectangular pieces together.

Thus, our real problem boils down to: why is the Divergence Theorem true for rectangular solids?

Suppose that we take the rectangular solid T where $0 \leq x \leq a$, $0 \leq y \leq b$, and $0 \leq z \leq c$ (a, b, and c are all positive). Let $\mathbf{F} = (P, Q, R)$ be a C^1 vector field on an open set containing T.

Archimedes supposedly discovered/realized this principle as he was getting in a bath tub. The story has it that he was so excited that he ran naked through the streets, shouting "eureka!" ("I have found it!")

This simple phrasing sometimes makes people yell "Eureka - I've found a counterexample to Archimedes' Principle!" They consider an object with a flat bottom, which is lying flush against the flat bottom of a container of water; there'd be no upward force. The unstated assumption in Archimedes' Principle is that the entire boundary surface is either in contact with the water, or is at, or above, the water level.

Let M_1 denote the face where $x = 0$, $0 \leq y \leq b$, and $0 \leq z \leq c$, and let M_2 denote the face where $x = a$, $0 \leq y \leq b$, and $0 \leq z \leq c$. Note that the outward-pointing unit normal vector to M_1 is $-\mathbf{i} = (-1, 0, 0)$, and the outward-pointing unit normal vector to M_2 is $\mathbf{i} = (1, 0, 0)$. Also, note that, on M_1 and M_2, $dS = dA = dz\,dy$.

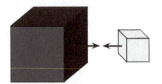

Figure 4.6.6: Canceling flux contributions from joining two rectangular solids.

Thus,

$$\iint_{M_1 \cup M_2} \mathbf{F} \cdot \mathbf{n}\, dS = \iint_{M_1} \mathbf{F} \cdot \mathbf{n}\, dS + \iint_{M_2} \mathbf{F} \cdot \mathbf{n}\, dS =$$

$$\int_0^b \int_0^c (P(0, y, z), *, *) \cdot (-1, 0, 0)\, dz\,dy \; + \; \int_0^b \int_0^c (P(a, y, z), *, *) \cdot (1, 0, 0)\, dz\,dy,$$

where, as usual, we have written $*$'s in vector field positions that will be multiplied by zero in the dot product.

Thus,

$$\iint_{M_1 \cup M_2} \mathbf{F} \cdot \mathbf{n}\, dS = \int_0^b \int_0^c \big(P(a, y, z) - P(0, y, z) \big)\, dz\,dy =$$

$$\int_0^b \int_0^c \left(\int_0^a \frac{\partial P}{\partial x}\, dx \right) dy\,dz = \iiint_T \frac{\partial P}{\partial x}\, dV.$$

By the symmetry of our set-up, if we calculate the flux integral over the two faces with fixed y values, we'll get $\iiint_T \frac{\partial Q}{\partial y}\, dV$, and if we calculate over the two faces with fixed z values, we'll get $\iiint_T \frac{\partial R}{\partial z}\, dV$.

Therefore,

$$\iint_{\partial T} \mathbf{F} \cdot \mathbf{n}\, dS = \iiint_T \left(\frac{\partial P}{\partial x} + \frac{\partial Q}{\partial y} + \frac{\partial R}{\partial z} \right) dV = \iiint_T \left(\vec{\nabla} \cdot \mathbf{F} \right) dV,$$

i.e., the Divergence Theorem holds on rectangular solids.

Finally, we should comment on what the Divergence Theorem, or the sketch of its proof, says about how you should think about the divergence of a vector field.

Suppose that we have a C^1 vector field on a region in $\mathbb{R}^3$ and we take a point $\mathbf{p}$ in our region. Is there an intuitive way of thinking about the divergence of $\mathbf{F}$ at $\mathbf{p}$? Yes.

Take a solid cube, C, centered at $\mathbf{p}$, of side-length Δ, where Δ is close to zero. Then,

$$\iint_{\partial C} \mathbf{F} \cdot \mathbf{n}\, dS = \iiint_C \left(\vec{\nabla} \cdot \mathbf{F} \right) dV \approx \left(\vec{\nabla} \cdot \mathbf{F} \right)\Big|_{\mathbf{p}} \Delta^3,$$

and so,

$$\left(\vec{\nabla} \cdot \mathbf{F} \right)\Big|_{\mathbf{p}} \approx \frac{\iint_{\partial C} \mathbf{F} \cdot \mathbf{n}\, dS}{\Delta^3} = \frac{\text{flux through the cube}}{\text{volume of the cube}}.$$

When you take the limit, you obtain equality:

$$\left(\vec{\nabla} \cdot \mathbf{F}\right)\Big|_{\mathbf{p}} = \lim_{\Delta \to 0} \frac{\text{flux through the cube}}{\text{volume of the cube}}.$$

In this sense, the divergence is the instantaneous flux per volume, at **p**, and so can be thought of as the *flux density* of **F** at **p**.

4.6.1 Exercises

Online answers to select exercises are here.

Basics:

In each of the following exercises, you are given a force field $\mathbf{F} = \mathbf{F}(x, y, z)$, in Newtons, where x, y, and z are in meters. You are also given a closed surface M. Use the Divergence Theorem to calculate the flux of **F** through M, in N-m^2, where you use the default outward-pointing almost-orientation for M.

1. $\mathbf{F}(x, y, z) = (x, y, z)$, and M is the cube, centered at the origin, where each side has length 2. ▶

2. $\mathbf{F}(x, y, z) = (z, x, y)$, and M is the cube, centered at the origin, where each side has length 2.

3. $\mathbf{F}(x, y, z) = (-y, x, z)$, and M is the right circular cylinder, with a top and bottom, centered around the z-axis, of radius 5, between $z = -7$ and $z = 7$.

4. $\mathbf{F}(x, y, z) = (0, 0, -z)$, and M is the bottom hemisphere of the sphere of radius 3, centered at the origin, with a disk on the top, closing the surface.

5. $\mathbf{F}(x, y, z) = (0, 0, -1)$, and M is the bottom hemisphere of the sphere of radius 3, centered at the origin, with a disk on the top, closing the surface.

6. $\mathbf{F}(x, y, z) = (2y, xz, -x)$, and M consists of the portion of the graph of $z = 4 - x^2 - y^2$, where $z \geq 0$, together with the closing disk in the xy-plane.

7. $\mathbf{F}(x, y, z) = (xy, yz, zx)$, and M is the cube which bounds the solid region where $0 \leq x \leq 3$, $0 \leq y \leq 3$, and $0 \leq z \leq 3$.

8. $\mathbf{F}(x, y, z) = (xy + e^z, yz + \sin x, zx + \tan^{-1} y)$, and M is the cube which bounds the solid region where $0 \leq x \leq 3$, $0 \leq y \leq 3$, and $0 \leq z \leq 3$. ▶

In each of the following exercises, you are given a velocity vector field $\mathbf{V} = \mathbf{V}(x, y, z)$, in m/s, of a flowing fluid, where x, y and z are in meters. You are also given an oriented non-closed surface M_1. (a) Find a convenient closing surface, i.e., an oriented surface M_2 so that $M = M_1 \cup M_2$ is a ccpr-surface with the standard outward-pointing orientation. **(b) Calculate the flux of V through M_2, using the definition of flux. (c) Calculate the flux of V through M, using the Divergence Theorem. (d) Combine parts (b) and (c) to calculate the flux of V through M_1.**

9. $\mathbf{V}(x,y,z) = (x,y,z)$, and M_1 is the surface of a cube, with no top, centered at the origin, where each edge has length 2, each face/side is parallel to a coordinate plane, and which is oriented outward from the filled-in cube.

10. $\mathbf{V}(x,y,z) = (2y, xz, -x)$, and M_1 consists of the portion of the graph of $z = 4 - x^2 - y^2$, where $z \geq 0$, oriented upward.

11. $\mathbf{V}(x,y,z) = (-y, x, z)$, and M_1 is the right circular cylinder, centered around the z-axis, of radius 5, between $z = -7$ and $z = 7$, oriented outward.   ▶

12. $\mathbf{V}(x,y,z) = (0, 0, -z)$, and M_1 is the bottom hemisphere of the sphere of radius 3, centered at the origin, oriented downward.

13. $\mathbf{V}(x,y,z) = (xy, yz, zx)$, and M_1 is the boundary of the solid tetrahedron in the first octant, below the plane where $x + y + z = 6$, except the base triangle in the xy-plane is missing. M_1 is oriented outward from the solid tetrahedron.

14. $\mathbf{V}(x,y,z) = (3, 5, 7)$, and M_1 is the right circular half-cone, whose vertex is at the origin, which is centered around the positive z-axis, which has height $H = 12$ and radius $R = 6$, and is oriented down.

More Depth:

In each of the following exercises, you are given a vector field F and a solid region T, such that the boundary ∂T consists of a finite number of non-intersecting ccpr-surfaces $M_1, \ldots, M_k$. Orient all pieces of ∂T outward from T, and calculate the flux $\iint_{\partial T} \mathbf{F} \cdot \mathbf{n}\, dS$ of F through ∂T.

15. $\mathbf{F}(x,y,z) = (-5x, 3y, 4z)$, and T is the solid region between the two spheres, centered at the origin, of radius 2 and radius 7.   ▶

16. $\mathbf{F} = (3y, 4z, -5x)$, and T is the solid region between the sphere, of radius 7, centered at the origin, and the ellipsoid where $2x^2 + 3y^2 + 5z^2 = 1$.

17. $\mathbf{F} = \left(1, 1, \dfrac{z}{\sqrt{x^2 + y^2}}\right)$, and T is the solid region between the two cylinders where $x^2 + y^2 = 9$ and $x^2 + y^2 = 25$, and $0 \leq z \leq 7$.

18. $\mathbf{F} = \left(1, 1, \dfrac{z}{\sqrt{x^2 + y^2 + z^2}}\right)$, and T is the solid region between the two spheres, centered at the origin, of radius 2 and radius 7.

19. Using our derivation of Gauss' Law, Theorem 4.6.8, as a reference, state and derive an analogous result for the gravitational flux from a collection of point-masses.

20. (**Dusseault's Problem**) It can be difficult to think of how to apply Archimedes' Principle, Theorem 4.6.10, in situations where part of the object is blocked from the water. In such cases, you may be able to calculate the net buoyancy force by integrating, and using the vector version of the Divergence Theorem, Theorem 4.6.9.

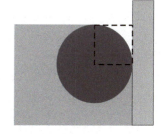

Figure 4.6.7: Side-view of a partially submerged log, wedged against a wall.

Consider a perfectly cylindrical log, of radius R and length L, of uniform density, which is wedged up against a wall by water, where the water rises just to the top of the log. See Figure 4.6.7. We assume that the log is stationary.

As before, we assume that air pressure is negligible, so that there are three forces acting on the log: the force $\mathbf{F}_w$ from the water pressure, the force of gravity, $\mathbf{F}_g$, and the normal force $\mathbf{F}_n$ from the wall. Gravity acts straight down. The normal force from the wall acts perpendicular to the wall. As we are assuming that the log is not moving, we must have $\mathbf{F}_w + \mathbf{F}_g + \mathbf{F}_n = \mathbf{0}$.

Set up a coordinate system so that Figure 4.6.7 is in the yz-plane, so that the x-axis runs along the center of the log, and $z = 0$ is at the top of the water, which is also at the top of the log. Let $\mathbf{F}_w = F_{wx}\mathbf{i} + F_{wy}\mathbf{j} + F_{wz}\mathbf{k}$.

(a) Without performing any calculations, explain why $F_{wx} = 0$.

(b) What is the relationship between $\mathbf{F}_n$ and $F_{wy}\mathbf{j}$?

(c) What is the relationship between $\mathbf{F}_g$ and $F_{wz}\mathbf{k}$?

(d) Consider the "solid" curved wedge of air, T, between the log and the wall, and below where $z = 0$. What is the volume of T? (Hint: the dotted square is in the figure in order to help you with this.)

(e) If T were full of water, what would Archimedes' Principle give you for the total buoyancy force on the log?

(f) Now, in order to produce the actual $\mathbf{F}_w$, subtract from your answer to part (e) the "missing" force $\mathbf{F}_C$ from water pressure on the curved side C (along the log) of T. Calculate $\mathbf{F}_C$ by considering the appropriate surface integral over the entire boundary of T, using the vector version of the Divergence Theorem, and calculating and eliminating the contributions from all of portions, other than C, of the surface of T. Note that these other portions consist of the flat ends, parallel to the yz-plane, a top rectangle where $z = 0$, and a side rectangle, parallel to the xz-plane, along the wall.

(g) Let ρ denote the density of water, and let $\rho_{\log}$ denote the density of the log. Show that $\rho_{\log} > \rho$.

(h) Can you think of a way to produce $F_{wz}\mathbf{k}$ by applying Archimedes' Principle in a clever way, so that you can avoid most of the above steps?

4.7   Stokes' Theorem

Stokes' Theorem is frequently referred to as *Green's Theorem in space,* for it tells us that the line integral of a vector field around a closed curve in space is equal to the integral of the curl of the vector field over a surface whose boundary is the given curve.

Basics:

Stokes' Theorem is a generalization of Green's Theorem to the case where the curve and the surface are in $\mathbb{R}^3$, instead of in $\mathbb{R}^2$.

In the previous section, we defined ccpr-surfaces, that is, compact, connected, piecewise-regular surfaces. In this section, we need *ccpr-surfaces with boundary.* Recall the definition of a topological submanifold with boundary from Definition 4.5.8.

> **Definition 4.7.1.** *A **ccpr-surface M with boundary** is a compact, connected, piecewise-regular, 2-dimensional topological submanifold with boundary in $\mathbb{R}^3$, where the boundary ∂M consists a finite number of non-intersecting simple, closed, piecewise-regular curves.*

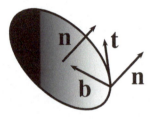

Figure 4.7.1: Compatible orientations on the surface and its boundary.

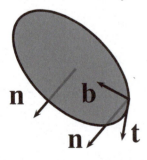

Figure 4.7.2: Compatible orientations on the surface and its boundary.

Before we can state Stokes' Theorem, we have to discuss orientations on a ccpr-surface with boundary and compatible orientations on the boundary curves.

Suppose that we have an **oriented** ccpr-surface M with boundary, so that we have selected a positive side/direction of M. We want to define a *compatible* or *positive* orientation on the boundary curves. The standard way of saying which direction is the positive/compatible direction on a boundary curve is to say that, if you walked along the curve in the positive direction, with your head pointing in the direction of the positive direction on the surface, then the surface is always to your left.

We can describe this in more technical mathematical terms. In Figure 4.7.1 and Figure 4.7.2, we have indicated compatible orientations on the surfaces and on the boundary; they are orientations such that, if **n** is a positive normal vector to the surface, **t** is a positive direction on the curve, and **b** is a vector which is tangent to the surface, but points in towards the surface, then the triple $(\mathbf{t}, \mathbf{b}, \mathbf{n})$ is a right-handed ordered triple.

Whenever we discuss an oriented ccpr-surface M with boundary ∂M, we assume, unless explicitly stated otherwise, that the boundary curves are given their compatible orientation.

Now we can state Stokes' Theorem.

Theorem 4.7.2. (Stokes' Theorem) *Suppose that M is an oriented ccpr-surface with boundary ∂M, which is given its compatible orientation. Let $\mathbf{F}$ be a continuously differentiable vector field on an open set in $\mathbb{R}^3$ which contains M.*

Then, the line integral of $\mathbf{F}$ along ∂M is equal to the flux integral of the curl $\vec{\nabla} \times \mathbf{F}$ over M, i.e.,

$$\int_{\partial M} \mathbf{F} \cdot d\mathbf{r} \;=\; \iint_M (\vec{\nabla} \times \mathbf{F}) \cdot \mathbf{n} \, dS.$$

For the proof, see [3], section 8.8. This is actually a proof of a much more general Stokes' Theorem for differential forms. To obtain our Stokes' Theorem, see formula 8.21 of [3].

Before we look at some specific examples, let's first see why Stokes' Theorem reduces to Green's Theorem, Theorem 4.4.6, when considering a vector field on a region in $\mathbb{R}^2$.

Let R be a compact region in $\mathbb{R}^2$, which is the closure of its interior, and whose boundary, ∂R, consists of a finite number of simple, closed, oriented, piecewise-regular curves, which are oriented so that the interior of R is always on the left as you move in the direction of the orientation. Let $\mathbf{F} = (P(x,y),\, Q(x,y))$ be a C^1 vector field on an open subset of $\mathbb{R}^2$ which contains R.

We consider R as a surface in $\mathbb{R}^3$ by considering each point in R as having z-coordinate 0; let M denote this "new" surface in $\mathbb{R}^3$. We give M an orientation by picking the positive unit normal vector $\mathbf{n}$ to be $\mathbf{k} = (0,0,1)$; with this orientation, the counterclockwise orientation on ∂R yields the compatible orientation on ∂M.

Let $\mathbf{F} = (P(x,y),\, Q(x,y))$ be a continuously differentiable vector field on an open subset of $\mathbb{R}^2$ which contains R. We "extend" the vector field $\mathbf{F}$ to $\mathbb{R}^3$ by defining the vector field $\hat{\mathbf{F}}$:

$$\hat{\mathbf{F}}(x,y,z) \;=\; \big(P(x,y),\, Q(x,y),\, 0\big).$$

Then, writing $*$'s in positions that will be multiplied by 0, we find

$$(\vec{\nabla} \times \hat{\mathbf{F}}) \cdot \mathbf{n} \;=\; \left(*,\, *,\, \frac{\partial Q}{\partial x} - \frac{\partial P}{\partial y}\right) \cdot (0,0,1) \;=\; \frac{\partial Q}{\partial x} - \frac{\partial P}{\partial y},$$

and

$$dS \;= dA.$$

Thus, Stokes' Theorem, applied to M and $\hat{\mathbf{F}}$, collapses to become Green's Theorem:

$$\int_{\partial R} \mathbf{F} \cdot d\mathbf{r} \;=\; \iint_R (Q_x - P_y)\, dA.$$

Now let's look at an example that's not in the xy-plane.

Example 4.7.3. Let C the circle, in the plane $y = z$, of radius 3, centered at the origin. Let M be the closed disk bounded by C, so that $C = \partial M$. Note that, as M is contained in the plane where $y - z = 0$, the unit normal vectors to M are $(0, 1, -1)/\sqrt{2}$ and $(0, -1, 1)/\sqrt{2}$. We orient M by selecting the "upward" normal vector $\mathbf{n} = (0, -1, 1)/\sqrt{2}$ as the positive unit normal vector, and we give ∂M the compatible orientation. See Figure 4.7.3.

Let $\mathbf{F} = (e^x + y,\ x + \sin y + 2z,\ 3x - 5y + z^2)$. We would like to calculate the line integral

$$\int_{\partial M} \mathbf{F} \cdot d\mathbf{r}.$$

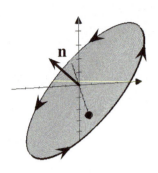

Figure 4.7.3: Compatible orientations on M and its boundary $C = \partial M$.

We **could** parameterize the circle and calculate the integral from the definition. However, if the curl of $\mathbf{F}$ turns out to be something simple, it might be easiest to use Stokes' Theorem, and to calculate the line integral by calculating the flux integral of $\vec{\nabla} \times \mathbf{F}$ over the disk.

We calculate

$$\vec{\nabla} \times \mathbf{F} = \begin{vmatrix} \mathbf{i} & \mathbf{j} & \mathbf{k} \\ \dfrac{\partial}{\partial x} & \dfrac{\partial}{\partial y} & \dfrac{\partial}{\partial z} \\ e^x + y & x + \sin y + 2z & 3x - 5y + z^2 \end{vmatrix} =$$

$$(-5 - 2)\mathbf{i} - (3 - 0)\mathbf{j} + (1 - 1)\mathbf{k} = (-7, -3, 0).$$

Thus, Stokes' Theorem tells us that

$$\int_{\partial M} \mathbf{F} \cdot d\mathbf{r} = \iint_M (\vec{\nabla} \times \mathbf{F}) \cdot \mathbf{n}\, dS = \iint_M (-7, -3, 0) \cdot \frac{(0, -1, 1)}{\sqrt{2}}\, dS =$$

$$\frac{3}{\sqrt{2}}\, (\text{surface area of } M) = \frac{3}{\sqrt{2}}\, \pi(3)^2 = \frac{27\pi}{\sqrt{2}}.$$

Example 4.7.4. Consider the vector field

$$\mathbf{F} = (e^x + y,\ x + \sin y + 2z,\ 3x - 5y + z^2)$$

from the previous example. We would like to calculate the flux integral of the curl of $\mathbf{F}$,

$$\iint_H (\vec{\nabla} \times \mathbf{F}) \cdot \mathbf{n}\, dS,$$

where H is the hemisphere, with boundary, of radius 3, centered at the origin, which is "above and to the left" of the plane given by $y = z$, oriented outward from the center; see Figure 4.7.4.

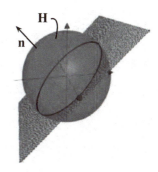

Of course, we calculated the curl of $\mathbf{F}$ in the previous example; we found

$$\vec{\nabla} \times \mathbf{F} = (-7, -3, 0).$$

However, even though the curl is a constant vector, it still doesn't look trivial to calculate

$$\iint_H (\vec{\nabla} \times \mathbf{F}) \cdot \mathbf{n}\, dS = \iint_H (-7, -3, 0) \cdot \mathbf{n}\, dS.$$

However, Stokes' Theorem tells us that, if we give ∂H the compatible orientation, then

$$\iint_H (\vec{\nabla} \times \mathbf{F}) \cdot \mathbf{n}\, dS = \int_{\partial H} \mathbf{F} \cdot d\mathbf{r}.$$

Figure 4.7.4: The hemisphere H, above and to the left of the plane where $y = z$.

But ∂H is precisely the circle $C = \partial M$ that we considered in the previous example, with the same orientation, and we calculated that

$$\int_{\partial M} \mathbf{F} \cdot d\mathbf{r} = \frac{27\pi}{\sqrt{2}}.$$

Therefore,

$$\iint_H (\vec{\nabla} \times \mathbf{F}) \cdot \mathbf{n}\, dS = \int_{\partial H} \mathbf{F} \cdot d\mathbf{r} = \int_{\partial M} \mathbf{F} \cdot d\mathbf{r} = \frac{27\pi}{\sqrt{2}}.$$

The last example gives us an important consequence of Stokes' Theorem:

if M_1 and M_2 are two oriented ccpr-surfaces with boundary, which are oriented in the "same" direction and which have the same boundary, then they yield the same flux integrals of the curl of a vector field F.

Here, when we write "same direction", we mean the direction so that the compatible orientations on the common boundary are the same. We can also describe the "same direction" in another important way.

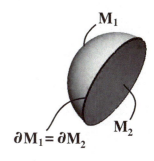

If M_1 and M_2 have the same boundary, then the union $M = M_1 \cup M_2$ is a ccpr-surface **without** boundary, for we eliminate the boundary when we "glue" M_1 and M_2 together along their common boundary. Thus, there is a solid region T bounded by the union of M_1 and M_2, and saying that M_1 and M_2 are oriented in the "same direction" means that we pick the positive direction to be away from T for one of M_1 or M_2, and, on the other surface, we pick the positive direction to be **into** T.

Figure 4.7.5: The union of two surfaces along their common boundary.

With these orientations, Stoke's Theorem yields what we wrote above:

$$\iint_{M_1} (\vec{\nabla} \times \mathbf{F}) \cdot \mathbf{n}\, dS \;=\; \int_{\partial M_1} \mathbf{F} \cdot d\mathbf{r} \;=\; \int_{\partial M_2} \mathbf{F} \cdot d\mathbf{r} \;=\; \iint_{M_2} (\vec{\nabla} \times \mathbf{F}) \cdot \mathbf{n}\, dS.$$

This has another interesting consequence. Suppose that we picked the positive directions to be out of T on M_1, and into T on M_2; we know then that

$$\iint_{M_1} (\vec{\nabla} \times \mathbf{F}) \cdot \mathbf{n}\, dS \;=\; \iint_{M_2} (\vec{\nabla} \times \mathbf{F}) \cdot \mathbf{n}\, dS.$$

If we reverse the orientation on M_2 to point outward from T, then we would have our usual orientation on $M = M_1 \cup M_2$: the orientation that always points outward from T. This was our default orientation on a ccpr-surface without boundary in Section 4.6. We write $M = M_1 - M_2$, to indicate that the orientation on M_2 is the negative of what it was.

But reversing the orientation on M_2 negates the integral:

$$\iint_{-M_2} (\vec{\nabla} \times \mathbf{F}) \cdot \mathbf{n}\, dS \;=\; -\iint_{M_2} (\vec{\nabla} \times \mathbf{F}) \cdot \mathbf{n}\, dS.$$

Thus,

$$\iint_{M} (\vec{\nabla} \times \mathbf{F}) \cdot \mathbf{n}\, dS \;=\; \iint_{M_1} (\vec{\nabla} \times \mathbf{F}) \cdot \mathbf{n}\, dS \;-\; \iint_{M_2} (\vec{\nabla} \times \mathbf{F}) \cdot \mathbf{n}\, dS \;=\; 0,$$

i.e., the flux of the curl of a vector field through a closed surface is always zero.

To conclude this from Stokes' Theorem, we actually need to know that every ccpr-surface without boundary can be formed by joining two ccpr-surface with boundary along their common boundary. This is true; start with the closed surface and cut out a small "disk" from a regular portion of the surface, then glue it back.

Corollary 4.7.5. *Suppose that M is a ccpr-surface* **without** *boundary. Let $\mathbf{F}$ be a continuously differentiable vector field on an open set in $\mathbb{R}^3$ which contains M.*
   Then,

$$\iint_{M} (\vec{\nabla} \times \mathbf{F}) \cdot \mathbf{n}\, dS \;=\; 0.$$

The corollary above may seem surprising, but it really shouldn't: if $\mathbf{F}$ is, in fact, a C^2 vector field which is defined on an open set containing the entire solid region T, enclosed by M, then the Divergence Theorem, Theorem 4.6.2, tells us that

$$\iint_{M} (\vec{\nabla} \times \mathbf{F}) \cdot \mathbf{n}\, dS \;=\; \iiint_{T} \vec{\nabla} \cdot (\vec{\nabla} \times \mathbf{F})\, dV.$$

Now, we use that the divergence of the curl, $\vec{\nabla} \cdot (\vec{\nabla} \times \mathbf{F})$, is always 0, as we calculated in Theorem 4.1.17.

However, Corollary 4.7.5 is true in cases where we can't use the Divergence Theorem, such as when $\mathbf{F}$ is C^1, but not C^2, or when $\mathbf{F}$ is not defined everywhere inside the enclosed solid region.

We would like to use Stokes' Theorem to interpret the curl of a vector field in terms of *rotation* produced by the vector field. In particular, we would like to explain why vector fields which have zero curl everywhere are called *irrotational*.

Suppose that we have a continuously differentiable vector field **F** on an open subset $\mathcal{U}$ of $\mathbb{R}^3$. For our discussion, we would like to think of **F** as a force field, because we want to discuss how this force field causes disks to rotate around axes; this is a matter of *torque*, force applied at a distance (recall Example 1.5.12). However, to aid with your intuition, it may be more helpful to imagine **F** as being the velocity vector field of a flowing fluid, and assume that our disks have a texture which is rough enough so that friction with the fluid produces a force which is proportional to the velocity. Thinking this way allows us to visualize **F** as "flowing", and still discuss the torque.

To specify a disk in space, we specify a point **p** that's the center of the disk, we specify a positive radius ϵ for the disk, and we specify an axis through **p** that's perpendicular to the disk. In fact, it's helpful to give ourselves a *directed axis* for the disk by specifying a non-zero vector **v** or, equivalently, by specifying the unit normal vector $\mathbf{n} = \mathbf{v}/|\mathbf{v}|$ in the direction of **v**. We want to investigate the rotation of the disk around the directed axis **v** or **n**, based at **p**. As we are interested in the "infinitesimal rotation produced at **p**", we will eventually let the radius ϵ approach 0.

In Figure 4.7.6, we have drawn a disk, perpendicular to **n**, centered at **p**. At one point on the boundary circle of the disk, we have included, as we did earlier in the section, an inward pointing unit vector **b** (which points towards the center of the circle), and a unit vector **t**, which is tangent to the circle. We have also included one force vector from our vector field **F**.

As before, we have chosen **t** to point in the *positive* direction with respect to the orientation of the disk given by **n**; this means that $(\mathbf{t}, \mathbf{b}, \mathbf{n})$ is a right-handed triple, or that, if you curl the fingers on your right hand around the circle, in the direction given by **t**, then your thumb points in the direction of **n**.

We are not interested in the components of **F** that would make the disk tilt, for this would shift the axis; we are fixing the axis, and want to look at rotation around it. Thus, we want to eliminate the component of **F** in the direction of **n**, for such a component would make the disk tilt. This means that we want to look at the torque produced not by **F**, but rather by the orthogonal projection of **F** into the plane of the disk, i.e., we look at $\mathbf{F}_{\text{disk}} = \mathbf{F} - (\mathbf{F} \cdot \mathbf{n})\mathbf{n}$.

Now, the radial vector from **p** out to our point on the disk of radius ϵ is $-\epsilon\mathbf{b}$, and so the magnitude of the torque at **p**, produced by $\mathbf{F}_{\text{disk}}$ is $|-\epsilon\mathbf{b} \times \mathbf{F}_{\text{disk}}|$. In the exercises, we have you show that there's an equality

$$|-\epsilon\mathbf{b} \times \mathbf{F}_{\text{disk}}| \;=\; \epsilon|\mathbf{F} \cdot \mathbf{t}|, \tag{4.3}$$

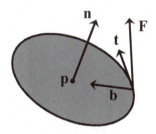

Figure 4.7.6: A disk, with a unit normal **n**, radial vector **b**, "positive" unit tangent **t**, and a single force vector.

i.e., that the number $\mathbf{F} \cdot \mathbf{t}$ measures the magnitude of the torque that we're after, but has a plus sign for rotation produced in the positive direction, and a minus sign for rotation produced in the negative direction.

This means that the line integral of $\mathbf{F}$ around the oriented boundary of the disk measures the tendency of the force field to produce rotation around the directed axis. Why? Because if we parametrize the oriented boundary circle by $\mathbf{r} = r(s)$, where s is arc length (recall Remark 1.6.16), then $\mathbf{r}'(s)$ is equal to the unit tangent vector $\mathbf{t}$, and $\mathbf{F} \cdot d\mathbf{r} = (\mathbf{F} \cdot \mathbf{t})\, ds$. Therefore, $\int_C \mathbf{F} \cdot d\mathbf{r}$ is a cumulative measure of the rotation produced by $\mathbf{F}$, taking into account the force acting at each point on the boundary of the disk.

Since the line integral measures the rotation produced by the vector field, we will use the line integral, and take a limit as ϵ approaches 0, to define the *rotational density* of $\mathbf{F}$ at $\mathbf{p}$, around each axis. We will then relate this to the curl of the vector field.

Before we do this, we should give you an important warning: the line integral of $\mathbf{F}$ around the boundary circle of a disk measures the rotation produced in the disk, but there need **not** be any noticeable rotation in the vector field itself, i.e., there does not have to be a circular, or even closed, flow in the vector field itself in order for the field to produce rotation in a disk. Consider Figure 4.7.7, which is a top-view of a disk in the plane $z = 0$, and part of the vector field $\mathbf{F}(x, y, z) = (0, x, 0)$, in a region in which all of the x-coordinates are positive.

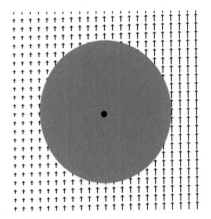

Figure 4.7.7: A "vertical" flow, which would make the disk rotate counterclockwise.

You don't see the vector field in Figure 4.7.7 circling around in any way; it's always pointing straight up. But if the center of the disk is fixed, and the disk is free to rotate around the center, then it should be clear that the disk would rotate counterclockwise, for the force vectors have greater magnitude on the right than on the left.

Of course, if the vector field itself were rotating, it would certainly produce rotation in a disk, but, the point is, vector fields can easily produce rotations even when the vector fields, themselves, don't appear to rotate.

With all of this discussion behind us, we are ready to relate rotation and the curl of a vector field.

We remind you that $\mathbf{F}$ is a continuously differentiable vector field on an open subset $\mathcal{U}$ of $\mathbb{R}^3$.

Let $\mathbf{n}$ be a unit vector, based at $\mathbf{p}$, and let M_ϵ denote the disk, of radius $\epsilon > 0$, centered at $\mathbf{p}$, in the plane which is perpendicular to $\mathbf{n}$. Note that, as $\mathcal{U}$ is an open set, for small enough ϵ, the disk M_ϵ will be contained in $\mathcal{U}$. In what follows, we assume that ϵ is always chosen small enough so that this true.

We pick the orientation on the surface M_ϵ to be in the direction given by $\mathbf{n}$. We give the boundary circle ∂M_ϵ its compatible orientation; this means that, if you point the thumb on your right hand in the direction of $\mathbf{n}$, then your fingers should curl in the direction of the positive orientation on ∂M_ϵ.

From our earlier discussion, we know that $\int_{\partial M_\epsilon} \mathbf{F} \cdot d\mathbf{r}$ is a measure of the rotation of M_ϵ produced by $\mathbf{F}$. A positive integral indicates that there's a net rotation produced by $\mathbf{F}$, around $\mathbf{n}$, in the direction given by the right-hand rule, and a negative integral indicates a net rotation in the opposite direction.

What does this have to do with the curl of the vector field $\mathbf{F}$?

If ϵ is close to zero, then Stokes' Theorem tells us that

$$\int_{\partial M_\epsilon} \mathbf{F} \cdot d\mathbf{r} \;=\; \iint_{M_\epsilon} (\vec{\nabla} \times \mathbf{F}) \cdot \mathbf{n}\, dS \;\approx\; \left[(\vec{\nabla} \times \mathbf{F})_{|_\mathbf{p}} \cdot \mathbf{n} \right] (\text{Area of } M_\epsilon),$$

and, in terms of limits,

$$\lim_{\epsilon \to 0^+} \frac{\int_{\partial M_\epsilon} \mathbf{F} \cdot d\mathbf{r}}{\text{Area of } M_\epsilon} \;=\; \lim_{\epsilon \to 0^+} \frac{\int_{\partial M_\epsilon} \mathbf{F} \cdot d\mathbf{r}}{\pi \epsilon^2} \;=\; (\vec{\nabla} \times \mathbf{F})_{|_\mathbf{p}} \cdot \mathbf{n}.$$

The quantity on the left, above, should be interpreted as the *rotational density of* $\mathbf{F}$ *around* $\mathbf{n}$ *at* $\mathbf{p}$, and so, the above equalities tell us that $(\vec{\nabla} \times \mathbf{F})_{|_\mathbf{p}} \cdot \mathbf{n}$ equals the rotational density.

Therefore, if the curl of $\mathbf{F}$ is zero everywhere, then the rotational density at every point, around every axis, is zero; naturally, the vector field is called irrotational.

Now suppose that $(\vec{\nabla} \times \mathbf{F})_{|_\mathbf{p}} \neq \mathbf{0}$. Let $\mathbf{n}$ be the unit vector in the direction of $(\vec{\nabla} \times \mathbf{F})_{|_\mathbf{p}}$, i.e., let

$$\mathbf{n} \;=\; \frac{1}{\left| (\vec{\nabla} \times \mathbf{F})_{|_\mathbf{p}} \right|} (\vec{\nabla} \times \mathbf{F})_{|_\mathbf{p}}.$$

Then, the rotational density of $\mathbf{F}$, around $\mathbf{n}$, or around $(\vec{\nabla} \times \mathbf{F})_{|_\mathbf{p}}$, at $\mathbf{p}$ is

$$(\vec{\nabla} \times \mathbf{F})_{|_\mathbf{p}} \cdot \mathbf{n} \;=\; (\vec{\nabla} \times \mathbf{F})_{|_\mathbf{p}} \cdot \frac{1}{\left| (\vec{\nabla} \times \mathbf{F})_{|_\mathbf{p}} \right|} (\vec{\nabla} \times \mathbf{F})_{|_\mathbf{p}} \;=\; \left| (\vec{\nabla} \times \mathbf{F})_{|_\mathbf{p}} \right| \;>\; 0.$$

Therefore, at a point where the curl is non-zero, the rotational density around the curl vector is positive, equal to the magnitude of the curl, and the vector field instantaneously rotates around the curl vector, in the direction specified by the right-hand rule.

Online answers to select exercises are here.

4.7.1   Exercises

Basics:

In each of the following exercises, you are given a vector field $\mathbf{F} = \mathbf{F}(x, y, z)$, and an oriented ccpr-surface M with boundary. Calculate $\iint_M (\vec{\nabla} \times \mathbf{F}) \cdot \mathbf{n} \, dS$ by calculating $\int_{\partial M} \mathbf{F} \cdot d\mathbf{r}$, with the appropriate orientation on ∂M, and applying Stokes' Theorem.

1. Let $\mathbf{F}(x, y, z) = (3 + ze^y, 5 + z \sin x, z^3)$, and let M be the upper-hemisphere of radius 4, centered at the origin, oriented upward.

2. Let $\mathbf{F}(x, y, z) = (e^y, e^z, e^x)$. Let T be the solid cube where $0 \leq x \leq 2$, $0 \leq y \leq 2$, and $0 \leq z \leq 2$. Let M be the boundary of T, minus the top filled-in square, oriented outward from T.

3. Let $\mathbf{F}(x, y, z) = (e^y, e^z, e^x)$. Let T be the solid tetrahedron, in the first octant, below the plane where $x + 2y + 3z = 6$. Let M be the boundary of T, minus the face in the xz-plane, oriented outward from T.

4. Let $\mathbf{F} = (y \sin z, xe^z, y \tan^{-1} x)$, and let M be the portion of the elliptical paraboloid where $y = 1 - \dfrac{x^2}{4} - \dfrac{z^2}{9}$, where $y \geq 0$, oriented in the direction of $\mathbf{j}$ (i.e., so that each normal vector has a non-negative $\mathbf{j}$ component).

In each of the following exercises, you are given a vector field $\mathbf{F} = \mathbf{F}(x, y, z)$, and an oriented curve C. Calculate $\int_C \mathbf{F} \cdot d\mathbf{r}$ by selecting an oriented ccpr-surface M such that $\partial M = C$, calculating $\iint_M (\vec{\nabla} \times \mathbf{F}) \cdot \mathbf{n} \, dS$, and applying Stokes' Theorem.

5. $\mathbf{F}(x, y, z) = (x, y, z)$, and C is the circle of radius 5, centered at the origin in the xz-plane, oriented from the positive x-axis towards the positive z-axis.

6. $\mathbf{F}(x, y, z) = (z + \sin x, x + e^y, y + z^3)$, and C is the oriented circle ∂M from Example 4.7.3.

7. $\mathbf{F}(x, y, z) = (z + \sin x,\ x + e^y,\ y + z^3)$, and C is the oriented square that starts at $(0, 0, 0)$, goes to $(0, 2, 0)$, then goes to $(0, 2, 2)$, then goes to $(0, 0, 2)$ and, finally, goes back to $(0, 0, 0)$.

8. $\mathbf{F} = (y \sin z,\ xe^z,\ y \tan^{-1} x)$, and C is the ellipse where $\dfrac{x^2}{4} + \dfrac{z^2}{9} = 1$, in the plane where $y = 0$, oriented counterclockwise in the xz-plane.

In each of the following exercises, you are given a continuously differentiable vector field $\mathbf{F} = \mathbf{F}(x, y, z)$, and two oriented ccpr-surfaces M_1 and M_2, with a common boundary. By Stokes' Theorem, the flux integrals of the curl of $\mathbf{F}$ over the two oriented surfaces are the same. Calculate this common value by calculating whichever flux integral seems easiest.

9. $\mathbf{F}(x, y, z) = (z \sin x,\ ze^y,\ xyz^2)$. M_1 is the disk where $z = 0$, and $x^2 + y^2 \le 9$. M_2 is the upper-hemisphere, of radius 3, centered at the origin. Both surfaces are oriented upward.

10. $\mathbf{F}(x, y, z) = (2y + \sin z + ye^x,\ e^x + 3z,\ x(7 + \cos z))$. M_1 is the oriented surface M from Exercise 2 (5 sides of a box, oriented outward from the solid box). M_2 is the "missing top" from Exercise 2, oriented downward.

11. $\mathbf{F}(x, y, z) = (2y + \sin z + ye^x,\ e^x + 3z,\ x(7 + \cos z))$. M_1 is the oriented surface M from Exercise 3 (3 sides of a tetrahedron, oriented outward from the solid tetrahedron). M_2 is the "missing face" from Exercise 3, oriented in the direction of $\mathbf{j}$.

12. $\mathbf{F} = (e^x + y,\ x + \sin y + 2z,\ 3x - 5y + z^2)$. M_1 is the upper-hemisphere, of radius 1, centered at the origin. M_2 is the portion of the circular paraboloid where $z = 1 - x^2 - y^2$, and $z \ge 0$. Both surfaces are oriented upward.

> **More Depth:**

13. Calculate the rotational density of $\mathbf{F}(x, y, z) = (yz,\ e^x,\ x^2 + y^2)$, at $(0, 1, 2)$, around the unit vector in the direction of $\mathbf{v} = (1, 1, 1)$.

14. In this exercise, we want to verify Formula 4.3. The vectors $\mathbf{F}$, $\mathbf{t}$, $\mathbf{b}$, and $\mathbf{n}$ here are the same as in the discussion before Formula 4.3.

    a. Explain why, at a fixed point on the boundary of the disk, there exist unique constants u, v, and w such that

$$\mathbf{F} = u\mathbf{t} + v\mathbf{b} + w\mathbf{n}.$$

    b. In terms of u, v, w, $\mathbf{t}$, $\mathbf{b}$, and $\mathbf{n}$, from above, what is $\mathbf{F}_{\text{disk}}$?

    c. Show that $-\mathbf{b} \times \mathbf{F}_{\text{disk}} = (\mathbf{F} \cdot \mathbf{t})\mathbf{n}$.

d. Show that $|-\epsilon \mathbf{b} \times \mathbf{F}_{\text{disk}}| = \epsilon |\mathbf{F} \cdot \mathbf{t}|$.

15. Suppose that $\left(\vec{\nabla} \times \mathbf{F}\right)\Big|_{(7,-1,2)} = (1, 5, -3)$. Find a unit vector $\mathbf{n}$ such that the rotational density of $\mathbf{F}$ around $\mathbf{n}$ at $(7, -1, 2)$ equals 0.

16. Suppose that $(\vec{\nabla} \times \mathbf{F})|_{\mathbf{p}} \neq \mathbf{0}$. For what unit vector $\mathbf{n}$ is the rotational density of $\mathbf{F}$ around $\mathbf{n}$ at $\mathbf{p}$ the greatest? Explain.

17. Write an essay in which you discuss the similarities and interrelationships between the Fundamental Theorem of Line Integrals, Theorem 4.3.3, Green's Theorem, Theorem 4.4.1, the Divergence Theorem, Theorem 4.6.2, and Stokes' Theorem, Theorem 4.7.2.

18. Recall the Möbius Strip from Exercise 23 in Section 3.11. The Möbius Strip M is the basic surface given by the parameterization

$$\mathbf{r}(u, v) = \left(\left(R - u \sin\left(\frac{v}{2}\right)\right) \cos v, \ \left(R - u \sin\left(\frac{v}{2}\right)\right) \sin v, \ u \cos\left(\frac{v}{2}\right)\right),$$

where L and R are constants such that $0 < L < R$, and $-L \leq u \leq L$ and $0 \leq v \leq 2\pi$.

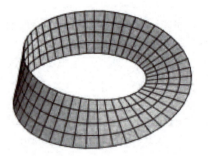

Figure 4.7.8: One view of the Möbius Strip

The boundary ∂M of M is parameterized by fixing $u = L$ and letting v "go around twice", i.e., ∂M has a parameterization given by

$$\mathbf{p}(t) = \left(\left(R - L \sin\left(\frac{t}{2}\right)\right) \cos t, \ \left(R - L \sin\left(\frac{t}{2}\right)\right) \sin t, \ L \cos\left(\frac{t}{2}\right)\right),$$

for $0 \leq t \leq 4\pi$.

The Möbius Strip is not orientable. In fact, we can observe its lack of orientability by noting the failure of Stokes' Theorem.

Let

$$\mathbf{F}(x, y, z) = \left(\frac{-y}{x^2 + y^2}, \frac{x}{x^2 + y^2}, 0\right).$$

This vector field is undefined along the z-axis, but M does not hit the z-axis.

(a) Verify that, using our parameterization, M does not intersect the z-axis.

(b) Show that $\vec{\nabla} \times \mathbf{F} = \mathbf{0}$, so that $\iint_M (\vec{\nabla} \times \mathbf{F}) \cdot \mathbf{n} \, dS = 0$.

(c) Show that the line integral $\int_{\partial M} \mathbf{F} \cdot d\mathbf{r}$ of $\mathbf{F}$ along ∂M, with the orientation given by the parameterization $\mathbf{p}$, is 4π.

Bibliography

[1] Bredon, G. *Topology and Geometry*. Graduate Texts in Mathematics. Springer-Verlag, 1993.

[2] Feynman, R., Leighton, R., and Sands, M. *The Feynman Lectures on Physics*. Addison-Wesley, 1963.

[3] Fleming, W. H. *Functions of Several Variables*. Addison-Wesley, 2nd edition, 1977.

[4] Massey, D. *Worldwide Differential Calculus*. Worldwide Center of Math., 1st edition, 2009.

[5] Massey, D. *Worldwide Integral Calculus, with infinite series*. Worldwide Center of Math., 1st edition, 2010.

[6] Milnor, J. *Morse Theory*, volume 51 of *Annals of Math. Studies*. Princeton Univ. Press, 1963. Based on lecture notes by Spivak, M. and Wells, R.

[7] Rudin, W. *Principles of Mathematical Analysis*. International Series in Pure and Applied Mathematics. McGraw-Hill, 1953.

[8] Trench, William F. *Introduction to Real Analysis*. Prentice Hall, 2003.

Index

About the Author:

David B. Massey was born in Jacksonville, Florida in 1959. He attended Duke University as an undergraduate mathematics major from 1977 to 1981, graduating *summa cum laude*. He remained at Duke as a graduate student from 1981 to 1986. He received his Ph.D. in mathematics in 1986 for his results in the area of complex analytic singularities.

Professor Massey taught for two years at Duke as a graduate student, and then for two years, 1986-1988, as a Visiting Assistant Professor at the University of Notre Dame. In 1988, he was awarded a National Science Foundation Postdoctoral Research Fellowship, and went to conduct research on singularities at Northeastern University. In 1991, he assumed a regular faculty position in the Mathematics Department at Northeastern. He has remained at Northeastern University ever since, where he is now a Full Professor.

Professor Massey has won awards for his teaching, both as a graduate student and as a faculty member at Northeastern. He has published over 30 research papers, and two research-level books. In addition, he was a chapter author of the national award-winning book on teaching: "Dear Jonas: What can I say?, Chalk Talk: E-advice from Jonas Chalk, Legendary College Teacher", edited by D. Qualters and M. Diamond, New Forums Press, (2004).

Professor Massey founded the Worldwide Center of Mathematics, LLC, in the fall of 2008, in order to give back to the mathematical community, by providing free or very low-cost materials and resources for students and researchers.